PLANT PATHOLOGY

PLANT PATHOLOGY

Editor

Mehmet Tilkic

scitus
academics

Plant Pathology

Edited by **Mehmet Tilkic**

This Book is Printed on Acid-Free Paper. ♾

ISBN: 9781681179445
Library of Congress Control Number: 2017949082

SCITUS Academics LLC,
616, Corporate Way, Suite 2, 4766,
Valley Cottage, NY 10989

www.scitusacademics.com

Printed in United States of America.

Preface

Plant diseases play a significant role on our everyday lives. The science that deals with the study of diseases of plants, their development and control is called Plant Pathology. Plant diseases are caused by biotic agents like fungi, bacteria, actinomycets, Mycoplasma, viruses, nematodes, and flowering parasites or by abiotic like unfavorable environmental conditions or nutritional deficiencies. Plants and their pathogens are engaged in continuous evolutionary battles and sustainable disease management requires novel systems to create environments conducive for short-term and long-term disease control. This work provides the knowledge of the fundamental factors that drive host–pathogen co-evolution in wild systems can provide new insights into disease development in agriculture. Such evolutionary principles can be used to guide the formulation of sustainable disease management strategies which can minimize disease epidemics while simultaneously reducing pressure on pathogens to evolve increased infectivity and aggressiveness.

This volume provides the latest information on plant pathogenic fungi, viruses, bacteria, phytoplasmas and spiroplasmas, including pathogen biology from genetics to geographical distributions. This book is an attempt in the process to provide readers with a relevant book which acquaints them with the current developments and advances in the field of plant pathology and biotechnology, particularly within the context of the widening scope of such arenas as biochemistry. In this book, expert scholars share their research knowledge and key literature which are vital toward the diagnosis of plant diseases across the globe, addressing traditional plant pathology techniques, as well as advanced molecular diagnostic approach.

Table of Contents

CHAPTER 1

Attenuated Plant Viruses: Preventing virus Diseases and Understanding the Molecular Mechanism

Masamichi Nishiguchi • Kappei Kobayashi

246-Ehime University MatsuyamaJapan

ABSTRACT

Attenuated viruses have been isolated and studied not only as a practical means of controlling virus diseases but also to gain a molecular understanding of viral virulence and cross protection. They have been isolated from crop fields and generated through high/low temperature treatment or by mutagens such as nitrous acid and ultraviolet irradiation. Some viruses have been beneficially used in fields and evaluated for one or more decades. Molecular genetic studies on attenuated viruses have revealed that amino acid substitutions are located in replicase and the movement protein in tobamovirus, protein 2b for cucumovirus, and P1 and HC-Pro for potyvirus. In most cases, with a few exceptions, symptom attenuation is positively correlated with a reduced level of RNA silencing suppression. Molecular mechanisms underlying virus attenuation and cross protection and the rationale for practical use of attenuated viruses for effective virus disease control are discussed.

Keywords: Attenuated virus Virulence Cross protection Interference RNA silencing Posttranscriptional gene silencing (PTGS) Viral silencing suppressor

INTRODUCTION

Attenuated or mild virus strains have been used to protect crops against severe strains of the same or related virus species. This method of crop protection is based on the phenomenon called cross protection/ interference, which was first reported by McKinney (1929), who showed that tobacco plants infected with a yellow strain of *Tobacco mosaic virus* (TMV) show resistance to a severe mosaic strain of TMV. Salaman (1933) reported that tobacco inoculated with a mild strain of *Potato virus X* (PVX) is resistant to severe strains of PVX. Holmes (1934) isolated a masked strain (M) of TMV from tomato stems that had been incubated at 34.5°C for 14 days after inoculation with a severe strain and elicited some resistance in tobacco and tomato against a severe strain. Since then, there have been a number of reports on vegetables and fruit trees including tomato, cucumber, cacao, papaya, and citrus (reviewed by Gal-On and Shiboleth 2006).

More recently, transgenic plants with a viral sequence have been proven to confer resistance to the virus from which the transgene was derived or to related viruses (Prins et al. 2008). One remarkable example is transgenic papaya; with the coat protein (CP) gene of *Papaya ringspot virus* (PRSV), it expresses extreme resistance to PRSV (Fuchs and Gonsalves 2008), although the resistance mechanism in this papaya appears to be CP-mediated. These papaya fruits have been sold in US and Canadian markets.

The mechanism of resistance in these transgenic plants has been interpreted by either of two distinct models: protein-mediated or RNA-mediated mechanisms (Baulcombe 1996). In the former model, pre-existing CP affects the virus infection process, i.e., inhibition of uncoating and replication as shown for TMV (Bendahmane et al. 2007). In the latter, the resistance results from RNA silencing by host plants (Ratcliff et al. 1999).

The phenomenon of RNA silencing has been thought to play a pivotal role in defending eukaryotes against invasive genes such as viruses, transposons and transgenes (Ding and Voinnet 2007). To counteract this host defense mechanism, viruses have evolved a means of counter-attacking host plants: virus encoded suppressor(s) of RNA silencing (Béclin et al. 1998; Brigneti et al. 1998; Kasschau and Carrington 1998). It is generally accepted that viral RNA silencing suppressors (VRSs) are more or less related to viral pathogenicity because the VRS represses silencing, that is, virus infection in host plants is more vigorous, which may virtually result in more severe symptoms (Díaz-Pendón and Ding 2008).

In this review, the present status of attenuated virus studies is described first in relation to a means of protecting crops from viral diseases. Then the mechanism of viral attenuation and cross protection is discussed with reference to RNA silencing. Finally, some characteristics of attenuated viruses that are needed for effective virus disease control are discussed in light of the molecular understanding of attenuated viruses.

SELECTION, ISOLATION AND INDUCTION OF ATTENUATED VIRUSES

The most commonly used attenuated virus strains have been isolated from naturally occurring variants that cause attenuated symptoms or by random or direct mutagenesis of wild, severe strains using nitrous acid or ultraviolet irradiation. Nitrous acid is a well-known mutagen that deaminates cytosine and adenine to produce uracil and hypoxanthine, respectively, and thus, induces transitional mutations. UV irradiation causes the formation of pyrimidine dimers in DNA as well as in RNA (Brégeon and Sarasin 2005) and likely induces any type of mutation, including base changes and insertion/deletions.

Some attenuated viruses have been isolated after cultivation at higher (35°C) or lower (15°C) temperatures than the optimal. The underlying mechanism of attenuated virus generation after temperature treatment is not yet clear. However, the concept of quasispecies supports that temperature treatment would facilitate the selection of mutated viruses from preexisting variation rather than that the treatment induces the mutation of a virus genome. A pseudorecombinant RNA virus is one choice for selecting an attenuated virus by the exchange of a viral RNA segment with the corresponding RNA from an attenuated strain in multisegmented viruses including *Cucumber mosaic virus* (CMV). Some attenuated viruses (strains/isolates) are listed in Table 1.

An attenuated strain of *Tomato mosaic virus* (ToMV), $L_{11}A$, has been in practical use for over 30 years to protect tomato in greenhouses against severe strain infection (Oshima 1981). Tomato cultivars carrying resistance genes (*Tm-1*, *Tm-2* and/or *Tm-2^a*) have been developed, and the frequency of pre-inoculation with the attenuated strain has decreased. However, when a resistant variety is not available, an attenuated strain is still required. In Europe and New Zealand, an attenuated strain of ToMV (MII-16), isolated by treatment with nitrous acid, was once used (Rast 1975). By the same method, an attenuated strain (ToMV-K) was produced and has been used in China for two decades (Yang et al. 2002). An attenuated strain (SH33b) of *Cucumber green mottle mosaic* *virus* (CGMMV) was isolated through nitrous acid and UV irradiation and has been used for muskmelon in greenhouses in Japan (Motoyoshi and Nishiguchi 1988). Some attenuated strains (Pa18 and C-1421) of *Pepper mild mottle virus* (PMMoV) have been used in pepper fields (Goto et al. 1984; Nagai 1987).

Table 1. Examples of attenuated viruses used for cross protection

Virus	Strain or isolate (method) (country)	Crop	
CGMMV	SH33b[b,d,e] (Japan)	Muskmelon	Motoyoshi and Nishiguchi (1988)
CMV	S51[a,f], S52[a,f] (China); KO3[a,f], CM95[a] (Japan)	Tomato, Tobacco, etc.	Kosaka and Fukunishi (1997), Sayama et al. (1993), Tien and Wu (1991)
CSSMV	Mild strains[a] (Ghana)	Cacao	Hughes and Ollennu (1994)
CTV	HM55[a], M-16A[b] (Japan), mild strains[a] (Brazil)	Citrus	Costa and Müller (1980), Ieki et al. (1997), Sasaki (1974)
PMMoV	Pa18[b], C-1421[b] (Japan)	Pepper	Goto et al. (1984), Nagai (1987)
PRSV	HA5-1[d] (USA, Taiwan)	Papaya	Yeh and Gonsalves (1984)
SMV	Ala15-M2[c] (Japan)	Soybean	Kosaka and Fukunishi (1993)
TMV	M[b] (USA)	Tobacco	Holmes (1934)
ToMV	$L_{11}A$[b] (Japan), MII-16[d] (Holland, UK, etc.), K[d] (China)	Tomato	Goto and Nemoto (1971), Rast (1975), Yang et al. (2002)
ZYMV	ZY95[c,d] (Japan)	Cucumber	Kosaka and Fukunishi (1997)

CGMMV, *Cucumber green mottle mosaic virus*; CMV, *Cucumber mosaic virus*; CSSMV, *Cocoa swollen shoot mosaic virus*; CTV, *Citrus tristeza virus*; PMMoV, *Pepper mild mottle virus*; PRSV, *Papaya ringspot virus*; SMV, *Soybean mosaic virus*; TMV, *Tobacco mosaic virus*; ToMV, *Tomato mosaic virus*; ZYMV, *Zucchini yellow mosaic virus*

[a] Natural selection

[b] High temperature

[c] Low temperature

[d] Nitrous acid

[e] Ultraviolet irradiation

[f] Satellite RNA

Tomato and tobacco diseases by CMV are also controlled by attenuated strains in Japan and China (Sayama et al. 1993; Tien and Wu 1991). In some cases, satellite RNA was added to field isolates of CMV to attenuate symptoms (Sayama et al. 1993). In Japan, this strategy has been applied not only to tomato but also to cucumber, petunia and Texas bluebell (*Eustoma grandiflorum*) (Sayama 1996).

Soybean has been preventively inoculated with an attenuated strain (Ala-15) of *Soybean mosaic virus*, which was isolated from soybean plants incubated at a lower temperature (15°C) (Kosaka and Fukunishi 1993). An attenuated strain (ZYMV-2002) of *Zucchini yellow mosaic virus* (ZYMV) isolated through low temperature treatment has also been applied to cucumber (Kosaka et al. 2006). Papaya in Hawaii and Taiwan was tested for cross protection by an attenuated strain of PRSV (HA-21), showing significant protection (Yeh et al. 1988). Some attenuated, field-selected strains of *Cacao swollen shoot virus* (CSSV) have been used for cross protection on cacao to prevent severe damage from CSSV (Hughes and Ollennu 1994).

SIDE EFFECTS OF PREVENTIVE INOCULATION WITH ATTENUATED VIRUS AND MONITORING ATTENUATED VIRUSES APPLIED TO FIELDS

The goal of preventive inoculation with an attenuated virus is to protect crops from infection with severe strains. The quantity and/or quality of the final harvested products are significantly important, and they should have a similar level of quality to virus-free control. In the case of CMV, the content of vitamin C was higher in tomato inoculated with attenuated CMV compared with that in virus-free tomato (Sayama 1996). This result is an unexpected advantage of using an attenuated virus.

In monitoring attenuated CMV in fields, necrotic spot, a serious disease that manifests itself on the surface of the fruits, was markedly reduced by application of the attenuated CMV. The longer the attenuated virus was used, the less frequent was disease occurrence, and there was no serious damage for 10 years (Sayama et al. 2001). When an attenuated strain of CGMMV was monitored in greenhouses after serial use of the attenuated strain, no substantial disease damage was found, implying that preventive inoculation reduces the chance of infection by a severe strain and that the attenuated strain is inactivated more effectively than the wild severe strain SH (Anonymous 1988).

MOLECULAR AND BIOLOGICAL CHARACTERIZATION OF ATTENUATED VIRUSES

Tobamovirus

Attenuated virus genomes have been reported for several tobamoviruses (Table 1; Figure 1a). In $L_{11}A$, 11 base substitutions were located in the genome when compared with the wild-type, severe strain L (Nishiguchi et al. 1985). Of those, only three at amino acid (aa) positions 348, 760 and 896 in the 130-kDa protein caused amino acid replacements. These three amino acid replacements were compared with those in L_{11}, from which $L_{11}A$ was derived (Goto and Nemoto 1971). Only one of them, the change from cysteine (C) in wild-type strains to tyrosine (Y) in attenuated strains at aa position 348 (at nucleotide position 1117), was conserved within these two attenuated virus strains ($L_{11}A$ and L_{11}), suggesting that the amino acid replacement at aa position 348 is important for virus attenuation (Nishiguchi et al. 1985). Kubota et al. (2003) later proved that the change of C–Y at aa position 348 in the infectious L cDNA clone rendered the mosaic symptom into an attenuated phenotype. That mutated clone is referred to as TLJ. It was also found to be the only nucleotide substitution between L_{11} and L over the genome (Kubota et al. 2003). They also showed that the wild-type 130-kDa protein had greater RNA silencing suppressor activity than that of the attenuated mutant TLJ, suggesting that the amino acid replacement from C to Y at aa position 348 in the 130-kDa protein plays an important role in reducing the RNA silencing suppression activity of the 130-kDa protein. More recently, Hagiwara-Komoda et al. (2008) reported that the 130-kDa protein of ToMV was detected in both a soluble and a crude membrane fraction. The latter contains the TOM1 protein, which is required for multiplication of tobamoviruses in *Arabidopsis* and tobacco (Asano et al. 2005; Yamanaka et al. 2000). In the *NtTOM1* (tobacco ortholog of *TOM1*)-overexpressing transgenic tobacco plants, the level of the 130-kDa protein in the soluble fraction was reduced because the overexpressed NtTOM1 recruited and anchored the 130-kDa protein to membranes. The reduced level of the soluble 130-kDa protein resulted in a reduction of RNA silencing suppression. The propagation of both L and L_{11} were inhibited in this NtTOM1-overexpressing line, implying that silencing suppression was apparently caused by the 130-kDa protein in the soluble fraction (Hagiwara-Komoda et al. 2008). The molecular mechanism underlying the silencing suppression by the tobamovirus 130-kDa protein (replicase protein) (including the 122-kDa protein of cr-TMV and the 126-kDa protein of TMV) is postulated to prevent formation of the RNA-induced silencing complex (RISC) after siRNA and miRNA syntheses (Csorba et al. 2007; Ding et al. 2004; Kubota et al. 2003; Vogler et al. 2007). The 130-kDa protein was found to bind to double-stranded RNA (dsRNA) but not to single-stranded RNA (ssRNA) (Hagiwara-Komoda et al. 2008). The dsRNA-binding activity and RNA-dependent RNA polymerase activity of the 130-kDa protein in the membrane fractions were not significantly different between L and TLJ. However, the 130-kDa protein level in the soluble fraction was reduced in the attenuated strain TLJ compared with that in the wild-type L, implying that the binding affinity of the attenuated 130-kDa protein to TOM1 or a TOM1-containing membrane is higher than that of the wild-type 130 kDa. Yamanaka et al. (2000) showed that the helicase domain of the 130 kDa interacts with TOM1. The mutation at aa position 348 of 130 kDa would alter the structure of the helicase domain, increase the TOM1-binding activity, and thus reduce the silencing suppressor activity. Replication protein (122 kDa) of TMV-Cg was found to disturb miRNA biogenesis by binding to small RNA duplexes and changing the miRNA profiles (Kurihara et al. 2007). It would be worth analyzing the difference in the effects of the wild-type and attenuated 130-kDa replication proteins on the miRNA profiles.

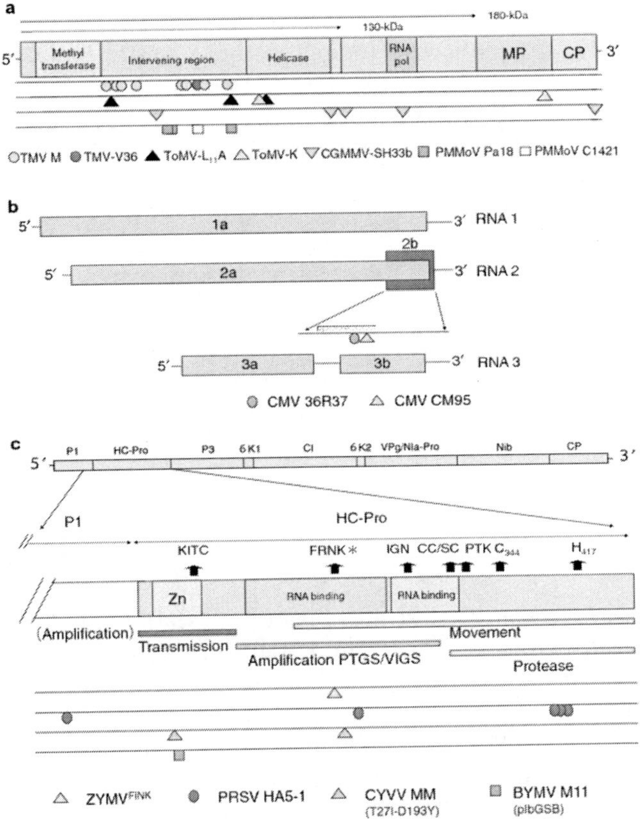

Figure 1. Location of amino acid substitutions among severe and attenuated strains in tobamovirus, cucumovirus and potyvirus. **A** Tobamovirus; **B** cucumovirus; **C** potyvirus (modified from Plisson et al. 2003). Only amino acid substitutions are described on the genome RNA. Three domains, methyltransferase, helicase and RNA polymerase (RNA pol) are shown in the 180-kDa ORF (Koonin 1991; Shintaku et al. 1996). MP and CP in **A** represent movement and coat proteins, respectively. *Dotted square* in **B** represents arginine-rich region in 2b protein (Goto et al. 2007). KITC, FRNK, IGN, CC/SC, C344 and H417 in **C** represent domains (single letter of amino acid and number of amino acids from N-terminal of HC-Pro) in HC-Pro (Plisson et al. 2003). *Asterisk* shows CDNQLD domain, 12 amino acids downstream from FRNK (Desbiez et al. 2010). *PTGS* posttranscriptional gene silencing, *VIGS* virus-induced gene silencing. *Tobacco mosaic virus* (TMV) M (Shintaku et al. 1996), TMV V-36 (Lewandowski and Dawson 1993), *Tomato mosaic virus* (ToMV-L_{11}A) (Nishiguchi et al. 1985), ToMV-K (Yang et al. 2002), *Cucumber green mottle mosaic virus* (CGMMV)-SH33b (Tan et al. 1997), *Pepper mild mottle virus* (PMMoV)-Pa18 (Ichiki et al. 2005), PMMoV-C1421 (Hagiwara et al. 2002), *Cucumber mosaic virus* (CMV) 36R37 (Kobori et al. 2005), CMV CM95 (Goto et al. 2007), *Zucchini yellow mosaic virus* (ZYMV)^{FINK} (Shiboleth et al. 2007), *Papaya ringspot virus* (PRSV) HA 5-1 (Chiang et al. 2007), *Clover yellow vein virus* (ClYVV) MM(T27I-D193Y) (Yambao et al. 2008), *Bean yellow mosaic virus* (BYMV) M11(pIbGSB) (Nakazono-Nagaoka et al. 2009)

Genomic characterization of an attenuated strain (ToMV-K) (Yang et al. 2002) showed that two mutations resulted in the production of UGA ochre and UAA opal nonsense codons in the 126-kDa replicase and MP genes, respectively, and that both mutations were shown to have roles in symptom attenuation by producing the truncated forms of the 126 kDa and MP. The attenuated phenotype of a masked strain (M) of TMV (Table 1) has been characterized in detail (Ding et al. 2004; Shintaku et al. 1996). The TMV-M 126-kDa protein is reported to have a reduced level of suppression of RNA silencing, while that of wild-type TMV-U1 has higher suppressive activity. TMV-M has eight amino acid substitutions in the 128-kDa protein. Each substitution more or less affects symptom attenuation (Ding et al. 2004).

An attenuated strain (SH33b) of CGMMV accumulated approximately two-thirds of the parental strain SH in melon plants. When the genome sequence of SH33b was compared with that of the parental SH, 14 nucleotide substitutions were found, plus three amino acid replacements in the 129-kDa replicase and one in the CP (Tan et al. 1997). Recently, the amino acid at position 479 (E–G) has been suggested to play a role in RNA silencing suppression activity (Chen et al. 2007).

The genomes of attenuated strains (Pa18 and C-1421) of PMMoV were also compared with a severe strain, PMMoV-J. As a result, valine (V) to alanine at aa position 649 in C-1421 and lysine to arginine (R) at aa position 549, V to threonine at aa position 556 and serine to leucine in Pa18 at aa position 760 were responsible for the attenuated phenotype (Hagiwara et al. 2002; Ichiki et al. 2005). Viral amplification of these mutants was 25% that of PMMoV-J. An artificial mutant that carried these four amino acid changes accumulated at a reduced level in infected plants in comparison to PMMoV-J. The attenuated strain of PMMoV engineered to contain the ToMV $L_{11}A$ type amino acid mutation Y or D at aa 348 or 760 in the corresponding position of the PMMoV 130-kDa protein was reported to induce attenuated symptoms (Tsuda et al. 2007), suggesting a common mechanism underlying symptom attenuation among tobamoviruses. However, symptoms were attenuated after infection with ToMV L that had a partial deletion in the 3′ noncoding region, but not in PMMoV with the similar deletion (Tsuda et al. 2007). These results indicate that mutations in the tobamovirus genome could have distinct effects on gene function, even if their genomes are highly conserved within the tobamoviruses.

Cucumovirus

CMV with a defective 2b protein caused mild symptoms (Ding et al. 1995). CMV 2b inhibits AGO1-mediated cleavage, implying that 2b is involved in biogenesis of the host's small RNA and affects symptom expression (Cillo et al. 2009; Lewsey et al. 2009; Zhang et al. 2006). A severe strain (CM95R) of CMV is a spontaneous mutant from an attenuated strain (CM95), which was selected through a series of local lesion isolations (Kosaka and Fukunishi 1997). The 2b protein encoded by CM95 differs in one amino acid residue from that of CM95R (Goto et al. 2007) (Figure 1b). The 2b protein is known to be involved in virus movement, virulence and posttranscriptional gene silencing (PTGS) suppressor (Brigneti et al. 1998; Shi et al. 2002; Soards et al. 2002). The 2b protein of CM95 showed weak PTGS suppressor activity compared with that of CM95R (unpublished results by Kobori and others in Goto et al. 2007) because of loss of the siRNA-binding property of 2b of CM95. The point mutation in the 2b of CR95 is responsible for binding to siRNA (Goto et al. 2007). Another attenuated strain of CMV, 36R37, also had a different amino acid substitution in 2b (Kobori et al. 2005) (Figure 1b). Other types of attenuated CMV strains harbor satellite RNA that affects viral pathogenesis. An attenuating satellite RNA has base substitutions and insertions in the region responsible for the necrotic response and an insertion/deletion in other regions (Sayama et al. 1993). The role of these base changes in attenuating symptoms remains to be resolved.

Potyvirus

As listed in Table 1 and Figure 1c, there are also reports of attenuated potyviruses (Chiang et al. 2007; Lin et al. 2007b; Yambao et al. 2008). When an attenuated strain (HA5-1) of PRSV was compared with its parent strain HA using an infectious cDNA clone, two amino acid substitutions among three in the C-terminal region of P1 at aa positions 309 and 481, and two of four in HC-Pro at aa positions 753 and 944 of HA5-1 appear to be critical in symptom attenuation in papaya (Chiang et al. 2007). In a comparison of attenuated and severe strains of *Clover yellow vein virus* (ClYVV), two amino acid substitutions in HC-Pro were found to be responsible for symptom attenuation (Yambao et al. 2008). Both mutations caused weaker RNA silencing suppression than the wild type did, suggesting a correlation between viral virulence and silencing suppression activity. In ZYMV, one amino acid substitution from R to isoleucine ($ZYMV^{FINK}$)was found in HC-Pro at aa position 180 in the FRNK motif of the attenuated strain AGII (Shiboleth et al. 2007). Other amino acid substitutions, phenylalanine to leucine and glutamic acid to asparagine at aa positions 205 and 396, respectively, also affected symptom expression in ZYMV (Lin et al. 2007b). A CDNQLD motif, located 12 amino acids downstream from the FRNK, also plays a role in symptom severity (Desbiez et al. 2010). The attenuated strain (GAC mutant) with two amino acid substitutions at aa positions 180 and 396 did not affect

RNA silencing suppression, implying that symptom attenuation does not always correlate with reduced RNA silencing suppression (Lin et al. 2007b). Although this strain conferred complete protection, the authors did not mention why. Because the viral accumulation kinetics for the attenuated strain AG were similar to those for the severe strain (Gal-On 2000). This similarity might be responsible for the cross protection ability. In studies with HC-Pro mutants of *Tobacco etch virus*, hyposuppression of RNA silencing was correlated with milder virulence, but hypersuppression did not correlate with stronger virulence than with the wild type (Torres-Barcelo et al. 2008). HC-Pro is known as a multifunctional protein: aphid transmissibility, viral cell-to-cell and long-distance movement, viral amplification, symptom expression, synergism with other viruses, interaction with 20S proteasome and cysteine type proteolysis, siRNA binding and silencing suppressor including miR-guided cleavage of target mRNAs (Maia et al. 1996; Plisson et al. 2003; Shiboleth et al. 2007). It is important to determine the structural requirements for RNA silencing suppressor activity, virus attenuation and cross protection.

MECHANISMS OF CROSS PROTECTION/INTERFERENCE

There are at least two mechanisms of cross protection: protein mediated and nonprotein mediated (Baulcombe 1996). In the former, the CP of a protecting virus present in cells may inhibit the uncoating step of an incoming virus. This case was first reported by Sherwood and Fulton (1982) who showed difficulty of infection by TMV in the yellow region of leaves with mosaic symptoms. Abel et al. (1986) showed delayed expression of symptoms after inoculation with TMV in CP-expressing transgenic tobacco. When this transgenic tobacco line was inoculated with TMV-RNA, no resistance was detected, indicating the involvement of CP in the uncoating step. There is a strong correlation between CP accumulation and resistance (Abel et al. 1986; Loesch-Fries et al. 1987) although the CP-mediated resistance (CPMR) mechanism is not yet fully understood. More recently, Bendahmane et al. (2007) suggested that there are at least two mechanisms for CPMR against TMV: interference by CP in the disassembly of an infecting virus and interference in the accumulation of viral replication complexes by CP. They stress that the stability of CP–CP interaction and the configuration of the quaternary structure would be important for conferring resistance.

Because a viroid, which has neither protein-coding capacity nor proteins (such as CP) that associate with the genome, can also confer cross protection/interference, there must be another mechanism or mechanisms than that mediated by the CP. One such mechanism is gene silencing by the host plant (Ratcliff et al. 1999). In addition to the dsRNA produced during the replication of RNA viruses, highly structured ssRNA is likely to trigger RNA silencing. A recent report, however, implied that theoretical highly structured RNA is not always a good substrate for small RNA synthesis (Qi et al. 2009). When plants are infected with a virus, RNA silencing will be induced in the host plants, and an incoming second virus, which is similar to the first virus, will be targeted for RNA degradation. However, such is not the case in PVX CPMR: the inhibition of RNA silencing did not affect resistance (Bazzini et al. 2006). Recently, Ziebell and Carr (2009) showed cross protection between CMV strains in RNA-silencing-impaired *Arabidopsis* mutants (*dcl2, 3* and *4*), suggesting that DCL2, 3 and 4 are not required for cross protection in this system. We cannot exclude the possibility, however, of the involvement of other gene/copy in this system.

Ziebell and Carr (2009) also found that the two CMV strains were mutually exclusive in the same infected region. Although the same phenomena have also been reported in *Potyvirus* and *Cheravirus*, the mechanism has not yet been elucidated (Dietrich and Maiss 2003; Takahashi et al. 2007). When these results are considered together, cross protection/interference is not likely due to a single mechanism but to multiple ones (Lin et al. 2007a). Which mechanism plays a major role in the protection would depend upon the particular system and conditions.

FROM MOLECULAR UNDERSTANDING TO PRACTICAL EVALUATION OF ATTENUATED-VIRUS-BASED CONTROL

Although several attenuated viruses have been used in disease control in commercial fields, their practical use is still under debate. For example, Hull (2002) points out that the use of a mild strain for protection is not recommended as a general practice because of (1) some levels of yield reduction, (2) risk as a virus reservoir for other crops, (3) acquisition of virulence by mutation, (4) serious damage from mixed infection with other viruses, (5) labor intensiveness and (6) risk of generating a new pathogenic virus by recombination with other viruses. Although these concerns are based on up-to-date knowledge from molecular biological studies, they are not always true for all virus systems. We emphasize that cross protection is still a practical means of choice for plant virus disease control, although it is recommended only under limited circumstances when no other protective means are available.

Nonetheless, the opposing view helps guide the development of new attenuated viruses having better practical properties. Addressing the issues raised by Hull (2002), we propose the following properties as criteria for a good attenuated virus strain for future development: (1) no symptoms or very mild symptoms are induced in any of the cultivated hosts, and the quality and quantity of the crop products are not reduced, (2) most host tissues are fully infected systemically, (3) they are highly stable genetically without mutating into a severe phenotype, (4) there is no vector transmission to other crops, (5) they protect against a wide range of viruses and strains, (6) the quality/quantity control of the inoculum and inoculations are easy and inexpensive. Of these criteria, genetic stability would be the most important factor for the practical use of attenuated viruses. Schneider and Roossinck (2001) reported that the diversity of RNA virus quasispecies depends on the host species where the viruses replicated. Therefore, the genetic stability of attenuated viruses should be examined in each of the hosts to which they will be applied. Because cross protection does not rely on harmful materials or chemicals such as pesticides and fungicides, it is suitable for sustainable agriculture, which is in high demand. Further molecular study and understanding of cross protection/interference should pave the way for broadening the practical applications of attenuated viruses.

ACKNOWLEDGMENTS

We wish to thank Dr. D. Murphy for his English correction. We are very grateful to Drs. K. Sayama, Y. Kosaka, T.U. Ichiki, C. Masuta and T. Yoneyama for their useful information and to Dr. M. Ishikawa for critical reading of part of the manuscript. This work was partially supported by the Ministry of Education, Culture, Sports, Science and Technology of Japan (MEXT), Grant-in-Aid for Scientific Research, 00465001, 2010 to M.N. and 21380032, 2010 to K.K.

REFERENCES

1. Abel PP, Nelson RS, De B, Hoffmann N, Rogers SG, Fraley RT, Beachy RN (1986) Delay of disease development in transgenic plants that express the tobacco mosaic virus coat protein gene. Science 232:738–743
2. Anonymous (1988) Control of melon mosaic disease by attenuated virus. Shizuoka Prefect Agric Fish Dep Rep. No. 1765, Shizuoka Pref. Gov. Shizuoka, pp 1–12
3. Asano M, Satoh R, Mochizuki A, Tsuda S, Yamanaka T, Nishiguchi M, Hirai K, Meshi T, Naito S, Ishikawa M (2005) Tobamovirus-resistant tobacco generated by RNA interference directed against host genes. FEBS Lett 579:4479–4484

4. Baulcombe DC (1996) Mechanisms of pathogen-derived resistance to viruses in transgenic plants. Plant Cell 8:1833–1844
5. Bazzini AA, Hopp HE, Beachy RN, Asurmendi S (2006) Posttranscriptional gene silencing does not play a significant role in Potato virus X coat protein-mediated resistance. Phytopathology 96:1175–1178
6. Béclin C, Berthomé R, Palauqui J-C, Tepfer M, Vaucheret H (1998) Infection of tobacco or Arabidopsis plants by CMV counteracts systemic post-transcriptional silencing of nonviral (trans)genes. Virology 252:313–317
7. Bendahmane M, Chen I, Asurmendi S, Bazzini AA, Szecsi J, Beachy RN (2007) Coat protein-mediated resistance to TMV infection of Nicotiana tabacum involves multiple modes of interference by coat protein. Virology 366:107–116
8. Brégeon D, Sarasin A (2005) Hypothetical role of RNA damage avoidance in preventing human disease. Mutat Res/Fundam Mol Mech Mutagen 577:293–302
9. Brigneti G, Voinnet O, Li W-X, Ji L-H, Ding S-W, Baulcombe DC (1998) Viral pathogenicity determinants are suppressors of transgene silencing in Nicotiana benthamiana. EMBO J 17:6739–6746
10. Chen H, Yamaoka N, Nishiguchi M (2007) Suppression of RNA silencing by 129K protein of Cucumber green mottle mosaic virus: comparison between severe and attenuated strains (abstract in Japanese). Jpn J Phytopathol 73:237
11. Chiang C-H, Lee C-Y, Wang C-H, Jan F-J, Lin S-S, Chen T-C, Raja JAJ, Yeh S-D (2007) Genetic analysis of an attenuated Papaya ringspot virus strain applied for cross-protection. Eur J Plant Pathol 118:333–348
12. Cillo F, Mascia T, Pasciuto MM, Gallitelli D (2009) Differential effects of mild and severe Cucumber mosaic virus strains in the perturbation of microRNA-regulated gene expression in tomato map to the 3′ sequence of RNA 2. Mol Plant Microbe Interact 22:1239–1249
13. Costa AS, Müller GW (1980) Tristeza control by cross protection: a US–Brazil cooperative success. Plant Dis 64:538–541
14. Csorba T, Bovi A, Dalmay T, Burgyán J (2007) The p122 subunit of Tobacco mosaic virus replicase is a potent silencing suppressor and compromises both small interfering RNA- and microRNA-mediated pathways. J Virol 81:11768–11780
15. Desbiez C, Girard M, Lecoq H (2010) A novel natural mutation in HC-Pro responsible for mild symptomatology of Zucchini yellow mosaic virus (ZYMV, Potyvirus) in cucurbits. Arch Virol 155:397–401
16. Díaz-Pendón JA, Ding SW (2008) Direct and indirect roles of viral suppressors of RNA silencing in pathogenesis. Annu Rev Phytopathol 46:303–326
17. Dietrich C, Maiss E (2003) Fluorescent labelling reveals spatial separation of potyvirus populations in mixed infected Nicotiana benthamiana plants. J Gen Virol 84:2871–2876
18. Ding S-W, Voinnet O (2007) Antiviral immunity directed by small RNAs. Cell 130:413–426
19. Ding S-W, Li W-X, Symons RH (1995) A novel naturally occurring hybrid gene encoded by a plant RNA virus facilitates long distance virus movement. EMBO J 14:5762–5772
20. Ding XS, Liu J, Cheng N-H, Folimonov A, Hou Y-M, Bao Y, Katagi C, Carter SA, Nelson RS (2004) The Tobacco mosaic virus 126-kDa protein associated with virus replication and movement suppresses RNA silencing. Mol Plant Microbe Interact 17:583–592
21. Fuchs M, Gonsalves D (2008) Safety of virus-resistant transgenic plants two decades after their introduction: lessons from realistic field risk assessment studies. Annu Rev Phytopathol 45:173–202
22. Gal-On A (2000) A point mutation in the FRNK motif of the potyvirus helper component-protease gene alters symptom expression in cucurbits and elicits protection against the severe homologous virus. Phytopathology 90:467–473
23. Gal-On A, Shiboleth YM (2006) Cross-protection. In: Loebenstein G, Carr JP (eds) Natural resistance mechanisms of plants to viruses. Springer, Berlin, pp 261–288

24. Goto T, Nemoto M (1971) Studies on control of plant virus diseases by interference of attenuated virus-selection of TMV attenuated strain and influence on various plants inoculated with the attenuated strain (in Japanese). Res Bull Hokkaido Natl Agric Exp Stn 99:67–76
25. Goto T, Iizuka N, Komochi S (1984) Selection and utilization of an attenuated isolate of pepper strain of tobacco mosaic virus (in Japanese with English summary). Ann Phytopathol Soc Jpn 50:221–228
26. Goto K, Kobori T, Kosaka Y, Natsuaki T, Masuta C (2007) Characterization of silencing suppressor 2b of cucumber mosaic virus based on examination of its small RNA-binding abilities. Plant Cell Physiol 48:1050–1060
27. Hagiwara K, Ichiki TU, Ogawa Y, Omura T, Tsuda S (2002) A single amino acid substitution in 126-kDa protein of Pepper mild mottle virus associates with symptom attenuation in pepper: the complete nucleotide sequence of an attenuated strain, C-1421. Arch Virol 147:833–840
28. Hagiwara-Komoda Y, Hirai K, Mochizuki A, Nishiguchi M, Meshi T, Ishikawa M (2008) Overexpression of a host factor TOM1 inhibits tomato mosaic virus propagation and suppression of RNA silencing. Virology 376:132–139
29. Holmes FO (1934) A masked strain of tobacco-mosaic virus. Phytopathology 24:845–873
30. Hughes Jd'A, Ollennu LAA (1994) Mild strain protection of cocoa in Ghana against cocoa swollen shoot virus—a review. Plant Pathol 43:442–457
31. Hull R (ed) (2002) Matthews' plant virology, 4th edn. San Diego, Academic Press
32. Ichiki TU, Nagaoka EN, Hagiwara K, Uchikawa K, Tsuda S, Omura T (2005) Integration of mutations responsible for the attenuated phenotype of Pepper mild mottle virus strains results in a symptomless cross-protecting strain. Arch Virol 150:2009–2020
33. Ieki H, Yamaguchi A, Kano T, Koizumi M, Iwanami T (1997) Control of stem pitting disease caused by citrus tristeza virus using protective mild strains in navel orange. Ann Phytopathol Soc Jpn 63:170–175
34. Kasschau KD, Carrington JC (1998) A counterdefensive strategy of plant viruses: suppression of posttranscriptional gene silencing. Cell 95:461–470
35. Kobori T, Ryang B-S, Natsuaki T, Kosaka Y (2005) A new technique to select mild strains of Cucumber mosaic virus. Plant Dis 89:879–882
36. Koonin EV (1991) The phylogeny of RNA-dependent RNA polymerases of positive-strand RNA viruses. J Gen Virol 72:2197–2206
37. Kosaka Y, Fukunishi T (1993) Attenuated isolates of soybean mosaic virus derived at a low temperature. Plant Dis 77:882–886
38. Kosaka Y, Fukunishi T (1997) Multiple inoculation with three attenuated viruses for the control of cucumber virus disease. Plant Dis 81:733–738
39. Kosaka Y, Ryang B-S, Kobori T, Shiomi H, Yasuhara H, Kataoka M (2006) Effectiveness of an attenuated Zucchini yellow mosaic virus isolate for cross-protecting cucumber. Plant Dis 90:67–72
40. Kubota K, Tsuda S, Tamai A, Meshi T (2003) Tomato mosaic virus replication protein suppresses virus-targeted posttranscriptional gene silencing. J Virol 77:11016–11026
41. Kurihara Y, Inaba N, Kutsuna N, Takeda A, Tagami Y, Watanabe Y (2007) Binding of tobamovirus replication protein with small RNA duplexes. J Gen Virol 88:2347–2352
42. Lewandowski DJ, Dawson WO (1993) A single amino acid change in tobacco mosaic virus replicase prevents symptom production. Mol Plant Microbe Interact 6:157–160
43. Lewsey M, Surette M, Robertson FC, Ziebell H, Choi SH, Ryu KH, Canto T, Palukaitis P, Payne T, Walsh JA, Carr JP (2009) The role of the Cucumber mosaic virus 2b protein in viral movement and symptom induction. Mol Plant Microbe Interact 22:642–654
44. Lin S-S, Henriques R, Wu H-W, Niu Q-W, Yeh S-D, Chua N-H (2007a) Strategies and mechanisms of plant virus resistance. Plant Biotechnol Rep 1:125–134

45. Lin S-S, Wu H-W, Jan F-J, Hou RF, Yeh S-D (2007b) Modifications of the helper component-protease of Zucchini yellow mosaic virus for generation of attenuated mutants for cross protection against severe infection. Phytopathology 97:287–296
46. Loesch-Fries LS, Merlo D, Zinnen T, Burhop L, Hill K, Krahn K, Jarvis N, Nelson S, Halk E (1987) Expression of alfalfa mosaic virus RNA 4 in transgenic plants confers virus resistance. EMBO J 6:1845–1851
47. Maia IG, Haenni A-L, Bernardi F (1996) Potyviral HC-Pro: a multifunctional protein. J Gen Virol 77:1335–1341
48. McKinney HH (1929) Mosaic diseases in the Canary Islands, West Africa and Gibraltar. J Agric Res 39:557–578
49. Motoyoshi F, Nishiguchi M (1988) Control of virus diseases by attenuated virus strains: comparison between attenuated strains of cucumber green mottle mosaic virus and tobacco mosaic virus. Gamma Field Symp Inst Radiat Breed Natl Inst Agrobiol Resour 27:91–109
50. Nagai Y (1987) Production of C-1421, an attenuated mutant of pepper strain of tobacco mosaic virus. Ann Phytopathol Soc Jpn 53:168–174
51. Nakazono-Nagaoka E, Takahashi T, Shimuzu T, Kosaka Y, Natsuaki T, Omura T, Sasaya T (2009) Cross-protection against Bean yellow mosaic virus (BYMV) and Clover yellow vein virus by attenuated BYMV isolate M11. Phytopathology 99:251–257
52. Nishiguchi M, Kikuchi S, Kiho Y, Ohno T, Meshi T, Okada Y (1985) Molecular basis of plant viral virulence: the complete nucleotide sequence of an attenuated strain of tobacco mosaic virus. Nucleic Acids Res 13:5585–5590
53. Oshima N (1981) Control of tomato mosaic disease by attenuated virus. Jpn Agric Res Q 14:222–228
54. Plisson C, Drucker M, Blanc S, German-Retana S, Le Gall O, Thomas D, Bron P (2003) Structural characterization of HC-Pro, a plant virus multifunctional protein. J Biol Chem 278:23753–23761
55. Prins M, Laimer M, Noris E, Schubert J, Wassenegger M, Tepfer M (2008) Strategies for antiviral resistance in transgenic plants. Mol Plant Pathol 9:73–83
56. Qi X, Bao FS, Xie Z (2009) Small RNA deep sequencing reveals role for Arabidopsis thaliana RNA-dependent RNA polymerases in viral siRNA biogenesis. PLoS ONE 4:e4971.
57. Rast ATB (1975) Variability of tobacco mosaic virus in relation to control of tomato mosaic in glasshouse tomato crops by resistance breeding and cross protection. Agric Res Rep 834:1–76
58. Ratcliff FG, MacFarlane SA, Baulcombe DC (1999) Gene silencing without DNA: RNA-mediated cross-protection between viruses. Plant Cell 11:1207–1215
59. Salaman RN (1933) Protective inoculation against a plant virus. Nature 131:468
60. Sasaki A (1974) Studies on Hassaku dwarf. Spec Bull Fruit Tree Exp Stn Hiroshima Pref 2:1–106
61. Sayama H (1996) Viral resistant tomato seedling production using attenuated cucumber mosaic virus (in Japanese). Plant Protect 50:20–25
62. Sayama H, Sato T, Kominato M, Natsuaki T, Kaper JM (1993) Field testing of a satellite-containing attenuated strain of cucumber mosaic virus for tomato protection in Japan. Phytopathology 83:405–410
63. Sayama H, Kominato M, Ubukata M, Sato T (2001) Three-year risk assessment of the practical application of cross-protection in processing tomato fields using an attenuated cucumber mosaic virus (CMV) strain containing an ameliorative satellite RNA. Acta Hortic 542:47–53
64. Schneider WL, Roossinck MJ (2001) Genetic diversity in RNA virus quasispecies is controlled by host–virus interactions. J Virol 75:6566–6571
65. Sherwood JL, Fulton RW (1982) The specific involvement of coat protein in tobacco mosaic virus cross protection. Virology 119:150–158

66. Shi B-J, Palukaitis P, Symons RH (2002) Differential virulence by strains of Cucumber mosaic virus is mediated by the 2b gene. Mol Plant Microbe Interact 15:947–955
67. Shiboleth YM, Haronsky E, Leibman D, Arazi T, Wassenegger M, Whitham SA, Gaba V, Gal-On A (2007) The conserved FRNK box in HC-Pro, a plant viral suppressor of gene silencing, is required for small RNA binding and mediates symptom development. J Virol 81:13135–13148
68. Shintaku MH, Carter SA, Bao Y, Nelson RS (1996) Mapping nucleotides in the 126-kDa protein gene that control the differential symptoms induced by two strains of tobacco mosaic virus. Virology 221:218–225
69. Soards AJ, Murphy AM, Palukaitis P, Carr JP (2002) Virulence and differential local and systemic spread of Cucumber mosaic virus in tobacco are affected by the CMV 2b protein. Mol Plant Microbe Interact 15:647–653
70. Takahashi T, Sugawara T, Yamatsuta T, Isogai M, Natsuaki T, Yoshikawa N (2007) Analysis of the spatial distribution of identical and two distinct virus populations differently labeled with cyan and yellow fluorescent proteins in coinfected plants. Phytopathology 97:1200–1206
71. Tan S-H, Nishiguchi M, Sakamoto W, Ogura Y, Murata M, Ugaki M, Tomiyama M, Motoyoshi F (1997) Molecular analysis of the genome of an attenuated strain of cucumber green mottle mosaic virus. Ann Phytopathol Soc Jpn 63:470–474
72. Tien P, Wu G (1991) Satellite RNA for the biocontrol of plant disease. Adv Virus Res 39:321–339
73. Torres-Barceló C, Martín S, Daròs J-A, Elena SF (2008) From hypo- to hypersuppression: effect of amino acid substitutions on the RNA-silencing suppressor activity of the Tobacco etch potyvirus HC-Pro. Genetics 180:1039–1049
74. Tsuda S, Kubota K, Kanda A, Ohki T, Meshi T (2007) Pathogenicity of Pepper mild mottle virus is controlled by the RNA silencing suppression activity of its replication protein but not the viral accumulation. Phytopathology 97:412–420
75. Vogler H, Akbergenov R, Shivaprasad PV, Dang V, Fasler M, Kwon M-O, Zhanybekova S, Hohn T, Heinlein M (2007) Modification of small RNAs associated with suppression of RNA silencing by tobamovirus replicase protein. J Virol 81:10379–10388
76. Yamanaka T, Ohta T, Takahashi M, Meshi T, Schmidt R, Dean C, Naito S, Ishikawa M (2000) TOM1, an Arabidopsis gene required for efficient multiplication of a tobamovirus, encodes a putative transmembrane protein. Proc Natl Acad Sci USA 97:10107–10112
77. Yambao MLM, Yagihashi H, Sekiguchi H, Sekiguchi T, Sasaki T, Sato M, Atsumi G, Tacahashi Y, Nakahara KS, Uyeda I (2008) Point mutations in helper component protease of Clover yellow vein virus are associated with the attenuation of RNA-silencing suppression activity and symptom expression in broad bean. Arch Virol 153:105–115
78. Yang G, Qiu B-S, Liu X-G, Li Y, Wang X-F (2002) Nonsense mutations of replicase and movement protein genes contribute to the attenuation of an avirulent tomato mosaic virus. Virus Res 87:119–128
79. Yeh S-D, Gonsalves D (1984) Evaluation of induced mutants of papaya ringspot virus for control by cross protection. Phytopathology 74:1086–1091
80. Yeh S-D, Gonsalves D, Wang H-L, Namba R, Chiu R-J (1988) Control of papaya ringspot virus by cross protection. Plant Dis 72:375–380
81. Zhang X, Yuan Y-R, Pei Y, Lin S-S, Tuschl T, Patel DJ, Chua N-H (2006) Cucumber mosaic virus-encoded 2b suppressor inhibits Arabidopsis Argonaute1 cleavage activity to counter plant defense. Genes Dev 20:3255–3268
82. Ziebell H, Carr JP (2009) Effects of dicer-like endoribonucleases 2 and 4 on infection of Arabidopsis thaliana by cucumber mosaic virus and a mutant virus lacking the 2b counter-defence protein gene. J Gen Virol 90:2288–2292

CHAPTER 2

Pathogen Infection Drives Patterns of Nutrient Resorption in Citrus Plants

Jirong Cao[1] , Chunzhen Cheng[2] , JunjieYang[1] & Qibing Wang[1]

[1] *State Key Laboratory of Vegetation and Environmental Change, Institute of Botany, Chinese Academy of Sciences, Beijing 100093, China.*

[2] *Institute of Horticultural Biotechnology, Fujian Agriculture and Forestry University, Fuzhou, Fujian 350002, China.*

ABSTRACT

Nutrient resorption processes in the plants infected by pathogen remain poorly understood. Huanglongbing (HLB) is a destructive disease of citrus. HLB-pathogen '*Candidatus* Liberibacter asiaticus' grows specifically in the phloem of hosts and may cause problems in the plant vascular system after infection. Therefore, it brings a great concern about the phloem nutrient transport and nutrient intra-cycling in HLB-affected plants. We investigated the effects of '*Ca.* L. asiaticus' infection on nitrogen (N) and phosphorus (P) concentrations and resorption in different citrus species (i.e. *Citrus reticulata, Citrus limon* and *Citrus maxima*). HLB-pathogen infection had distinctive impacts on nutrient resorption in different species. P resorption efficiency substantially decreased in infected *C. reticulata* plants relative to the healthy plants in summer, which may account for the marked decrease in the average fruit yield. P resorption was more efficient in infected *C. limon* plants than in the healthy plants. However, for *C. maxima* plants, HLB had no significant effects on N:P ratio in live leaves and resorption efficiency as well as on fruit yield. Keeping efficient internal nutrient cycling can be a strategy of citrus species being tolerant to HLB.

INTRODUCTION

Citrus Huanglongbing (HLB) disease is one of the most devastating diseases of citrus and is threatening citrus industry worldwide[1,2]. HLB-associated pathogens are the phloem-limited, Gram-negative bacteria named '*Candidatus* Liberibacter spp.', including the three species of '*Ca.* L. asiaticus', '*Ca.* L. americanus' and '*Ca.* L. africanus'[3]. These bacteria are transmitted chiefly by psyllids *Diaphorina citri* in Asia and America and by *Trioza erytreae* in Africa[4]. Timely monitoring of citrus trees for HLB symptoms is critical for early detection and management of the disease, but there are no effective measures to control HLB now. Removal of infected trees and treatments with insecticide against psyllids are the most common ways to control the spread of HLB pathogens locally[1,5].

One of the most distinct symptoms of HLB is the development of yellow shoots on infected trees[6]. HLB also induces other characteristic symptoms, such as blotchy mottle leaves and lopsided fruits with color inversion and aborted seeds[7]. However, it is difficult to diagnose HLB by symptoms. Some HLB symptoms can be masked by the symptoms of other diseases. In particular, the leaf symptom of '*Ca.* L. asiaticus'-infected plants is easy to be confused with the symptom caused by nutrient deficiency, especially in an early infected stage[8]. Therefore, much of research has attempted to find out the relations of nutrients and HLB symptom expression[9,10,11,12].

Nutrients are important factors in disease control, because nutrients affect plant, pathogen and microbial growth and their interactions. Dordas[13] reviewed literatures on the relationships between mineral nutrients and plant disease and concluded that most of essential nutrients influenced the severity of plant disease. It seems plausible that nutrient supplements could reduce the severity of disease symptom, yet there are many findings on both sides of this debate[13,14]. For instance, the severity of crown and root rot of tomato was reduced by supplements of nitrate-nitrogen and copper, but increased by ammonium-nitrogen, zinc and manganese[15]. Likewise, for HLB, some studies showed that the application of phosphorus (P) to infected citrus trees improved plant growth and yield[10,16]. However, Gottwald, Graham[12] revealed that the nutritional supplements (P and micronutrients) had no effects on growth, fruit quality and yield of '*Ca.* L. asiaticus'-infected trees. Furthermore, Zambrosi *et al.*[17] found that the great proportion of P in the flowers and fruits were from the remobilized P in citrus plants, and suggested that P resorption might be more important than P uptake to meet the demands of new vegetative and reproductive growth of citrus plants, even with adequate nutrient supply. Therefore, P resorption could play a vital role in the interaction between citrus plant and HLB-pathogen. However, little is known about the nutrient resorption in the plants infected by HLB pathogen. Nutrient resorption is a key strategy for conserving nutrients in plants. In this process, plants remove nutrients from senescing tissues and transport them through the phloem to other living tissues[18,19,20]. Phosphorus is readily phloem-mobile in plants[21]. It has been reported that callose deposition in phloem sieve tubes can restrict phloem transport in citrus plants infected with '*Ca.* L. asiaticus'[22,23]. Moreover, it was reported that '*Ca.* L. asiaticus' infection reduced P concentration by 35% in '*Ca.* L. asiaticus'-infected citrus plants compared with the healthy plants[10]. In this context, we hypothesized that (1) P would be a limiting nutrient in the citrus plants infected with '*Ca.* L. asiaticus', and (2) the HLB-pathogen infection would cause the change in P resorption in the host plant.

Few studies have focused on the responses of different citrus species to HLB to date. Besides, the effects of HLB on nitrogen (N) in the host plants are rarely reported[11]. In this study, we investigated the effects of '*Ca.* L. asiaticus' infection on the nutrient resorption in citrus plants among different species. We were interested in understanding how N and P concentrations in leaves changed responding to '*Ca.* L. asiaticus' infection, how the infection affected N and P resorption in citrus plants, and whether there were convergent responses of nutrient resorption to HLB among different citrus species.

RESULTS

Nutrients in live and senesced leaves

The '*Ca.* L. asiaticus' infection had significant effects on N and P concentrations in live leaves of citrus plants in our study. Even so, most of the N and P concentrations in live leaves of the '*Ca.* L. asiaticus'-infected plants still remained at optimum or high levels, according to the criteria shown in Table 1. There were two exceptions. Leaf N and P concentrations in infected *C. reticulata* plants in June dropped to the deficient and low levels. N concentrations in *C. limon* plants, which were deficient in the healthy trees as an unusual case, reached to the low level in the infected trees. Repeated measure-ANOVA based on mixed models shows that different species and sampling dates also have significant effects on concentrations of N and P in live leaves, and that there are interactive effects of species and health status on live-leaf N and P concentrations ($P < 0.0001$; Table 2). For example, in June, the concentrations of live-leaf N and P in *C. reticulata* plants significantly decreased in response to the HLB-pathogen infection, whereas those in *C. limon* plants remained unchanged (Figs 1a and 2a). In October, the P concentration of live leaves in infected *C. reticulata* plants recorded a marked increase ($P < 0.001$), while concentrations of live-leaf N and P in infected *C. limon* plants showed no significant changes, compared with those in the healthy plants (Figs 1b and 2b).

Table 1. Leaf N and P levels in healthy and '*Candidatus* Liberibacter asiaticus'-infected citrus plants.

Species	N		P	
	Healthy	Infected	Healthy	Infected
Jun.				
C. reticulata	Optimum	Deficient	Optimum	Low
C. limon	Deficient	Low	High	High
C. maxima	Optimum	Optimum	High	High
Oct.				
C. reticulata	Optimum	Optimum	Optimum	High
C. limon	Deficient	Low	High	High
C. maxima	Optimum	Optimum	High	High
Criteria*				
Deficient (%)	<2.2		<0.09	
Low (%)	2.2–2.4		0.09–0.11	
Optimum (%)	2.5–2.7		0.12–0.16	
High (%)	2.8–3.0		0.17–0.30	
Excess (%)	>3.0		>0.30	

Table 2. *P* values of repeated measures ANOVA for leaf nutrient variables and resorption parameters as dependent on species identity, health status and their interactions.

	Nliv	Nsen	NRE	Pliv	Psen	PRE	N:Pliv	N:Psen
Species	<0.0001	<0.0001	<0.0001	<0.0001	<0.0001	<0.0001	<0.0001	<0.0001
Status	<0.0001	<0.0001	<0.0001	0.0141	<0.0001	0.001	0.0259	0.9185
Species*Status	<0.0001	<0.0001	<0.0001	<0.0001	<0.0001	<0.0001	<0.0001	0.7334
Date	<0.0001	<0.0001	<0.0001	<0.0001	<0.0001	<0.0001	<0.0001	0.003
Species*Date	<0.0001	<0.0001	<0.0001	<0.0001	<0.0001	<0.0001	<0.0001	<0.0001
Status*Date	<0.0001	0.5720	<0.0001	<0.0001	0.0623	<0.0001	0.0020	0.0999
Species*Status*Date	<0.0001	0.0591	<0.0001	<0.0001	0.1727	<0.0001	<0.0001	0.6414

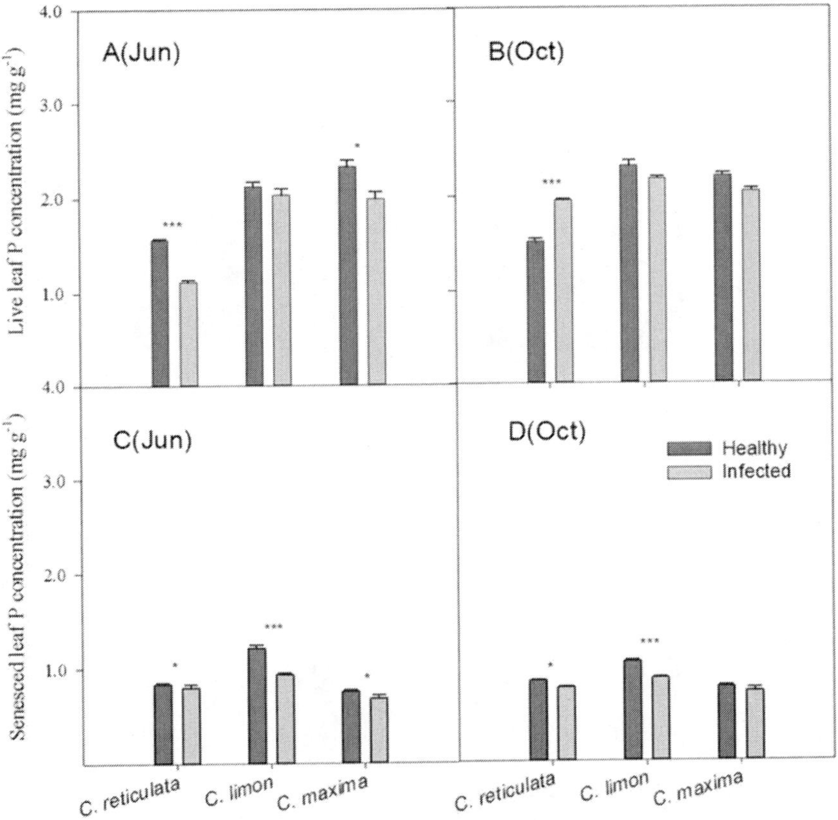

Figure 1. P concentrations in live- and senesced-leaves of different citrus species. Values are mean± Se, n=5. *, ** and *** above bars indicate significant differences at $P < 0.05$, 0.01 and 0.001, respectively, derived from the results of paired *t*-tests.

Concentrations of N and P in senesced leaves declined in infected *C. reticulata, C. limon* and *C.maxima* plants, compared with those in the healthy plants (Figs 1c,d and 2c,d). The effects of species and the interactive effects of species and health status on concentrations of N and P in senesced leaves were also significant ($P < 0.0001$; Table 2). In addition, there were no interactive effects of species, health status and sampling dates on N and P concentrations in senesced leaves (Table 2).

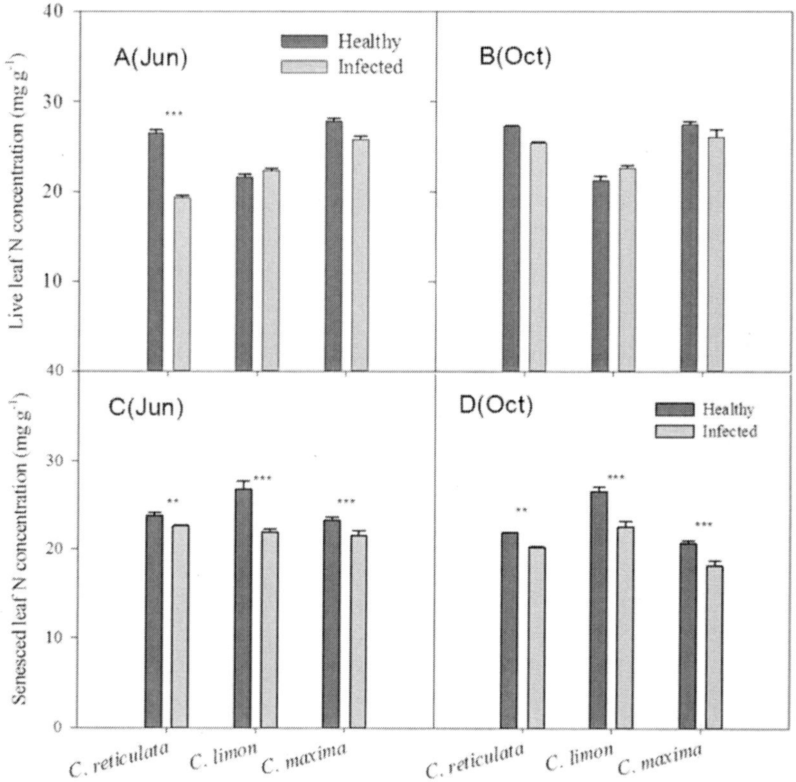

Figure 2: N concentrations in live- and senesced-leaves of different citrus species. Values are mean ± Se, n = 5. *, ** and *** above bars indicate significant differences at $P < 0.05$, 0.01 and 0.001, respectively, derived from the results of paired *t*-tests.

NUTRIENT RESORPTION EFFICIENCY

P resorption efficiency was significantly greater than N resorption efficiency across the three citrus species, regardless of '*Ca*. L. asiaticus' infection (matched pairs analysis $P < 0.0001$; Figure 3). The HLB-pathogen infection had marked impacts on P resorption efficiencies among different citrus species ($P = 0.001$ for HLB effects, $P < 0.0001$ for species effects, $P < 0.0001$ for their interactive effects; Table 2). P resorption efficiency recorded an overall increase in '*Ca*. L. asiaticus'-infected citrus plants compared with healthy plants ($P < 0.001$). Phosphorus resorption was more efficient in infected *C. limon* plants than in the healthy plants

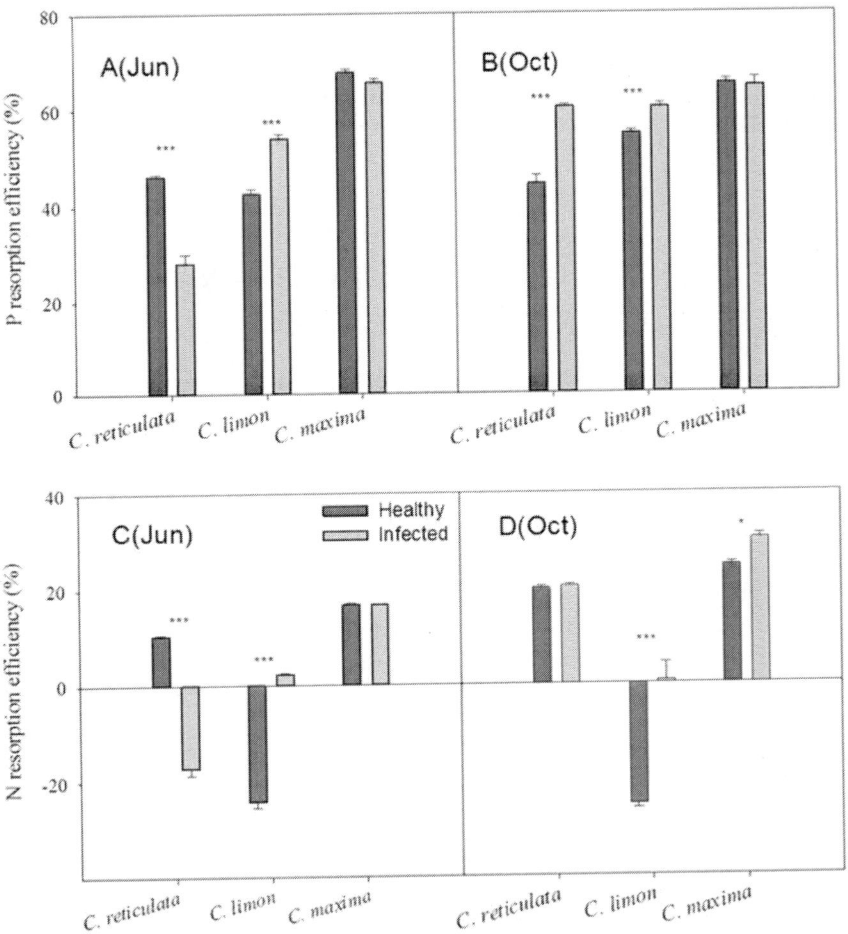

Figure 3. P (a,b) and N (c,d) resorption efficiencies of different citrus species. Values are mean±Se, n = 5. *, ** and *** above bars indicate significant differences at $P < 0.05$, 0.01 and 0.001, respectively, derived from the results of paired *t*-tests.

($P < 0.001$; Figure 3a,b). The P resorption efficiency in *C. maxima* plants showed no significant changes in response to '*Ca*. L. asiaticus' infection in summer and autumn. The fruit production of *C. maxima* plants was not affected by HLB (209.7 kg/tree in the healthy trees and 202.8 kg/tree in the infected trees). However, there were distinct seasonal patterns in P resorption efficiency in infected *C. reticulata* plants. Compared with the healthy plants, P resorption efficiency in HLB-positive *C. reticulata* plants substantially decreased in June, while it increased in October ($P < 0.001$; Figure 3a,b). N resorption efficiency was also significantly affected by the HLB-pathogen infection ($P < 0.0001$; Table 2) and presented similar patterns to P resorption efficiency (Figure 3c,d). Especially, N in infected *C. reticulata* plants showed accumulation in the senesced leaves in June. The average *C. reticulata* fruit yield dropped from 21.7 kg/tree in the healthy plants to 4.5 kg/tree in the infected plants.

CORRELATIONS OF NUTRIENT CONCENTRATION AND RESORPTION EFFICIENCY

P concentrations in live leaves were positively correlated with P resorption efficiencies ($r^2 = 0.44$ for healthy plants and $r^2 = 0.81$ for '*Ca.* L. asiaticus'-infected plants, $P < 0.0001$; Table 3). Differently, the negative correlation between P concentration in senesced leaves and P efficiency was found in healthy plants ($P < 0.0001$); however, this correlation was not significant in infected plants ($P = 0.137$). N concentrations in live leaves were also significantly correlated with N resorption efficiencies. Significant correlations between N concentrations in senesced leaves and N resorption efficiencies were observed in both healthy and infected plants.

Table 3. Pearson correlation coefficients of the measured variables (*, ** and *** denote the significant levels at $P < 0.05$, 0.01 and 0.001, respectively).

	Nsen		NRE		Pliv		Psen		PRE		N:Pliv		N:Psen	
	Healthy	Infected	Healthy	Infected	Healthy	Infected	Healthy	Infected	Healthy	Infected	Healthy	Infected	Healthy	Infected
Nliv	−0.744***	−0.528**	0.955***	0.930***	−0.314	0.659***	−0.880***	−0.407*	0.405*	0.845***	0.660***	−0.352	0.674***	0.058
Nsen			−0.895***	−0.758***	0.269	−0.265	0.748***	0.440*	−0.374*	−0.462*	−0.515**	0.105	−0.253	0.263
NRE					−0.339	0.634***	−0.894***	−0.443*	0.404*	0.832***	0.658***	−0.368*	0.555**	−0.075
Pliv							0.21	0.128	0.667***	0.898***	−0.913***	−0.931***	−0.024	−0.305
Psen									−0.572***	−0.278	−0.528**	−0.344	−0.817***	−0.738***
PRE											−0.361*	−0.730***	0.556**	−0.014
N:Pliv													0.295	0.414*

Resorption efficiencies of N and P were significantly correlated with one another ($r = 0.40$, $P = 0.027$ for healthy plants and $r = 0.83$, $P < 0.0001$ for infected plants; Table 3); so were concentrations of N and P in senesced leaves ($r = 0.75$, $P < 0.0001$ for the healthy and $r = 0.44$, $P = 0.015$ for the infected). A significant correlation between N and P concentrations in live leaves was detected in infected plants ($r = 0.66$, $P < 0.0001$), but was not in healthy plants ($r = -0.31$, $P = 0.091$). There were opposing effects of '*Ca.* L. asiaticus' infection on N:P ratios in live leaves, with a decrease in *C. reticulata* plants and an increase in *C. limon* plants ($P = 0.03$ and $P = 0.01$, respectively; Figure 4). However, the effects ofthe infection on N:P ratios in senesced leaves were not significant (Table 2). The N:P ratio in live leaves was significantly correlated with P resorption efficiency ($r = -0.36$, $P = 0.049$ for healthy plants and $r = -0.73$, $P < 0.0001$ for infected plants). The correlation of N:P in senesced leaves and P resorption efficiency was significant only in healthy plants (Table 3).

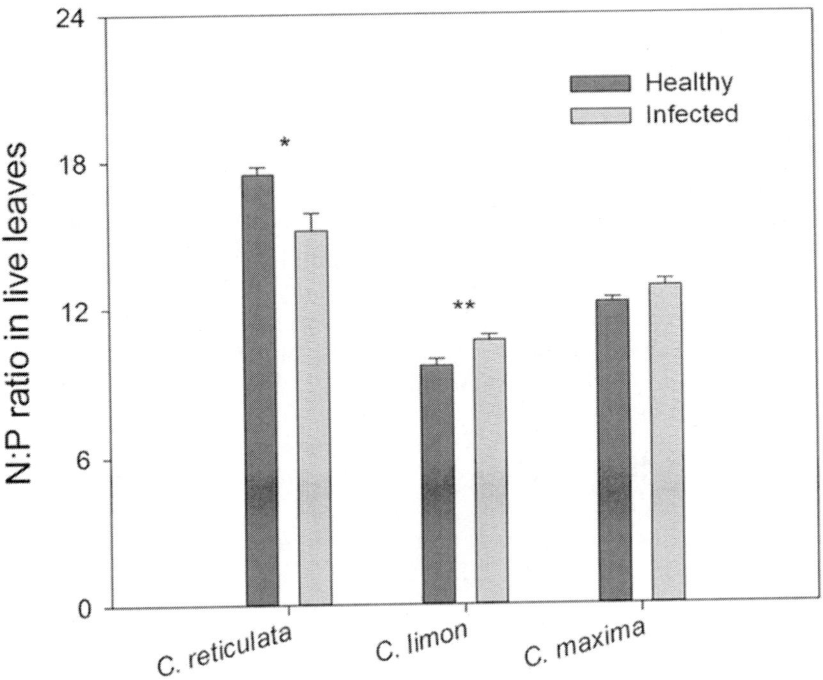

Figure 4. N:P ratios in live leaves of different citrus species. There were no significant differences in N:P ratio between sampling dates for all species, therefore the N:P ratios in June and October were compiled to calculate means ± Se (n = 10) for each species. * and ** above bars indicate significant differences at $P < 0.05$ and 0.01, respectively, derived from the results of paired *t*-tests.

DISCUSSION

HLB is a highly destructive citrus disease associated with the phloem-limited fastidious *'Ca.* Liberibacter spp.', bringing an unprecedented challenge to citrus fruit production throughout the world. There is no effective method available to control HLB at present. Therefore, it is urgent to understand the interaction of citrus and '*Ca.* L. asiaticus' to unravel the pathogenic processes and develop innovative strategies of disease control. To our knowledge, the study presented here is the first analysis of nutrient resorption in host plants responding to '*Ca.* L. asiaticus' infection.

Would citrus plants be P limited under HLB-pathogen infection?

A study associated with HLB found a substantial reduction of leaf P in '*Ca.* L. asiaticus'-positive citrus plants compared with healthy plants (*C. sinensis*)10. Accordingly, we hypothesized that P would be a limiting nutrient in '*Ca.* L. asiaticus'-infected citrus species (i.e. *C. reticulata*, *C. limon* and *C. maxima*). Our results were not completely in agreement with this hypothesis. For species *C. limon* and *C. maxima*, P concentrations in live leaves remain at optimum or even high levels in the plants infected with the HLB pathogen (Table 1). However, *C. reticulata* plants were P limited under the HLB-pathogen infection in summer. Mann *et al.*24 reported that P was deficient in '*Ca.* L. asiaticus'-infected citrus plants (*C. sinensis*). Our data also show that species, health status and their interaction have significant effects on the P in live leaves ($P < 0.0001$; Table 2). Therefore, the hypothesis was not suitable for all citrus species, at least not for *C. limon* and *C. maxima* in the present study. The differences in the responses of different citrus species to HLB, for instance, being sensitive, tolerant or resistant, might be the main reason for the inconsistency of the hypothesis and the results. Meanwhile, other factors such as seasons should be taken into account as well. The results from Folimonova *et al.*25 indicate that different species show their tolerance to HLB differently, and generally support our observation on the tolerance of the three citrus species. Our results suggest that the maintenance of optimum or higher P concentrations within the citrus plants can be a defensive strategy against the HLB-bacterial infection.

Would HLB-pathogen infection cause the change in P resorption in the host plant?

The HLB pathogen inhabits in a specialized niche, that is, the phloem sieve tubes, in host plants6. Electron microscope evidence illustrated the deposition of callose in the sieve tubes in citrus plants infected with '*Ca.* L.

asiaticus'[23,26]. The synthesis of pathogen-induced callose is a defense response of the plant immunity that is controlled by signaling pathways[27]. However, the excessive formation of callose in phloem sieve tubes can lead to the restriction of phloem transport in '*Ca.* L. asiaticus'-infected citrus plants (e.g. *C. reticulata, C. sinensis* and *C. paradisi*)[22,23,28]. Built upon these previous findings, a focus in our study was the effects of '*Ca.* L. asiaticus' infection on the nutrient resorption through phloem transport in citrus plants. Zhao, Sun[10] discovered that '*Ca.* L. asiaticus' infection in citrus plants induced miR399, a miRNA highly related to P starvation. Thus, we hypothesized that HLB-pathogen infection would cause the change in P resorption efficiency. Our results show that '*Ca.* L. asiaticus' infection has a marked impact on P resorption efficiency, with an overall increase in infected plants relative to healthy plants. Nutrient resorption is the process in which plants reallocate phloem-mobile nutrients from senescing tissues to storage sites[19,29]. Efficiency, as a measure of resorption, is uniquely featured by creating a linkage between nutrient demand (live-leaf nutrient concentration) and nutrient withdrawal (senesced-leaf nutrient concentration)[30]. Intuitively, the putative restriction in the phloem induced by '*Ca.* L. asiaticus' infection would hinder the resorption process, thus making the resorption less efficient. However, the interaction between plant immune system and pathogen is very complex[31], as was shown in our study.

In the study, we observed a significant negative correlation between P concentration of senesced leaf and P resorption efficiency in healthy plants (Table 3). This result indicates that the trend of P resorption proficiency (lower nutrient concentration in senesced leaves indicates higher nutrient proficiency) is consistent with that of P resorption efficiency in healthy plants. However, this linkage disappeared in '*Ca.* L. asiaticus'-infected plants (Table 3). It suggests that under '*Ca.* L. asiaticus' infection, the level to which infected plants reduced P in senesced leaves (measured as proficiency) can not predict the efficiency of reallocating P from senesced leaves to live leaves through phloem transport. Therefore, taking the complementary nature of resorption efficiency and proficiency into consideration as suggested by Killingbeck[29], we would possibly find a clue to know whether the P phloem transport was normal under the HLB-pathogen infection. The infection had distinctive impacts on the P resorption in the different species. The most remarkable was the pattern in *C. reticulata*. The P resorption efficiency decreased, whereas the P proficiency increased, in the infected *C. reticulata* plants compared with the healthy plants in summer. It turned out that the changes in resorption efficiency and proficiency were contradictory. The increase in P resorption proficiency in infected *C. reticulata* plants indicated that the removal of P from the senesced leaves actually happened. Coupled with the significant decrease in P concentration of live leaf in infected *C. reticulata* plants, it might be a cue revealing that P transport in phloem was abnormal and consequently the P withdrawn from the senesced leaves did not relocate to the live leaves. In fact, the process of nutrient resorption can be divided into two steps, that is, the removal of nutrients from senescing tissues and the transport of these nutrients to storage tissues in perennial plants[29]. Therefore, both steps are equally important. The induction of genes involved in plant defense against the HLB pathogen in *C. reticulata* plants indicates that *C. reticulata* is not completely vulnerable to the infection, though this species is considered to be susceptible to '*Ca*. L. asiaticus' infection by Albrecht and Bowman[32]. From physiological aspect, the results in our study agree on this count. In autumn, a substantial increase in the P resorption efficiency was observed in infected *C. reticulata* plants, relative to the healthy plants. These results to some extent can attribute to the favorable climate for better growth of plants. Our data show that season and the interaction of season and health status have significant effects on P concentration and resorption efficiency (Table 2). After the early summer (the first sampling), the increases of temperature and rainfall can be beneficial for the infected *C. reticulata* plants to overcome the temporary disorder in the P transport and to return to a greater P resorption[33,34,35]. The P resorption process in the infected *C. reticulata* plants reflected the dynamic interaction between P and the plant-pathogen system. Future study to elucidate the molecular mechanism underlying the HLB-affected phloem transport of P resorption is warranted.

Distinguishing from the pattern in *C. reticulata* plants, the P resorption efficiency in *C. maxima* plants was not affected by the HLB-pathogen infection. The resorption efficiency in infected *C. limon* plants were greater than that in the healthy plants in both seasons. These results suggest that P resorption may be an adaptive trait with great flexibility in some citrus species under the disease stress. Nutrients can affect disease susceptibility, tolerance and resistance in plants[14]. Disease resistance is the ability of a host plant to restrict or suppress the development and reproduction of the pathogen[36]. The tolerance of a host plant is regarded as the ability to maintain its growth and yield under pathogen infection[13]. The transcriptional profiling proves that many genes are differentially regulated in citrus following '*Ca*. L. asiaticus' infection[31,37], demonstrating that the host plant initiates its defense system to cope with the detrimental effects of the pathogen, instead of being passive in the plant-pathogen interaction. However, none of these genes are substantially expressed to suppress the pathogen development during the early stage of the infection[37]. To the best knowledge of authors, complete HLB resistance in citrus species has not been identified yet. However, there are some citrus species considered to be tolerant to HLB[25,38]. Disease tolerance is genetically controlled, and thus different species present variations in the tolerance to disease[39]. The results on *C. maxima* and *C. limon* provide a hint that some citrus species could balance tissue nutrient levels by regulating nutrient resorption to improve the tolerance of plant itself to HLB.

The significance of remobilized nutrients for fruit product

It has been shown that N and P can restrict the pathological effects induced by pathogen infection13. The changes in nutrient resorption efficiency will make impacts on other processes such as shoot growth and reproductive output, since nutrient resorption is one of the most important strategies employed by plants to conserve nutrients18. The results in our study imply that more efficient resorption of N and P may promote the growth of infected plants to counteract the destructive effects of HLB-pathogen infection. The process of nutrient resorption conserves nutrients in plants and reduces the dependence of plants on source in the soil29. The relative importance of nutrient uptake and resorption for growth varies in different plants. For citrus plants, P remobilization may be dominant over P uptake to meet the demands of new vegetative and fruit growth17. There is no doubt that fruit productivity is a major concern for citrus growers upon '*Ca.* L. asiaticus'-infection of citrus plants. The production of fruits requires prominent reallocation of nutrients from senescing leaves to reproductive tissues40, because fruits contain greater N and P concentrations than vegetative tissues41. Particularly, the remobilized P in citrus plants accounts for 72–97% of the P in flowers and fruits, even when P supply is sufficient17. In our study, the HLB-pathogen infection affected P resorption with important feedback on the plant productivity. The substantial decrease in P resorption efficiency in infected *C. reticulata* plants relative to the healthy plants in summer, may account for the marked decrease in the average fruit yield in autumn (21.7 kg/tree vs. 4.5 kg/tree in the healthy vs. the infected trees, respectively). However, for *C. maxima* plants, in which the HLB-pathogen infection did not significantly change the resorption efficiency, the fruit yield was not affected by HLB. These results can be understood as a strategy of species *C. maxima* being tolerant to HLB, i.e. keeping efficient internal nutrient cycling to sustain the fruit growth of the infected plants.

Effects of other factors on nutrient resorption

As discussed above, HLB-pathogen infection significantly drove the patterns in nutrient resorption in citrus plants. The effects in species and interactions of species and health status contributed to these patterns as well. In addition, live-leaf N:P ratio was significantly correlated with P resorption efficiency, accounting for 53% of the variance in P resorption efficiency in '*Ca.* L. asiaticus'-infected plants. Importantly, the balanced nutrition in plants can facilitate the expression of disease resistance or tolerance14. Güsewell42 reported that high N:P ratios could promote the tolerance of plant species under stress. In our study, the decreases in both live-leaf N:P and fruit yield in infected *C. reticulata* plants relative to the healthy plants indicate that species *C. reticulata* is sensitive to the HLB-pathogen infection. However, species *C. maxima* showed the tolerance to HLB, since HLB did not make an impact on live-leaf N:P and fruit yield in *C. maxima* plants. These results suggest that a delicately balanced nutrient system is critical for the host to encounter HLB-pathogen invasion.

In conclusion, to our knowledge, this study provides the first estimate of the nutrient resorption process under the specific circumstance, i.e. the HLB-bacterial infection. The findings in the study can help us to better understand the responses of different citrus species to HLB. Different citrus species exhibited the different abilities to tolerate the HLB-pathogen infection. Therefore, it is quite necessary for HLB research to focus on different citrus species. In our study, the citrus species displayed plastic responses to the infection in terms of nutrient resorption. The molecular mechanisms of the resorption process driven by HLB-pathogen infection deserve a further study.

METHODS

Study sites

This study was conducted in three groves growing with mandarin (*Citrus reticulata* Blanco cv. Shatangju), Eureka lemon (*Citrus limon* (L.) Burm. f.) and pamelo (*Citrus maxima* (Burm.) Merr. cv. Shatian Yu), respectively. The locations and general information of the groves are presented in Table 4. These citrus plants are the evergreen species that flush in early spring, summer and autumn, and produce ripened fruits in autumn.

Table 4. Description of the three sampling sites of this study in Guangdong, China.

Site	Species	Latitude	Longitude	Altitude (m)	MAT (°C)	MAP (mm)	pH	SOC (mg g^{-1})	TN (mg g^{-1})	AN (mg kg^{-1})	AP (mg kg^{-1})
Huizhou	*Citrus reticulata*	23°16.613′N	114°15.737′E	63	22.3	1897.0	4.62 (0.14)	10.17 (1.40)	0.930 (0.13)	23.82 (4.90)	18.73 (3.40)
Heyuan	*Citrus limon*	24°05.947′N	114°45.396′E	141	21.0	1742.0	4.38 (0.09)	11.49 (2.09)	0.927 (0.07)	28.77 (4.98)	4.27 (1.07)
Meizhou	*Citrus maxima*	24°26.070′N	116°05.853′E	178	21.3	1528.6	4.76 (0.57)	9.92 (2.35)	1.272 (0.22)	17.40 (2.89)	56.58 (6.13)

Sampling and laboratory analyses

In June 2013, the time of summer shoots immerging in the study areas, we visually identified five '*Ca.* L. asiaticus'-infected trees according to leaf symptoms and nearby five healthy trees as controls in each of the three groves. With the help of a skilled colleague, we carefully selected the healthy and HLB-pathogen infected trees following the method by Trivedi *et al.*43 to make sure that all the infected trees were in the similar stages of infection. The infected and healthy trees were further confirmed by leaf '*Ca.* L. asiaticus' detection using PCR analysis.

For leaf nutrient analysis, 30 fully expanded live leaves (full sun) and 30 freshly senesced leaves were collected randomly from the canopy of each tree and were separately packed into a brown-paper envelope. Soil samples (0–10 cm depth) were also collected under each tree, 30–50 cm away from the trunk. An auger of 3 cm in diameter was used to collect three cores per tree. The three cores were mixed into a single composite sample *in situ* for each tree, and packed into sealed plastic bags. All samples were kept in a cooler until transported to the laboratory.

In October 2013, the time when autumn shoots grew, we revisited the same trees that were sampled in June 2013, and collected leaf and soil samples for the same processing and analyzing as the summer samples (including PCR confirmation of '*Ca.* L. asiaticus'). At the sampling time, the fruits were ready to be harvested. We counted the number of all fruits on the sampling tree, and randomly picked 30 fruits (10 for *C. maxima*) and weighed them, in order to estimate the average fruit yield per tree (i.e. the average weight per fruit multiplies by the total number of fruits on each sampling tree). The fruit yield of *C. limon* trees were not recorded because of the harvest prior to the sampling.

The PCR analysis further confirmed that all the sampled healthy-trees were '*Ca.* L. asiaticus'-negative and all the sampled infected-trees were '*Ca.* L. asiaticus'-positive both in the summer and in the autumn.

Chemical analyses of leaf nutrients and soil property

All samples were transported to the laboratory at the day of collecting. The leaf samples were carefully cleaned using deionized water, dried at 65 °C for 48 h, and finely ground with a ball mill (NM200, Retsch, Haan, Germany). Total N in leaf was measured by dry combustion on a CN analyzer (vario El III, Elementar Analysensysteme GmbH, Germany). Total foliar P was analyzed using a concentrated sulfuric acid/hydrogen peroxide digest and an ascorbic acid molybdate colorimetric analysis on a Flow Injection Analysis (AA3, SEAL Analytical GmbH, Germany)44.

Field-moist soil samples were sieved through a 2 mm mesh size in order to remove plant materials, small insects and stones. A subsample of field-moist soil was stored in 4 °C refrigerator for determining soil water content, pH and available nutrients within a week. Soil water content was determined by drying for 24 h at 105 °C. Soil pH was determined in a 1:2 soil-to-water (w/v) ratio in deionized water. Available N (ammonium and nitrate) was extracted with 2 M KCl and measured by the AA345. Available P (Olsen's P) was extracted with 0.5 M $NaHCO_3$ (pH = 8.5) and analyzed on the AA3. The rest of soil samples were air-dried and ground to pass through a 100-mesh sieve. Soil organic C and total N concentrations were determined using the method for leaf N.

Leaf and soil nutrient concentrations were expressed on a dry mass basis.

Nutrient resorption calculations

Nutrient resorption is examined as resorption efficiency and resorption proficiency, following the descriptions from Killingbeck29 and Kobe *et al.*46. Resorption efficiency is calculated using the two equations below46.

$$[nutrient]_{sen} = A([nutrient]_{live})^B \qquad (1)$$

$$Res\ orption\ efficiency = 1 - \frac{A[nutrient]_{live}^B}{[nutrient]_{live}} \times 100\% \qquad (2)$$

where A and B are parameters estimated from the measured $[nutrient]_{live}$ and $[nutrient]_{sen}$ by regressions, $B \neq 1$, and $[nutrient]_{live}$ and $[nutrient]_{sen}$ represent nutrient concentrations in live leaves and senesced leaves, respectively46. Resorption proficiency is the level to which nutrients have been reduced in senesced leaves and is quantified using the senesced-leaf nutrient concentrations. Lower senesced-leaf nutrient concentration indicates higher nutrient resorption proficiency29.

Live leaves and senesced leaves were sampled with the consideration of similarity in order to avoid biasing caused by the possible reduction in mass and size during leaf senescence47,48. We estimated leaf weight by drying and leaf area using portable area meter (LI-3000A), and found that there was no significant difference in specific leaf area between live leaves and senesced leaves in healthy and *'Ca.* L. asiaticus'-infected citrus species. Thus, we analyzed nutrient resorption on a mass basis.

Statistical analysis

All data were tested for normality and homoscedasticity; if either assumption was violated, data were log transformed before analysis. We performed repeated measures analyses of variance based on mixed models to detect significant differences in each response variable through citrus health status, species and sampling dates. The significant differences in variables between *'Ca.* L. asiaticus'-infected and healthy citrus plants were analyzed using paired *t*-tests. Means were separated by Duncan's multiple comparison among the species. Pearson correlation was performed to examine the relationships between the variables measured. All statistical analyses were performed using SAS system for Windows V9.0 (SAS Institute, Cary, North Carolina, USA).

ADDITIONAL INFORMATION

How to cite this article: Cao, J. *et al.* Pathogen infection drives patterns of nutrient resorption in citrus plants. *Sci. Rep.* **5**, 14675;

REFERENCES

1. Callaway, E. Bioterror: the green menace. Nature 452, 148–150 (2008).
2. Wang, N. & Trivedi, P. Citrus Huanglongbing: a newly relevant disease presents unprecedented challenges. Phytopathol. 103, 652–665 (2013).
3. Sechler, A. et al. Cultivation of 'Candidatus Liberibacter asiaticus', 'Ca. L. africanus', and 'Ca. L. americanus' associated with Huanglongbing. Phytopathol. 99, 480–486 (2009).
4. Halbert, S. E. & Manjunath, K. L. Asian citrus psyllids (Sternorrhyncha: Psyllidae) and greening disease of citrus: a literature review and assessment of risk in Florida. Fla. Entomol. 87, 330–353 (2004).
5. Mishra, A., Karimi, D., Ehsani, R. & Albrigo, L. G. Evaluation of an active optical sensor for detection of Huanglongbing (HLB) disease. Biosyst. Eng. 110, 302–309 (2011).
6. Bove, J. M. Huanglongbing: a destructive, newly-emerging, century-old disease of citrus. J. Plant Pathol. 88, 7–37 (2006).
7. Bassanezi, R. B., Montesino, L. H. & Stuchi, E. S. Effects of Huanglongbing on fruit quality of sweet orange cultivars in Brazil. Eur. J. Plant Pathol. 125, 565–572 (2009).

8. Tian, S. et al. Spatial imaging of Zn and other elements in Huanglongbing-affected grapefruit by synchrotron-based micro X-ray fluorescence investigation. J. Exp. Bot. 65, 953–964 (2014).
9. Razi, M., Khan, I. A. & Jaskani, M. J. Citrus plant nutritional profile in relation to Huanglongbing prevalence in Pakistan. Pak. J. Agric. Sci. 48, 299–304 (2011).
10. Zhao, H. et al. Small RNA profiling reveals phosphorus deficiency as a contributing factor in symptom expression for citrus Huanglongbing disease. Mol. Plant 6, 301–310 (2013).
11. Nwugo, C. C., Lin, H., Duan, Y. & Civerolo, E. L. The effect of 'Candidatus Liberibacter asiaticus' infection on the proteomic profiles and nutritional status of pre-symptomatic and symptomatic grapefruit (Citrus paradisi) plants. BMC Plant Biol. 13, 59–82 (2013).
12. Gottwald, T. R., Graham, J. H., Irey, M. S., McCollum, T. G. & Wood, B. W. Inconsequential effect of nutritional treatments on Huanglongbing control, fruit quality, bacterial titer and disease progress. Crop Prot. 36, 73–82 (2012).
13. Dordas, C. Role of nutrients in controlling plant diseases in sustainable agriculture: a review. Sustain. Agric. 28, 33–46 (2008).
14. Huber, D. & Jones, J. The role of magnesium in plant disease. Plant Soil 368, 73–85 (2013).
15. Duffy, B. K. & Défago, G. Macro- and microelement fertilizers influence the severity of Fusarium crown and root rot of tomato in a soilless production system. Hortscience 34, 287–291 (1999).
16. Pustika, A. B. et al. Interactions between plant nutrition and symptom expression in mandarin trees infected with the disease Huanglongbing. Aust. Plant Dis. Notes 3, 112–115 (2008).
17. Zambrosi, F. B. et al. Contribution of phosphorus (32P) absorption and remobilization for citrus growth. Plant Soil 355, 353–362 (2012).
18. Chapin I., F. S. The Mineral Nutrition of Wild Plants. Annu. Rev. Ecol. Syst. 11, 233–260 (1980).
19. Aerts, R. Nutrient resorption from senescing leaves of perennials: Are there general patterns? J. Ecol. 84, 597–608 (1996).
20. Yuan, Z. & Chen, H. Y. H. Global trends in senesced-leaf nitrogen and phosphorus. Global Ecol. Biogeogr. 18, 532–542 (2009).
21. Jeschke, W. D. & Pate, J. S. Mineral nutrition and transport in xylem and phloem of Banksia prionotes (Proteaceae), a tree with dimorphic root morphology. J. Exp. Bot. 46, 895–905 (1995).
22. Kim, J. S., Sagaram, U. S., Burns, J. K., Li, J. L. & Wang, N. Response of sweet orange (Citrus sinensis) to 'Candidatus Liberibacter asiaticus' infection: microscopy and microarray analyses. Phytopathol. 99, 50–57 (2009).
23. Koh, E. J. et al. Callose deposition in the phloem plasmodesmata and inhibition of phloem transport in citrus leaves infected with "Candidatus Liberibacter asiaticus". Protoplasma 249, 687–697 (2012).
24. Mann, R. S. et al. Induced release of a plant-defense volatile 'deceptively' attracts insect vectors to plants infected with a bacterial pathogen. Plos Pathog. 8, 1–13 (2012).
25. Folimonova, S. Y., Robertson, C. J., Garnsey, S. M., Gowda, S. & Dawson, W. O. Examination of the responses of different genotypes of citrus to Huanglongbing (citrus greening) under different conditions. Phytopathol. 99, 1346–1354 (2009).
26. Etxeberria, E., Gonzalez, P., Achor, D. & Albrigo, G. Anatomical distribution of abnormally high levels of starch in HLB-affected Valencia orange trees. Physiol. Mol. Plant Pathol. 74, 76–83 (2009).
27. Luna, E. et al. Callose deposition: a multifaceted plant defense response. Mol. Plant-Microbe Interac. 24, 183–193 (2011).
28. Xu, M. et al. Preliminary research on soil conditioner mediated citrus Huanglongbing mitigation in the field in Guangdong, China. Eur. J. Plant Pathol. 135, 1–11 (2013).

29. Killingbeck, K. T. & Whitford, W. G. Nutrient resorption in shrubs growing by design, and by default in Chihuahuan Desert arroyos. Oecologia 128, 351–359 (2001).
30. Killingbeck, K. T. Nutrients in senesced leaves: keys to the search for potential resorption and resorption proficiency. Ecology 77, 1716–1727 (1996).
31. Alexander, H. M. & Holt, R. D. The interaction between plant competition and disease. Perspect. Plant Ecol. 1, 206–220 (1998).
32. Albrecht, U. & Bowman, K. D. Transcriptional response of susceptible and tolerant citrus to infection with Candidatus Liberibacter asiaticus. Plant Sci. 185, 118–130 (2012).
33. Coakley, S. M., Scherm, H. & Chakraborty, S. Climate change and plant disease management. Annu. Rev. Phytopathol. 37, 399–426 (1999).
34. Walters, D., Walsh, D., Newton, A. & Lyon, G. Induced resistance for plant disease control: maximizing the efficacy of resistance elicitors. Phytopathol. 95, 1368–1373 (2005).
35. Huang, J., Yu, H., Li, L., Yuan, Z. & Bartels, S. Water supply changes N and P conservation in a perennial grass Leymus chinensis. J. Integr. Plant Biol. 51, 1050–1056 (2009).
36. Graham, R. & Webb, M. Micronutrients and disease resistance and tolerance in plants in Micronutrients in agriculture 2nd edn (eds Mortvedt, J. et al.) 329–370 (Soil Sci Soc America, 1991).
37. Albrecht, U. & Bowman, K. D. Gene expression in Citrus sinensis (L.) Osbeck following infection with the bacterial pathogen Candidatus Liberibacter asiaticus causing Huanglongbing in Florida. Plant Sci. 175, 291–306 (2008).
38. Albrecht, U. & Bowman, K. D. Tolerance of the trifoliate citrus hybrid US-897 (Citrus reticulata Blanco × Poncirus trifoliata L. Raf.) to Huanglongbing. Hortscience 46, 16–22 (2011).
39. Agrios, N. G. Plant Pathology. 5th edn. (Elsevier, 2005).
40. Tully, K. L., Wood, T. E., Schwantes, A. M. & Lawrence, D. Soil nutrient availability and reproductive effort drive patterns in nutrient resorption in Pentaclethra macroloba. Ecology 94, 930–940 (2013).
41. Abrahamson, W. G. & Caswell, H. On the comparative allocations of biomass, energy, and nutrients in plants. Ecology 63, 982–991 (1982).
42. Güsewell, S. N. P ratios in terrestrial plants: variation and functional significance. New Phytol. 164, 243–266 (2004).
43. Trivedi, P. et al. Huanglongbing alters the structure and functional diversity of microbial communities associated with citrus rhizosphere. ISME J. 6, 363–383 (2012).
44. Robertson, G. P. et al. Soil carbon and nitrogen availability: nitrogen mineralization, nitrification, and soil respiration potentials in Standard Soil Methods for Long-term Ecological Research (eds Robertson, G. P. et al.) Ch. 13, 258–290 (Oxoford University Press, 1999).
45. Johnson, N. C. Resource stoichiometry elucidates the structure and function of arbuscular mycorrhizas across scales. New Phytol. 185, 631–647 (2010).
46. Kobe, R. K., Lepczyk, C. A. & Iyer, M. Resorption efficiency decreases with increasing green leaf nutrients in a global data set. Ecology 86, 2780–2792 (2005).
47. Van Heerwaarden, L. M., Toet, S. & Aerts, R. Current measures of nutrient resorption efficiency lead to a substantial underestimation of real resorption efficiency: facts and solutions. Oikos 101, 664–669 (2003).
48. Lü, X. et al. Convergent responses of nitrogen and phosphorus resorption to nitrogen inputs in a semiarid grassland. Global Change Biol. 19, 2775–2784 (2013).

CHAPTER 3

Geographical Distribution and Incidence of Cassava Bacterial Blight (Manihot esculenta Crantz) Caused by Xanthomonas axonopodis pv. manihotis in Two Agro-ecological Zones of Cote d'Ivoire

Affery Arthur Martin, Abo Kouabenan, Tuo Seydou, N'Zué Boni and Koné Daouda

[1] *Laboratory of Plant Physiology, Department of Biosciences, University Félix Houphouët-Boigny of Cocody-Abidjan, 22 B.P. 582, 22 Abidjan, Côte d'Ivoire*

[2] *Laboratory of Plant Pathology and Plant Biology, Department of Training and Research in Agriculture and Animal Resources, Institute National Polytechnique Félix Houphouët-Boigny, B.P. 1313, Yamoussoukro, Côte d'Ivoire*

[3] *National Center for Agronomic Research (CNRA), 01 B.P. 1740, 01 Abidjan, Côte d'Ivoire*

ABSTRACT

Background and Objective: Cassava (Manihot esculenta Crantz) is the second largest food crop in Côte d'Ivoire after yam. It is cultivated for its fleshy roots rich in starch and for the richness of its leaves in minerals (calcium, phosphorus and iron). However, this crop is subject to numerous biotic constraints of which one of the most formidable is bacterial blight caused by Xanthomonas axonopodis pv. manihotis. In Côte d'Ivoire, no concrete study has been conducted on this emerging disease with the aim of reducing its impact on the yield of tuberous roots of traditional and improved varieties. The objective of this study is to study the distribution of bacterial blight on the one hand and to assess the sensitivity of 8 varieties through an epidemiological study in two agro-ecological zones of Côte d'Ivoire on the other hand. Materials and Methods: Survey missions conducted in 2013 and 2014 have helped establish the health map of the different cassava production zones, followed by an epidemiological study of the disease with 8 cassava varieties of which four are improved, in two cassava production zones (Yamoussoukro and Ferkessedougou). Results: The results showed that bacterial blight was observed in all agro-ecological zones with an incidence ranging from 2.17-66.67%. As for the epidemiological study, it showed a contrast between the two areas of study. The severity of the disease was very low in the dry season (December-March) corresponding to the 7th until the 10th month of this study. Moreover, the rainy season contributed to an expansion of the disease. Diarrassouba and Yace (traditional varieties) and Bocou 1 (improved variety) were the most susceptible to the disease. Conclusion: A fight against this disease must be envisaged before it spreads to all cassava cultivation zones in Côte d'Ivoire.

INTRODUCTION

Cassava (*Manihot esculenta* Crantz, Euphorbiaceae) is an important crop in Africa, where it contributes to food security in many rural and urban households[1,2]. It can be used in a wide spectrum of industrial activities, including food industry, pharmaceutical laboratories, textile, paper and glue-making industry, as well as raw material for the production of biofuel ethanol[3]. It is also consumed in several forms[4,5]. In the early 1980s, following the fall in prices of agricultural raw materials such as cocoa and coffee on the international market, many farmers in the tropical zone engaged in cassava production not only for self-consumption of producing households, but also for marketing[6]. Cassava cultivation is less demanding in terms of input. Its cultivation is suitable for poor soils and arid climates.

This hardiness, combined with its multiple food uses makes it an important food crop for food security[7,8].

In 2013, the annual national yield of cassava in Côte d'Ivoire, which was estimated at 2,436,496 t, rose to 4,239,303 t in 2014 with 500,000 ha of cultivated area[9]. However, diseases and pests are the main causes of decline in yields of this crop. Thus, the damage caused by virus diseases and *Xanthomonas axonopodis* pv. *manihotis* bacterial blight can be considered as the main yield-limiting factors in all cassava cultivation zones[10,11]. Cassava bacterial blight caused by *Xanthomonas axonopodis* pv. *manihotis*, appears as the most important restrictive bacteria for this crop[10,12]. Bacterial blight was reported for the first time in Latin America in 1912 and appeared in Nigeria and Africa, in 1972 and is now extending to many countries of the African continent[13]. In most severe cases, bacterial blight can cause a total loss in tuberous roots and foliage, a reduction in starch content and a destruction of planting material. Decrease in yield vary[14] from 20-100%. This bacterial disease, however could have significant economic consequences if preventive measures are not adopted by farmers. The use and exchange of plant material between producers may be an important source of disease spreading. Faced with this situation, studies are needed so as to prevent or control the spread of this bacterial disease in Côte d'Ivoire. Thus, the assessment of the distribution and incidence of bacterial blight in farmers' fields Côte d'Ivoire are all preliminary work to be carried out. This study therefore focused on the assessment of the distribution and incidence of bacterial blight in cassava cultivation zones in Côte d'Ivoire. It consisted in assessing the behavior of 8 cassava varieties on-farm vis-à-vis *Xanthomonas axonopodis* pv. *manihotis* in two distinct agro-ecological zones of the country. In that respect, two experimental plots were established in 2014 in the center in Yamoussoukro and in the North in Ferkéssédougou respectively.

Materials and Methods

Two studies were carried out, on the one hand by performing a survey aiming at quantifying the importance of infections by bacterial blight in the main cassava cultivation zones of Côte d'Ivoire while identifying the heavily infested zones that can serve on the other hand as sites for cassava variety screening tests. Thus, two agro-ecological zones were selected basing on infection rate and ease of access.

Localization and description of experimental sites on-farm: The experiments were conducted in two agro-ecological zones namely the agro-ecological zones (AEZ) IV and VI. One of the trials was set up on the plot with the following coordinates: 005°135.54' North latitude, 06°479.42' West longitude and 232 m altitude. This test is located in central Cote d'Ivoire, in the department of Yamoussoukro in AEZ IV. As for the second test, its set up occurred in Koutiénédougou at 005°159.17' North latitude, 09°265.28' West longitude and 380 m altitude, in the department of Ferkéssédougou in Northern Côte d'Ivoire in AEZ VI. These test plots were chosen basing on the level of infection by bacterial blight, land availability, accessibility, topography, climate and **SOIL TYPE**. Both trials were conducted in farmers' fields from 2014-2015. The agro-ecological zones (AEZ) where these behavioral tests were conducted are in the dense humid and semi-deciduous forest and the humid tropical Savannah respectively. The AEZ IV has an altitude ranging from 0-200 m, with a rainfall regime showing a bimodal shape and an annual average oscillating between 1300 and 1750 mm while that of AEZ VI varies from 300-500 mm, with an annual rainfall ranging[15] between 1150 and 1350 mm.

The study area of Yamoussoukro is located in the center of Côte d'Ivoire. The climate of the region corresponds to the attenuated transitional equatorial regime (Baouléen climate) characterized by four seasons of which two rainy seasons (long rainy season from March-June and short rainy season from September-October) and two dry seasons (short dry season from July-August and very marked long dry season from November-February, however with some isolated rainfalls). The interannual average temperature of the Yamoussoukro region is about 26°C. The relative humidity varies from 75-85% with falls at 40% during the harmattan period. The region is marked by a succession of several **SOIL TYPE**s ranging from reddish soils that are more or less gravelly, having fine to medium texture and well-draining to yellow-brownish or brownish soils that are more or less gravelly having medium texture and quick-draining[16].

The department of Ferkéssédougou is characterized by a tropical sub-humid or sub-Sudanese transition climate, between the equatorial and hot desert type[17]. It has two seasons: A rainy one, extending from March-October and a dry one, from November-February. Between mid-November and February, the harmattan, a hot dry wind of Saharan origin from the Northern sector, blows and influences the climate of the region. The average daily temperature reaches 27°C with generally low thermal shifts except during the period marked by the harmattan (15-20°C) which extends from mid-November to February[18]. In this region, there are gallery forests along streams, consisting of wooded and shrub Savannah vegetation[19,20]. It consists mainly of remolded ferralitic soils, strongly to moderately desaturated. Very heterogeneous in space, these soils are characterized by a strong tendency to induration and sensitivity to compaction[18].

Plant material: In order to study the distribution and incidence of bacterial blight, the surveys focused on cassava plants (*Manihot esculenta*) of cultivated local and improved varieties. As for the study of the varietal behavior of cassava plants on-farm, three improved varieties originating from the National Center for Agronomic Reserach (CNRA), one improved variety originating from the Swiss Center for Scientific Research (CSRS) in Côte d'Ivoire and four other traditional (local) ones were under on-farm trial in two distinct agro-ecological zones. These included Bocou 1, Bocou 2, Bocou 3 and Yavo for improved varieties and Yacé, Akama, Diarrassouba and Dankwa and for local varieties (Table 1).

Survey and sampling method: Farmers' cassava fields were visited between July and August, 2013 and 2014, in six of the seven agro-ecological zones of Côte d'Ivoire. The observations were carried out during the rainy season of the year, corresponding to the most favorable period for the observation of symptoms. Only agro-ecological zone VII (AEZ VII) corresponding to the Northeast, where cassava cultivation is low, could not be visited during these surveys. Agro-ecological zones were defined by Halle and Bruzon[15]depending on weather conditions, **SOIL TYPE**, importance of the predominant ecosystem and major biotic and abiotic constraints of the crop. In each agro-ecological zone, several cities were visited. Several fields were observed in each city. The number of plots visited in a city was based on the phenological stages of cassava plants. The first field of each city was visited. Then the other prospection were carried out each time a distance of at least 30 km was covered. The observations were made along the main roads linking cities from North-South and from East-West. The surveys concerned only the seedlings of 3-6 months and 8-11 months after planting and dominant varieties of the visited plot. In each parcel visited, observations were made according to diagonals. The parameters rated for each plot were the name of the city of the prospected plot, the geographical coordinates and altitude, the age of the plants, the estimation of the plot surface area and the rating of phytosanitary parameters. Geographical coordinates and altitude were taken using a GARMIN GPS.

For the rating of phytosanitary parameters, 15 plants were randomly selected and rated on a scale of 1-5 according to the method recommended by Boher and Agbobli[21] and IITA[22].

During these surveys, leaves or stems of cassava plants showing symptoms of the disease were removed and placed into khaki envelopes and then stored at room temperature before being sent to the laboratory for isolation of the pathogen. Confirmation of the presence of the pathogen is obtained by isolation, inoculation and observation of symptoms.

Table 1: Characteristics of four traditional varieties of Côte d'Ivoire and four improved varieties disseminated

Varieties	Cycle (months)	Average yield (t ha⁻¹)	Characteristics	Common usage
Yace	11-18	20	Variety sensitive to mosaic, mites and scale insects, Dry matter content 40% Good for processing into attiéké	Attiéké, placali
Diarrassouba	*	*	Sensitive to mosaic, mites, bacterial blight and scale insects	Attiéké, placali, foutou
Dankwa	*	*	Study in progress	Attiéké, placali
Akama	*	*	Sensitive to mosaic, mites, bacterial blight and scale insects	Attiéké, placali, foutou
Bocou 1 (CNRA)	12-20	25	Variety having a very good vegetation cover Sensitive to mites High yield Dry matter content 39%	Attiéké, placali, foutou
Bocou 2 (CNRA)	11-16	25	Variety having a very good vegetation cover Sensitive to mites Sensitive to root rot Easy harvesting High yield Dry matter content 38%	Attiéké, placali, foutou
Bocou 3 (CNRA)	12-16	25	Variety having a very good vegetation cover Virus disease-tolerant Sensitive to root rot High yield Dry matter content 37%	Attiéké, placali
Yavo (CSRS)	*	*	Variety having a good vegetation cover	Attiéké, placali, foutou

N'Zue *et al.*[14] modified, *Undetermined

Distribution and incidence of bacterial blight in Côte d'Ivoire: The study of the distribution of bacterial blight was conducted on the basis of the disease symptoms (bacterial blight) observed in the plots visited. It consisted in rating the presence of bacterial blight and disease severity.

In order to determine the distribution of cassava bacterial blight in the production areas on a map, a classification of the agro-ecological zones prospected was made from medium impact data. Thus, the agro-ecological zones were grouped into four classes: incidence (1: 0-15%, 2: 15-35%, 3: 35-55% and 4: >55%).

Preparation of plots and planting material: In view of their size which does not require the use of agricultural machinery and wishing to remain in producers' conditions, the experimental plots were plowed manually using a hoe. Cuttings measuring 25 cm long having 4-6 germ nodes were used as seed. On each experimental plot, the different cassava varieties were planted the same day. The plots were weeded each using a hoe 4-5 times throughout the duration of the study.

Experimental designs: The **EXPERIMENTAL DESIGN** consisted of a three-block Fisher design (3 repetitions) spaced 2 m from each other. The trials were set up during the crop year 2014 in Yamoussoukro and Ferkéssédougou, respectively. The **EXPERIMENTAL DESIGN**s were bi-factorial of three blocks of which the city represented the first factor and cassava varieties the second factor. The main plot (Block) represented respectively an area of 8000 m^2 in Ferkéssédougou and Yamoussoukro. It was subdivided into three sub-blocks of 2500 m^2. The sub-blocks were separated by a distance of 2 m used as path. Each sub-block was subdivided into 8 plots (elementary) of 300 m^2, which were separated by an interval of 1.5 m and each of them contained a variety of cassava. In each elementary plot, 10 lines of 30 feet were noted and cassava plants were arranged at 1×1 m with 224 feet in the usable area. This usable area did not include plants that were placed within the border lines of each elementary plot. Each variety was represented once in each of the three sub-blocks. No phytosanitary treatment was applied throughout this study. The cuttings were obliquely planted on mounds in Ferkessédougou and on flat land in Yamoussoukro.

Symptomatology of cassava bacterial blight: Two months after planting the cuttings, 30 cassava plants of each elementary plot were selected and marked with tags. From that period, the trials were monitored so as to observe the possible presence of typical symptoms of bacterial blight. The observations were made monthly in two crop cycles on both sites. The symptoms were photographed and then described. Bacterial blight development monitoring was made according to a visual diagnosis based on the observation of characteristic symptoms appearing on the leaves and stems of cassava plants.

Assessment of the incidence of bacterial blight on two successive crops cycles: Two months after planting, the percentage of symptomatic plants was calculated by counting the plants showing symptoms of bacterial blight in relation to the total number of plants in the plot, according to the equation of Bansal *et al.*[23]:

$$\text{Disease incidence (DI\%)} = \frac{\text{No. of diseased plants}}{\text{Total No. of planrts}} \times 100$$

The calculation of the incidence enabled to know the number of cassava plants attacked and not attacked.

Assessment of the severity of bacterial blight symptoms on cassava plants: In order to assess the severity within the trial plots, 30 plants were randomly observed on two diagonals of each farm. The degree of severity of foliar symptoms began 30 days after the planting of cuttings on a monthly basis. This assessment was made on the underside of leaves of field plants with the rating scale established by the International Institute of Tropical Agriculture[23] which indicates the severity of bacterial blight symptoms.

This scale consists of five degrees or ratings ranging from 1-5. Thus, the following ratings were defined 1 = no symptoms, 2 = presence of angular leaf spots only, 3 = limited leaf burn, wilt, defoliation, the presence of a gummy exudate on the stems and petioles, 4 = widespread foliar burns, wilt, defoliation and necrosis of extremities (die-back) and 5 = complete defoliation and cauline necrosis, stunting and necrosis of side branches with vegetative stop.

The disease Severity Index (SI%) was calculated using the equation adapted by Song *et al.*[24]:

$$\text{Severity index (SI\%)} = \frac{\sum(\text{Scale} \times \text{No. of plants infected})}{(\text{Highest scale} \times \text{Total No. of plants})} \times 100$$

Statistical analysis of data: The data were collected and recorded with the Excel spreadsheet (2007), before being subjected to analysis of variance (ANOVA) using Statistica software version 7.1 on data from both crop cycles.

A post ANOVA analysis was performed in each case for the classification of averages at 5% threshold using the Newman-Keuls test.

RESULTS

Distribution and incidence of cassava bacterial blight in Côte d'Ivoire: The surveys conducted in the six agro-ecological zones have helped detect the presence of bacterial diseases caused by plant **PATHOGENIC BACTERIA** on cassava plants. Indeed, the infected plants were young ones and adult ones. The infection reached the leaves as well as the stems, making the plant more vulnerable. The typical symptoms of bacterial blight are expressed in leaves in the form of small brown spots of irregular shape surrounded by a yellow halo of variable diameter up to 3 mm. These spots are less frequent on the stems. Over 160 plots visited during the surveys, cassava bacterial blight was observed in 15% of the plots. Symptoms of bacterial blight were observed in all the agro-ecological zones prospected (Figure 1). However, a variable incidence was recorded from one plot to another, from one city to another and from one agro-ecological zone to another. In the East and Southwest, bacterial blight was not observed in the visited farms. In cities where it was observed, the infection rates were between 0 and 67%. Thus, the attacks of the bacteria are stronger in AEZ V and VI with the highest incidence values (60 and 66.67% respectively). Agro-ecological zones I and II, showed the lowest values (2.17 and 2.70%, respectively). As for agro-ecological zones III and IV, they had respective intermediate values of 24 and 13.33% (Figure 2). In AEZ VI wilt strewn with die-back were observed on the plants. In the other AEZs, the disease was overall expressed by angular spots without significant wilt.

Symptoms of bacterial blight observed on two experimental sites: During the two crop cycles on both experimental sites, different types of symptoms of cassava bacterial blight were observed on all 8 cassava varieties tested, traditional varieties (Yacé, Akama, Diarrassouba and Dankwa) as well improved varieties (Bocou 1, Bocou 2, Bocou 3 and Yavo). But the symptoms were stronger on local varieties (traditional ones). During the whole period of observation, bacterial blight symptoms were more visible in the rainy than in the dry season. The disease was more severe on young plants than on old plants.

Overall, such symptoms appeared in the beginning on leaves in the form of necrotic spots with moist appearance (lesions). These lesions appeared between leaf veins and were more visible on the underside of leaves (Figure 3a). They were small in size of angular shapes. Then these angular lesions merged and became progressively bigger spots that destroyed the leaf blade. The latter turned brown later on and carried at the front end of the brown spot a soaked spot (Figure 3b). These attacks lead to leaf "burns" that wilt (Figure 3c), die and fall, causing defoliation and die-back (Figure 3d) or the final death of the shoot. The leaf "burn" first appeared on the blade, then gradually reached the petiole. The latter takes a horizontal position before the fall of the leaf (Figure 3d). Drops of a brownish gum were also observed on leaves, petioles and stems of infected plants.

Incidence and severity of cassava bacterial blight in the two study sites: The results of Figure 4 and 5 and Table 2 show that the 8 cassava varieties tested proved sensitive to bacterial blight, with nuances in the cities of study. These varieties showed low to very high incidences that ranged from 1-89% depending on the city (Figure 4, 5). Overall, the highest incidence values were obtained on both study sites between the 3rd and 6th months after planting the cuttings (Figure 4, 5) with peaks in the 4th month. The traditional varieties (Diarrassouba and Yacé) and the improved variety (Bocou 1) were the most sensitive with higher attack rates. In contrast, the local varieties Akama and Dankwa and the improved ones Yavo, Bocou 2 and Bocou 3 showed a relatively low incidence of bacterial blight over the same period. Overall, the lowest incidence was recorded in the 2nd month and the highest one in the 4th month. In that period, there was a general increase of the disease on both sites.

Thus, in the city of Yamoussoukro (Figure 4), the average incidents observed on plants of 2-10 months were inferior to 60%. In this city, in the 4th month, corresponding to the period when the strongest attacks were recorded, the incidence of bacterial blight symptoms ranged from 20-56.67%. The highest incidences were observed on traditional varieties Diarrassouba (56.67%), Yacé (42.22%) and Akama (35.56%), while the lowest ones were observed on improved varieties Yavo (20%) and Bocou 2 (21.11%).

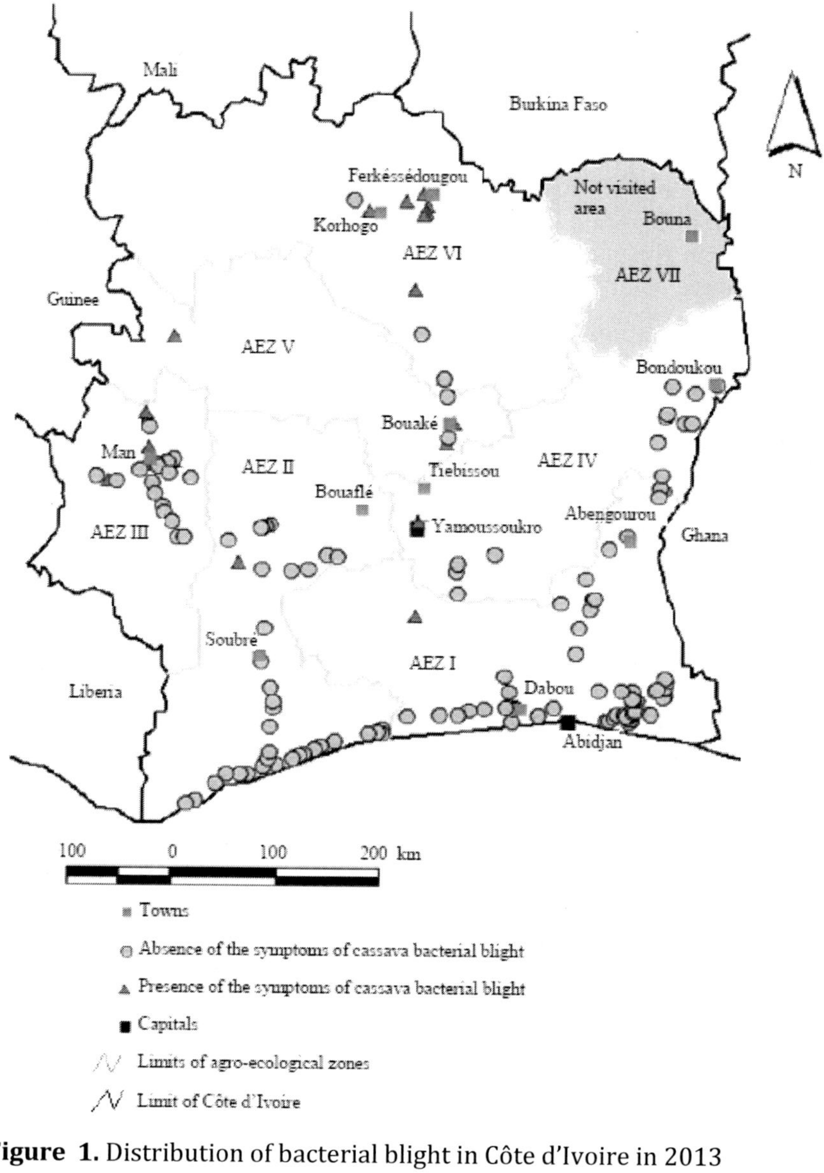

Figure 1. Distribution of bacterial blight in Côte d'Ivoire in 2013

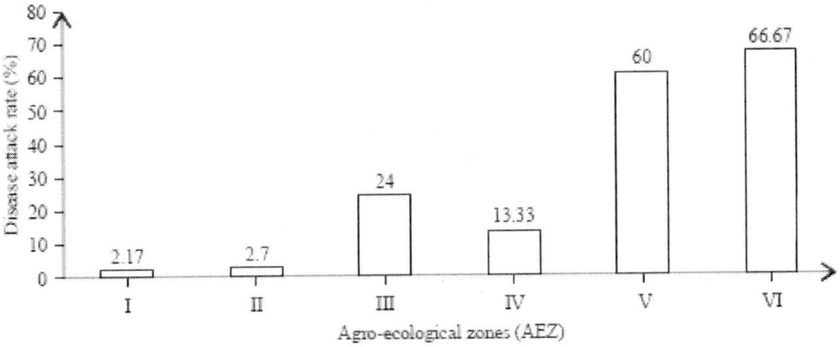

Figure 2. Incidence of bacterial blight on cassava plants depending on the agro-ecological zones (AEZ)

Figure 3(a-d). Symptoms of cassava bacterial blight seen on the leaves of local varieties and improved varieties on the study sites: (a, b) Wet aspect lesions on the leaves of the variety Diarrassouba in Ferkéssédougou, (c) Angular spots on leaves of the variety Yacé in Yamoussoukro and (d) Die-back on a young plant of the variety Bocou 1 in Ferkéssédougou

As for the city of Ferkessédougou (Figure 5), the highest average incidences observed on the plants in the 4th month after the planting the cuttings ranged between 24.44 and 88.89%. The traditional varieties Diarrassouba and Yacé recorded the highest values of average incidences of 88.89 and 48.89% respectively.

Table 2. Level of severity of bacterial blight on the different varieties of cassava on the two study sites

		Severity of infection (%) in different months								
City	Varieties	2	3	4	5	6	7	8	9	10
Yamoussoukro	Yacé	21.11^{b}	26.22^{ab}	32.22^{ab}	31.78^{a}	27.78^{ab}	23.11^{ab}	23.78^{a}	22.00^{a}	23.78^{a}
	Akama	20.22^{a}	28.67^{a}	30.22^{b}	26.89^{b}	22.67^{c}	22.22^{ab}	22.22^{ab}	20.00^{b}	20.00^{b}
	Dankwa	20.22^{a}	23.33^{b}	26.22^{bc}	25.33^{bc}	23.56^{bc}	20.44^{b}	22.22^{ab}	21.33^{ab}	20.88^{b}
	Diarrassouba	20.67^{a}	21.33^{c}	36.67^{a}	27.33^{b}	29.78^{a}	24.22^{a}	22.00^{ab}	21.78^{ab}	21.56^{ab}
	Yavo	20.22^{a}	21.56^{c}	24.67^{c}	22.22^{c}	21.77^{c}	20.67^{b}	20.67^{b}	20.88^{b}	20.22^{b}
	Bocou 1	20.67^{a}	26.44^{ab}	28.00^{b}	28.00^{ab}	27.56^{ab}	23.11^{ab}	22.22^{ab}	20.44^{b}	20.22^{b}
	Bocou 2	20.44^{a}	24.67^{b}	25.33^{bc}	27.33^{b}	27.11^{ab}	22.44^{ab}	22.67^{ab}	21.11^{ab}	20.67^{b}
	Bocou 3	20.44^{a}	22.22^{bc}	26.88^{bc}	25.56^{bc}	25.56^{b}	23.33^{ab}	22.22^{ab}	20.89^{b}	21.11^{ab}
Overall average		20.50	24.31	28.78	26.81	25.72	22.44	22.25	21.05	21.06
CV (%)		1.51	10.90	14.13	10.12	10.96	5.84	3.83	3.13	5.78
Ferkéssédougou	Yacé	21.11^{a}	32.00^{ab}	34.89^{b}	33.78^{a}	30.22^{a}	23.56^{ab}	21.33^{bc}	21.11^{b}	20.67^{b}
	Akama	20.22^{b}	30.44^{b}	27.56^{c}	30.89^{ab}	26.00^{b}	22.00^{b}	21.78^{b}	21.11^{b}	21.33^{a}
	Dankwa	20.67^{b}	26.44^{bc}	29.33^{bc}	27.33^{bc}	23.33^{bc}	21.56^{bc}	21.56^{bc}	22.00^{ab}	21.33^{a}
	Diarrassouba	20.44^{b}	31.56^{ab}	48.67^{a}	33.11^{a}	28.44^{a}	22.00^{b}	24.00^{a}	24.44^{a}	21.33^{a}
	Yavo	20.22^{b}	23.78^{c}	26.22^{c}	24.22^{c}	22.22^{c}	20.89^{c}	20.89^{bc}	20.44^{c}	20.22^{b}
	Bocou 1	21.78^{a}	36.00^{a}	32.89^{ab}	31.78^{ab}	27.56^{ab}	24.44^{a}	22.44^{b}	21.33^{b}	21.11^{a}
	Bocou 2	20.44^{b}	25.78^{bc}	26.67^{c}	28.22^{b}	26.67^{ab}	21.11^{bc}	20.44^{c}	20.67^{c}	20.44^{b}
	Bocou 3	21.33^{a}	23.11^{c}	28.89^{bc}	29.33^{b}	27.56^{ab}	24.22^{a}	23.78^{ab}	20.67^{c}	20.44^{b}
Overall average	20.78	28.64	31.89	29.83	26.5	22.47	22.03	21.47	20.86	
CV (%)	2.75	15.86	23.26	10.74	9.93	6.24	5.87	6.03	2.24	

For the same city and in the same column, the averages followed by the same letter are not statistically different at 5% threshold of the Newman-Keuls test

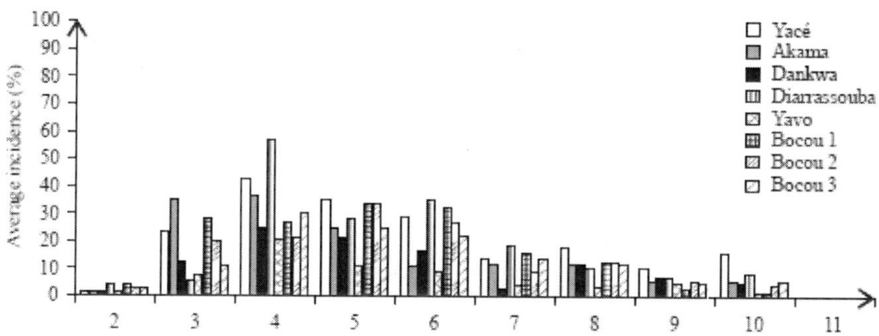

Figure 4. Incidence of cassava bacterial blight in Yamoussoukro depending on cassava varieties

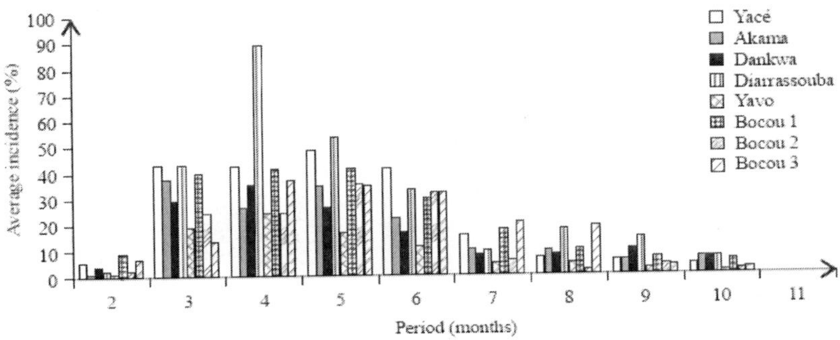

Figure 5. Incidence of cassava bacterial blight in Ferkéssédougou depending on cassava varieties

In contrast, the lowest values were observed on improved varieties Yavo and Bocou 2 by 24.44%, respectively.

Like the incidence of cassava bacterial blight, regardless of the phenological stage of the plants, the statistical analyses showed a significant to very significant difference between the level of average severity of symptoms in all varieties studied (Table 2). The most significant damage was observed in Ferkéssédougou with die-back effect on variety Diarrassouba. In Yamoussoukro, the levels of average severity of symptoms ranged from 20.22-36.67% with peaks in the 4th month (Table 2). The degrees of severity observed in that period were 36.67, 32.22 and 30.22% respectively for traditional varieties (Diarrassouba, Yacé and Akama). The disease was more severe on all the traditional varieties expressed by angular spots with leaf wilt.

Regarding the city of Ferkessédougou, bacterial blight severity levels ranged from 20.22-48.67%. Four months after planting cuttings, corresponding to the period of higher severity of the disease (Table 2), the level of severity of symptoms varied from 26.22-48.67%. The disease was more severe on local varieties Diarrassouba (48.67%) and Yacé (34.89%) as well as the improved variety Bocou 1 (32.89%). Bacterial blight was stronger in this city with attacks on stems as well as leaves and young small branches.

DISCUSSION

The results of this study reveal the presence of cassava bacterial blight caused by *Xanthomonas axonopodis* pv. *manihotis* in all the agro-ecological zones visited in Côte d'Ivoire with varying incidence. The symptoms observed on the leaves and stems were similar to those described in the literature in other African and Latin American countries[25]. Indeed, the symptoms observed during these surveys were expressed in the aspect of translucent spots characteristic of bacterial attacks described in the aspect of angular spots and leaf burn. The endemic nature of bacterial blight as vascular disease has already been noted in other countries such as South Africa, with no mention of neither the areas of prevalence nor the varieties attacked[26]. This observation could be related to the presence of different strains of the parasite having variable aggressiveness levels. The disease attacks improved varieties as well as traditional (local) varieties with attack levels. Indeed, in the same AEZ several cassava varieties were encountered, some were more sensitive and others were less sensitive. This could cause a change in the incidence of disease from one variety to another[26]. This disease is more rampant in agro-ecological zone VI (AEZ VI), which is a Savannah zone, in the departments of Korhogo and Ferkéssédougou with an incidence of 66.67%. It is less present in AEZ I (2.17%), which is a forest zone. These observations corroborate those of Banito *et al.*[27] which in the ecological conditions of Togo found that bacterial blight was absent in the forest zone and more severe in the Savannah zone and the forest-Savannah transition area. Several studies in various African countries confirm this assertion with generally highest incidences in the Savannah than in the transition forest zones and rarely in forest zones[28-30]. According to the study of Boher and Agbobli[22], some improved varieties express few symptoms and thus escape the visual health check. However, according to these researchers, such varieties can thereafter be an important source of inoculum. In Côte d'Ivoire, cassava bacterial blight has been increasing significantly in recent years. This situation might be the result of the intervention of several factors: Increased cassava cultivation, seed exchange, lack of resistant varieties, poor agricultural practices and inadequate control measures. Indeed, the climatic conditions of the country are very favorable to the emergence of bacterial blight. In Côte d'Ivoire, there is an alternation of rainy and dry season, maximum moisture and significant daytime and night temperature differences. Now, according to Lozano[31], the variation in temperature between day and night helps favoring the disease. The high incidence observed in the AEZ VI would thus be linked to the subsistence of high humidity and a significant daytime and night temperature difference. These results confirm those obtained by Lamouroux[32] in Central African Republic in 1979, indicating that the equatorial or transition climate is conducive to the development of cassava bacterial blight. Our surveys were conducted in the rainy season and confirm the detection of diseases in this period, namely bacterial blight which parasitic phase process is characterized by epiphytic multiplication of the bacteria, its penetration and proliferation in tissues of the host, unlike the dry season, which is a phase of survival. However, the almost continuous presence of the plant in the landscape could encourage rapid dissemination of the parasite either naturally or with rain runoff and certain insects, without omitting soil depletion[22]. For the establishment of a new plot, cassava producers tend to exchange plant material (cuttings) each other. According to Verdier *et al.*[33], the low availability of plant material explains the frequent exchanges of cuttings between farmers. Indeed, the use of contaminated cuttings is usually the cause of the occurrence of the disease in new plantations where it was initially absent[10,22]. A deep analysis is needed to see if there is morphotype variability of the causative agent of the disease observed. This analysis shall begin with the isolation of the pathogen associated with brown spots of cassava from infected tissues.

This study in natural conditions of infection by *Xanthomonas axonopodis* pv. *manihotis*, which took place in two distinct agro-ecological zones (AEZ V and AEZ VI) with 8 cassava varieties of which four improved varieties and four other local ones has helped highlight the sensitivity or tolerance of these varieties.

In these experimental conditions, different types of symptoms of bacterial blight were observed on improved varieties as well as local varieties in the cities of Ferkéssédougou and Yamoussoukro and whatever the phenological stage of the plant. These symptoms varied depending on the varieties tested and the city. Local varieties were more sensitive to bacterial blight in both cities with incidences ranging from 1-89%. Indeed, 2 months after the planting of cuttings, the plants of local varieties already showed a high level of infection. This could be due to the fact that these cuttings were contaminated before they had been planted. As for improved varieties, symptoms were less expressed on them. The change in sensitivity observed among varieties can be explained by the genetic characteristics of each variety and the existence of Xam races[34]. Also, the sensitivity of varieties to bacterial diseases would depend on environmental conditions and especially their genetic heritage. The level of infection of improved varieties disseminated is lower compared to that of local varieties, which is consistent with the results of Zinsou *et al.*[34] where the 7 varieties of cotton showed different levels of symptoms with 3 isolates of *Xanthomonas axonopodis* pv. *malvacearum*. This spatial variation in the incidence of bacterial blight could be explained by the difference in vegetation, which is denser in Yamoussoukro (forest-Savannah transition zone), which would contribute less to the development of bacterial blight, than in Ferkéssédougou which is located in the Savannah zone. According to

Onyeka *et al.*[30], the incidence of bacterial blight may depend on various environmental conditions influencing more or less the success of transmissions. The results of this study highlight the climate parameters that influence the evolution of the disease. Indeed, the period from the third to the 6th month corresponding respectively to the months from August-November is the period of the rainy season in the study areas and it would favor the development of the disease in the different varieties.

CONCLUSION

This study has highlighted the existence of cassava bacterial blight in all the agro-ecological zones of Côte d'Ivoire. This disease attacks traditional (local) varieties as well as improved varieties but with a higher incidence in traditional varieties. The disease is more rampant in the Savannah zone than in the forest zone.

In natural condition, the varieties of cassava studied are sensitive to bacterial blight with variations from one city to another. Local varieties Akama and Dankwa were less tolerant to the disease than the improved variety Bocou 1. The most important bacterial attack occurred during the rainy season and resulted in mortalities in the most sensitive varieties. The data reveal an interest for the use of traditional varieties Akama and Dankwa and improved varieties Yavo, Bocou 2 and Bocou 3 in the areas of high pressure by cassava bacterial blight.

ACKNOWLEDGMENTS

The authors wish to thank the Plant Physiology Laboratory, Department of Biosciences and University of Félix Houphouët-Boigny of Cocody-Abidjan for providing the laboratory facilities.

REFERENCES

1. Ambang, Z., A. Amougou, B. Ndongo, J. Nantia and G.M. Chewachong, 2009. [Resistance to mosaic virus disease of manihot glaziovii by grafting on M. esculenta]. Tropicultura, 27: 8-14.
2. Ameu, M., J. Hire, D. Kamalandua, G. Khisa, C.K. Nzeza, C.N. Ndokib and W.O. Abangi, 2014. Champs-ecoles paysans sur le manioc: Ressources a l'intention des facilitateurs d'Afrique sub-Saharienne. Organisation des Nations-Unies pour l'Alimentation et l'Agriculture (FAO), Rome, pp: 233.

3. Banito, A., K.E. Kpemoua, B. Bissang and K. Wydra, 2010. Assessment of Cassava root and stem rots in ecozones of Togo and evaluation of the pathogen virulence. Pak. J. Bot., 42: 2059-2068.
4. Banito, A., V. Verdier, K.E. Kpemoua and K. Wydra, 2007. Assessment of major cassava diseases in Togo in relation to agronomic and environmental characteristics in a systems approach. Afr. J. Agric. Res., 2: 418-428.
5. Bansal, V.K., P.D. Kharbanda, G.R. Stringam, M.R. Thiagarajah and J.P. Tewari, 1994. A comparison of greenhouse and field screening methods for blackleg resistance in doubled haploid lines of Brassica napus. Plant Dis., 78: 276-281.
6. Boher, B. and C.A. Agbobli, 1992. Cassava bacterial blight in Togo: Description of the pathogen, geographic distribution and varietal sensibility. Agronomie Tropicale, 46: 131-136.
7. Daniel, J.F., B. Boher and F. Kohler, 1981. [Bacterial pathogens of Cassava in the people's republic of the Congo and the central African republic: Etiology, epidemiology]. Agronomie EDP Sci., 1: 751-756.
8. Daniel, J.F., B. Boher, P. N'Dongo and L. Makoundou, 1985. Some aspects of survival of Xanthomonas campestris pv. manihotis, the causal agent of cassava bacterial blight. Agronomie, 5: 339-346.
9. FAO., 2015. FAOSTAT database. Food and Agriculture Organization of the United Nations, Geneva. http://faostat3.fao.org/home/index_fr.html?l ocale=fr# DOWNLOAD.
10. Fagbemissi, R.C., O. Coulibaly, R. Hanna and D. Endamana, 2002. [Adoption of cassava varieties and sustainable biological control of cassava green mite in Benin]. IITA-Benin, pp: 1-17.
11. Guillaumet, J.L. and E. Adjanohoun, 1971. La vegetation. In Le Milieu Naturel de la Côte d'Ivoire. Mem. Orstom No. 50. Paris, France, Orstom, pp: 157-263.
12. Halle, B. and V. Bruzon, 2006. Profil environnemental de la Côte d'Ivoire. Rapport Final. Consortium AFC, AGRIFOR Consult, pp: 133.
13. IITA., 1997. The cassava mosaic disease in east Africa. International Institute of Tropical Agriculture (IITA), Ouganda, pp: 1-20.
14. Kouakou, K.E., B.T.A. Goula and A.M. Kouassi, 2012. Analyze of climate variability and change impacts on hydro-climate parameters: Case study of Côte d'Ivoire. Int. J. Sci. Eng. Res., 3: 1-8.
15. Lamouroux, M., 1979. Notice explicative No. 34. Carte Pedologique du Togo au 1/1000000. Bondy, ORSTOM, pp: 84.
16. Lozano, J.C., 1986. Cassava bacterial blight: A manageable disease. Plant Dis., 70: 1089-1093.
17. Mamba-Mbayi, G., P. Tshilenge-Djim, K.K. Nkongolo and A. Kalonji-Mbuyi, 2014. Characterization of congolese strains of Xanthomonas axonopodis pv. manihotis associated with cassava bacterial blight. Am. J. Plant Sci., 5: 1191-1201.
18. N'Guessan, K.A., A.M. Kouassi, R. Gnaboa, K.S. Traore and P.V. Houenou, 2014. Analyse de phenomenes hydrologiques dans un bassin versant urbanise: Cas de la ville de Yamoussoukro (Centre de la Côte d'Ivoire). Larhyss J., 17: 135-154.
19. N'Guessan, K.A., N. Diarrassouba, K.A. Alui, K.Y. Nangha, I.J. Fofana and A. Yao-Kouame, 2015. [Physical indicators of soil degradation in Northern Côte d'Ivoire: Case of Boundiali and Ferkessedougou]. Afrique Sci., 11: 115-128.
20. N'Zue, B., 2007. Caracterisation morphologique, selection varietale et amelioration du taux de multiplication vegetative chez le manioc (Manihot esculenta Crantz). Ph.D. Thesis, Universite de Cocody, Abidjan.
21. N'Zue, B., G.P. Zohouri, C. Djedji and O. Tahouo, 2013. Bien cultiver le manioc en Côte d'Ivoire. Fiche Technique du Centre National de Recherche Agronomique (CNRA), pp: 4.
22. Ntawuruhunga, P., G. Okao-Okuja, A. Bembe, M. Obambi, J.C.A. Mvila and J.P. Legg, 2007. Incidence and severity of Cassava mosaic disease in the Republic of Congo. Afr. Crop Sci. J., 15: 1-9.

23. Ntawuruhunga, P., G. Ssemakula, H. Ojulong, A. Bua, P. Ragama, C. Kanobe and J. Whyte, 2006. Evaluation of advanced cassava genotypes in Uganda. Afr. Crop Sci. J., 14: 17-25.
24. Ogunjobi, A.A., A.G.O.E. Fagade, O. Dixon and R. Bandyopadhyay, 2010. Assessment of large population of cassava accessions for resistant to cassava bacterial blight infection in the screen house environment. J. Agric. Biotechnol. Sust. Dev., 2: 87-91.
25. Onyeka, T.J., A.G.O. Dixon, R. Bandyopadhy, R.U. Okechukwu and B. Bamkefa, 2004. Distribution and current status of bacterial blight and fungal diseases of cassava in Nigeria. International Institute of Tropical Agriculture (IITA), Ibadan, Nigeria, pp: 37.
26. Pene, C.B. and M. Kehe, 2005. [Performances of three late-season sugarcane varieties under soil water deficit at the yield formation stage in Northern Ivory Coast]. Agronomie Africaine, 17: 7-18.
27. Song, W., L. Zhou, C. Yang, X. Cao, L. Zhang and X. Liu, 2004. Tomato Fusarium wilt and its chemical control strategies in a hydroponic system. Crop Protect., 23: 243-247.
28. Verdier, V., 1998. [Studies on the variability of Xanthomonas axonopodis pv. manihotis, the causal agent of the bacterial blight disease of cassava (Manihot esculenta Crantz)]. Ph.D. Thesis, University of Paris-Sud, France.
29. Verdier, V., G. Mosquera and K. Assigbetse, 1998. Detection of the cassava bacterial blight pathogen, Xanthomonas axonopodis pv. manihotis, by polymerase chain reaction. Plant Dis., 82: 79-83.
30. Wydra, K. and V. Verdier, 2002. Occurrence of cassava diseases in relation to environmental, agronomic and plant characteristics. Agric. Ecosyst. Environ., 93: 211-226.
31. Wydra, K. and W. Msikita, 1998. Overview of present situation of cassava diseases in West Africa. Proceedings of the 6th Triennial Symposium of International Society of Tropical Root Crops-Africa Branch, October 22-28, 1998, Lilongwe, Malawi, pp: 198-206.
32. Zinga, I., C.R. Nguimalet, D.P. Lakouetene, G. Konate, K.E. Kosh and S. Semballa, 2008. [The impacts of African cassava mosaic in Central African Republic]. Geo-Eco-Trop, 32: 47-60.
33. Zinga, I., R.D. Longue, E.K. Komba, C. Beaumont and S. Semballa, 2016. [Assessment of protein and chlorophyll contents in leaves of five local varieties of Cassava infected by African mosaic virus in central African republic]. Tropicultura, 34: 3-9.
34. Zinsou, V.A., L.A.C. Afouda, A. Hougni, S.S. Affomasse and E. Sekloka, 2015. [Distribution of cotton bacterial blight in benin and susceptibility of local varieties to its attacks]. Tropicultura, 33: 55-62.

CHAPTER 4

Exploring the Phytoplasmas, Plant Pathogenic Bacteria

Kensaku Maejima • Kenro Oshima • Shigetou Namba

Department of Agricultural and Environmental Biology, Graduate School of Agricultural and Life SciencesThe University of TokyoTokyoJapan

ABSTRACT

Phytoplasmas are plant pathogenic bacteria associated with devastating damage to over 700 plant species worldwide. It is agriculturally important to identify factors involved in their pathogenicity and to discover effective measures to control phytoplasma diseases. Despite their economic importance, phytoplasmas remain the most poorly characterized plant pathogens, primarily because efforts at in vitro culture, gene delivery, and mutagenesis have been unsuccessful. However, recent molecular studies have revealed unique biological features of phytoplasmas. This review summarizes the history and recent progress in phytoplasma research, focusing on (1) the discovery of phytoplasmas, (2) molecular classification of phytoplasmas, (3) diagnosis of phytoplasma diseases, (4) reductive evolution of the genomes, (5) characteristic features of the plasmids, (6) molecular mechanisms of insect transmissibility, and (7) virulence factors involved in their unique symptoms.

Keywords: Diagnosis Genome Phytoplasma Insect transmission Molecular classification Virulence factors

INTRODUCTION

Phytoplasmas are plant pathogenic bacteria in the class Mollicutes and are formally called mycoplasma-like organisms (MLOs) (Doi et al. 1967). They are transmitted by insect vectors (leafhoppers, planthoppers, and psyllids) and infect hundreds of plant species worldwide, including many economically important crops, fruit trees, and ornamental plants (Hogenhout et al. 2008; Oshima et al. 2013). Infected plants show a wide range of symptoms including stunting, yellowing, witches' broom (development of numerous tiny shoot branches with small leaves), phyllody (formation of leaf-like tissues instead of flowers), virescence (greening of floral organs), proliferation (growth of shoots from floral organs), purple top (reddening of leaves and stems), and phloem necrosis (Figure 1). Phytoplasma infection is often fatal and causes devastating damage to global agricultural production. Due not only to difficulties in culturing phytoplasmas in vitro and in manipulating them genetically but also to their recalcitrant biological properties such as restriction to phloem cells, nontransmissibility by manual inoculation, and high vector specificity, these organisms are one of the most poorly characterized groups of plant pathogens. However, recent molecular studies have revealed interesting features of phytoplasmas. This review summarizes the history and recent progress in phytoplasma research, especially in Japan.

PHYTOPLASMAL DISEASES AND DISCOVERY OF THE PATHOGEN

Phytoplasmal diseases have been observed globally. In Japan, mulberry dwarf disease has caused severe damage to mulberry plants, the sole source of food for silkworms, since the Tokugawa period (1603–1868) (Okuda 1972). In 1890, Shirai (the first president of the Phytopathological Society of Japan and the first professor of plant pathology at the University of Tokyo) reported that this disease was transmissible by grafting, although the causal agent remained unknown (Shirai 1890). Since the early twentieth century, other phytoplasmal diseases, including paulownia witches' broom disease (Kawakami 1902) and rice yellow dwarf disease (Anonymous 1919) in Japan, as well as aster yellows disease (Kunkel 1926) in the United States, have been reported. These diseases were initially attributed to plant viruses because of their insect transmissibility and virus-like symptoms. However, Doi et al. (1967) discovered small pleomorphic bodies that resembled mycoplasmas (bacterial pathogens of humans and animals) in ultrathin sections of the phloem of plants affected by diseases such as mulberry dwarf, paulownia witches' broom, and aster yellows. The associated agents were named mycoplasma-like organisms (MLOs) because of their morphological similarity to mycoplasmas, as well as their sensitivity to tetracycline antibiotics (Ishiie et al. 1967). Discovery of these novel plant pathogens, MLOs, was confirmed in subsequent studies (Granados et al. 1968; Hull et al. 1969).

Figure 1. Unique symptoms caused by phytoplasma infection. **A** Phyllody symptoms on hydrangea induced by infection with '*Candidatus* Phytoplasma japonicum' JHP phytoplasma (*right*). **B** Witches' broom symptoms on *Ziziphus jujuba* induced by infection with '*Ca*. P. ziziphi' (*left*). **C** Witches' broom symptoms on *Paulownia tomentosa* induced by infection with '*Ca*. P. asteris' (*left*). Photo courtesy of Dr. Norio Nishimura (Koibuchi College of Agriculture and Nutrition)

MOLECULAR CLASSIFICATION OF PHYTOPLASMAS

More than four decades since their discovery, MLOs remain the most poorly characterized phytopathogens from every viewpoint, including biology and taxonomy, because they are difficult to culture in vitro.

Moreover, since classification of MLOs was not possible, each MLO was named based on its natural host and symptoms. According to this nomenclature, as of the late 1980s, several hundred MLOs had been reported globally, and 63 MLOs were reported in Japan (Kishi 1987). In the early 1990s, the 16S rRNA gene sequences of MLOs were compared with each other, as well as with *Acholeplasma laidlawii*, *Spiroplasma citri*, and several mycoplasmas (Kuske and Kirkpatrick 1992; Namba et al. 1993b). These analyses revealed that MLOs were a monophyletic group within the class Mollicutes, but were more closely related to *Acholeplasma* spp. than to *Spiroplasma* spp. and animal mycoplasmas. In 1994, the name "phytoplasmas" was adopted by the Phytoplasma Working Team at the 10th Congress of the International Organization of Mycoplasmology to collectively denote MLOs. In 2004, phytoplasmas were proposed to be within a novel genus '*Candidatus* Phytoplasma' (IRPCM 2004). Approximately 40 phytoplasmal species have been reported worldwide (Hogenhout et al. 2008; Jung et al. 2002, 2003b, c; Sawayanagi et al. 1999). In Japan, approximately 70 strains of phytoplasmas (PSJ NIAS 2012) have been reclassified into nine species (Figure 2).

DIAGNOSIS OF PHYTOPLASMA DISEASES

Phytoplasmas cause devastating damage to many plant species worldwide. For example, in 2001, a phytoplasma outbreak in apple trees caused losses of about €100 million in Italy and €25 million in Germany (Strauss 2009). Lethal yellowing of palm has killed millions of coconut palm trees in the Caribbean over the past 40 years (Brown et al. 2006). In Japan, phytoplasmas also cause serious economic losses to agricultural production.

'*Candidatus* Phytoplasma asteris' PaWB strain (PaWB) is the causal agent of paulownia witches' broom disease, an important paulownia tree disease. Once infected with PaWB, tree vigor is reduced, occasionally followed by dieback. Until 1960, more than 1 million PaWB-infected trees were found in Japan, excluding Hokkaido Prefecture and the Tohoku region (Ito 1960), and the disease later spread to almost every plantation in the Tohoku region (Nakamura et al. 1998). Paulownia witches' broom disease also occurs in China (Hiruki 1999), where it has affected 880,000 ha of plantations (Yue et al. 2008). '*Candidatus* Phytoplasma oryzae' RYD strain (RYD) is the causal agent of rice yellow dwarf disease, a serious rice disease in many regions of Asia. Early infection with RYD results in a nearly 80 % yield loss. In the 1960s, rice yellow dwarf disease occurred in 10–50 % of rice growing areas of Japan and caused serious economic losses in many prefectures (Komori 1966; Sameshima 1967). In other cases, sweet potato witches' broom disease threatened the food supply in the Ryukyu Islands in the 1950s and 1960s (Nakamori and Maezato 1968; Shinkai 1964), and hydrangea phyllody disease remains a destructive disease in hydrangea cultivation (Kanehira et al. 1996; Sawayanagi et al. 1999; Takinami et al. 2013).

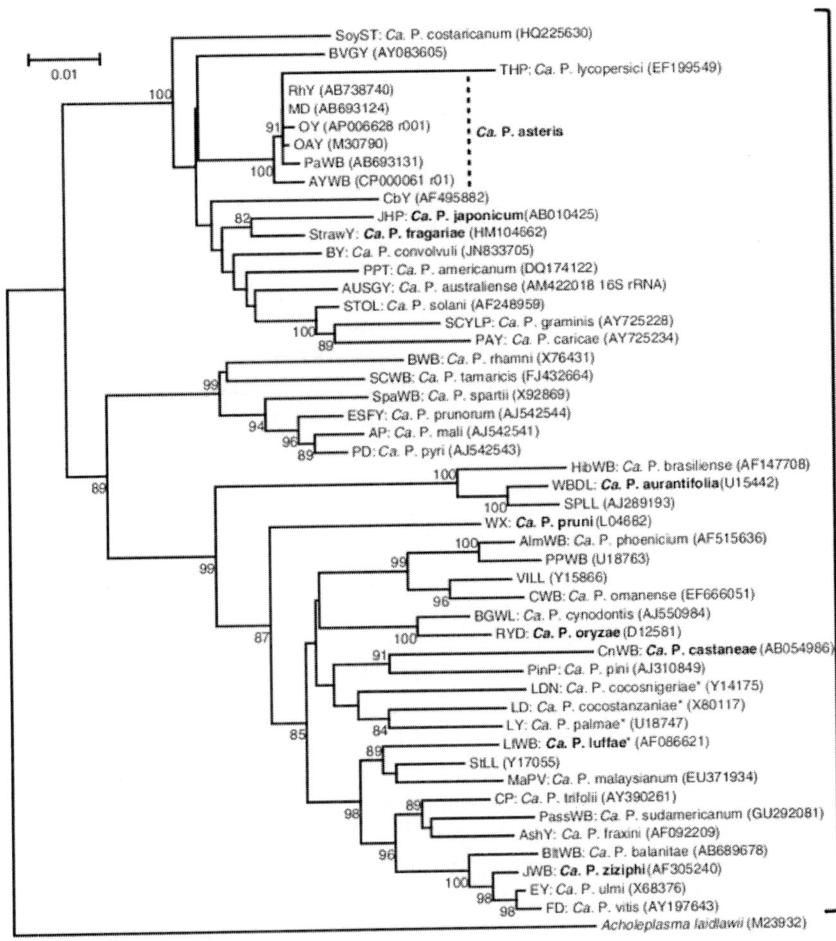

Figure 2. Phylogenetic relationship of phytoplasmas. A phylogenetic tree was constructed using the neighbor-joining method (Saitou and Nei 1987) with 16S rRNA gene sequences from phytoplasmas and acholeplasma (outgroup). Numbers on the branches are the bootstrap values (only values >80 % are shown). GenBank accession numbers are in parentheses. *Asterisks* indicate the proposed provisional names (IRPCM 2004). *Scale bar* indicates the number of nucleotide substitutions

per site. Species found in Japan are in *bold type*. *AlmWB* almond witches' broom, *AP* apple proliferation, *AshY* ash yellows, *AUSGY* Australian grapevine yellows, *AY-WB* aster yellows phytoplasma strain witches' broom, *BGWL* Bermuda grass white leaf, *BltWB* Balanites triflora witches' broom, *BVGY* Buckland Valley grapevine yellows, *BWB* buckthorn witches' broom, *BY* bindweed yellows, *Ca. P.Candidatus* Phytoplasma, *CbY* chinaberry yellows, *CnWB* chestnut witches' broom, *CP* clover proliferation, *CWB* cassia witches' broom, *ESFY* European stone fruit yellows, *EY* elm yellows, *FD* flavescence dorée of grapevine, *HibWB* hibiscus witches' broom, *JHP* Japanese hydrangea phyllody, *JWB* jujube witches' broom, *LD* coconut lethal yellowing, substrain Tanzanian lethal disease, *LDN* coconut lethal yellowing, substrain Nigerian Awka disease, *LY* coconut lethal yellowing, *LfWB* loofah witches' broom, *MaPV* Malaysian periwinkle virescence, *MD* mulberry dwarf, *OAY* oenothera aster yellows, *OY* onion yellows, *PassWB* passion fruit witches' broom, *PaWB* paulownia witches' broom, *PAY* papaya, *PD* pear decline, *PinP* pine phytoplasma, *PPT* potato purple top wilt, *PPWB* Caribbean pigeon pea witches' broom, *RhY* rhus yellows, *RYD* rice yellow dwarf, *SCWB* salt cedar witches' broom, *SCYLP* sugarcane yellow leaf syndrome, *SoyST* soybean stunt, *SpaWB* spartium witches' broom, *SPLL* sweet potato little leaf, *STOL* stolbur, *StrawY* strawberry yellows, *THP* 'Hoja de perejil' (parsley leaf) of tomato, *ViLL* vigna little leaf, *WBDL* witches' broom disease of lime, *WX* western X-disease

There is no effective way to control phytoplasma diseases; thus, early detection and removal of infected plants are critical to prevent the spread of disease. Until the early 1980s, phytoplasma diseases were diagnosed by transmission electron microscopic (TEM) observation because of their small size and the difficulty of culturing phytoplasmas in vitro. However, that approach requires expensive TEM equipment and time to prepare ultrathin sections of phloem tissue. In the 1980s, simple diagnostic techniques for phytoplasma diseases, such as direct fluorescence detection (DFD) (Namba et al. 1981) and DAPI staining (Hiruki and da Rocha 1986), were developed, both of which utilize florescent microscopy. The DFD method detects autofluorescence of necrotic phloem cells, and DAPI detects phytoplasmal DNA. Enzyme-linked immunosorbent assay (ELISA), the most prevalent method to diagnose viral diseases, was rarely used to detect phytoplasmal diseases in the 1980s, except for a few successful cases (Lin and Chen 1985), because it was difficult to prepare a phytoplasma-specific antibody at that time due to difficulties in purifying phytoplasmal cells.

Around 1990, advances in molecular biology enabled direct detection of phytoplasmal DNA by DNA–DNA hybridization (Kirkpatrick et al. 1987)

and the polymerase chain reaction (PCR) (Deng and Hiruki 1991). PCR amplification of the 16S rRNA genes of phytoplasmas (Lee et al. 1993; Namba et al. 1993a; Schneider et al. 1995) has become the gold standard in the diagnosis of phytoplasmal diseases. PCR has also been applied to analyze phytoplasmal localization and dynamics in planta (Nakashima and Hayashi 1995b; Sahashi et al. 1995; Wei et al. 2004b). Recently, loop-mediated isothermal amplification (LAMP) (Notomi et al. 2000) has also been used to detect several phytoplasmas and is expected to become a rapid and reliable field-diagnostic system for phytoplasma diseases (Sugawara et al. 2012; Tomlinson et al. 2010). LAMP is more sensitive and rapid than PCR amplification and does not require DNA purification or special equipment such as a thermal cycler. The first LAMP-based detection kit for phytoplasmas has been commercially available since 2011 in Japan (http://nippongene-analysis.com/phytoplasma-fs.htm). Application of the kit for on-site diagnosis will aid in the early control of phytoplasmal diseases.

REDUCTIVE EVOLUTION OF THE GENOMES

Biological and molecular characteristics of phytoplasmas are still quite unclear because of the difficulty in culturing them in vitro. However, in the late 1990s, several genome projects were initiated globally to explore the genomic features of phytoplasmas. At the same time, several mild-pathogenicity mutants (Shiomi et al. 1998; Uehara et al. 1999) and a non-insect-transmissible mutant (Oshima et al. 2001b) were identified in Japan. Among these mutant strains, the '*Ca.* P. asteris' OY-M strain (OY-M) was thought to be suitable for whole genome analysis because the chromosome size of OY-M (ca. 870 kbp) was much smaller than that of the '*Ca.* P. asteris' OY-W strain (OY-W; ca. 1,000 kbp) (Oshima et al. 2001b). In parallel with the OY-M genome project, the draft genomic sequences of OY-W were obtained (Oshima et al. 2002; Miyata et al. 2002b), revealing phytoplasma codon usage (Miyata et al. 2002a), catalytic activity assays of phytoplasmal proteins (Miyata et al. 2003), and the presence of two rRNA operons (Jung et al. 2003a). In 2004, the complete genomic sequence of OY-M was determined (Oshima et al. 2004). To date, the complete genomic sequences of three other phytoplasmas have been reported (Bai et al. 2006; Tran-Nguyen et al. 2008; Kube et al. 2008).

Although phytoplasma genomes contain genes for basic cellular functions such as DNA replication, transcription, translation, and protein translocation, they lack genes for amino acid biosynthesis, fatty acid biosynthesis, the tricarboxylic acid cycle, and oxidative phosphorylation, just as in the mycoplasma genome (Razin et al. 1998). Phytoplasma genomes, however, encode even fewer metabolically functional proteins than do mycoplasma genomes, which were previously thought to have the minimum possible gene set (Mushegian and Koonin 1996). For example, phytoplasma genomes lack the pentose phosphate pathway genes and genes encoding F_1F_o-type ATP synthase. Therefore, ATP synthesis in phytoplasmas is strongly dependent on glycolysis instead of ATP synthase (Oshima et al. 2007). Interestingly, phytoplasmas harbor multiple copies of transporter-related genes not found in mycoplasmas. These genomic features suggest that phytoplasmas are highly dependent on metabolic compounds from their hosts (Oshima et al. 2013). Phytoplasma genomes contain the *sodA* gene, which encodes superoxide dismutase that can inactivate reactive oxygen species (ROS) (Miura et al. 2012). Plants deploy a broad range of defenses during infection by various pathogens. The oxidative burst, which produces ROS, is one of the earliest events in the plant defense response. Since the genomes of mycoplasmas do not contain this gene, the presence of sodA may defend phytoplasmas against the unique threat of ROS released by the plant cell. Phytoplasma genomes also contain clusters of repeated gene sequences called potential mobile units (PMUs) (Bai et al. 2006), which consist of similar genes organized in a conserved order (Arashida et al. 2008a; Jomantiene and Davis 2006). A PMU in '*Ca.* P asteris' AY-WB strain (AY-WB) exists as both linear chromosomal and circular extrachromosomal elements (Toruño et al. 2010), suggesting the ability to transpose within the genome.

CHARACTERISTIC FEATURES OF THE PLASMIDS

In general, a phytoplasma genome consists of one chromosome and several small plasmids (Firrao et al. 2007). Phytoplasmal plasmids were cloned in the late 1980s (Davis et al. 1988; Nakashima et al. 1991) and have often been used as targets of DNA hybridization to compare phytoplasma strains (Nakashima and Hayashi 1995a). Complete nucleotide sequence analyses of the plasmids revealed that they may be products of DNA recombination between the phytoplasmal plasmids (Nishigawa et al. 2002a) and that each plasmid encodes a replication initiation protein (Rep) involved in rolling-circle replication, as well as several other unknown proteins (Nakashima and Hayashi 1997; Kuboyama et al. 1998; Nishigawa et al. 2003). Interestingly, the phytoplasmal Reps are similar to the Reps encoded by bacterial plasmids and to geminivirus (Nakashima and Hayashi 1997; Nishigawa et al. 2001) and circovirus Reps (Oshima et al. 2001a), suggestive of evolutionary relationships between phytoplasmas and viruses.

The functions of other genes in the phytoplasmal plasmids remain unknown. However, some genes are apparently related to adaptation to the insect host. For example, '*Ca.* P. asteris' OY-NIM strain (OY-NIM), a non-insect-transmissible derivative line of OY-M, lacked the ability to express the ORF3 gene encoded in plasmids (Nishigawa et al. 2002b; Ishii et al. 2009a). Sequence analysis of the plasmids over 10 years revealed that one of the plasmids was gradually lost from OY-NIM and finally disappeared during the maintenance of OY-NIM in plant tissue culture (Ishii et al. 2009b). These results suggested that the plasmid is not essential for phytoplasmal viability in a plant host, but may be a key element for adaption to an insect host.

MOLECULAR MECHANISMS OF INSECT TRANSMISSIBILITY

Phytoplasmas are transmitted by specific insect species in a persistent manner. It is important to understand the molecular mechanisms underlying these strict pathogen–insect specificities in order to control phytoplasmal diseases. Many insect vectors involved in phytoplasma diseases have been identified in Japan (Table 1). Building on knowledge about the insect vectors, recent studies on phytoplasmal localization in nonvector insects revealed that the phytoplasma–insect specificity is determined by the ability to pass through several physical barriers, including the midgut and salivary gland (Nakajima et al. 2002, 2009).

Since phytoplasmas are endocellular parasites that lack a cell wall, their membrane proteins and secreted proteins function directly in the host cell. In many phytoplasmas, a subset of membrane proteins (usually referred to as immunodominant membrane proteins) accounts for a major portion of the total cellular membrane proteins (Kakizawa et al. 2006b). Immunodominant membrane proteins were classified into three types: immunodominant membrane protein (Imp) (Kakizawa et al. 2009), antigenic membrane protein (Amp) (Kakizawa et al. 2004), and immunodominant membrane protein A (IdpA) (Neriya et al. 2011), which are not orthologues. In both OY-W and OY-M, Amp was a major surface membrane protein (Kakizawa et al. 2009). Cloning of *amp* genes from several phytoplasma strains showed that Amp proteins are under positive selection, and positively selected amino acids can be found in the central hydrophilic domain (Kakizawa et al. 2006a). In addition, the Amp protein forms a complex with insect microfilaments (Suzuki et al. 2006). Interestingly, the formation of Amp–microfilament complexes was correlated with the phytoplasma-transmitting capability of leafhoppers, suggesting that the interaction between Amp and insect microfilaments may play a major role in determining the transmissibility of phytoplasmas (Suzuki et al. 2006). The interaction between Amp and actin (a component of microfilament), as well as the ATP synthase β subunit of insect vectors, has also been observed for the CYP strain of '*Ca*. P. asteris' (Galetto et al. 2011).

Table 1. Summary of insect vectors of phytoplasma diseases in Japan

Disease name	Host plant	Causal phytoplasma species	Insect vector	References
Anemone witches' broom	*Anemone coronaria*	*Ca*. P. asteris	*Macrosteles striifrons*	Namba (1996)
Buplever yellows	*Bupleurum falcatum*	*Ca*. P. asteris	*Macrosteles striifrons*	Namba (1996)
Carrot yellows	*Daucus carota*	*Ca*. P. pruni	*Ophiola flavopicta*	Namba (1996)
Celery yellows	*Apium graveolens*	*Ca*. P. pruni	*Ophiola flavopicta*	Namba (1996)
China aster yellows	*Callistephus chinensis*	*Ca*. P. pruni	*Ophiola flavopicta*	Namba (1996)
Cineraria witches' broom	*Senecio* × *hybridus*	*Ca*. P. asteris	*Macrosteles striifrons*	Namba (1996)
Cosmos yellows	*Cosmos bipinnatus*	*Ca*. P. asteris	*Macrosteles striifrons*	Wakibe et al. (1995)
Cryptotaenia japonica witches' broom	*Cryptotaenia japonica*	*Ca*. P. asteris	*Macrosteles striifrons*	Namba (1996)
			Hishimonus sellatus	Nishimura et al. (1998)
			Hishimonoides sellatiformis	Nishimura et al. (2004)
Garland chrysanthemum witches' broom	*Chrysanthemum coronarium* var. *spatiosum*	*Ca*. P. asteris	*Macrosteles striifrons*	Shiomi and Sugiura (1983)
Gentian witches' broom	*Gentiana scabra* var. *buergeri*	*Ca*. P. pruni	*Ophiola flavopicta*	Namba (1996)
		Ca. P. asteris	*Macrosteles striifrons*	Tanaka et al. (2006)
Geranium witches' broom	*Pelargonium* × *hortorum*	*Ca*. P. pruni	*Ophiola flavopicta*	Namba (1996)
Glaucidium witches' broom	*Glaucidium palmatum*	*Ca*. P. pruni	*Ophiola flavopicta*	Tanaka et al. (1998)
Hovenia witches' broom	*Hovenia tomentella*	*Ca*. P. asteris	*Hishimonus sellatus*	Kusunoki et al. (2002)
Iceland poppy yellows	*Papaver nudicaule*	*Ca*. P. asteris	*Macrosteles striifrons*	Namba (1996)
Jujube witches' broom	*Zizyphus jujuba*	*Ca*. P. ziziphi	*Hishimonus sellatus*	Kusunoki et al. (2002)
Legume witches' broom	(plants in the family Fabaceae)	*Ca*. P. aurantifolia	*Orosius orientalis*	Namba (1996)
Marguerite yellows	*Argyranthemum frutescens*	*Ca*. P. asteris	*Macrosteles striifrons*	Namba (1996)
Mulberry dwarf	*Morus* spp.	*Ca*. P. asteris	*Hishimonus sellatus*	Namba (1996)
			Hishimonoides sellatiformis	Namba (1996)
			Tautoneura mori	Jiang et al. (2005)
Onion yellows	*Allium cepa*	*Ca*. P. asteris	*Macrosteles striifrons*	Namba (1996)
			Hishimonus sellatus	Nishimura et al. (1998)
			Hishimonoides sellatiformis	Nishimura et al. (2004)
Potato witches' broom	*Solanum tuberosum*	*Ca*. P. pruni	*Ophiola flavopicta*	Namba (1996)
Rhus yellows	*Rhus javanica*	*Ca*. P. asteris	*Hishimonus sellatus*	Tanaka et al. (2000)
Rice yellow dwarf	*Oryza sativa*	*Ca*. P. oryzae	*Nephotettix cincticeps*	Namba (1996)
			Nephotettix nigropictus	Namba (1996)
			Nephotettix virescens	Namba (1996)
Rocket larkspur witches' broom	*Consolida ambigua*	*Ca*. P. asteris	*Macrosteles striifrons*	Tanaka et al. (2007)
Statice witches' broom	*Limonium* spp.	*Ca*. P. asteris	*Macrosteles striifrons*	Shiomi et al. (1999)
Strawberry witches' broom	*Fragaria* × *ananassa*	*Ca*. P. asteris	*Macrosteles striifrons*	Namba (1996)
Sweet potato witches' broom	*Ipomoea batatas*	*Ca*. P. aurantifolia	*Orosius ryukyuensis*	Namba (1996)
Tomato yellows	*Lycopersicon esculentum*	*Ca*. P. pruni	*Ophiola flavopicta*	Namba (1996)
		Ca. P. asteris	*Macrosteles striifrons*	Kato et al. (1988)
Tsuwabuki witches' broom	*Farfugium japonicum*	*Ca*. P. pruni	*Ophiola flavopicta*	Namba (1996)
Udo dwarf	*Aralia cordata*	*Ca*. P. pruni	*Ophiola flavopicta*	Namba (1996)
Water dropwort yellows	*Oenanthe javanica*	*Ca*. P. asteris	*Macrosteles striifrons*	Shiomi and Sugiura (1983)

Moreover, a phytoplasmal microarray analysis of OY-M revealed that expression of approximately one-third of the genes is affected during host switching between plant and insect, suggesting that the phytoplasma alters its gene expression in response to its host (Figure 3) and may use transporters, membrane proteins, secreted proteins, and metabolic enzymes in a host-specific manner (Oshima et al. 2011).

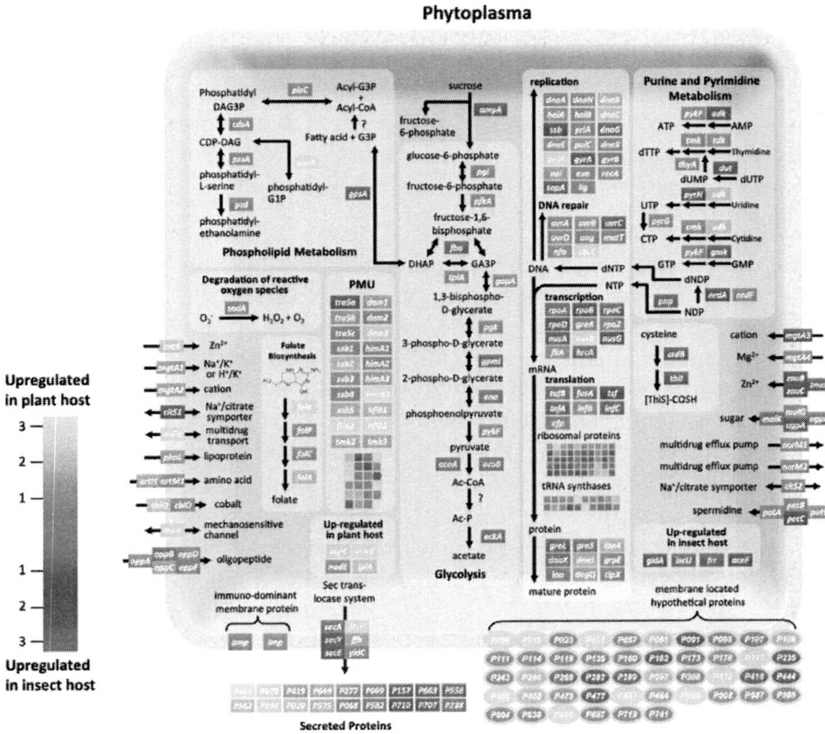

Figure 3. Global patterns of gene regulation in phytoplasma associated with host switching between plant and insect (modified from Oshima et al. 2011). Genes upregulated in the plant host are in green; those upregulated in the insect host are in *red*

VIRULENCE FACTORS INVOLVED IN UNIQUE SYMPTOMS

Phytoplasmas cause dramatic changes in plant development (Arashida et al. 2008b); however, the molecular mechanisms underlying their pathogenicity remain unclear. The phytoplasma genome lacks homologs of the type III secretion system, which are essential for the virulence of most phytopathogenic bacteria (Abramovitch et al. 2006), and for toxins produced by *Pseudomonas* spp., cell wall maceration enzymes of *Pectobacterium* spp., and the type IV secretion system of *Rhizobium* (*Agrobacterium*) spp. (Oshima et al. 2004). On the other hand, phytoplasmas have no cell walls and reside within host cells, and their secreted proteins are believed to directly function in the host plant or insect cells. The genes encoding SecA, SecY, and SecE, required for protein translocation in *Escherichia coli* (Economou 1999), were identified in strain OY-M of '*Ca*. P. asteris' (Kakizawa et al. 2001, 2004) and in three other phytoplasma genomes (Bai et al. 2006; Kube et al. 2008; Tran-Nguyen et al. 2008). In addition, SecA protein was expressed in plants infected with five respective phytoplasma species (Wei et al. 2004a). These results suggest that the Sec system is widely conserved among phytoplasmas.

Homologs of the virulence genes of other phytopathogenic bacteria were not identified in the phytoplasma genomes. Therefore, phytoplasmal proteins secreted via the Sec system are candidate virulence factors that induce morphological changes. In 2009, a secreted protein (TENGU) was identified from OY-M as the first phytoplasmal virulence factor (Hoshi et al. 2009). TENGU induces witches' broom and dwarfism, which are typical of phytoplasma infection in *Nicotiana benthamiana* and *Arabidopsis thaliana*. TENGU homologs from several phytoplasma strains induce similar symptoms (Sugawara et al. 2013). Microarray analysis of TENGU-transgenic *Arabidopsis* plants revealed that TENGU inhibits auxin-related pathways, thereby affecting plant development (Figure 4) (Hoshi et al. 2009). More recently, TENGU was found to be processed in planta into a small functional peptide, similar to the proteolytic processing of plant endogenous peptide signals (Sugawara et al. 2013). As for phyllody symptoms, a previous study has shown the involvement of the floral homeotic MADS-domain transcription factors of the floral ABCE model (floral quartet model) (Himeno et al. 2011). Recently, PHYL1 and SAP54 were found to be homologous secreted proteins that induce phyllody in floral organs of *Arabidopsis thaliana* (MacLean et al. 2011; Maejima et al. 2014). PHYL1 actually interacts with and degrades these MADS-domain transcription factors (Maejima et al. 2014). *PHYL1* gene was genetically and functionally conserved among other phytoplasma strains and species. Therefore, *PHYL1* and its homologs were designated as members of the phyllody-inducing gene family "phyllogen." SAP11, a secreted protein from AY-WB, contains eukaryotic nuclear localization signals and localizes in plant cell nuclei (Bai et al. 2009). SAP11 downregulates JA synthesis and increases the fecundity of insect vectors (Sugio et al. 2011). In addition to these secreted proteins, the consumption of plant metabolites by phytoplasmas may be associated with virulence. The duplicated glycolytic genes in OY-W are likely

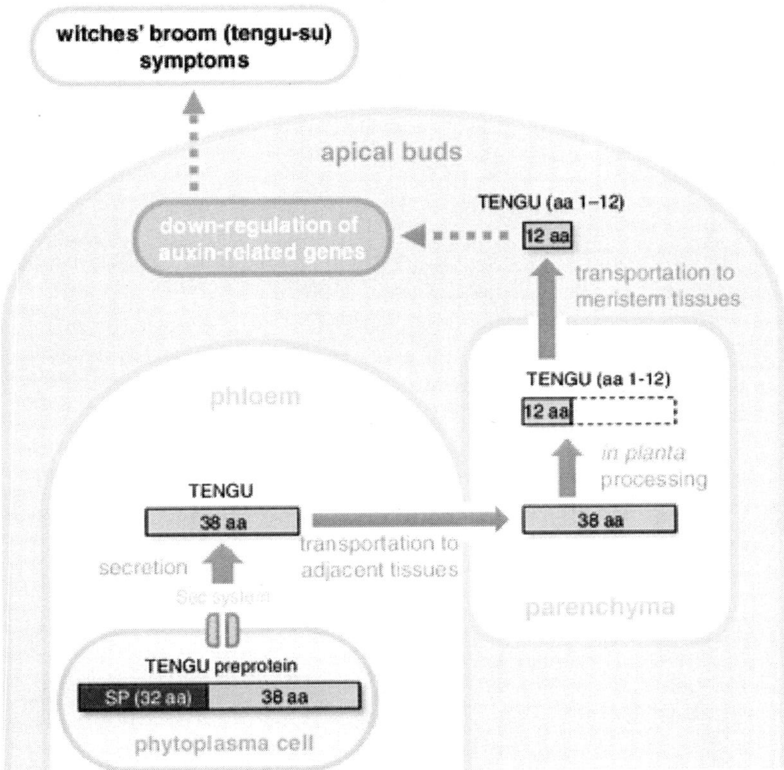

Figure 4. Hypothetical mechanism of witches' broom (tengu-su) symptoms induced by a phytoplasmal virulence factor, TENGU. TENGU, a small peptide secreted from the phytoplasma, is translated as a 70-amino acid preprotein with a signal peptide (SP) of 32 amino acids (aa) at its N terminus. The C-terminal 38 aa of TENGU preprotein are secreted into plant phloem via the Sec system, a bacterial conserved protein translocation system on the cellular membrane, where the N-terminal signal peptide is cleaved. TENGU is transported from phloem into parenchyma and apical buds. During the transportation, TENGU is proteolytically processed into the active form consisting of N-terminus 12 aa. The active form of TENGU inhibits auxin-related pathways, resulting in induction of witches' broom (tengu-su) symptoms

responsible for the severe symptoms (Oshima et al. 2007). Moreover, in the most recent study, activation of the host anthocyanin biosynthesis pathway in response to phytoplasma infection is responsible for purple top symptoms and is associated with a reduction of leaf cell death (Himeno et al. 2014). Through all these studies, the molecular mechanisms of phytoplasma symptoms are becoming clear. On the other hand, the biological significance of the unique symptoms, which is also an interesting topic, remains largely unknown, and awaits further study.

CONCLUDING REMARKS

Although phytoplasmas remain the most poorly characterized phytopathogens, recent studies have identified virulence factors that induce typical phytoplasma disease symptoms and have characterized the unique reductive evolution of the genome. Phytoplasma-related diseases are expected to increase because the warming global climate is advantageous to the cold-sensitive vectors of the phytoplasmas. Therefore, pest control and detection of phytoplasmas are important. Further analysis of phytoplasmas at the molecular level will increase our understanding of these economically important and biologically fascinating microorganisms.

REFERENCES

1. Abramovitch RB, Anderson JC, Martin GB (2006) Bacterial elicitation and evasion of plant innate immunity. Nat Rev Mol Cell Biol 7:601–611
2. Anonymous (1919) Annu Rep Kochi Agric Res Inst: 62
3. Arashida R, Kakizawa S, Hoshi A, Ishii Y, Jung H-Y, Kagiwada S, Yamaji Y, Oshima K, Namba S (2008a) Heterogeneic dynamics of the structures of multiple gene clusters in two pathogenetically different lines originating from the same phytoplasma. DNA Cell Biol 27:209–217
4. Arashida R, Kakizawa S, Ishii Y, Hoshi A, Jung H-Y, Kagiwada S, Yamaji Y, Oshima K, Namba S (2008b) Cloning and characterization of the antigenic membrane protein (Amp) gene and in situ detection of Amp from malformed flowers infected with Japanese hydrangea phyllody phytoplasma. Phytopathology 98:769–775
5. Bai X, Zhang J, Ewing A, Miller SA, Radek A, Schevchenko DV, Tsukerman K, Walunas T, Lapidus A, Campbell JW, Hogenhout SA (2006) Living with genome instability: the adaptation of phytoplasmas to diverse environments of their insect and plant hosts. J Bacteriol 188:3682–3696
6. Bai X, Correa VR, Toruño TY, Ammar E-D, Kamoun S, Hogenhout SA (2009) AY-WB phytoplasma secretes a protein that targets plant cell nuclei. Mol Plant Microbe Interact 22:18–30
7. Brown SE, Been BO, McLaughlin WA (2006) Detection and variability of the lethal yellowing group (16Sr IV) phytoplasmas in the *Cedusa* sp. (Hemiptera: Auchenorrhyncha: Derbidae) in Jamaica. Ann Appl Biol 149:53–62
8. Davis MJ, Tsai JH, Cox RL, McDaniel LL, Harrison NA (1988) Cloning of chromosomal and extrachromosomal DNA of the mycoplasmalike organism that causes maize bushy stunt disease. Mol Plant Microbe Interact 1:295–302
9. Deng S, Hiruki C (1991) Amplification of 16S rRNA genes from culturable and nonculturable mollicutes. J Microbiol Methods 14:53–61

10. Doi Y, Teranaka M, Yora K, Asuyama H (1967) Mycoplasma- or PLT group-like microorganisms found in the phloem elements of plants infected with mulberry dwarf, potato witches' broom, aster yellows or paulownia witches' broom (in Japanese with English summary). Ann Phytopath Soc Japan 33:259–266
11. Economou A (1999) Following the leader: bacterial protein export through the Sec pathway. Trends Microbiol 7:315–320
12. Firrao G, Garcia-Chapa M, Marzachì C (2007) Phytoplasmas: genetics, diagnosis and relationships with the plant and insect host. Front Biosci 12:1353–1375
13. Galetto L, Bosco D, Balestrini R, Genre A, Fletcher J, Marzachì C (2011) The major antigenic membrane protein of "*Candidatus* Phytoplasma asteris" selectively interacts with ATP synthase and actin of leafhopper vectors. PLoS ONE 6:e22571
14. Granados RR, Maramorosch K, Shikata E (1968) Mycoplasma: suspected etiologic agent of corn stunt. Proc Natl Acad Sci USA 60:841–844
15. Himeno M, Neriya Y, Minato N, Miura C, Sugawara K, Ishii Y, Yamaji Y, Kakizawa S, Oshima K, Namba S (2011) Unique morphological changes in plant pathogenic phytoplasma-infected petunia flowers are related to transcriptional regulation of floral homeotic genes in an organ-specific manner. Plant J 67:971–979
16. Himeno M, Kitazawa Y, Yoshida T, Maejima K, Yamaji Y, Oshima K, Namba S (2014) Purple top symptoms are associated with reduction of leaf cell death in phytoplasma-infected plants. Sci Rep 4:4111
17. Hiruki C (1999) Paulownia witches'-broom disease important in East Asia. Acta Hort 496:63–68
18. Hiruki C, da Rocha A (1986) Histochemical diagnosis of mycoplasma infections in *Catharanthus roseus* by means of a fluorescent DNA-binding agent, 4,6-diamidino-2-phenylindole-2HCl (DAPI). Can J Plant Pathol 8:185–188
19. Hogenhout SA, Oshima K, Ammar E-D, Kakizawa S, Kingdom HN, Namba S (2008) Phytoplasmas: bacteria that manipulate plants and insects. Mol Plant Pathol 9:403–423
20. Hoshi A, Oshima K, Kakizawa S, Ishii Y, Ozeki J, Hashimoto M, Komatsu K, Kagiwada S, Yamaji Y, Namba S (2009) A unique virulence factor for proliferation and dwarfism in plants identified from a phytopathogenic bacterium. Proc Natl Acad Sci USA 106:6416–6421
21. Hull R, Home RW, Nayar RM (1969) Mycoplasma-like bodies associated with sandal spike disease. Nature 224:1121–1122
22. IRPCM (2004) '*Candidatus* Phytoplasma', a taxon for the wall-less, non-helical prokaryotes that colonize plant phloem and insects. Int J Syst Evol Microbiol 54:1243–1255
23. Ishii Y, Kakizawa S, Hoshi A, Maejima K, Kagiwada S, Yamaji Y, Oshima K, Namba S (2009a) In the non-insect-transmissible line of onion yellows phytoplasma (OY-NIM), the plasmid-encoded transmembrane protein ORF3 lacks the major promoter region. Microbiology 155:2058–2067
24. Ishii Y, Oshima K, Kakizawa S, Hoshi A, Maejima K, Kagiwada S, Yamaji Y, Namba S (2009b) Process of reductive evolution during 10 years in plasmids of a non-insect-transmissible phytoplasma. Gene 446:51–57
25. Ishiie T, Doi Y, Yora K, Asuyama H (1967) Suppressive effects of antibiotics of tetracycline group on symptom development of mulberry dwarf disease. Ann Phytopath Soc Japan 33:267–275
26. Ito K (1960) Witches' broom of paulownia trees (in Japanese). Forest Pests News 9:221–225
27. Jiang H, Saiki T, Watanabe K, Kawakita H, Sato M (2005) Possible vector insect of mulberry dwarf phytoplasma, *Tautoneura mori* Matsumura. J Gen Plant Pathol 71:370–372
28. Jomantiene R, Davis RE (2006) Clusters of diverse genes existing as multiple, sequence-variable mosaics in a phytoplasma genome. FEMS Microbiol Lett 255:59–65

29. Jung H-Y, Sawayanagi T, Kakizawa S, Nishigawa H, Miyata S, Oshima K, Ugaki M, Lee J-T, Hibi T, Namba S (2002) '*Candidatus* Phytoplasma castaneae', a novel phytoplasma taxon associated with chestnut witches' broom disease. Int J Syst Evol Microbiol 52:1543–1549
30. Jung H-Y, Miyata S, Oshima K, Kakizawa S, Nishigawa H, Wei W, Suzuki S, Ugaki M, Hibi T, Namba S (2003a) First complete nucleotide sequence and heterologous gene organization of the two rRNA operons in the phytoplasma genome. DNA Cell Biol 22:209–215
31. Jung H-Y, Sawayanagi T, Kakizawa S, Nishigawa H, Wei W, Oshima K, Miyata S, Ugaki M, Hibi T, Namba S (2003b) '*Candidatus* Phytoplasma ziziphi', a novel phytoplasma taxon associated with jujube witches'-broom disease. Int J Syst Evol Microbiol 53:1037–1041
32. Jung H-Y, Sawayanagi T, Wongkaew P, Kakizawa S, Nishigawa H, Wei W, Oshima K, Miyata S, Ugaki M, Hibi T, Namba S (2003c) '*Candidatus* Phytoplasma oryzae', a novel phytoplasma taxon associated with rice yellow dwarf disease. Int J Syst Evol Microbiol 53:1925–1929
33. Kakizawa S, Oshima K, Kuboyama T, Nishigawa H, Jung H-Y, Sawayanagi T, Tsuchizaki T, Miyata S, Ugaki M, Namba S (2001) Cloning and expression analysis of *Phytoplasma* protein translocation genes. Mol Plant Microbe Interact 14:1043–1050
34. Kakizawa S, Oshima K, Nishigawa H, Jung H-Y, Wei W, Suzuki S, Tanaka M, Miyata S, Ugaki M, Namba S (2004) Secretion of immunodominant membrane protein from onion yellows phytoplasma through the Sec protein-translocation system in *Escherichia coli*. Microbiology 150:135–142
35. Kakizawa S, Oshima K, Jung H-Y, Suzuki S, Nishigawa H, Arashida R, Miyata S, Ugaki M, Kishino H, Namba S (2006a) Positive selection acting on a surface membrane protein of the plant pathogenic phytoplasmas. J Bacteriol 188:3424–3428
36. Kakizawa S, Oshima K, Namba S (2006b) Diversity and functional importance of phytoplasma membrane proteins. Trends Microbiol 14:254–256
37. Kakizawa S, Oshima K, Ishii Y, Hoshi A, Maejima K, Jung H-Y, Yamaji Y, Namba S (2009) Cloning of immunodominant membrane protein genes of phytoplasmas and their *in planta* expression. FEMS Microbiol Lett 293:92–101
38. Kanehira T, Horikoshi N, Yamakita Y, Shinohara M (1996) Occurrence of hydrangea phyllody in Japan and detection of the causal phytoplasma (in Japanese with English summary). Ann Phytopathol Soc Jpn 62:537–540
39. Kato S, Shiomi T, Wakibe H, Iwanami S (1988) Tomato yellows transmitted by the leafhopper vector, *Macrosteles orientalis* Virbaste (in Japanese with English summary). Ann Phytopath Soc Japan 54:220–223
40. Kawakami T (1902) On the hexenbesen of *Paulownia tomentosa* (in Japanese). Shokwabo, Tokyo
41. Kirkpatrick BC, Stenger DC, Morris TJ, Purcell AH (1987) Cloning and detection of DNA from a nonculturable plant pathogenic mycoplasma-like organism. Science 238:197–200
42. Kishi K (1987) The world of loupe in plant pathology (in Japanese). Ann Phytopath Soc Japan 53:275–278
43. Komori N (1966) Occurrence and control of rice yellow dwarf disease in Ibaraki Prefecture (in Japanese). Plant Prot 20:285–288
44. Kube M, Schneider B, Kuhl H, Dandekar T, Heitmann K, Migdoll AM, Reinhardt R, Seemüller E (2008) The linear chromosome of the plant-pathogenic mycoplasma '*Candidatus* Phytoplasma mali'. BMC Genom 9:306
45. Kuboyama T, Huang C–C, Lu X, Sawayanagi T, Kanazawa T, Kagami T, Matsuda I, Tsuchizaki T, Namba S (1998) A plasmid isolated from phytopathogenic onion yellows phytoplasma and its heterogeneity in the pathogenic phytoplasma mutant. Mol Plant Microbe Interact 11:1031–1037
46. Kunkel LO (1926) Studies on aster yellows. Am J Bot 13:646–705

47. Kuske CR, Kirkpatrick BC (1992) Phylogenetic relationships between the western aster yellows mycoplasmalike organism and other prokaryotes established by 16S rRNA gene sequence. Int J Syst Bacteriol 42:226–233
48. Kusunoki M, Shiomi T, Kobayashi M, Okudaira T, Ohashi A, Nohira T (2002) A leafhopper (*Hishimonus sellatus*) transmits phylogenetically distant phytoplasmas: rhus yellows and hovenia witches' broom phytoplasma. J Gen Plant Pathol 68:147–154

1. Lee I-M, Hammond RW, Davis RE, Gundersen DE (1993) Universal amplification and analysis of pathogen 16S rDNA for classification and identification of mycoplasmalike organisms. Phytopathology 83:834–842
2. Lin CP, Chen TA (1985) Monoclonal antibodies against the aster yellows agent. Science 227:1233–1235
3. MacLean AM, Sugio A, Makarova OV, Findlay KC, Grieve VM, Toth R, Nicolaisen M, Hogenhout SA (2011) Phytoplasma effector SAP54 induces indeterminate leaf-like flower development in Arabidopsis plants. Plant Physiol 157:831–841
4. Maejima K, Iwai R, Himeno M, Komatsu K, Kitazawa Y, Fujita N, Ishikawa K, Fukuoka M, Minato N, Yamaji Y, Oshima K, Namba S (2014) Recognition of floral homeotic MADS-domain transcription factors by a phytoplasmal effector, phyllogen, induces phyllody. Plant J.
5. Miura C, Sugawara K, Neriya Y, Minato N, Keima T, Himeno M, Maejima K, Komatsu K, Yamaji Y, Oshima K, Namba S (2012) Functional characterization and gene expression profiling of superoxide dismutase from plant pathogenic phytoplasma. Gene 510:107–112
6. Miyata S, Furuki K, Oshima K, Sawayanagi T, Nishigawa H, Kakizawa S, Jung H-Y, Ugaki M, Namba S (2002a) Complete nucleotide sequence of the S10-spc operon of phytoplasma: gene organization and genetic code resemble those of Bacillus subtilis. DNA Cell Biol 21:527–534
7. Miyata S, Furuki K, Sawayanagi T, Oshima K, Kuboyama T, Tsuchizaki T, Ugaki M, Namba S (2002b) Gene arrangement and sequence of str operon of phytoplasma resemble those of Bacillus more than those of Mycoplasma. J Gen Plant Pathol 68:62–67
8. Miyata S, Oshima K, Kakizawa S, Nishigawa H, Jung H-Y, Kuboyama T, Ugaki M, Namba S (2003) Two different thymidylate kinase gene homologues, including one that has catalytic activity, are encoded in the onion yellows phytoplasma genome. Microbiology 149:2243–2250
9. Mushegian AR, Koonin EV (1996) A minimal gene set for cellular life derived by comparison of complete bacterial genomes. Proc Natl Acad Sci USA 93:10268–10273
10. Nakajima S, Nishimura N, Matsuda I, Shiomi T, Namba S, Tsuchizaki T (2002) Detection of mulberry dwarf and onion yellows phytoplasmas by PCR from vector insects and nonvector insects (in Japanese with English summary). Jpn J Phytopathol 68:39–42
11. Nakajima S, Nishimura N, Jung H-Y, Kakizawa S, Fujisawa I, Namba S, Tsuchizaki T (2009) Movement of onion yellows phytoplasma and Cryptotaenia japonica witches' broom phytoplasma in the nonvector insect Nephotettix cincticeps (in Japanese with English summary). Jpn J Phtopathol 75:29–34
12. Nakamori K, Maezato T (1968) Control of sweet potato witches' broom disease in Okinawa (in Japanese). Plant Prot 22:19–24
13. Nakamura H, Ohgake S, Sahashi N, Yoshioka N, Kubono T, Takahashi T (1998) Seasonal variation of paulownia witches'-broom phytoplasma in paulownia trees and distribution of the disease in the Tohoku district of Japan. J Forest Res 3:39–42
14. Nakashima K, Hayashi T (1995a) Extrachromosomal DNAs of rice yellow dwarf and sugarcane white leaf phytoplasmas. Ann Phytopathol Soc Jpn 61:456–462

15. Nakashima K, Hayashi T (1995b) Multiplication and distribution of rice yellow dwarf phytoplasma in infected tissues of rice and green rice leafhopper Nephotettix cincticeps. Ann Phytopathol Soc Jpn 61:451–455
16. Nakashima K, Hayashi T (1997) Sequence analysis of extrachromosomal DNA of sugarcane white leaf phytoplasma. Ann Phytopathol Soc Jpn 63:21–25
17. Nakashima K, Kato S, Iwanami S, Murata N (1991) Cloning and detection of chromosomal and extrachromosomal DNA from mycoplasmalike organisms that cause yellow dwarf disease of rice. Appl Environ Microbiol 57:3570–3575
18. Namba S (1996) Taxonomy of phytoplasmas (in Japanese). Plant Prot 50:152–156
19. Namba S, Yamashita S, Doi Y, Yora K (1981) Direct fluorescence detection method (DFD method) for diagnosing yellow-type virus diseases and mycoplasma diseases of plants (in Japanese with English summary). Ann Phytopath Soc Japan 47:258–263
20. Namba S, Kato S, Iwanami S, Oyaizu H, Shiozawa H, Tsuchizaki T (1993a) Detection and differentiation of plant-pathogenic mycoplasmalike organisms using polymerase chain reaction. Phytopathology 83:786–791
21. Namba S, Oyaizu H, Kato S, Iwanami S, Tsuchizaki T (1993b) Phylogenetic diversity of phytopathogenic mycoplasmalike organisms. Int J Syst Bacteriol 43:461–467
22. Neriya Y, Sugawara K, Maejima K, Hashimoto M, Komatsu K, Minato N, Miura C, Kakizawa S, Yamaji Y, Oshima K, Namba S (2011) Cloning, expression analysis, and sequence diversity of genes encoding two different immunodominant membrane proteins in poinsettia branch-inducing phytoplasma (PoiBI). FEMS Microbiol Lett 324:38–47
23. Nishigawa H, Miyata S, Oshima K, Sawayanagi T, Komoto A, Kuboyama T, Matsuda I, Tsuchizaki T, Namba S (2001) In planta expression of a protein encoded by the extrachromosomal DNA of a phytoplasma and related to geminivirus replication proteins. Microbiology 147:507–513
24. Nishigawa H, Oshima K, Kakizawa S, Jung H-Y, Kuboyama T, Miyata S, Ugaki M, Namba S (2002a) Evidence of intermolecular recombination between extrachromosomal DNAs in phytoplasma: a trigger for the biological diversity of phytoplasma? Microbiology 148:1389–1396
25. Nishigawa H, Oshima K, Kakizawa S, Jung H-Y, Kuboyama T, Miyata S, Ugaki M, Namba S (2002b) A plasmid from a non-insect-transmissible line of a phytoplasma lacks two open reading frames that exist in the plasmid from the wild-type line. Gene 298:195–201
26. Nishigawa H, Oshima K, Miyata S, Ugaki M, Namba S (2003) Complete set of extrachromosomal DNAs from three pathogenic lines of onion yellows phytoplasma and use of PCR to differentiate each line. J Gen Plant Pathol 69:194–198
27. Nishimura N, Nakajima S, Sawayanagi T, Namba S, Shiomi T, Matsuda I, Tsuchizaki T (1998) Transmission of Cryptotaenia japonica witches' broom and onion yellows phytoplasmas by Hishimonus sellatus Uhler. Ann Phytopathol Soc Jpn 64:474–477
28. Nishimura N, Nakajima S, Kawakita H, Sato M, Namba S, Fujisawa I, Tsuchizaki T (2004) Transmission of Cryptotaenia japonica witches' broom and onion yellows by Hishimonoides sellatiformis (in Japanese with English summary). Jpn J Phytopathol 70:22–25
29. Notomi T, Okayama H, Masubuchi H, Yonekawa T, Watanabe K, Amino N, Hase T (2000) Loop-mediated isothermal amplification of DNA. Nucleic Acids Res 28:e63
30. Okuda S (1972) Phytoplasmal diseases in Japan (in Japanese). Plant Prot 26:180–183
31. Oshima K, Kakizawa S, Nishigawa H, Kuboyama T, Miyata S, Ugaki M, Namba S (2001a) A plasmid of phytoplasma encodes a unique replication protein having both plasmid- and virus-like domains: clue to viral ancestry or result of virus/plasmid recombination? Virology 285:270–277
32. Oshima K, Shiomi T, Kuboyama T, Sawayanagi T, Nishigawa H, Kakizawa S, Miyata S, Ugaki M, Namba S (2001b) Isolation and characterization of derivative lines of the onion yellows phytoplasma that do not cause stunting or phloem hyperplasia. Phytopathology 91:1024–1029

33. Oshima K, Miyata S, Sawayanagi T, Kakizawa S, Nishigawa H, Jung H-Y, Furuki K, Yanazaki M, Suzuki S, Wei W, Kuboyama T, Ugaki M, Namba S (2002) Minimal set of metabolic pathways suggested from the genome of onion yellows phytoplasma. J Gen Plant Pathol 68:225–236
34. Oshima K, Kakizawa S, Nishigawa H, Jung H-Y, Wei W, Suzuki S, Arashida R, Nakata D, Miyata S, Ugaki M, Namba S (2004) Reductive evolution suggested from the complete genome sequence of a plant-pathogenic phytoplasma. Nat Genet 36:27–29
35. Oshima K, Kakizawa S, Arashida R, Ishii Y, Hoshi A, Hayashi Y, Kagiwada S, Namba S (2007) Presence of two glycolytic gene clusters in a severe pathogenic line of Candidatus Phytoplasma asteris. Mol Plant Pathol 8:481–489
36. Oshima K, Ishii Y, Kakizawa S, Sugawara K, Neriya Y, Himeno M, Minato N, Miura C, Shiraishi T, Yamaji Y, Namba S (2011) Dramatic transcriptional changes in an intracellular parasite enable host switching between plant and insect. PLoS ONE 6:e23242
37. Oshima K, Maejima K, Namba S (2013) Genomic and evolutionary aspects of phytoplasmas. Front Microbiol 4:230
38. PSJ NIAS (ed) (2012) Common names of plant diseases in Japan, 2nd edn (in Japanese). Phase out, Aichi
39. Razin S, Yogev D, Naot Y (1998) Molecular biology and pathogenicity of mycoplasmas. Microbiol Mol Biol Rev 62:1094–1156
40. Sahashi N, Nakamura H, Yoshikawa N, Kubono T, Shoji T, Takahashi T (1995) Distribution and seasonal variation in detection of phytoplasma in bark phloem tissues of single paulownia trees infected with witches' broom. Ann Phytopathol Soc Jpn 61:481–484
41. Saitou N, Nei M (1987) The neighbor-joining method: a new method for reconstructing phylogenetic trees. Mol Biol Evol 4:406–425
42. Sameshima T (1967) Occurrence and control of rice yellow dwarf disease in Miyazaki Prefecture (in Japanese). Plant Prot 21:47–50
43. Sawayanagi T, Horikoshi N, Kanehira T, Shinohara M, Bertaccini A, Cousin M-T, Hiruki C, Namba S (1999) 'Candidatus Phytoplasma japonicum', a new phytoplasma taxon associated with Japanese hydrangea phyllody. Int J Syst Bacteriol 49:1275–1285
44. Schneider B, Seemüller E, Smart CD, Kirkpatrick BC (1995) Phylogenetic classification of plant pathogenic mycoplasma-like organisms or phytoplasmas. In: Razin S, Tully JG (eds) Molecular and diagnostic procedures in mycoplasmology. Academic Press, San Diego, pp 369–380
45. Shinkai A (1964) Transmission of sweet potato witches' broom disease by Orosius ryukyuensis (in Japanese). Plant Prot 18:259–262
46. Shiomi T, Sugiura M (1983) Water dropwort yellows and chrysanthemum witches' broom occurred in Ishikawa Prefecture (in Japanese with English summary). Ann Phytopath Soc Japan 49:367–370
47. Shiomi T, Tanaka M, Wakiya H, Zenbayashi R (1996) Occurrence of welsh onion yellows (in Japanese with English summary). Ann Phytopathol Soc Jpn 62:258–260
48. Shiomi T, Tanaka M, Sawayanagi T, Yamamoto S, Tsuchizaki T, Namba S (1998) A symptomatic mutant of onion yellows phytoplasma derived from a greenhouse-maintained isolate (in Japanese with English summary). Ann Phytopathol Soc Jpn 64:501–505
49. Shiomi T, Tanaka M, Uematsu S, Wakibe H, Nakamura H (1999) Statice witches' broom, a Macrosteles striifrons-borne phytoplasma disease (in Japanese with English summary). Ann Phytopathol Soc Jpn 65:87–90
50. Shirai M (1890) The cause of mulberry dwarf disease (in Japanese). Jpn Silk Culturist 32:1–2
51. Strauss E (2009) Phytoplasma research begins to bloom. Science 325:388–390
52. Sugawara K, Himeno M, Keima T, Kitazawa Y, Maejima K, Oshima K, Namba S (2012) Rapid and reliable detection of phytoplasma by loop-mediated isothermal amplification targeting a house-keeping gene. J Gen Plant Pathol 78:389–397

53. Sugawara K, Honma Y, Komatsu K, Himeno M, Oshima K, Namba S (2013) The alteration of plant morphology by small peptides released from the proteolytic processing of the bacterial peptide TENGU. Plant Physiol 162:2005–2014
54. Sugio A, Kingdom HN, MacLean AM, Grieve VM, Hogenhout SA (2011) Phytoplasma protein effector SAP11 enhances insect vector reproduction by manipulating plant development and defense hormone biosynthesis. Proc Natl Acad Sci USA 108:E1254–E1263
55. Suzuki S, Oshima K, Kakizawa S, Arashida R, Jung H-Y, Yamaji Y, Nishigawa H, Ugaki M, Namba S (2006) Interaction between the membrane protein of a pathogen and insect microfilament complex determines insect–vector specificity. Proc Natl Acad Sci USA 103:4252–4257
56. Takinami Y, Maejima K, Takahashi A, Keima T, Shiraishi T, Okano Y, Komatsu K, Oshima K, Namba S (2013) First report of 'Candidatus Phytoplasma asteris' infecting hydrangea showing phyllody in Japan. J Gen Plant Pathol 79:209–213
57. Tanaka M, Shiomi T, Tairako K, Uehara T, Matsuda I (1998) Occurrence of glaucidium witches' broom (in Japanese with English summary). Ann Phytopathol Soc Jpn 64:205–207
58. Tanaka M, Osada S, Matsuda I (2000) Transmission of rhus (Rhus javanica L.) yellows by Hishimonus sellatus and host range of the causal phytoplasma. J Gen Plant Pathol 66:323–326
59. Tanaka M, Kodama K, Chiba K, Iwadate Y, Usugi T (2006) Occurrence of gentian witches' broom caused by 'Candidatus Phytoplasma asteris' (in Japanese with English summary). Jpn J Phytopathol 72:191–194
60. Tanaka M, Tanina K, Kasuyama S, Usugi T (2007) Occurrence of rocket larkspur witches' broom caused by "Candidatus Phytoplasma asteris" in Japan. J Gen Plant Pathol 73:286–289
61. Tomlinson JA, Boonham N, Dickinson M (2010) Development and evaluation of a one-hour DNA extraction and loop-mediated isothermal amplification assay for rapid detection of phytoplasmas. Plant Pathol 59:465–471
62. Toruño TY, Musić MS, Simi S, Nicolaisen M, Hogenhout SA (2010) Phytoplasma PMU1 exists as linear chromosomal and circular extrachromosomal elements and has enhanced expression in insect vectors compared with plant hosts. Mol Microbiol 77:1406–1415
63. Tran-Nguyen LTT, Kube M, Schneider B, Reinhardt R, Gibb KS (2008) Comparative genome analysis of "Candidatus Phytoplasma australiense" (subgroup tuf-Australia I; rp-A) and "Ca. Phytoplasma asteris" strains OY-M and AY-WB. J Bacteriol 190:3979–3991
64. Uehara T, Tanaka M, Shiomi T, Namba S, Tsuchizaki T, Matsuda I (1999) Histopathological studies on two symptom types of phytoplasma associated with lettuce yellows. Ann Phytopathol Soc Jpn 65:465–469
65. Wakibe H, Shiomi T, Tanaka M (1995) Cosmos yellows disease transmitted by Macrosteles striifrons in Saga Prefecture (in Japanese with English summary). Proc Assoc Plant Prot Kyushu 41:51–54
66. Wei W, Kakizawa S, Jung H-Y, Suzuki S, Tanaka M, Nishigawa H, Miyata S, Oshima K, Ugaki M, Hibi T, Namba S (2004a) An antibody against the SecA membrane protein of one phytoplasma reacts with those of phylogenetically different phytoplasmas. Phytopathology 94:683–686
67. Wei W, Kakizawa S, Suzuki S, Jung H-Y, Nishigawa H, Miyata S, Oshima K, Ugaki M, Hibi T, Namba S (2004b) In planta dynamic analysis of onion yellows phytoplasma using localized inoculation by insect transmission. Phytopathology 94:244–250
68. Yue HN, Wu YF, Shi YZ, Wu KK, Li YR (2008) First report of paulownia witches'-broom phytoplasma in China. Plant Dis 92:1134

CHAPTER 5

Keeping Control: The Role of Senescence and Development in Plant Pathogenesis and Defense

Eva Häffner [1,*], Sandra Konietzki [1] and Elke Diederichsen [1,2]

[1]*Freie Universität Berlin, Fachbereich Biologie, Chemie, Pharmazie, Institut für Biologie, Dahlem Centre of Plant Sciences, Angewandte Genetik, Albrecht-Thaer-Weg 6, 14195 Berlin, Germany*

[2]*Norddeutsche Pflanzenzucht H.G. Lembke KG, Hohenlieth, D-24363 Holtsee, Germany*

ABSTRACT

Many plant pathogens show interactions with host development. Pathogens may modify plant development according to their nutritional demands. Conversely, plant development influences pathogen growth. Biotrophic pathogens often delay senescence to keep host cells alive, and resistance is achieved by senescence-like processes in the host. Necrotrophic pathogens promote senescence in the host, and preventing early senescence is a resistance strategy of plants. For hemibiotrophic pathogens both patterns may apply. Most signaling pathways are involved in both developmental and defense reactions. Increasing knowledge about the molecular components allows to distinguish signaling branches, cross-talk and regulatory nodes that may influence the outcome of an infection. In this review, recent reports on major molecular players and their role in senescence and in pathogen response are reviewed. Examples of pathosystems with strong developmental implications illustrate the molecular basis of selected control strategies. A study of gene expression in the interaction between the hemibiotrophic vascular pathogen *Verticillium longisporum* and its cruciferous hosts shows processes that are fine-tuned to counteract early senescence and to achieve resistance. The complexity of the processes involved reflects the complex genetic control of quantitative disease resistance, and understanding the relationship between disease, development and resistance will support resistance breeding.

Keywords: senescence; development; pathogen; resistance; biotroph; necrotroph; hemibiotroph; *Verticillium*; signaling; gene expression

1. INTRODUCTION

Plant senescence, the developmental program leading to nutrient remobilization and finally the degradation and death of tissue, can be triggered by many factors, both intrinsic and extrinsic [1]. Pathogen infections interact with developmental processes in a complex way. The outcome of this interaction may greatly influence plant productivity as well as resistance and susceptibility. Pathogenesis and development mutually influence each other. On the one hand, developmental conditions of the host plant may determine the outcome of pathogen infection. On the other hand, pathogen infection can change the developmental program of the host [2]. Symptoms of senescence often accompany the progression of disease. In other cases, senescence is delayed in response to pathogen infection. The lifestyle of the pathogen plays an important role for the developmental response of the host [3,4]. Necrotrophic pathogens derive nutrients from dead cells, whereas biotrophic pathogens need living cells to meet their nutritional demands. Between both extremes, a third nutritional type is recognized: the hemibiotrophs. Hemibiotrophs adopt both lifestyles, switching from a biotrophic phase at the beginning of the infection to a necrotrophic lifestyle as pathogenesis proceeds [5]. Generally, senescence and cell death are regarded as favorable to the necrotrophic lifestyle, and necrotrophic pathogens have adopted strategies to induce senescence by manipulating the respective signaling pathways in plants [4]. Consequently, a defense strategy against necrotrophs would consist of the host's ability to prevent senescence-like processes. Conversely, biotrophic pathogens benefit from living cells and change signaling processes in the host to delay senescence. The most efficient defense strategy against biotrophs is therefore the controlled induction of cell death during the hypersensitive response (HR), thus starving the pathogen in confined regions. It can easily be imagined that the situation is especially complex for hemibiotrophs. Resistance against necrotrophs and hemibiotrophs is often quantitative, *i.e.*, restricting, but not eliminating the pathogen. Figure 1 gives an overview on developmental modifications of hosts by pathogens with different lifestyles and the respective resistance reactions of the host.

The constantly growing knowledge about the molecular mechanisms controlling development and pathogen response provides us with the opportunity to recognize patterns and strategies that promote resistance and increase productivity of the host plant. Intriguingly, the same signaling molecules often control both senescence and pathogen defense. Especially in quantitative disease resistance, molecular responses to pathogen infection are often difficult to interpret: They may be induced by the pathogen for its own benefit, or they may be defense reactions of the host. Studies using resistant and susceptible genotypes as well as mutants and transgenes contribute to disentangling the complex network of defense and development. It becomes more and more obvious that the respective signaling pathways that are triggered by relatively few and simple molecules have multiple functions, that they are cross-linked in multiple ways and that individual branches can be switched on or off by both host and pathogen. Major interest lies on components that represent regulatory nodes and points of convergence, since they may be decisive for resistance or susceptibility.

In the present article, the interrelation between pathogenesis and senescence is treated on two levels: In Chapter 2, current knowledge about major players in the control of senescence and pathogen defense is reviewed. Signaling pathways and their role in senescence and defense against pathogens are described and regulatory nodes and pathway integrators are highlighted. Chapter 3 gives examples of host-pathogen-interactions with strong developmental implications—to see "the players in action" and learn about control strategies of plants and pathogens. Focus is laid on pathogenesis caused by necrotrophic and hemibiotrophic pathogens and its interaction with plant senescence. Complex mutual influence of pathogenesis and development is observed in the interaction between the hemibiotrophic vascular pathogen *Verticillium longisporum* and its cruciferous hosts. The most obvious developmental implication, premature senescence, is at the same time the most problematic disease symptom caused by *V. longisporum* in oilseed rape crops, leading to massive yield losses [6]. Over the past years, we have collected data giving a precise description of the interaction between development and pathogenesis/resistance in *Brassica* spp. and in *Arabidopsis thaliana*. We have used *A. thaliana* to address the underlying molecular mechanisms. Some of our published and unpublished data on the development-pathogenesis-defense-interplay are included in the present review.

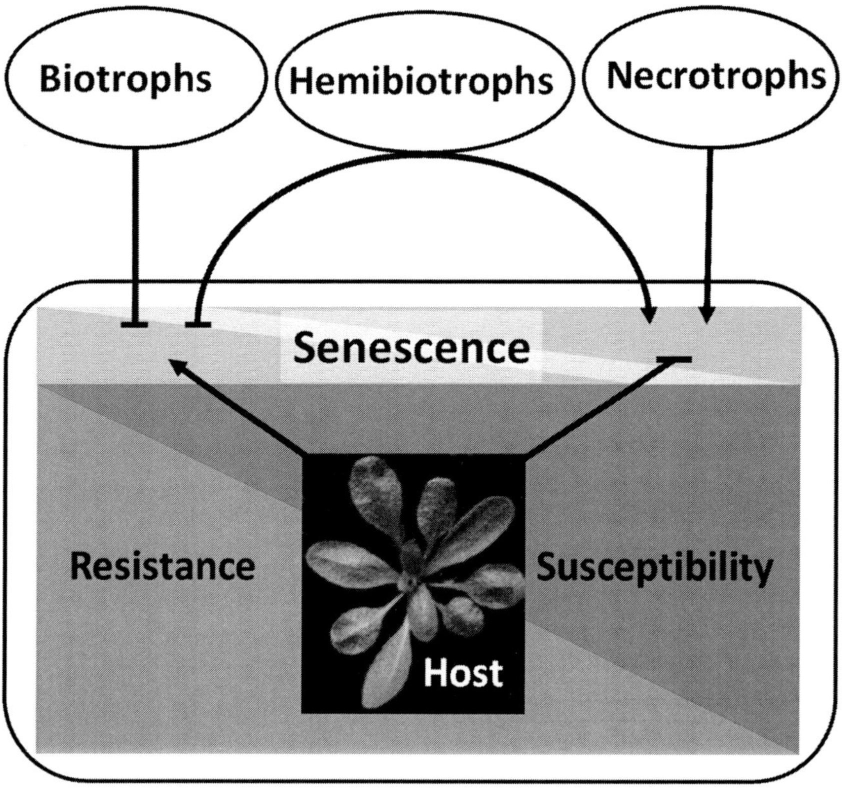

Figure 1. Relationship between senescence and resistance/susceptibility in necrotrophic and biotrophic host-pathogen interactions. Biotrophic pathogens and hemibiotrophs in their biotrophic stage inhibit senescence to increase susceptibility. The host can control pathogen growth and promote resistance by activating senescence-like processes. Necrotrophic pathogens and hemibiotrophs in their necrotrophic stage induce senescence to increase susceptibility. The host can control pathogen growth and promote resistance by inhibiting senescence-like processes. Hemibiotrophs switch from a biotrophic to a necrotrophic lifestyle in the course of their development. Arrows stand for activation/induction, bars for inhibition.

2. MAJOR PLAYERS IN THE CONTROL OF SENESCENCE AND PATHOGEN RESPONSE

Among the most important players in the control of senescence and pathogen defense are phytohormones like ethylene (ET), jasmonic acid (JA), salicylic acid (SA) and abscisic acid (ABA), and to a minor extent brassinosteroids (BR) as hormones that promote senescence. Cytokinins, auxins, and gibberellins can be regarded as counteracting hormones that delay senescence. Transcription factors play a pivotal role in all signaling networks as they target the responsive genes. They often act as regulatory nodes between signaling branches and thus contribute to the fine-tuning of developmental or defense reactions. Reactive oxygen species (ROS) control senescence and cell death as well as the early events after pathogen recognition. Since the signaling networks behind these key players are each highly complex and cross-linked among each other, mainly those modules relevant for senescence and pathogen response and nodes of interaction and integration will be pointed out. If not mentioned otherwise, all reviewed studies used the model plant *Arabidopsis thaliana*.

2.1. Phytohormones

2.1.1. Ethylene

The gaseous plant hormone ET controls many developmental processes as well as responses to environmental stimuli [7]. Its role in the interaction with necrotrophic and hemibiotrophic pathogens is characterized by two opposite effects: ET primarily promotes senescence [8,9] and is involved in programmed cell death during HR [10]. Cell ageing and death is beneficial to necrotrophic pathogens. Accordingly, inducing or synthesizing ET is a virulence strategy applied by many necrotrophic or hemibiotrophic pathogens (e.g., [11,12,13]). However, ET interacts synergistically with JA in the activation of defense against necrotrophic pathogens, thereby promoting resistance [14].

The rate-limiting step of ET biosynthesis is catalyzed by 1-AMINOPROPANE-1-CARBOXYLATE (ACC-) synthases (ACS). In *A. thaliana*, nine ACS isoforms exist that are differentially regulated to adjust ET levels to changing needs [15,16]. ET is perceived in *A. thaliana* via five different receptors located in the endoplasmic reticulum [17,18]. In the absence of ET, they activate the serine-threonine-kinase CONSTITUTIVE TRIPLE RESPONSE 1 (CTR1), which in turn represses the activity of ETHYLENE-INSENSITIVE 2 (EIN2) by phosphorylation [19]. Upon the binding of ET to the receptors, CTR1 is inactivated and EIN2 mediates the stabilization of the transcription factors EIN3 and EIN3-LIKE 1 (EIL1) in the nucleus [20,21]. EIN3 and EIL1 activate a great diversity of primary responsive genes like ET response factors (*Erf*). ERF transcription factors induce secondary responsive genes, thus modulating a great variety of developmental and stress-induced responses [22].

The senescence-promoting effect of ET has long been known [23,24]. ET is neither required nor sufficient to induce senescence, but it can accelerate the process depending on different genetic and environmental conditions. Some of the genes modulating the plant's receptiveness to ET-mediated induction of senescence have been identified [9,25]. A mechanism by which ET accelerates leaf senescence has been elucidated only recently [26]: The central transcription factor EIN3 has been shown to repress the transcription of micro-RNA miR164, thereby allowing the accumulation of its target gene product, the ARABIDOPSIS NAC DOMAIN CONTAINING PROTEIN 92 (ANAC092)/ORESARA 1 (ORE1), a transcription factor that positively regulates leaf senescence [27,28].

The activation of defense responses against necrotrophic pathogens, which is characterized by the production of PLANT DEFENSIN 1.2 (PDF1.2), requires JA and ET. Both hormones together induce transcriptional activation of the ET response factors ERF1 and OCTADECANOID-RESPONSIVE ARABIDOPSIS AP2/ERF 59 (ORA59) via EIN3 and EIL1 [29,30]. Generally, ET contributes to resistance against necrotrophic pathogens, and mutation of central signaling components results in enhanced susceptibility [10], but there are exceptions (e.g., [31]). In some cases, disrupting different components of the ET signaling pathway had opposite effects on the outcome of the infection. Even different receptor mutants can show very different pathogen responses: While *ein4* mutants of *A. thaliana* showed increased susceptibility to the hemibiotrophic fungus *Verticillium longisporum* [32], *etr1* mutants were more resistant than wild type [32,33]. Initially, the ET receptors were regarded as functionally redundant [34]; however, recent evidence supports the hypothesis that different receptor combinations can produce different outputs depending on the tissue and the ET concentration [35].

2.1.2. Jasmonic Acid

Much like ET, JA plays a prominent role in the control of senescence, cell death and in resistance to necrotrophic pathogens. JA is generated by lipid peroxidation in chloroplasts and peroxisomes. The physiologically active compound, the amino-acid conjugate jasmonoyl-isoleucine (JA-Ile) acts by de-repressing its target genes [36]. The JA-Ile receptor is a complex consisting of a JASMONATE ZIM DOMAIN (JAZ)-protein and the F-box-protein CORONATINE INSENSITIVE 1 (COI1). In the absence of JA-Ile, JAZ-proteins act as repressors of target genes by binding transcription factors such as MYC2 or EIN3. Upon recruitment by COI1, JAZ proteins are ubiquitinated by the SKP1-CULLIN-F-BOX- (SCF)-complex and targeted for degradation by the 26S proteasome. Transcription factors are released and target genes are transcribed [37,38,39].

The induction of leaf senescence was the first effect on plant development that was described for JA [40]. Exogenous application of methyl jasmonate can induce senescence in *A. thaliana* leaves and JA-levels rise during senescence. The facts that several JA biosynthesis genes are up-regulated during senescence and that the *coi1*-mutant was not responding to JA-induced senescence confirmed a regulatory role of the hormone in senescence [41]. Recent research has revealed some of the underlying molecular mechanisms. JA levels have been shown to be developmentally controlled by the transcription factors TCP2, 4 and 10. They exerted a direct effect on JA biosynthesis via transcriptional control of LIPOXYGENASE 2 (LOX2) [42]. JA signaling acts synergistically with ET as EIN3 and EIL1 are interactors of JAZ [30]. Release of JAZ repression of EIN3 and EIL1 allows downstream processes to occur (see above). The final proof whether this JA/ET synergism during defense reactions also contributes to the induction of senescence is still missing [43]. Another key factor in the JA-mediated induction of senescence has been shown to be RUBISCO ACTIVASE (RCA), which is transcriptionally down-regulated in a COI1-dependent manner [44]. Typical markers for senescence, like chlorophyll degradation and the expression of senescence-associated genes (SAGs) were associated with RCA-down-regulation. *rca*-mutants showed an early senescence phenotype and reduced transcription of JA-regulated genes [44]. Furthermore, JA has been shown to promote programmed cell death by inducing the production of ROS in mitochondria [45]. In another context, JA acted as a negative regulator of senescence: JA has been shown to induce the expression of the ETIOSPECIFYING SENESCENCE REGULATOR/ETIOSPECIFYER PROTEIN (ESR/ESP). This protein had an inhibitory effect on the senescence-promoting transcription factor WRKY53, which in turn was induced by SA. In the absence of WRKY53, ESR/ESP mediated the conversion of glucosinolates to nitriles, leading to enhanced resistance against *Alternaria brassicicola and Pseudomonas syringae* [46].

Synergism between the defense activation pathways controlled by JA and ET is mainly characterized by the induction of *Pdf1.2* via ERF1 and ORA59 [29]. This pathway is antagonized by two other major defense responses: One of them is triggered upon insect attack and wounding and requires synergism between JA and ABA. In this branch of JA signaling, transcription factor MYC2 is activated, which leads to expression of defense genes against insect attack as represented by the marker gene *Vegetative storage protein 2* (*Vsp2*). Antagonism takes place at the phytohormone level (ABA *vs.* ET) and at the transcription factor level (MYC2 *vs.* ERF1 and ORA59) [47]. The second signaling cascade repressing the ERF/ORA59-branch of JA-signaling is typically triggered by biotrophic pathogens and is based on the induction of SA upon recognition of pathogen-associated molecular patterns (PAMPs) or pathogen effectors (see below). Multiple molecular components mediate the antagonistic crosstalk between JA- and SA- activated defense pathways to achieve a high degree of flexibility in the fine-tuning of adequate defense reactions [47]. Besides this antagonism, JA has an important role in the establishment of systemic acquired resistance (SAR) [48].

2.1.3. Salicylic Acid

SA is primarily known as a defense hormone [49], but it has multiple roles in development as well [50]. SA has been shown to promote developmental senescence via *npr1* and *phytoalexin-deficient 4* (*pad4*) [51]. Furthermore, it has been shown to induce the transcription factor WRKY53, an important positive regulator of senescence and pathogen defense [52]. The signaling pathway leading to PATHOGENESIS-RELATED (PR) gene induction and SAR during PAMP-triggered immunity (PTI) and effector-triggered immunity (ETI) is the backbone of SA-mediated defense responses [49]. Activation of the transcriptional co-activator NONEXPRESSOR OF PR-GENES (NPR1) leads to the expression of a large set of defense-related genes such as PR1 and several WRKY transcription factors, and to the establishment of SAR [47]. Independent of NPR1, SA is also required for HR [53], during which controlled localized cell death occurs with participation of ROS [49]. During HR, biotrophic pathogens are encapsulated at the infection site and prevented from spreading to healthy parts of the plant.

2.1.4. Abscisic Acid

ABA has perhaps the most multi-faceted role in senescence regulation and in the control of defense against pathogens. Whereas ABA is generally regarded as a senescence-promoting hormone, its role in pathogen defense is ambiguous for biotrophic as well as necrotrophic pathogens [54]. Crosstalk with many other signaling pathways and cellular processes such as JA-, ET-, SA-, cytokinin- and sugar signaling, ROS production and signaling as well as autophagy exists [47,55,56,57,58].

ABA is perceived by receptor complexes consisting of an RCAR (REGULATORY COMPONENTS OF ABA RECEPTORS)/PYR1 (PYRABACTIN RESISTANT 1)/PYL (PYR-LIKE) protein [59,60] and a PROTEIN PHOSPHATASE 2C (PP2C), like ABA INSENSITIVE 1 (ABI1) and ABI2. In *A. thaliana*, 14 RCAR isoforms and nine RCAR-interacting PP2C-isoforms have been identified that can interact with each other in a combinatorial way, possibly addressing different downstream signaling components [61,62]. Downstream signaling involving SNF1-RELATED KINASEs (SnRKs) like OPEN STOMATA 1 (OST1) leads to the targeting of ion channels in the guard cells and to transcriptional activation or repression of target genes via transcription factors like ABI5 [62,63].

ABA plays a crucial role in developmental as well as stress-induced senescence [64]. During developmental senescence, ABA levels rise (e.g., [65,66]), and ABA biosynthesis genes are up-regulated [67]. ABA-inducible RECEPTOR PROTEIN KINASE 1 (RPK1) has been shown to play an important role in developmental leaf senescence [68]. ABA furthermore mediates the connection between senescence and carbohydrate metabolism: By down-regulating extracellular invertase via INHIBITOR OF CELL WALL INVERTASE (INVINH), ABA reduces sink strength of tissues. INVINH induction proved to be essential for ABA-induced senescence in tomato [69]. Cytokinins counteract ABA in the modulation of sink-source-relationships by inducing extracellular invertase and thus delaying senescence [70]. Extracellular invertase plays a crucial role in phloem unloading and providing sink tissues with carbohydrates [71]. ABA-responsive transcription factors that control senescence have also been identified: ATAF1 was shown to integrate H_2O_2- and ABA-induced senescence and acted upstream of ANAC092 [72]. Interestingly, the NAC transcription factor VND-INTERACTING 2 (VNI2) was activated by ABA under salinity stress and negatively regulated senescence. As a bifunctional transcription factor, it has been proposed to act as an integration point between senescence and stress regulatory pathways [73]. Cross-talk of ABA and SA in inducing developmental senescence is suggested by experiments with the *A. thaliana* mutant *saul1* that showed early developmental senescence in low light. The U-BOX ARMADILLO E3 UBIQUITIN LIGASE 1 (SAUL1) presumably targets the ABA-degrading enzyme AAO3 for degradation by the 26S-proteasome [74]. SA and PAD4- but not NPR1-dependent signaling were required for the early-senescence phenotype of *saul1* [75]. The PAD4-pathway is also required for cell death during HR.

ABA can have positive and negative effects on defense responses, which depends mainly on the pathogen. Mostly, ABA negatively affects resistance [76,77,78,79,80]. Accordingly, the hormone is produced by many pathogens, both biotrophic and necrotrophic, as a virulence factor [54]. ABA suppresses defense against biotrophs by antagonizing SA-production and the NPR1-branch of SA-signaling [81,82]. Potentially, also the JA/ET branch of JA-signaling can be antagonized by promoting the antagonistic JA-ABA branch [47]. Resistance-promoting effects of ABA were mainly observed in interactions with necrotrophic pathogens [83]. A positive effect of ABA on defense responses was exerted by priming for callose deposition [84]. Furthermore, ABA has been shown to play an important role in H_2O_2-homeostasis by controlling CATALASE 1- (CAT1) induction and H_2O_2-production [85]. As described below, ROS like H_2O_2 play an important role in early defense responses under controlled conditions, but become highly toxic at higher concentrations [3].

2.1.5. Brassinosteroids

BR are important plant growth hormones and have a senescence-promoting effect [86]; plants defective in BR biosynthesis like the *A. thaliana* mutant *de-etiolated 2* (*det2*) or the BR receptor mutant *brassinosteroid insensitive 1* (*bri1*) showed delayed senescence [87,88]. BR induced resistance against the (hemi-)biotrophic pathogens tobacco mosaic virus (TMV), *Oidium* and *Pseudomonas syringae* in tobacco and *Xanthomonas oryzae* and *Magnaporthe grisea* in rice [89]. BR have been shown to induce PR1 in an NPR1-independent way [90]. Furthermore, potato plants sprayed with BR became more resistant to *Phytophthora infestans* and showed increased ABA- and ET levels [91]. The important BR signaling components BRI1-ASSOCIATED RECEPTOR KINASE 1 (BAK1) and BAK-LIKE 1 (BKK1) play a role in cell death regulation, however, in a way that was independent of BRI1 and BR [87].

2.1.6. Cytokinins, Auxins, Gibberellins

Whereas all the aforementioned phytohormones generally have senescence-promoting effects, cytokinins, auxins and gibberellins are regarded as more or less senescence-retarding hormones [55]. While few studies addressed the role of gibberellins in senescence (e.g., [92,93]), the evidence for **cytokinin**-induced delay of senescence is paramount [55,86,94]. Tobacco plants transformed with the *Isopentenyl-phosphotransferase* (*Ipt*-) gene, which encodes the rate-limiting enzyme for cytokinin biosynthesis, under the promoter of *Senescence-associated-gene 12* (*Sag12*) displayed a functional stay-green phenotype with prolonged photosynthetic activity [95]. Functional stay-green genotypes with an enhanced cytokinin status were also more resistant to drought [94,96]. Surprisingly, transgenic tobacco plants over-expressing *Cytokinin oxidase* (*Ckx*)-genes with a lower endogenous cytokinin content did not show early senescence [97,98]. It was concluded that cytokinin does not control the onset of senescence but prevents its execution [86].

Auxins, mainly represented by indole acetic acid (IAA), control many aspects of plant growth and development [99]. Auxins are synthesized in young tissues from tryptophan via two different pathways, highlighting their exceptional importance [100]. Another distinct feature is the polar auxin transport by auxin efflux carriers with polar distribution in cells. It is a prerequisite for the fine auxin gradients shaping plant architecture [99]. Auxin signaling shows many parallels to JA signaling: Upon auxin binding, the F-box protein TIR forms a complex with AUX/IAA repressors, targeting them for ubiquitination and subsequent degradation [101,102]. AUXIN RESPONSE FACTORs (ARF) are released and initiate the transcription of target genes [99]. Auxins hold a complex role in senescence. While in some studies a senescence-retarding effect of auxins was found, others suggest a role in its induction [86]. Exogenously applied auxin was shown to delay senescence [103], and over-expression of the auxin biosynthesis gene YUCCA6 caused elevated auxin levels and extreme longevity [104]. Furthermore, *arf2* mutants that are defective in the auxin signaling repressor ARF2 showed delayed senescence [105]. Antagonistic cross-talk of auxin- and JA-signaling has been discovered by Jiang *et al.*, with WRKY57 acting as a regulatory node which was competitively up-regulated by auxin and down-regulated by JA. WRKY57 delayed senescence [106]. While these findings support a senescence-retarding role of auxin, it is unexpected that auxin biosynthesis genes were up-regulated and contents of free auxin rose during senescence [107,108]. In addition, auxin has been found to act synergistically with ET in the induction of senescence via SENESCENCE-ASSOCIATED RECEPTOR-LIKE KINASE (SARK) in soybean and in *A. thaliana* [109]. Likewise, over-expression of the auxin responsive *Small auxin-up RNA gene 36* (*Saur36*) led to an early-senescence phenotype of *A. thaliana*, suggesting a positive regulation of senescence by auxin [110].

The role of the growth-promoting hormones in pathogen defense has long remained cryptic. During the last decade many surprising roles of gibberellins, cytokinins and auxins in biotic stress response have been described, often in the context of synergism or antagonism with known stress signaling pathways [47,111,112]. Gibberellins target DELLA repressors for degradation [113]—a signaling pathway that is analogous to the one for JA and auxin signaling. Several DELLA proteins compete with MYC2 in binding to JAZ repressors [114], thereby facilitating JA-signaling. GA-mediated DELLA degradation thus leads to repression of JA-signaling. Recently, GbWRKY1 transcription factor has been shown to dampen JA-response in the course of *Verticillium dahliae* infection in cotton, thereby de-repressing GA-signaling [115]. The mutual repression of JA- and GA-signaling reflects the antagonism between growth and defense that controls resource allocation [116,117]. Through the JA-SA-antagonism, gibberellin signaling acts synergistically with SA, promoting resistance to biotrophs [118]. Likewise, cytokinins have been shown to act synergistically with SA via ARABIDOPSIS RESPONSE REGULATOR 2 (ARR2), which is recruited by SA-controlled transcription factors to enhance expression of PR1 [111].

Auxin has been shown to interact antagonistically with SA, thereby increasing susceptibility to biotrophic pathogens [112]. SA has been found to repress the auxin receptor TIR1, thus stabilizing the AUX/IAA-repressors of auxin signaling [119]. This is consistent with the finding that PR1 is hyperinducible by SA in auxin receptor mutants [120]. Furthermore, SA has been shown to induce WES1/GH3.5, an enzyme that conjugates IAA to amino acids, thus reducing free IAA levels [121]. *A. thaliana* plants over-expressing WES1 were more resistant to *P. syringae* and showed elevated PR1 levels [121]. A similar observation was made in rice, where reduction of free IAA by over-expression of the conjugating enzyme OsGH3.8 resulted in enhanced resistance against *Xanthomonas oryzae* pv. *oryzae*, which was, however, SA-independent. It was hypothesized that increased resistance was due to inhibition of expansins [122]. Strikingly, a different study found a direct disease-promoting effect of an IAA-aspartate conjugate (IAA-Asp) via induction of microbial virulence factors [123]: Exogenously applied IAA-Asp promoted disease symptoms caused by *Botrytis cinerea* and *P. syringae* DC3000, and the enzyme GH3.2, which conjugates IAA to aspartate, was up-regulated by infection with these pathogens. Accordingly, *gh3.2* mutants were more resistant against both pathogens. Other IAA-conjugating GH3-enzymes showed minor or no effects on pathogenesis in this study. While the disease-promoting role of auxin in interactions with biotrophic pathogens is relatively clear, its role in hemibiotrophic and necrotrophic interactions is ambiguous. It was shown that auxin signaling is required for quantitative resistance against the necrotrophs *B. cinerea* and *Plectosphaerella cucumerina*, as *auxin resistant* mutants *axr1*, *axr2* and *axr6* were more susceptible to these pathogens [124]. Interestingly, *axr6* also has an important role in JA-signaling, since it encodes CULLIN1, a component of the SCF complex interacting with COI1 [125]. However, *axr1* and *axr2* showed increased resistance to *Fusarium oxysporum*, suggesting a disease-promoting role of auxin signaling for this pathogen [126]. In the same study, auxin transport was shown to mediate susceptibiliy.

2.2. Transcription Factors

The vital importance of transcription factors in regulating developmental processes and stress responses has already been emphasized in the context of the signaling pathways described above. The most relevant groups in the control of senescence and defense comprise WRKY transcription factors [127], APETALA 2/ETHYLENE RESPONSIVE ELEMENT BINDING PROTEINS (AP2/EREBP) [128], NAM-ATAF1,2-CUC2 (NAC) [129,130], R2R3-MYB [131], basic HELIX LOOP HELIX proteins (bHLH) [132] and C2H2-type zinc finger transcription factors [133]. In this section, the focus will be on selected WRKY transcription factors, because they are often not attributed to individual phytohormone signaling pathways. Instead, many of them have an integrative function for diverse incoming signals. Hence, some WRKY transcription factors play a pivotal role as nodes of convergence in the regulation of both senescence and defense against pathogens [52,134]. WRKY transcription factors constitute a gene family comprising 75 members in *A. thaliana* with diverse functions in development and responses to biotic and abiotic stress [127,135].

WRKY transcription factors with a confirmed role in senescence include WRKY6, WRKY53, WRKY70 and WRKY30 [136,137]. WRKY6 expression increased during senescence and it strongly induced SENESCENCE-INDUCED RECEPTOR-LIKE KINASE/FLG22-INDUCED RECEPTOR-LIKE KINASE (SIRK/FRK1), a gene induced during senescence and also during PAMP-triggered immunity [137]. WRKY53 was up-regulated during senescence and underlay a negative feedback loop. WRKY53 over-expression accelerated senescence, and knockout lines showed delayed senescence [138]. WRKY30 was expressed during developmental leaf senescence and both WRY53 and WRKY30 were shown to be responsive to ROS [136]. WRKY70 is a negative regulator of senescence. Despite their contrasting role in regulating senescence, WRKY53 and WKY70 both were shown to be positive regulators of defense with a role in SAR [134,136]. WRKY38 and 62, on the other hand, acted as negative regulators of SA- and NPR1- mediated basal defense [139]. Another regulatory node is formed by WRKY18, 40 and 60. They were shown to form homo- and heterocomplexes with each other to establish a highly plastic system to respond to different microbial pathogens [140]. Triple knockout mutants showed enhanced resistance to biotrophic *P. syringae*, but increased susceptibility to necrotrophic *B. cinerea*. The mutant phenotype correlated with increased *PR1*-transcript levels and decreased *Pdf1.2* transcript levels, emphasizing the role of these three transcription factors in mediating JA/ET-induced defenses. All three factors are also negative regulators of ABA-signaling. Especially WRKY40 repressed the transcription factor *Abi5* [141]. Conversely, WRKY8 is described as a positive regulator of ABA- and a negative regulator of ET signaling [142].

2.3. Reactive Oxygen Species, Programmed Cell Death and Autophagy

ROS like hydrogen peroxide, the superoxide anion or the hydroxyl ion function as signaling molecules during senescence and pathogen response, but also as antimicrobial compounds and as toxins leading to programmed cell death and necrosis [143,144,145]. Programmed cell death and autophagy, key processes in senescence and defense against pathogens, are both responsive to ROS [144,145]. It was suggested that three levels of ROS activity can be distinguished in plant cells depending on the concentration: At low levels, ROS can be controlled by the antioxidant machinery of the plant. At medium levels, ROS change the redox status of the cell in a way that leads to programmed cell death, while at high, toxic concentrations, cells are killed in an uncontrolled way, leading to necrosis [145]. ROS originate from different sources in the plant: They are a by-product of both photosynthesis and respiration and their concentration is usually controlled by antioxidant enzymes such as catalases, superoxide dismutases, peroxidases and glutathione-S-transferases, and antioxidant compounds like ascorbic acid and glutathione [146,147]. As signaling molecules, ROS are produced in a controlled way by the RESPIRATORY BURST OXIDASE HOMOLOGUE (RBOH)-NADPH-oxidases [148,149,150].

ROS signaling is closely linked to developmental events. During senescence, the antioxidant capacity of cells decreases and ROS concentrations increase [151,152]. Conversely, ROS have been shown to induce leaf senescence [153,154]. Important senescence-promoting transcription factors like ANAC092/ORE1 [155], ATAF1 [72], ORE1-SISTER 1 (ORS1) [156] or WRKY53 [138] were found to be up-regulated by H_2O_2. Recently, the NAC transcription factor ATAF1 has been identified as an integration point cross-linking H_2O_2- and ABA-signaling in the control of leaf senescence [72].

Rapid biphasic ROS production follows pathogen recognition during PTI and ETI and is termed oxidative burst [157]. Defense reactions triggered by ROS include direct effects like oxidative reinforcement of cell walls by cell-wall peroxidases [158]. More importantly, multiple signaling processes leading for example to the production of antimicrobial compounds are triggered by ROS [159]. Interestingly, ANAC042/JUB1, a NAC transcription factor which is induced by H_2O_2, had a negative regulatory effect on senescence [154] and has been shown to be involved in camalexin induction [159]. It may have a role in the fine-tuning between senescence- and defense signaling by H_2O_2. Ca^{2+} is not only required to trigger ROS production by RBOHs [160,161], but is also involved in downstream signaling processes [162]. MITOGEN-ACTIVATED PROTEIN KINASE-(MAPK-) signaling cascades are activated by ROS [161,163], such as MPK6, which is a major integrator of senescence and defense regulatory pathways [36,49,55,164]. SA, JA and ET are induced by ROS [143] and the respective signaling pathways are tightly cross-linked. A prominent example is the redox sensitivity of NPR1, which is reduced after a compensatory rise of glutathione levels following the oxidative burst. Reduction leads to monomerization of NPR1, thus facilitating SA-signaling [165]. ROS furthermore triggered the ABA-mediated closure of guard cells [162], and ABA has been shown to induce H_2O_2 via MAPK-signaling [58]. While the aforementioned hormone signaling pathways are mainly reinforced by ROS, an inhibitory effect of ROS on auxin signaling has been reported, leading to the so-called "stress-induced morphogenetic response" [166]. ROS are highly mobile signals and are thus an essential component of systemic responses like SAR [148,163].

HR involving programmed cell death is one of the most efficient defense reactions initiated by ROS [149,167]. However, the benefit of HR for the host strongly depends on the lifestyle of the pathogen [3,168]: HR is a particularly efficient strategy against biotrophic pathogens, as it leads to confined lesions at the sites of pathogen entry. Being dependent on living cells, the pathogen is prevented from spreading. A strategy of biotrophic pathogens therefore consists in the induction of antioxidants, leading to ROS scavenging and the inhibition of HR [169]. HR can also be efficient against necrotrophs when it is associated with the deposition of antimicrobial metabolites such as callose or lignin at the entry sites of the pathogen. However, being a pro-death strategy, cell death and HR can also facilitate the growth of necrotrophic pathogens, and ROS induction leading to cell death in the host can be a virulence strategy [170]. In this case, the capacity of the host to produce antioxidants would promote resistance. Programmed cell death is closely linked to autophagy, a process by which cell constituents are enclosed in vesicles (autophagosomes) that fuse with the vacuole. Recent studies support the view that autophagy, though induced upon senescence, is a pro-survival strategy of cells which is necessary for controlled nutrient recycling during senescence and stress [171]. *A. thaliana atg* mutants, which are deficient in autophagy, show premature senescence, are hypersensitive to oxidative stress and hypersusceptible to necrotrophic pathogens [172]. The hypersusceptibility to *B. cinerea* infection could be explained by an interaction of ATG18a with WRKY33, which mediates defense against necrotrophic pathogens [173]. There are indications that autophagy negatively regulates HR-PCD [174]. Accelerated cell death in *atg* mutants depends on SA signaling via NPR1 [175]. In other studies, a cell death-promoting effect of autophagy has been found [176]. Interestingly, the role of autophagy in HR-PCD depended on the type of R gene-signaling during ETI: While HR induction via TIR-NBS-LRR receptors depended on autophagic processes, HR mediated by CC-NBS-LRR was largely independent of autophagy [176].

Taken together, the evidence suggests a positive role of ROS in defense against necrotrophic pathogens as long as their production is tightly controlled by redox buffering and signaling processes that prevent cell death and necrosis.

3. DEVELOPMENTAL IMPLICATIONS OF HOST-PATHOGEN-INTERACTIONS

3.1. Host Development Affects Pathogenesis

The following examples demonstrate that developmental conditions or developmental stages of the host plant can affect the outcome of the host-pathogen interaction. Examples of developmentally conditioned susceptibility and resistance are presented for pathogens with different lifestyles.

3.1.1. Senescence-Like Processes Confer Quantitative Resistance against Biotrophic Pathogens in Cereals

Three examples from cereals demonstrate that senescence-like processes can confer resistance against devastating biotrophic pathogens.

Lr34, a gene belonging to the pleiotropic drug resistance family of ABC-transporters, is a resistance gene from wheat that confers broad-spectrum-, durable and quantitative resistance to the biotrophic pathogens *Puccinia triticina*, *P. striiformis* and *Blumeria graminis* [177]. The resistance conferred by *Lr34* has strong developmental implications: It works only in adult plants during the grain-filling phase and it is connected with senescence-like processes. Resistance was highly correlated with a morphological trait called leaf-tip necrosis, which was most pronounced in the uppermost leaf, the flag leaf [177]. Genes up-regulated in the presence of the *Lr34*-gene comprised ABA-inducible genes, genes associated with cold and drought stress and with seed maturation. Typical defense genes were not differentially expressed [178]. It was suggested that senescence-like processes are the cause for increased resistance [177].

A second gene conferring quantitative resistance against *P. striiformis* in wheat is *Yr36*/WHEAT KINASE START 1 (WKS1) which is present in some wild wheat species, but not in hexaploid bread wheat [179]. WKS1 contains a lipid-binding START domain and a kinase domain. The protein exists in six splice variants, only one of which is full-length, while the others have truncated START domains. The full-length transcript is up-regulated by *P. striiformis*-infection, while the other transcripts are not [179]. Full-length WKS1 was shown to phosphorylate THYLAKOID-ASSOCIATED ASCORBATE PEROXIDASE (tAPX), thus reducing its capacity to detoxify H_2O_2 [180]. Transgenic wheat containing multiple copies of *Wks1* showed accelerated senescence even without infection. It was concluded that WKS1 facilitates cell death that is slower than HR cell death, thus allowing restricted pathogen growth [180].

In the third example, *Mlo* from barley, coding for a membrane protein with seven transmembrane domains [181] delayed senescence and conferred susceptibility to biotrophic *B. graminis* f. sp. *hordei* (Bgh), but increased resistance against the hemibiotroph *Magnaporthe grisea* and the necrotroph *Bipolaris sorokiniana*. *mlo* mutants showed accelerated leaf senescence, were more susceptible against *M. grisea*, but highly resistant against Bgh. This resistance was associated with rapid necrosis formation and H_2O_2- production at the infection sites [182]. *Mlo* is an inducible gene that is up-regulated by a number of pathogen and stress treatments [182] and controls basal resistance [183].

3.1.2. Late-Senescing Genotypes Are More Resistant to Necrotrophic and Hemibiotrophic Pathogens

A late-senescing genotype may confer quantitative resistance to necrotrophic and some hemibiotrophic pathogens. However, these effects are mostly multigenic and individual genes are hard to identify. In cowpea (*Vigna unguiculata*), resistance against the necrotrophic pathogen *Macrophomina phaseolina* was associated with late maturity, and two *M. phaseolina* resistance QTL co-localized with QTL mediating late maturation [184]. It was shown that *M. phaseolina* infection in the legume model plant *Medicago truncatula* induced genes that lead to inactivation of auxin and suppression of auxin signaling [185].

Exogenous application of auxin could alleviate disease symptoms. To date it is not known whether the identified resistance QTL [184] are involved in auxin homeostasis or signaling.

Crucial events of pathogenesis often depend on developmental stages of the host. For example, mature-green tomatoes are more resistant to infection by the necrotrophic pathogen *B.cinerea* than red ripe fruits [186]. Gene expression analysis showed how developmental and pathogenesis effects were intricately interwoven for each of the involved hormone signaling networks controlled by ET, SA, ABA and JA [186]. Generally, ET seemed to have a disease-promoting effect in tomatoes by accelerating maturity. Down-regulation of the central signaling components LeEIL3 and LeEIL4 is interpreted as a plant strategy to dampen the disease-promoting effect of ET. LeERF1, which is a crucial component of the PDF1.2-branch of JA/ET-signaling, was up-regulated during infection. In contrast, the regulation of ABA synthesis and signaling components pointed to a clear disease-promoting role of this hormone in the process of ripening. More resistant, mature green tomatoes showed up-regulation of the ABA-response inhibitor RECEPTOR FOR ACTIVATED C KINASE 1 (RACK1) [186].

In the interaction between *V. longisporum* and its cruciferous hosts, the onset of flowering is a key event affecting pathogenesis. In susceptible *Brassica* genotypes, the phase of systemic colonization through the xylem was triggered by the onset of flowering [187]. The time-course of *V. longisporum*-colonization in the susceptible *A. thaliana* ecotype Landsberg *erecta* (L*er*) showed the same pattern (Figure 2, [188]). Systemic colonization of the shoot system, which can be monitored by re-isolating the pathogen from apical shoot segments plated on malt agar, started with the onset of flowering. Extensive fungal proliferation, which was detected by quantifying fungal DNA-contents via qPCR, started at the onset of silique maturity. Figure 2 also shows that the resistant *A. thaliana* ecotype Burren (Bur) was barely colonized in the course of its development.

Resistance against *V. longisporum* in *A. thaliana* as well as in *Brassica* was also often associated with a late-flowering developmental type (Figure 3A,B). In *B. alboglabra*, the late-flowering accession "99" showed almost no systemic colonization of apical shoot segments in an experiment with root-dip inoculation with *V. longisporum* isolate 43 (Figure 3C). The same was true for the late-flowering *A. thaliana* ecotype Bur (Figure 3D). The early flowering genotypes *B. alboglabra* "94" and *A. thaliana* ecotype L*er* were extensively colonized (Figure 3C,D). In *B. alboglabra*, resistance to systemic colonization was paralleled by a slower disease progression based on the symptoms of stunting and chlorosis (Figure 3E). In *A. thaliana*, stunting resistance as determined by the performance height

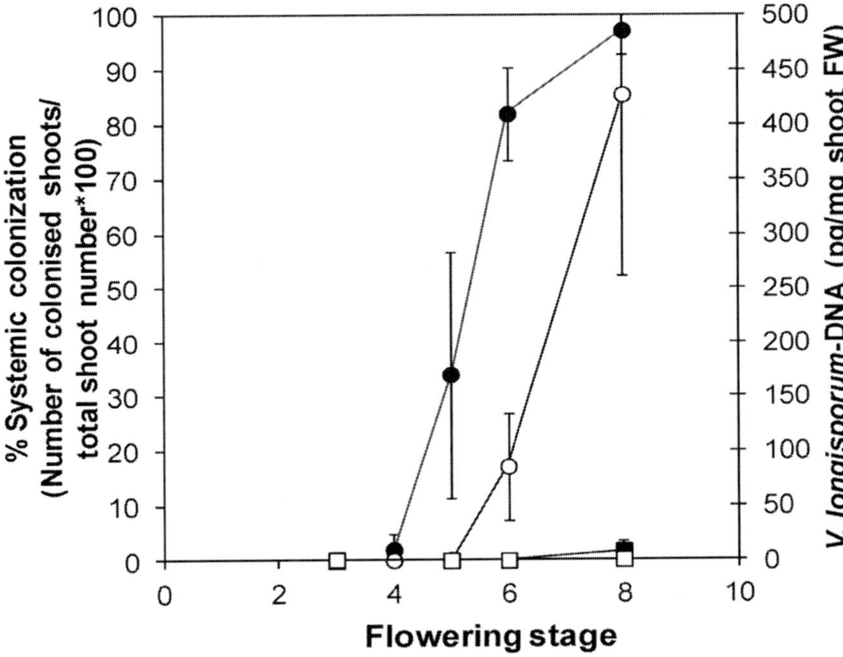

Figure 2. Systemic colonization of L*er* and Bur by *V. longisporum* in the course of plant development. Percentage of colonized apical shoot segments in a plating assay (closed symbols) and fungal DNA-content determined by qPCR in apical shoot segments (open symbols) in L*er* (circles) and Bur (squares) at different developmental stages. Developmental stages: 3 = 1–3 flowers open, 4 = 4–10 flowers open, 5 = more than 10 flowers open, 6 = up to 3 siliques mature, 7 = 4–6 siliques mature, 8 = up to 50% of siliques on main shoot mature. *N* = 4 (Bur) or 5 (L*er*), one replicate corresponded to a batch of 30 plants. Error bars denote standard deviations (data from [188]).

[79] was increased in late-flowering ecotype Bur only during summer when light conditions in the greenhouse were optimal (Figure 3F). Several colonization resistance QTL were found in the vicinity of QTL controlling development time in *A. thaliana*, which explained the correlation of both traits [188]. A similar correlation was found in connection with the *V. longisporum* resistance QTL *Vet1* [189]. that also delayed development. To date it is not known whether both effects are controlled by the same or by closely linked gene(s). However, the correlation between longevity and resistance against some types of stress seems to be a general principle. Long-lived, late-flowering *A. thaliana* mutants have been shown to be more tolerant to oxidative

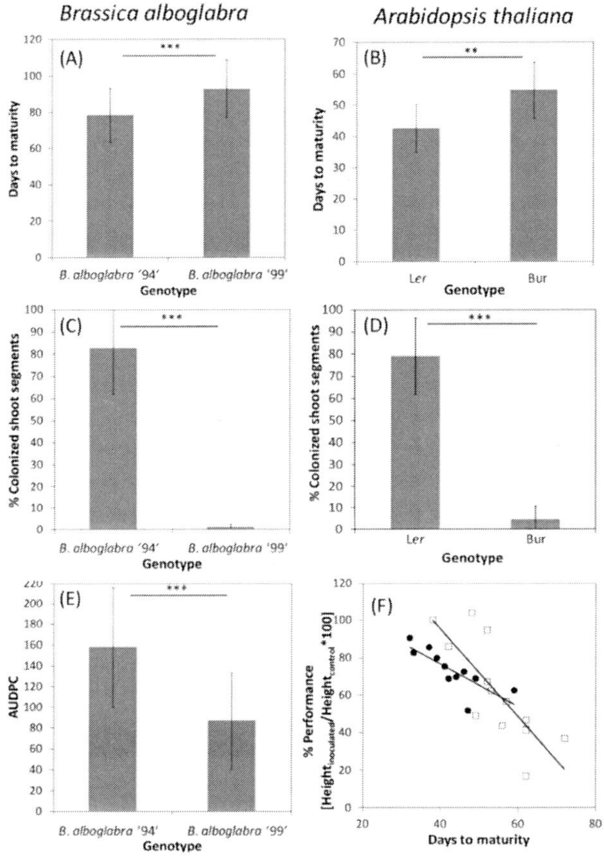

Figure 3. Developmental type and *V. longisporum* disease parameters of two accessions of *B. alboglabra* (left side) and *A. thaliana* (right side). (**A**,**B**) Development time of mock-inoculated control plants from inoculation to maturity (*B. alboglabra*) or germination to maturity (*A. thaliana*). (**A**) Single-plant data pooled from three independent experiments, $N = 58$ (94), $N = 59$ (99); (**B**) Average values from means of 9 and 11 independent experiments comprising at least 15 plants each; (**C**,**D**) Systemic colonization by *V. longisporum*. (**C**) $N = 5$; (**D**) $N = 9$ (L*er*), $N = 11$ (Bur); (**E**) Area under the disease progress curve (AUDPC, [191]) of *V. longisporum*-inoculated plants. High values stand for fast disease progression. Single-plant values from six independent experiments, $N = 243$ (94), $N = 253$ (99); (**F**) Performance of L*er* (circles) and Bur (squares) after *V. longisporum* infection depending on development time (data from [188]). Development was prolonged in winter experiments under lower light intensities. Performance (%) = $\text{Height}_{\text{inoculated}}/\text{Height}_{\text{control}}$ *100. Each data point represents a separate experiment comprising at least 30 plants. Experiments were performed at different seasons over three years under greenhouse conditions. Correlations: L*er*: $r = -0.79$, $p = 0.004$; Bur: $r = -0.79$, $p < 0.001$. Error bars denote standard deviations, asterisks denote significant differences below the 0.01 (**) and 0.001 (***) significance levels respectively (Student's *t*-test).

stress caused by the herbicide paraquat, suggesting that longevity and oxidative stress tolerance are tightly linked [190]. Evidence for the role of oxidative stress in *V. longisporum* pathogenesis will be presented below (see Section 3.2.3.).

3.2. Pathogenesis Affects Host Development

3.2.1. Pathogens Manipulate Host Development for Their Own Benefit

The most commonly described developmental implication of pathogenesis is the induction and exploitation of signaling pathways that benefit the pathogen, but are deleterious to the host [4,54,192]. Hormones or toxins produced by the pathogen may delay or promote senescence, depending on the lifestyle of the pathogen. The following examples illustrate how developmental pathways are exploited by pathogens and how plants can counter-steer during quantitative disease resistance.

Biotrophic and some hemibiotrophic pathogens preferentially produce or induce growth-promoting plant hormones, such as auxins or cytokinins, or change the host's response to these hormones. The AvrRpt2 Type III effector of *P. syringae* DC3000 has been shown to increase auxin levels and auxin sensitivity in the host to promote disease in *A. thaliana* [193]. Upon infection by *Plasmodiophora brassicae*, a biotrophic protist causing clubroot disease in crucifers, cytokinins and auxins contribute to hypertrophy of root gall cells [194]. *P. brassicae* can produce cytokinins by itself [195] and induce transcriptomic changes in the host that lead to increased auxin and cytokinin contents in root galls [196]. Cytokinins also participate in the formation of green islands—green areas at the infection sites of obligate biotrophs like *P. graminis* on otherwise senescent host tissue [197]. While effectors of biotrophic pathogens promote disease in compatible interactions, host resistance in incompatible interactions is achieved by effector recognition and subsequent ETI based on HR cell death. ETI often results in complete resistance, which can, however, easily be overcome by mutations in the pathogen [198].

Necrotrophic and some hemibiotrophic pathogens promote senescence by targeting the respective signaling pathways. Two independent studies showed that some hemibiotrophic vascular pathogens can induce senescence-like processes by targeting the F-box protein COI1, which is part of the JA-binding JAZ-COI1-co-receptor [199,200]. *Fusarium oxysporum* as well as *V. longisporum* needed functional *Coi1* in the root to induce senescence symptoms during infection within the shoot of *A.*

thaliana. *coi1* mutants showed a high level of quantitative resistance against both pathogens as inferred from disease symptoms, although host colonization still occurred. Strikingly, *coi1*-mediated resistance was independent of JA biosynthesis and JA-induced defense pathways as well as from SA-mediated defense [199,200]. Other pathogens targeting the JA-signaling pathway inhibit JA-mediated defenses via the JA-SA antagonistic crosstalk [201]. This seemed not to be the case in *coi1*-mediated resistance. COI1 was not a target of potential *V. longisporum*-produced JA-mimicks, since JA-mediated defenses were absent in *V. longisporum*-infected *delayed dehiscence 2* (*dde2*)-mutants that are impaired in JA biosynthesis [199]. Due to the suppression of pathogen-induced senescence, *coi1*-mutants were less colonized during the late stages of infection in both pathosystems [199,200].

NECROSIS AND ETHYLENE INDUCING PEPTIDE 1 (NEP1)-LIKE PROTEINs (NLPs) are widespread microbial toxins produced by necrotrophic and hemibiotrophic pathogens to induce cell death [202]. Hemibiotrophic *Moniliophthora perniciosa*, the causal pathogen of witches' broom disease in cacao, initiated its necrotrophic phase and the production of NLP MpNEP2 upon carbon starvation in the apoplast [203]. In the preceding biotrophic phase, the pathogen increased sink strength in infected tissue possibly by producing cytokinins. This led to increased concentrations of extracellular hexoses in infected tissues and a quicker rise of intracellular sugars. Subsequent early senescence was accompanied by repression of photosynthesis. It was hypothesized that this strategy allows the pathogen to acquire soluble apoplastic carbohydrates during its biotrophic phase and to gain prime access to intracellular carbohydrates by killing senescent cells during its necrotrophic phase [203]. This example shows how the spatial and temporal control of signals can manipulate a host for the maximum benefit of the pathogen.

3.2.2. "Developmental Buffering" Confers Quantitative Resistance to *V. longisporum* in *A. thaliana*

With very few exceptions, ETI is not an efficient resistance mechanism against necrotrophic or hemibiotrophic pathogens [4]. In contrast, toxins may elicit ETI and HR for the benefit of the necrotroph [204]—a process that has been termed effector-triggered susceptibility [205]. Resistance against hemibiotrophs and necrotrophs therefore relies on diverse other mechanisms that are often complex and not fully understood. In the following, the developmental effects of *V. longisporum* on its cruciferous hosts are described and signaling processes modulating susceptibility and quantitative resistance are presented.

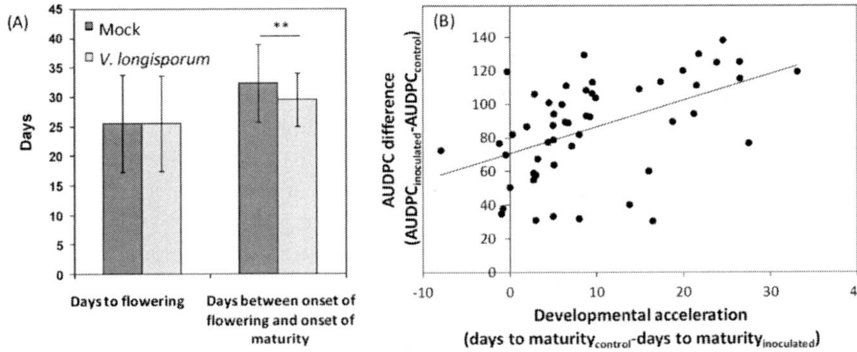

Figure 4. Developmental acceleration in *A. thaliana* and *B. alboglabra* after *V. longisporum* infection. **(A)** Flowering time and maturation time in a collection of 100 *A. thaliana* (Bur × *Ler*)-recombinant inbred lines (RILs) after mock- and *V. longisporum* inoculation respectively. $N = 94$ to 102, each replicate corresponded to a batch of at least 15 plants. Asterisks denote a significant difference of $p < 0.01$ (t-test), error bars denote standard deviations. **(B)** Correlation between developmental acceleration by *V. longisporum* and disease progression, expressed as AUDPC difference between inoculated and mock-inoculated plants in *B. alboglabra* (94 × 99)-F3-families. $N = 52$, $r = 0.49$, $p < 0.01$. Each data point represents one F3-family, differences were calculated from means of 38 inoculated and 11 mock-inoculated plants in each F3-family.

V. longisporum induced premature senescence in *A. thaliana* and *B. alboglabra* in greenhouse experiments (Figure 4). This finding is in accordance with observations on oilseed rape in the field [206]. Premature senescence in *A. thaliana* was not due to an acceleration of flowering time, but reduced the time from flowering to maturity (Figure 4A). Acceleration of maturity was observed in a population of (Bur × *Ler*)-recombinant inbred lines (RILs). QTL controlling resistance against premature senescence, however, could not be identified. In *B. alboglabra* we showed that acceleration of maturity was positively correlated with disease progression (Figure 4B). These results suggest, in accordance with theories for necrotrophic pathogens, that acceleration of maturity was mainly exerted during the necrotrophic, systemic phase of infection and that premature senescence is a disease symptom and not a result of defense signaling.

Comparing a resistant and a susceptible genotype with respect to individual signaling pathways can give information about components involved in resistance and susceptibility as well as in developmental processes. For this purpose, phytohormone analyses and gene expression studies were performed in near-isogenic lines originating from the *A. thaliana* ecotypes Bur and L*er*. The lines were polymorphic for a genomic region harboring the major resistance QTL *vec1* and differed in resistance to systemic colonization by *V. longisporum.* Previous studies showed a disease-promoting effect for ABA and JA at the stage of silique maturity as inferred from phytohormone contents of Bur, L*er*, and two NILs representing the respective alleles in *vec1* (Figure 5, [79]). NIL5, carrying alleles of the resistant parent Bur in a ~3 Megabase region of *vec1*, showed less severe ABA-induction and suppression of JA-production after *V. longisporum* infection in shoots of maturing plants [79].

3.2.3. Transcriptional Response to *V. longisporum* at Maturity in a Susceptible and a Resistant *A. thaliana* Line

A microarray analysis was performed to monitor genome-wide transcriptomic changes in response to *V. longisporum* controlled by *vec1*. For this purpose, a tailor-made near-isogenic line (tmNIL130) has been selected that contained a max. 532 kb- introgression of Bur-alleles surrounded by L*er*-alleles in *vec1* (Figure 6A). The introgression was designed to be as small as possible while still controlling a significant level of resistance (Figure 6C). The transcriptional response to *V. longisporum* in tmNIL130 was compared to the one in NIL9 that carried L*er*-alleles in the indicated region. In all other regions, the two lines were isogenic at the genotyped marker loci except for a region at the top of chromosome 4 (Figure 6A). Differential gene expression was studied in samples containing the hypocotyl and the shoot basis, since colonization resistance has been reported to be effective in the hypocotyl of *Brassica* spp. [207]. Thus, local and systemic responses were to be determined. Samples were collected during the two crucial stages of *V. longisporum* colonization (Figure 6B): shortly after the onset of flowering when the shoot started elongation and systemic colonization was shown to start (time point 1, 13 days post inoculation, dpi), and upon maturation of the first silique when extensive proliferation was shown to occur in susceptible genotypes (time point 2, 27–29 dpi). The complete data set and the metadata containing all experimental descriptions have been released to Gene Expression Omnibus (GEO, [208]) repository at the National Center for Biotechnology Information (NCBI) under the reference number GSE70021 [209].

Generally, partially resistant tmNIL130 was characterized by a much more pronounced transcriptional response to *V. longisporum* infection compared to NIL9 (Figure 6B). At the onset of flowering, only 18

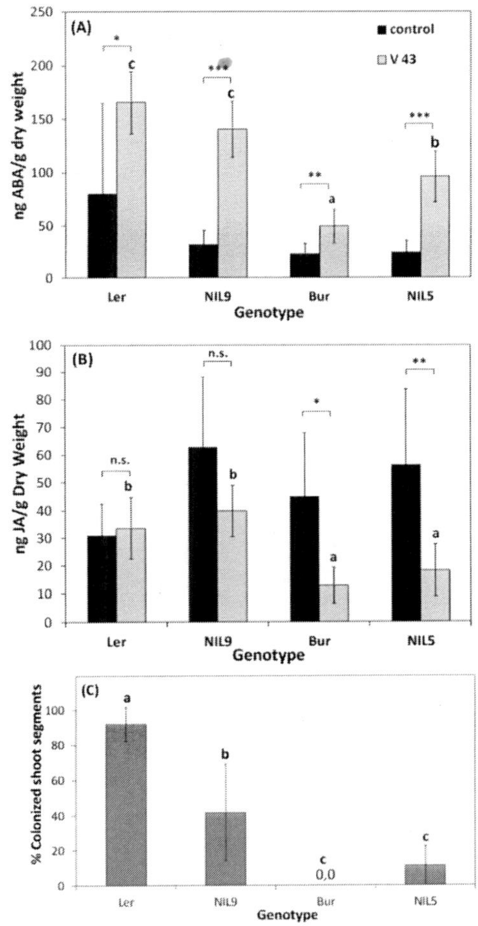

Figure 5. Hormone contents and systemic colonization by *V. longisporum* in *A. thaliana* ecotypes L*er*, Bur, and near-isogenic lines differing in a region containing the colonization resistance QTL *vec1* on chromosome 2. Contents of (**A**) abscisic acid (ABA) and (**B**) jasmonic acid (JA) in Bur, L*er* and two (Bur × L*er*)-near-isogenic lines (NILs): NIL9 (L*er* alleles in the variable region) and NIL5 (Bur alleles). Asterisks refer to the significance level of differences between mock-inoculated and *V. longisporum*-treated plants within one genotype (*t*-test). Mean contents of inoculated plants marked with different letters differed significantly at $p < 0.05$. Mean phytohormone contents of controls did not differ significantly (one-way ANOVA and post-hoc Tukey test). $N = 6–7$, each replicate corresponded to a batch of 15 plants. Vertical bars denote standard deviations (data from [79]). (C) Systemic colonization by *V. longisporum* in the same plants that have been used for phytohormone measurements. Means marked with different letters differed at $p < 0.05$ (one-way ANOVA and post-hoc Tukey test). $N = 6–7$, each replicate corresponded to a batch of 15 plants. Vertical bars denote standard deviations.

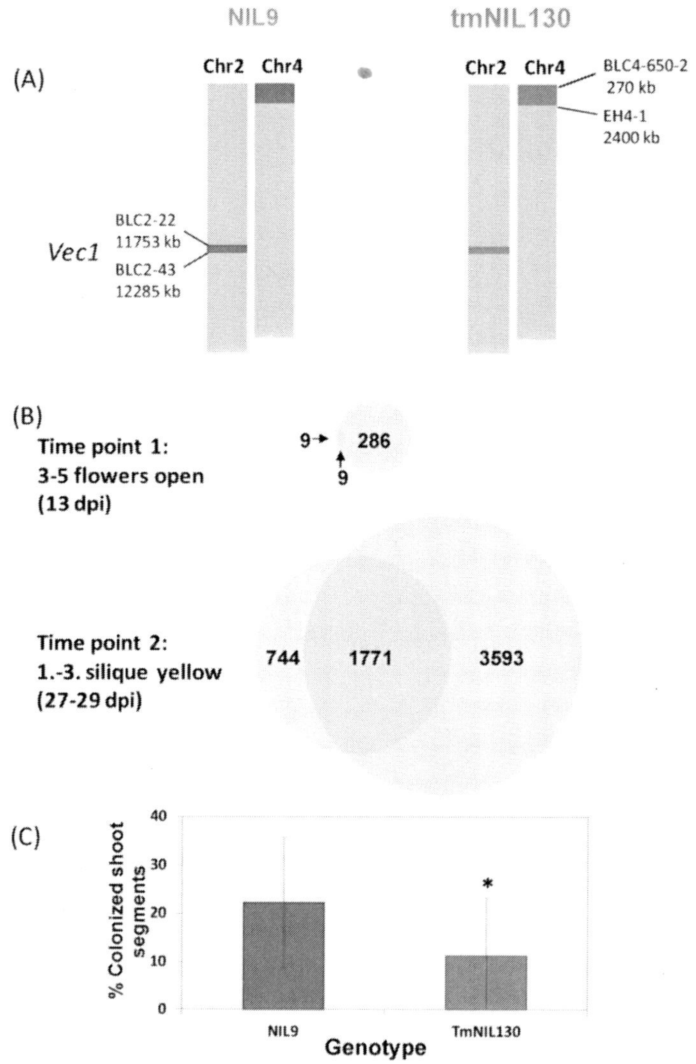

Figure 6. Differential gene expression and systemic colonization after *V. longisporum* infection in two (Bur × L*er*)-NILs differing in a region within the colonization resistance QTL *vec1*. (**A**) Genotype of NIL9 and tmNIL130 in the variable regions. Red parts stand for L*er* alleles, green parts for Bur alleles in the variable regions. Grey parts are isogenic with respect to the tested marker loci (for marker grid in isogenic regions see [79]). Names and physical positions in kilobases (kb) of markers delimiting variable regions are given next to the bars representing chromosomes. For primer information see Supplementary File Table S1; (**B**) Genes differentially expressed by *V. longisporum* infection in the hypocotyl-shoot-transition zone of NIL9 and tmNIL130 at two time points after infection. For growth-, inoculation- and RNA extraction protocol see GEO

accession GSE70021. Hybridization to Affymetrix AraGene-1_1-st was performed by the Nottingham Arabidopsis Stock Centre (NASC) Affymetrix service [212]. Data were rma- (robust multichip average) normalized using the Bioconductor package affy [213], and differential expression was calculated using linear models with the Bioconductor package limma [214] by pairwise comparisons. Annotations were provided by the Bioconductor pd.aragene.1.1.st annotation package [215]. Only expression differences with a significance level of $p < 0.05$ were regarded at the probeset level. If more than one probeset represented a transcript cluster (≙ mRNA), transcript cluster means were calculated from values for individual probesets; (**C**) Systemic colonization of NIL9 and tmNIL130 at time point 2 (onset of silique maturity). $N = 12$, *t*-test. Samples were taken from 30 plants per replicate, among which were also the plants sampled for microarray analysis.

transcripts were differentially expressed after *V. longisporum* challenge in NIL9, but 295 transcripts were mostly up-regulated by *V. longisporum* in tmNIL130. Among them were many defense-related genes like the transcription factor WRKY33, which was shown to mediate JA/ET-dependent defense responses against necrotrophs [210] with 2.6-fold up-regulation, or *Frk1*, a PTI marker gene [211] with 8.6-fold up-regulation after infection. At the onset of maturity, massive transcriptional reprogramming has occurred in both lines, but in tmNIL130 still more than twice as many genes were differentially expressed compared to NIL9.

To identify processes involved in defense and/or senescence, MAPMAN [216] was applied to sort the genes that were differentially expressed at the onset of maturity on the basis of their annotation (Table 1 and Table 2). Up- and down-regulated genes in both lines with roles in phytohormone signaling, the response to oxidative stress as well as WRKY transcription factors were compiled. Four major trends were identified:

(1) Ethylene-associated transcripts showed a shift towards defense-relevant branches in tmNIL130.

Genes involved in ET synthesis, signaling and response were predominantly up-regulated after *V. longisporum* infection in both lines. Expression changes were more pronounced in tmNIL130. ACC-synthases (ACS) were more strongly up-regulated in tmNIL130 and interestingly, the catalytically inactive isoform ACS1 was only up-regulated in tmNIL130. ACS1-expression was previously associated with young tissues only [217]. ET receptors were induced only in tmNIL130. It must be kept in mind that owing to the inhibitory nature of the ET receptors, increasing their density decreases sensitivity to the hormone.

Transcription factor ERF14, which acts in the ERF1/ORA59 defense signaling branch of JA and ET [218], was strongly up-regulated in tmNIL130. ORA59, however, was down-regulated in NIL9, not in tmNIL130 (Table 1). These findings suggest that the defense branch of ET signaling was strengthened after infection in tmNIL130, while it was weakened in NIL9.

(2) Many genes acting in auxin synthesis, homeostasis, transport and response were differentially expressed upon infection in tmNIL130.

From the sheer number of down-regulated genes in tmNIL130 it could be concluded that auxin played a negative role in *V. longisporum* resistance (Table 1). However, the auxin biosynthesis gene NIT2 was 16-fold up-regulated in tmNIL130. Auxin-conjugating enzymes were also up-regulated to a considerable degree, among them GH3.2, which has been shown to mediate susceptibility to *B. cinerea* and *P. syringae* [123]. However, with tmNIL130 being the resistant line, this mechanism did not seem to apply to *V. longisporum* infection. Auxin induction and tight control at the same time point to a sophisticated and specialized role of auxin signaling in mediating *V. longisporum* resistance during the late stage of infection at the onset of maturity.

(3) Several defense-related WRKY transcription factors were exclusively up-regulated in tmNIL130.

Several WRKY transcription factors with sometimes seemingly contrasting roles were exclusively up-regulated in tmNIL130 by *V. longisporum* infection (Table 1). Among the most strongly up-regulated WRKY transcription factors were WRKY40 and WRKY60 as positive regulators of defense against necrotrophs and negative regulators of ABA signaling [140]. However, the 6-fold up-regulated WRKY63 has been reported to positively regulate some ABA-responses [219]. Highly induced WRKY62 as well as WRKY70 and WRKY53 have been shown to positively regulate basal defense [136,139]. WRKY53 and WRKY6 are also positive regulators of senescence. Interestingly, moderately up-regulated WRKY57 was shown to mediate JA-auxin cross-talk, inhibiting JA-induced senescence and promoting auxin-signaling [106]. This would be in accordance with the finding that JA played a disease-promoting role during the late stage of infection as concluded from phytohormone analyses (Figure 5). Considering their emerging role as molecular switches between pathways, WRKY transcription factors possibly contributed to a much more elaborate control of the defense—senescence signaling network in tmNIL130 at the onset of maturity compared to NIL9.

Table 1. Regulatory pathway genes differentially expressed by *V. longisporum* infection in the hypocotyl and basal shoot of two *A. thaliana* (Bur × L*er*) near-isogenic lines (NILs) at the onset of silique maturity in a microarray experiment.

NIL9		tmNIL130			
Gene	**FC**	**Gene**	**FC**	**Gene**	**FC**
Ethylene biosynthesis, -signaling and -response					
at2g25450	6.94	at5g20400	2.13	at5g44210 (ERF9)	2.59
at4g37770 (ACS8)	4.07	at5g59540	2.48	at3g23150 (ETR2)	2.46
at5g65800 (ACS5)	3.90	at5g43440	2.18	at1g04310 (ERS2)	2.84
at1g01480(ACS2)	3.59	at1g03400	2.10	at3g50260 (CEJ1)	2.84
at4g26200 (ACS7)	3.18	at5g43450	2.13	at2g40940 (ERS1)	2.40
at5g44210 (ERF9)	2.94	at2g25450	12.06	at5g25190	3.67
at5g25190	5.09	at4g37770 (ACS8)	4.99	at3g11930	2.25
at4g37580 (HLS1)	2.07	at5g65800(ACS5)	3.88	at1g04370 (ATERF14)	4.98
at3g11930	3.67	at1g01480 (ACS2)	11.90	at1g55150	3.49
at1g55150	2.10	at3g61510 (ACS1)	5.00	at1g09740	2.32
at3g20640	2.40	at4g26200 (ACS7)	2.17		
at1g77330	0.28	at2g31230 (ATERF15)	0.40		
at5g07580	0.38	at2g20100	0.45		
at2g31230 (ATERF15)	0.42	at1g27660	0.32		
at1g06160 (ORA59)	0.33	at1g49830	0.33		
at5g44350	0.21	at4g29100	0.48		
at1g27660	0.37				
Jasmonic acid biosynthesis, signaling and response					
		at1g18020	2.89		
		at1g52070	2.64		
		at5g48180 (NSP5)	2.25		
at1g55020 (LOX1)	0.44	at1g09400	0.35		
at1g76690 (OPR2)	0.39	at3g16470 (JR1)	0.24		
at3g16470 (JR1)	0.34	at3g16450	0.40		
at3g16450	0.35				
Salicylic acid biosynthesis, signaling and response					
at1g66720	6.23	at1g66720	12.03		
at5g04370 (NAMT1)	0.18	at5g38020	0.20	at4g26420 (GAMT1)	0.35
at5g37990	0.09	at1g05670	0.38	at5g37990	0.08
		at5g04370 (NAMT1)	0.45		
Abscisic acid biosynthesis, signaling and response					
at4g18350 (NCED2)	4.26	at4g18350 (NCED2)	10.11	at5g23350	2.39
at5g50720 (ATHVA22E)	2.03	at1g30100 (NCED5)	3.32	at5g50720 (ATHVA22E)	2.31
at1g74520 (ATHVA22A)	2.84	at3g24650 (ABI3)	8.23		
at4g32810 (CCD8)	0.41	at2g36020 (HVA22J)	0.20	at2g17770 (ATBZIP27)	0.28
at2g44990 (CCD7)	0.15	at4g32810 (CCD8)	0.28	at5g62490 (ATHVA22B)	0.49
at3g63210 (MARD1)	0.50	at2g44990 (CCD7)	0.16	at1g45249 (ABF2)	0.39
at2g17770 (ATBZIP27)	0.32	at3g43600 (AAO2)	0.26		
at5g23370	0.46	at1g52920 (GCR2)	0.48		
at1g45249 (ABF2)	0.36	at3g63210 (MARD1)	0.45		
Brassinosteroid synthesis, signaling and response					
at3g50660 (DWF4)	4.09	at3g50660 (DWF4)	2.99	at2g22830 (SQE2)	2.28
		at1g76090 (SMT3)	2.30	at4g36780	2.16
		at4g37760 (SQE3)	0.34		

Table 1. *Cont.*

NIL9		tmNIL130			
Gene	**FC**	**Gene**	**FC**	**Gene**	**FC**
Cytokinin biosynthesis, signaling and response					
at5g05860 (UGT76C2)	2.34	at5g05860 (UGT76C2)	2.06	at1g22400 (UGT85A1)	2.72
at2g17820 (AHK1)	2.10	at4g29740 (CKX4)	5.18		
at2g47430 (CKI1)	0.34	at5g35750 (AHK2)	0.45	at2g01830 (AHK4)	0.44
at2g01830 (AHK4)	0.44	at2g25180 (ARR12)	0.32		
Gibberellic acid biosynthesis, signaling and response					
at5g59845	6.82	at1g02400 (ATGA2OX4)	3.05		
		at1g75750 (GASA1)	10.66		
		at5g59845	5.97		
at1g52820	0.24	at4g23340	0.33	at5g51810 (GA20OX2)	0.22
at4g25420 GA5	0.22	at4g25420 (GA5)	0.33	at1g14920 (GAI)	0.34
at1g14920 (GAI)	0.42				
Auxin biosynthesis, signaling and response					
at5g13360	2.46	at3g44300 (NIT2)	16.35	at5g54510 (GH3.6)	3.00
at1g10810	4.09	at5g55250 (IAMT1)	2.62	at4g36110	4.11
at5g13320 (PBS3)	4.09	at3g15450	6.61	at1g17345	2.57
at5g54510 GH3.6	2.78	at2g23170 (GH3.3)	15.67	at4g37390 (YDK1/GH3.2)	5.77
at1g17345	2.48	at1g10810	3.26	at2g04850	2.06
at2g04850	2.69	at5g13320 (PBS3)	6.08	at4g12410	2.09
at4g12410	3.37	at1g60730	2.48	at5g13370	4.57
		at5g01100	2.93	at4g27450	3.94
at1g77690 (LAX3)	0.37	at3g10870 (MES17)	0.49	at3g12955	0.34
at5g47530	0.29	at1g05560 (UGT1)	0.45	at3g02250	0.45
at4g17280	0.17	at1g77690 (LAX3)	0.42	at4g00880	0.30
at1g16510	0.48	at5g16530 (PIN5)	0.42	at1g19830	0.26
at2g14960	0.24	at1g73590 (PIN1)	0.47	at3g61900	0.45
at3g02250	0.46	at5g47530	0.21	at1g14970	0.31
at4g00880	0.46	at4g17280	0.21	at3g61750	0.25
at3g61750	0.34	at3g43120	0.34	at5g65470	0.41
at1g28130 (GH3.17)	0.27	at3g47620 (AtTCP14)	0.40	at4g34750 (SAUR_E)	0.31
		at3g25290	0.31	at2g04852	0.49
		at2g34680 (AIR9)	0.42	at4g12980	0.39
WRKY transcription factors					
at4g31800 (WRKY18)	2.38	at1g69810 (WRKY36)	2.05	at1g69310 (WRKY57)	1.69
at5g46350 (WRKY8)	3.85	at4g31800 (WRKY18)	2.76	at3g56400 (WRKY70)	2.48
at2g25000 (WRKY60)	4.09	at5g46350 (WRKY8)	4.19	at5g64810 (WRKY51)	4.65
at1g80590 (WRKY66)	4.13	at1g66600 (WRKY63)	6.33	at1g80590 (WRKY66)	3.18
at2g47260 (WRKY23)	4.09	at4g23810 (WRKY53)	3.01	at5g26170 (ATWRKY50)	2.96
at4g22070 (WRKY31)	3.19	at2g23320 (WRKY15)	2.31	at5g49520 (WRKY48)	1.90
		at4g23550 (WRKY29)	2.83	at2g47260 (WRKY23)	6.37
		at2g25000 (WRKY60)	3.91	at4g22070 (WRKY31)	4.67
		at1g18860 (WRKY61)	4.38	at1g62300 (WRKY6)	3.73
		at5g13080 (WRKY75)	3.47	at5g24110 (WRKY30)	3.15
		at5g01900 (WRKY62)	5.49	at1g80840 (WRKY40)	6.85
at5g52830 (WRKY27)	0.45	at4g39410 (WRKY13)	0.50	at2g37260 (WRKY44)	0.34
at3g58720	0.48	at5g52830 (WRKY27)	0.34	at2g44745	0.47
at2g34830 (WRKY35)	0.46	at2g34830 (WRKY35)	0.44		

Classification of transcripts with MAPMAN [216]. In each class, differentially expressed genes in the susceptible NIL9 are displayed on the left side, those in resistant tmNIL130 on the right side. Up-regulated genes are at the top of each group, down-regulated genes at the bottom. Only those transcripts were included whose fold change (FC) was over 2 or below 0.5. Means for transcripts (represented by transcript clusters in AraGene-1_1-st microarrays) were calculated from individual probeset values, for which the cutoff was set at $p < 0.05$.

Table 2. Genes involved in the response to oxidative stress differentially expressed by *V. longisporum* infection in the hypocotyl and basal shoot of two *A. thaliana* (Bur × Ler) near-isogenic lines (NILs) at the onset of silique maturity in a microarray experiment.

NIL9		tmNIL130			
Gene	FC	Gene	FC	Gene	FC
Redox homeostasis					
at1g69880 (ATH8)	5.95	at1g69880 (ATH8)	7.52	at3g02870 (VTC4)	2.60
at5g61440 (ACHT5)	2.79	at1g19730 (ATH4)	3.45	at1g63460	2.04
at4g39830	4.77	at5g61440 (ACHT5)	3.45	at2g31570 (ATGPX2)	2.45
at1g20630	2.82	at5g16400 (TRXF2)	2.46	at2g16060 (GLB1)	3.65
		at1g21750 (ATPDIL1-1)	2.58	at1g03850	5.63
		at3g19010	2.40	at4g15700	4.01
		at2g43350 (ATGPX3)	2.24	at1g32760	2.01
		at4g39830	5.01	at1g20630 (CAT1)	2.62
		at4g09010 (APX4)	3.47	at5g17770 (CBR1)	2.30
		at4g08390 (SAPX)	2.54		
at4g33040	0.44	at3g20560 (ATPDIL5-3)	0.48	at4g33040	0.31
at5g11930	0.33	at4g18260	0.16	at5g11930	0.26
		at3g19000	0.44	at2g47870	0.41
		at5g21100	0.37	at4g25100 (FSD1)	0.23
		at2g47880	0.24	at1g31170 (SRX)	0.45
Peroxidases					
at3g49120 (ATPERX34)	2.51	at5g19880	2.53	at1g49570	5.56
at2g37130 (PER21)	4.44	at5g19890	4.38	at2g34060	2.20
at1g71695 (PER12)	2.29	at3g49120 (ATPERX34)	2.36	at5g64110	4.40
at5g15180	6.40	at2g37130 (PER21)	3.78	at5g64100	6.12
at5g64110	3.77	at1g71695 (PER12)	3.00	at5g05340	3.91
at5g51890	24.95	at4g37520 (PER50)	2.98	at4g33420	3.14
		at5g64120	11.16	at5g51890	18.62
		at5g39580	6.27	at1g05240	2.48
		at5g15180	4.10		
at5g67400	0.09	at1g68850	0.26		
at5g14130	0.31	at5g42180 (PER64)	0.21		
at5g06720	0.46				
at1g30870	0.33				
at5g42180 (PER64)	0.26				
at1g44970	0.11				
at1g05250	0.15				
at1g05240	0.13				
Glutathione-S-transferases					
at5g02780	7.93	at1g02920 (ATGSTF7)	4.38	at1g14550	5.27
at2g29440 (ATGSTU6)	6.41	at2g29470 (ATGSTU3)	8.22	at1g14540	6.54
at3g09270: (ATGSTU8)	3.21	at4g19880	2.23	at3g09270 (ATGSTU8)	3.94
at1g17180 (ATGSTU25)	3.02	at1g17170 (ATGSTU24)	2.26	at1g17180 (ATGSTU25)	2.19
at1g69930 (ATGSTU11)	2.05	at5g02780	3.33	at1g69930: (ATGSTU11)	6.67
		at2g29440 (ATGSTU6)	8.74	at1g02930 (ATGSTF6)	5.20
		at4g02520 (ATGSTF2)	2.74	at2g29480 (ATGSTU2)	4.03
		at2g02930 (ATGSTF3)	4.33	at2g29450 (ATGSTU5)	2.83
		at1g69920 (ATGSTU12)	5.13		
		at1g74590 (ATGSTU10)	4.16		
at1g49860 (ATGSTF14)	0.19	at3g03190 (ATGSTF11)	0.27		
at1g78370 (ATGSTU20)	0.20	at1g78370 (ATGSTU20)	0.24		
at1g10360 (ATGSTU18)	0.44				
at5g62480: (ATGSTU9)	0.39				
at5g17220: (ATGSTF12)	0.10				

Classification of transcripts with MAPMAN [216]. In each class, differentially expressed genes in the susceptible NIL9 are displayed on the left side, those in resistant tmNIL130 on the right side. Up-regulated genes are at the top of each group, down-regulated genes at the bottom. Only those transcripts were included whose fold change (FC) was over 2 or below 0.5. Means for transcripts (represented by transcript clusters in AraGene-1_1-st microarrays) were calculated from individual probeset values, for which the cutoff was set at $p < 0.05$.

(4) tmNIL130 showed a much stronger response to oxidative stress.

The differential expression of genes involved in oxidative stress response was perhaps the most conspicuous difference between NIL9 and tmNIL130 (Table 2). While in NIL9, only few ROS-detoxifying enzymes responded to infection, most of which were down-regulated, tmNIL130 showed up-regulation of a large portfolio of enzymes involved in redox homeostasis and ROS metabolism like catalase and peroxidases. A much greater diversity of glutathione-S-transferases, which are generally thought to detoxify oxidation products in the cell [220], were up-regulated in tmNIL130. Keeping in mind that these processes took place during silique maturity and the late phase of *V. longisporum* infection, this reaction clearly points to a stringent control of cell death and senescence in tmNIL130, obviously restricting pathogen growth.

4. CONCLUDING REMARKS

Plant senescence and host response to pathogen infection are both associated with numerous signaling events and major transcriptional changes. Considering the fact that signaling pathways overlap and often the same molecules are involved in both senescence and response to biotic stress, one might be inclined to conclude that some developmental changes inevitably occur upon pathogen infection. Recent work has greatly increased our knowledge of signal integration, crosstalk between signaling pathways and the fine-tuning of stress responses and developmental processes. The emerging picture suggests that in most cases distinct branches of a signaling pathway control different developmental and defense responses. Some signaling components act as regulatory nodes to channel the plant's response into a certain direction. The capacity of the plant to control such responses may decide over disease or resistance especially in quantitative disease resistance, which is the more common type of resistance to necrotrophic and hemibiotrophic pathogens. The interaction of necrotrophic and hemibiotrophic pathogens with their hosts is characterized by extensive transcriptional changes and the response of the majority of signaling pathways available to plants. Often these changes lead to the induction of senescence and cell death, which is favorable to a necrotrophic lifestyle based on the killing of cells. While pathogen-induced senescence relies on the induction of the "wrong" signaling components, the capacity of the plant to buffer these interferences and channel the signaling pathways into the "right" responses seems to account for a large part of quantitative and developmentally influenced resistance. This capacity

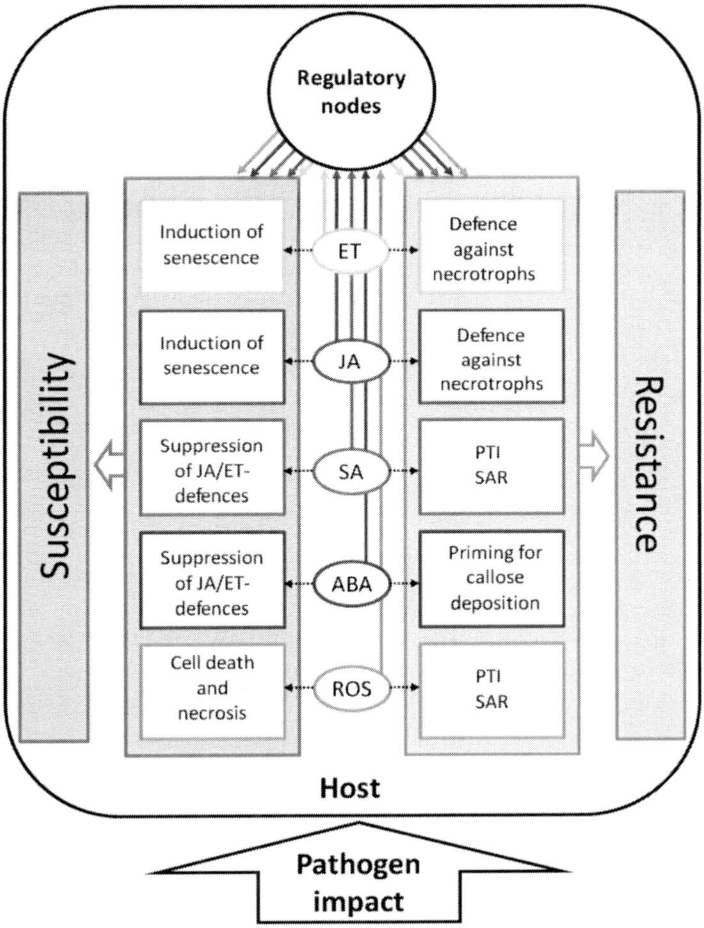

Figure 7. A model for the roles of potentially senescence-inducing signaling molecules during necrotrophic host-pathogen-interactions. Ethylene (ET), jasmonic acid (JA), salicylic acid (SA), abscisic acid (ABA) and reactive oxygen species (ROS) are usually induced during necrotrophic interactions. If the senescence-promoting effects prevail, host susceptibility will be increased. If the host succeeds to restrict signaling events to defence-related branches, resistance can be achieved. Regulatory nodes, such as transcription factors, may act as molecular switches between signaling branches. Colored arrows symbolize signaling pathways triggered by different signaling molecules. Dashed arrows symbolize potential responses. PTI = PAMP-triggered immunity, SAR = systemic acquired resistance, PAMP = pathogen-associated molecular pattern.

may be pre-formed: A certain developmental type or stage is *per se* less susceptible to pathogen manipulations because the targets are not available or the cellular environment is otherwise hostile to the pathogen. During infection, much depends on induced responses in the host: Resistance can be achieved by counteracting senescence and cell death and by activating defense responses (Figure 7).

Multiple mechanisms may underlie an adequate control of signaling events: The timing and intensity of the response, the concentration of signaling molecules or the tissue-specific expression of genes for hormone synthesis, receptors, transporters or responsive genes. It is a challenge for the future to identify regulators that control developmental homeostasis and enable the plant to stay on the narrow path of an adequate response to pathogens.

ACKNOWLEDGMENTS

We thank Sophia Harrand and Rebecca Werner for excellent technical assistance and Wolfram Brenner for advice concerning the microarray experiment. We are grateful to Karen Zeise (Berlin, Germany) for providing the fungal strain. Financial support from Deutsche Forschungsgemeinschaft (DFG) Bonn, Germany (grants DI1502/1-1 and DI1502/3-1), Norddeutsche Pflanzenzucht (NPZ), Germany and Federal State of Berlin, Germany (Nafög-scholarship) is gratefully acknowledged. We thank Petr Karlovsky for our long-standing and productive collaboration and Thomas Schmülling and Reinhard Kunze for supporting and hosting the project.

AUTHOR CONTRIBUTIONS

EH planned, supervised and analyzed all experiments with *A. thaliana* and wrote the manuscript. SK planned, supervised and analyzed all experiments with *B. alboglabra*. ED conceived the study and helped to write the manuscript.

CONFLICTS OF INTEREST

The authors declare no conflict of interest. The funding sponsors had no role in the design of the study; in the collection, analyses, or interpretation of data; in the writing of the manuscript, and in the decision to publish the results. All previously published figures have been published under the terms of the Creative Commons Attribution License (http://creativecommons.org/licenses/by/2.0), which permits unrestricted use, distribution, and reproduction in any medium, provided the original work is properly cited.

REFERENCES

1. Guo, Y.; Gan, S.S. Convergence and divergence in gene expression profiles induced by leaf senescence and 27 senescence-promoting hormonal, pathological and environmental stress treatments. Plant Cell Environ. 2012, 35, 644–655.
2. Grant, M.R.; Jones, J.D.G. Hormone (dis)harmony moulds plant health and disease. Science 2009, 324, 750–752.
3. Barna, B.; Fodor, J.; Harrach, B.; Pogány, M.; Király, Z. The janus face of reactive oxygen species in resistance and susceptibility of plants to necrotrophic and biotrophic pathogens. Plant Physiol. Biochem. 2012, 59, 37–43.
4. Mengiste, T. Plant immunity to necrotrophs. Ann. Rev. Phytopathol. 2012, 50, 267–294.
5. Newton, A.C.; Fitt, B.D.; Atkins, S.D.; Walters, D.R.; Daniell, T.J. Pathogenesis, parasitism and mutualism in the trophic space of microbe-plant interactions. Trends Microbiol. 2010, 18, 365–373.
6. Gladders, P. Relevance of Verticillium wilt (Verticillium longisporum) in winter oilseed rape in the UK. HCGA Res. Rev.2009, 72, 1–39.
7. Abeles, F.B.; Morgan, P.W.; Saltveit, M.E., Jr. Ethylene in Plant Biology; Academic Press: San Diego, CA, USA, 2012.
8. Grbić, V.; Bleecker, A.B. Ethylene regulates the timing of leaf senescence in Arabidopsis. Plant J. 1995, 8, 595–602.
9. Jing, H.-C.; Schippers, J.H.; Hille, J.; Dijkwel, P.P. Ethylene-induced leaf senescence depends on age-related changes and old genes in Arabidopsis. J. Exp. Bot. 2005, 56, 2915–2923.
10. Van Loon, L.C.; Geraats, B.P.; Linthorst, H.J. Ethylene as a modulator of disease resistance in plants. Trends Plant Sci.2006, 11, 184–191.
11. Bailey, B.A. Purification of a protein from culture filtrates of Fusarium oxysporum that induces ethylene and necrosis in leaves of Erythroxylum coca. Phytopathology 1995, 85, 1250–1255.
12. Cristescu, S.M.; de Martinis, D.; te Lintel Hekkert, S.; Parker, D.H.; Harren, F.J. Ethylene production by Botrytis cinerea in vitro and in tomatoes. Appl. Environ. Microbiol. 2002, 68, 5342–5350.
13. Weingart, H.; Völksch, B. Ethylene production by Pseudomonas syringae pathovars in vitro and in planta. Appl. Environ. Microbiol. 1997, 63, 156–161.
14. Penninckx, I.A.; Thomma, B.P.; Buchala, A.; Métraux, J.-P.; Broekaert, W.F. Concomitant activation of jasmonate and ethylene response pathways is required for induction of a plant defensin gene in arabidopsis. Plant Cell 1998, 10, 2103–2113.
15. Argueso, C.T.; Hansen, M.; Kieber, J.J. Regulation of ethylene biosynthesis. J. Plant Growth Regul. 2007, 26, 92–105.
16. Wang, K.L.-C.; Li, H.; Ecker, J.R. Ethylene biosynthesis and signaling networks. Plant Cell 2002, 14, S131–S151.
17. Hua, J.; Meyerowitz, E.M. Ethylene responses are negatively regulated by a receptor gene family in Arabidopsis thaliana. Cell 1998, 94, 261–271.

18. Stepanova, A.N.; Alonso, J.M. Ethylene signaling and response: Where different regulatory modules meet. Curr. Opin. Plant Biol. 2009, 12, 548–555.
19. Ju, C.; Yoon, G.M.; Shemansky, J.M.; Lin, D.Y.; Ying, Z.I.; Chang, J.; Garrett, W.M.; Kessenbrock, M.; Groth, G.; Tucker, M.L. CTR1 phosphorylates the central regulator EIN2 to control ethylene hormone signaling from the ER membrane to the nucleus in Arabidopsis. Proc. Natl. Acad. Sci. USA 2012, 109, 19486–19491.
20. An, F.; Zhao, Q.; Ji, Y.; Li, W.; Jiang, Z.; Yu, X.; Zhang, C.; Han, Y.; He, W.; Liu, Y. Ethylene-induced stabilization of ETHYLENE INSENSITIVE 3 and EIN3-LIKE 1 is mediated by proteasomal degradation of EIN3 binding F-box 1 and 2 that requires EIN2 in Arabidopsis. Plant Cell 2010, 22, 2384–2401.
21. Wen, X.; Zhang, C.; Ji, Y.; Zhao, Q.; He, W.; An, F.; Jiang, L.; Guo, H. Activation of ethylene signaling is mediated by nuclear translocation of the cleaved EIN2 carboxyl terminus. Cell Res. 2012, 22, 1613–1616.
22. Yoo, S.-D.; Cho, Y.; Sheen, J. Emerging connections in the ethylene signaling network. Trends Plant Sci. 2009, 14, 270–279.
23. Doubt, S.L. The Response of Plants to Illuminating Gas; University of Chicago: Chicago, IL, USA, 1917; Volume 227.
24. Zimmerman, P.; Hitchcock, A.; Crocker, W. The effect of ethylene and illuminating gas on roses. Contrib. Boyce Thompson Inst. 1931, 3, 459–481.
25. Jing, H.-C.; Anderson, L.; Sturre, M.J.; Hille, J.; Dijkwel, P.P. Arabidopsis cpr5 is a senescence-regulatory gene with pleiotropic functions as predicted by the evolutionary theory of senescence. J. Exp. Bot. 2007, 58, 3885–3894.
26. Li, Z.; Peng, J.; Wen, X.; Guo, H. ETHYLENE-INSENSITIVE 3 is a senescence-associated gene that accelerates age-dependent leaf senescence by directly repressing miR164 transcription in Arabidopsis. Plant Cell 2013, 25, 3311–3328.
27. Balazadeh, S.; Siddiqui, H.; Allu, A.D.; Matallana-Ramirez, L.P.; Caldana, C.; Mehrnia, M.; Zanor, M.I.; Köhler, B.; Mueller-Roeber, B. A gene regulatory network controlled by the NAC transcription factor ANAC092/atNAC2/ORE1 during salt-promoted senescence. Plant J. 2010, 62, 250–264.
28. Kim, J.H.; Woo, H.R.; Kim, J.; Lim, P.O.; Lee, I.C.; Choi, S.H.; Hwang, D.; Nam, H.G. Trifurcate feed-forward regulation of age-dependent cell death involving miR164 in Arabidopsis. Science 2009, 323, 1053–1057.
29. Pré, M.; Atallah, M.; Champion, A.; de Vos, M.; Pieterse, C.M.; Memelink, J. The AP2/ERF domain transcription factor ORA59 integrates jasmonic acid and ethylene signals in plant defense. Plant Physiol. 2008, 147, 1347–1357.
30. Zhu, Z.; An, F.; Feng, Y.; Li, P.; Xue, L.; Mu, A.; Jiang, Z.; Kim, J.-M.; To, T.K.; Li, W. Derepression of ethylene-stabilized transcription factors (EIN3/EIL1) mediates jasmonate and ethylene signaling synergy in Arabidopsis. Proc. Natl. Acad. Sci. USA 2011, 108, 12539–12544.
31. Hirsch, J.; Deslandes, L.; Feng, D.X.; Balagué, C.; Marco, Y. Delayed symptom development in ein2–1, an Arabidopsis ethylene-insensitive mutant, in response to bacterial wilt caused by Ralstonia solanacearum. Phytopathology 2002, 92, 1142–1148.
32. Johansson, A.; Staal, J.; Dixelius, C. Early responses in the Arabidopsis-Verticillium longisporum pathosystem are dependent on NDR1, JA- and ET-associated signals via cytosolic NPR1 and RFO1. Mol. Plant Microbe Interact. 2006, 19, 958–969.
33. Pantelides, I.S.; Tjamos, S.E.; Paplomatas, E.J. Ethylene perception via ETR1 is required in Arabidopsis infection by Verticillium dahliae. Mol. Plant Pathol. 2010, 11, 191–202.
34. Bleecker, A.B.; Kende, H. Ethylene: A gaseous signal molecule in plants. Ann. Rev. Cell Dev. Biol. 2000, 16, 1–18.
35. Liu, Q.; Wen, C.-K. Arabidopsis ETR1 and ERS1 differentially repress the ethylene response in combination with other ethylene receptor genes. Plant Physiol. 2012, 158, 1193–1207.

36. Wasternack, C.; Hause, B. Jasmonates: Biosynthesis, perception, signal transduction and action in plant stress response, growth and development. An update to the 2007 review in Annals of Botany. Ann. Bot. 2013, 111, 1021–1058.
37. Chini, A.; Fonseca, S.; Chico, J.M.; Fernández-Calvo, P.; Solano, R. The ZIM domain mediates homo- and heteromeric interactions between Arabidopsis JAZ proteins. Plant J. 2009, 59, 77–87.
38. Thines, B.; Katsir, L.; Melotto, M.; Niu, Y.; Mandaokar, A.; Liu, G.; Nomura, K.; He, S.Y.; Howe, G.A.; Browse, J. JAZ repressor proteins are targets of the SCF-COI1 complex during jasmonate signalling. Nature 2007, 448, 661–665.
39. Yan, J.; Zhang, C.; Gu, M.; Bai, Z.; Zhang, W.; Qi, T.; Cheng, Z.; Peng, W.; Luo, H.; Nan, F. The Arabidopsis CORONATINE INSENSITIVE 1 protein is a jasmonate receptor. Plant Cell 2009, 21, 2220–2236.
40. Ueda, J.; Kato, J. Isolation and identification of a senescence-promoting substance from wormwood (Artemisia absinthium L.). Plant Physiol. 1980, 66, 246–249.
41. He, Y.; Fukushige, H.; Hildebrand, D.F.; Gan, S. Evidence supporting a role of jasmonic acid in Arabidopsis leaf senescence. Plant Physiol. 2002, 128, 876–884.
42. Schommer, C.; Palatnik, J.F.; Aggarwal, P.; Chételat, A.; Cubas, P.; Farmer, E.E.; Nath, U.; Weigel, D. Control of jasmonate biosynthesis and senescence by miR319 targets. PLoS Biol. 2008, 6, e230.
43. Kim, J.; Chang, C.; Tucker, M.L. To grow old: Regulatory role of ethylene and jasmonic acid in senescence. Front. Plant Sci. 2015, 6, 20.
44. Shan, X.; Wang, J.; Chua, L.; Jiang, D.; Peng, W.; Xie, D. The role of Arabidopsis RUBISCO ACTIVASE in jasmonate-induced leaf senescence. Plant Physiol. 2011, 155, 751–764.
45. Zhang, L.; Xing, D. Methyl jasmonate induces production of reactive oxygen species and alterations in mitochondrial dynamics that precede photosynthetic dysfunction and subsequent cell death. Plant Cell Physiol. 2008, 49, 1092–1111.
46. Miao, Y.; Zentgraf, U. The antagonist function of Arabidopsis WRKY53 and ESR/ESP in leaf senescence is modulated by the jasmonic and salicylic acid equilibrium. Plant Cell 2007, 19, 819–830.
47. Pieterse, C.M.; van der Does, D.; Zamioudis, C.; Leon-Reyes, A.; van Wees, S.C. Hormonal modulation of plant immunity. Ann. Rev. Cell Dev. Biol. 2012, 28, 489–521.
48. Truman, W.; Bennett, M.H.; Kubigsteltig, I.; Turnbull, C.; Grant, M. Arabidopsis systemic immunity uses conserved defense signaling pathways and is mediated by jasmonates. Proc. Natl. Acad. Sci. USA 2007, 104, 1075–1080.
49. Vlot, A.C.; Dempsey, D.M.A.; Klessig, D.F. Salicylic acid, a multifaceted hormone to combat disease. Ann. Rev. Phytopathol. 2009, 47, 177–206.
50. Rivas-San Vicente, M.; Plasencia, J. Salicylic acid beyond defence: Its role in plant growth and development. J. Exp. Bot. 2011, 62, 3321–3338.
51. Morris, K.; Mackerness, S.A.H.; Page, T.; John, C.F.; Murphy, A.M.; Carr, J.P.; Buchanan-Wollaston, V. Salicylic acid has a role in regulating gene expression during leaf senescence. Plant J. 2000, 23, 677–685.
52. Zentgraf, U.; Laun, T.; Miao, Y. The complex regulation of WRKY53 during leaf senescence of Arabidopsis thaliana. Eur. J. Cell Biol. 2010, 89, 133–137.
53. Raffaele, S.; Rivas, S.; Roby, D. An essential role for salicylic acid in atMYB30-mediated control of the hypersensitive cell death program in Arabidopsis. FEBS Lett. 2006, 580, 3498–3504.
54. Robert-Seilaniantz, A.; Grant, M.; Jones, J.D. Hormone crosstalk in plant disease and defense: More than just jasmonate-salicylate antagonism. Ann. Rev. Phytopathol. 2011, 49, 317–343.
55. Fischer, A.M. The complex regulation of senescence. Crit. Rev. Plant Sci. 2012, 31, 124–147.
56. Gazzarrini, S.; McCourt, P. Genetic interactions between ABA, ethylene and sugar signaling pathways. Curr. Opin. Plant Biol. 2001, 4, 387–391.

57. Vanhee, C.; Batoko, H. Autophagy involvement in responses to abscisic acid by plant cells. Autophagy 2011, 7, 655–656.
58. Zhang, A.; Jiang, M.; Zhang, J.; Tan, M.; Hu, X. Mitogen-activated protein kinase is involved in abscisic acid-induced antioxidant defense and acts downstream of reactive oxygen species production in leaves of maize plants. Plant Physiol. 2006, 141, 475–487.
59. Ma, Y.; Szostkiewicz, I.; Korte, A.; Moes, D.; Yang, Y.; Christmann, A.; Grill, E. Regulators of PP2C phosphatase activity function as abscisic acid sensors. Science 2009, 324, 1064–1068.
60. Park, S.-Y.; Fung, P.; Nishimura, N.; Jensen, D.R.; Fujii, H.; Zhao, Y.; Lumba, S.; Santiago, J.; Rodrigues, A.; Tsz-fung, F.C. Abscisic acid inhibits type 2C protein phosphatases via the PYR/PYL family of start proteins. Science 2009, 324, 1068–1071.
61. Fuchs, S.; Tischer, S.V.; Wunschel, C.; Christmann, A.; Grill, E. Abscisic acid sensor RCAR7/PYL13, specific regulator of protein phosphatase coreceptors. Proc. Natl. Acad. Sci. USA 2014, 111, 5741–5746.
62. Raghavendra, A.S.; Gonugunta, V.K.; Christmann, A.; Grill, E. ABA perception and signalling. Trends Plant Sci. 2010, 15, 395–401.
63. Hubbard, K.E.; Nishimura, N.; Hitomi, K.; Getzoff, E.D.; Schroeder, J.I. Early abscisic acid signal transduction mechanisms: Newly discovered components and newly emerging questions. Genes Dev. 2010, 24, 1695–1708.
64. Zhang, H.; Zhou, C. Signal transduction in leaf senescence. Plant Mol. Biol. 2013, 82, 539–545.
65. Gepstein, S.; Thimann, K.V. Changes in the abscisic acid content of oat leaves during senescence. Proc. Natl. Acad. Sci. USA 1980, 77, 2050–2053.
66. Weaver, L.M.; Gan, S.; Quirino, B.; Amasino, R.M. A comparison of the expression patterns of several senescence-associated genes in response to stress and hormone treatment. Plant Mol. Biol. 1998, 37, 455–469.
67. Buchanan-Wollaston, V.; Page, T.; Harrison, E.; Breeze, E.; Lim, P.O.; Nam, H.G.; Lin, J.F.; Wu, S.H.; Swidzinski, J.; Ishizaki, K. Comparative transcriptome analysis reveals significant differences in gene expression and signalling pathways between developmental and dark/starvation-induced senescence in Arabidopsis. Plant J. 2005, 42, 567–585.
68. Lee, I.C.; Hong, S.W.; Whang, S.S.; Lim, P.O.; Nam, H.G.; Koo, J.C. Age-dependent action of an ABA-inducible receptor kinase, RPK1, as a positive regulator of senescence in Arabidopsis leaves. Plant Cell Physiol. 2011, 52, 651–662.
69. Jin, Y.; Ni, D.-A.; Ruan, Y.-L. Posttranslational elevation of cell wall invertase activity by silencing its inhibitor in tomato delays leaf senescence and increases seed weight and fruit hexose level. Plant Cell 2009, 21, 2072–2089.
70. Balibrea Lara, M.E.; Gonzalez Garcia, M.-C.; Fatima, T.; Ehneß, R.; Lee, T.K.; Proels, R.; Tanner, W.; Roitsch, T. Extracellular invertase is an essential component of cytokinin-mediated delay of senescence. Plant Cell 2004, 16, 1276–1287.
71. Godt, D.E.; Roitsch, T. Regulation and tissue-specific distribution of mRNAs for three extracellular invertase isoenzymes of tomato suggests an important function in establishing and maintaining sink metabolism. Plant Physiol.1997, 115, 273–282.
72. Garapati, P.; Xue, G.-P.; Munné-Bosch, S.; Balazadeh, S. Transcription factor ATAF1 in Arabidopsis promotes senescence by direct regulation of key chloroplast maintenance and senescence transcriptional cascades. Plant Physiol. 2015.
73. Yang, S.-D.; Seo, P.J.; Yoon, H.-K.; Park, C.-M. The Arabidopsis NAC transcription factor VNI2 integrates abscisic acid signals into leaf senescence via the COR/RD genes. Plant Cell 2011, 23, 2155–2168.
74. Raab, S.; Drechsel, G.; Zarepour, M.; Hartung, W.; Koshiba, T.; Bittner, F.; Hoth, S. Identification of a novel E3 ubiquitin ligase that is required for suppression of premature senescence in Arabidopsis. Plant J. 2009, 59, 39–51.
75. Vogelmann, K.; Drechsel, G.; Bergler, J.; Subert, C.; Philippar, K.; Soll, J.; Engelmann, J.C.; Engelsdorf, T.; Voll, L.M.; Hoth, S. Early senescence and cell death in Arabidopsis

saul1 mutants involves the PAD4-dependent salicylic acid pathway. Plant Physiol. 2012, 159, 1477–1487.

76. Anderson, J.P.; Badruzsaufari, E.; Schenk, P.M.; Manners, J.M.; Desmond, O.J.; Ehlert, C.; Maclean, D.J.; Ebert, P.R.; Kazan, K. Antagonistic interaction between abscisic acid and jasmonate-ethylene signaling pathways modulates defense gene expression and disease resistance in Arabidopsis. Plant Cell 2004, 16, 3460–3479.

77. Audenaert, K.; de Meyer, G.B.; Höfte, M.M. Abscisic acid determines basal susceptibility of tomato to Botrytis cinerea and suppresses salicylic acid-dependent signaling mechanisms. Plant Physiol. 2002, 128, 491–501.

78. De Torres-Zabala, M.; Truman, W.; Bennett, M.H.; Lafforgue, G.; Mansfield, J.W.; Rodriguez Egea, P.; Bögre, L.; Grant, M. Pseudomonas syringae pv. tomato hijacks the Arabidopsis abscisic acid signalling pathway to cause disease. EMBO J.2007, 26, 1434–1443.

79. Häffner, E.; Karlovsky, P.; Splivallo, R.; Traczewska, A.; Diederichsen, E. ERECTA, salicylic acid, abscisic acid and jasmonic acid modulate quantitative disease resistance of Arabidopsis thaliana to Vverticillium longisporum. BMC Plant Biol. 2014, 14, 85.

80. Ulferts, S.; Delventhal, R.; Splivallo, R.; Karlovsky, P.; Schaffrath, U. Abscisic acid negatively interferes with basal defence of barley against Magnaporthe oryzae. BMC Plant Biol. 2015, 15, 7.

81. De Torres-Zabala, M.; Bennett, M.H.; Truman, W.H.; Grant, M.R. Antagonism between salicylic and abscisic acid reflects early host–pathogen conflict and moulds plant defence responses. Plant J. 2009, 59, 375–386.

82. Jiang, C.-J.; Shimono, M.; Sugano, S.; Kojima, M.; Yazawa, K.; Yoshida, R.; Inoue, H.; Hayashi, N.; Sakakibara, H.; Takatsuji, H. Abscisic acid interacts antagonistically with salicylic acid signaling pathway in rice-Magnaporthe grisea interaction. Mol. Plant Microbe Interact. 2010, 23, 791–798.

83. Adie, B.; Pérez-Pérez, J.; Pérez-Pérez, M.; Godoy, M.; Sánchez-Serrano, J.-J.; Schmelz, E.; Solano, R. ABA is an essential signal for plant resistance to pathogens affecting ja biosynthesis and the activation of defenses in Arabidopsis. Plant Cell 2007, 19, 1665–1681.

84. Ton, J.; Flors, V.; Mauch-Mani, B. The multifaceted role of ABA in disease resistance. Trends Plant Sci. 2009, 14, 310–317.

85. Xing, Y.; Jia, W.; Zhang, J. AtMKK1 mediates ABA-induced CAT1 expression and H2O2 production via atMPK6-coupled signaling in Arabidopsis. Plant J. 2008, 54, 440–451.

86. Jibran, R.; Hunter, D.A.; Dijkwel, P.P. Hormonal regulation of leaf senescence through integration of developmental and stress signals. Plant Mol. Biol. 2013, 82, 547–561.

87. He, K.; Gou, X.; Yuan, T.; Lin, H.; Asami, T.; Yoshida, S.; Russell, S.D.; Li, J. BAK1 and BKK1 regulate brassinosteroid-dependent growth and brassinosteroid-independent cell-death pathways. Curr. Biol. 2007, 17, 1109–1115.

88. Wang, Z.-Y.; Seto, H.; Fujioka, S.; Yoshida, S.; Chory, J. BRI1 is a critical component of a plasma-membrane receptor for plant steroids. Nature 2001, 410, 380–383.

89. Nakashita, H.; Yasuda, M.; Nitta, T.; Asami, T.; Fujioka, S.; Arai, Y.; Sekimata, K.; Takatsuto, S.; Yamaguchi, I.; Yoshida, S. Brassinosteroid functions in a broad range of disease resistance in tobacco and rice. Plant J. 2003, 33, 887–898.

90. Divi, U.K.; Rahman, T.; Krishna, P. Brassinosteroid-mediated stress tolerance in Arabidopsis shows interactions with abscisic acid, ethylene and salicylic acid pathways. BMC Plant Biol. 2010, 10, 151.

91. Krishna, P. Brassinosteroid-mediated stress responses. J. Plant Growth Regul. 2003, 22, 289–297.

92. Li, J.; Yu, K.; Wei, J.; Ma, Q.; Wang, B.; Yu, D. Gibberellin retards chlorophyll degradation during senescence of Paris polyphylla. Biol. Plant. 2010, 54, 395–399.

93. Yu, K.; Wang, Y.; Wei, J.; Ma, Q.; Yu, D.; Li, J. Improving rhizome yield and quality of Paris polyphylla through gibberellic acid-induced retardation of senescence of aerial parts. Plant Signal. Behav. 2009, 4, 413–415.

94. Gregersen, P.L.; Culetic, A.; Boschian, L.; Krupinska, K. Plant senescence and crop productivity. Plant Mol. Biol. 2013, 82, 603–622.
95. Gan, S.; Amasino, R.M. Inhibition of leaf senescence by autoregulated production of cytokinin. Science 1995, 270, 1986–1988.
96. Merewitz, E.B.; Gianfagna, T.; Huang, B. Photosynthesis, water use, and root viability under water stress as affected by expression of SAG12-IPT controlling cytokinin synthesis in Agrostis stolonifera. J. Exp. Bot. 2011, 62, 383–395.
97. Schmülling, T.; Werner, T.; Riefler, M.; Krupková, E.; Bartrina y Manns, I. Structure and function of cytokinin oxidase/dehydrogenase genes of maize, rice, Arabidopsis and other species. J. Plant Res. 2003, 116, 241–252.
98. Werner, T.; Motyka, V.; Strnad, M.; Schmülling, T. Regulation of plant growth by cytokinin. Proc. Natl. Acad. Sci. USA2001, 98, 10487–10492.
99. Teale, W.D.; Paponov, I.A.; Palme, K. Auxin in action: Signalling, transport and the control of plant growth and development. Nat. Rev. Mol. Cell Biol. 2006, 7, 847–859.
100. Cohen, J.D.; Slovin, J.P.; Hendrickson, A.M. Two genetically discrete pathways convert tryptophan to auxin: More redundancy in auxin biosynthesis. Trends Plant Sci. 2003, 8, 197–199.
101. Dharmasiri, N.; Dharmasiri, S.; Estelle, M. The F-box protein TIR1 is an auxin receptor. Nature 2005, 435, 441–445.
102. Kepinski, S.; Leyser, O. The Arabidopsis F-box protein TIR1 is an auxin receptor. Nature 2005, 435, 446–451.
103. Noh, Y.-S.; Amasino, R.M. Identification of a promoter region responsible for the senescence-specific expression of SAG12. Plant Mol. Biol. 1999, 41, 181–194.
104. Kim, J-I.; Murphy, A.S.; Baek, D.; Lee, S.-W.; Yun, D.-J.; Bressan, R.A.; Narasimhan, M.L. YUCCA6 over-expression demonstrates auxin function in delaying leaf senescence in Arabidopsis thaliana. J. Exp. Bot. 2011.
105. Lim, P.O.; Lee, I.C.; Kim, J.; Kim, H.J.; Ryu, J.S.; Woo, H.R.; Nam, H.G. Auxin response factor 2 (ARF2) plays a major role in regulating auxin-mediated leaf longevity. J. Exp. Bot. 2010.
106. Jiang, Y.; Liang, G.; Yang, S.; Yu, D. Arabidopsis WRKY57 functions as a node of convergence for jasmonic acid–and auxin-mediated signaling in jasmonic acid-induced leaf senescence. Plant Cell 2014, 26, 230–245.
107. Quirino, B.F.; Normanly, J.; Amasino, R.M. Diverse range of gene activity during Arabidopsis thaliana leaf senescence includes pathogen-independent induction of defense-related genes. Plant Mol. Biol. 1999, 40, 267–278.
108. Van der Graaff, E.; Schwacke, R.; Schneider, A.; Desimone, M.; Flügge, U.-I.; Kunze, R. Transcription analysis of Arabidopsis membrane transporters and hormone pathways during developmental and induced leaf senescence. Plant Physiol. 2006, 141, 776–792.
109. Xu, F.; Meng, T.; Li, P.; Yu, Y.; Cui, Y.; Wang, Y.; Gong, Q.; Wang, N.N. A soybean dual-specificity kinase, gmSARK, and its Arabidopsis homolog, atSARK, regulate leaf senescence through synergistic actions of auxin and ethylene. Plant Physiol. 2011, 157, 2131–2153.
110. Hou, K.; Wu, W.; Gan, S.-S. Saur36, a small auxin up RNA gene, is involved in the promotion of leaf senescence in Arabidopsis. Plant Physiol. 2013, 161, 1002–1009.
111. Choi, J.; Choi, D.; Lee, S.; Ryu, C.-M.; Hwang, I. Cytokinins and plant immunity: Old foes or new friends? Trends Plant Sci. 2011, 16, 388–394.
112. Kazan, K.; Manners, J.M. Linking development to defense: Auxin in plant–pathogen interactions. Trends Plant Sci.2009, 14, 373–382.
113. Sun, T.-P. The molecular mechanism and evolution of the GA–GID1–DELLA signaling module in plants. Curr. Biol.2011, 21, R338–R345.
114. Kazan, K.; Manners, J.M. JAZ repressors and the orchestration of phytohormone crosstalk. Trends Plant Sci. 2012, 17, 22–31.
115. Li, C.; He, X.; Luo, X.; Xu, L.; Liu, L.; Min, L.; Jin, L.; Zhu, L.; Zhang, X. Cotton WRKY1 mediates the plant defense-to-development transition during infection of cotton by

Verticillium dahliae by activating JASMONATE-ZIM-DOMAIN1 expression. Plant Physiol. 2014, 166, 2179–2194.

116. Hou, X.; Lee, L.Y.C.; Xia, K.; Yan, Y.; Yu, H. DELLAs modulate jasmonate signaling via competitive binding to JAZs. Dev. Cell 2010, 19, 884–894.

117. Yang, D.-L.; Yao, J.; Mei, C.-S.; Tong, X.-H.; Zeng, L.-J.; Li, Q.; Xiao, L.-T.; Sun, T.-P.; Li, J.; Deng, X.-W. Plant hormone jasmonate prioritizes defense over growth by interfering with gibberellin signaling cascade. Proc. Natl. Acad. Sci. USA 2012, 109, E1192–E1200.

118. Navarro, L.; Bari, R.; Achard, P.; Lisón, P.; Nemri, A.; Harberd, N.P.; Jones, J.D. DELLAs control plant immune responses by modulating the balance of jasmonic acid and salicylic acid signaling. Curr. Biol. 2008, 18, 650–655.

119. Wang, D.; Pajerowska-Mukhtar, K.; Culler, A.H.; Dong, X. Salicylic acid inhibits pathogen growth in plants through repression of the auxin signaling pathway. Curr. Biol. 2007, 17, 1784–1790.

120. Iglesias, M.J.; Terrile, M.C.; Casalongué, C.A. Auxin and salicylic acid signalings counteract the regulation of adaptive responses to stress. Plant Signal. Behav. 2011, 6, 452–454.

121. Park, J.-E.; Park, J.-Y.; Kim, Y.-S.; Staswick, P.E.; Jeon, J.; Yun, J.; Kim, S.-Y.; Kim, J.; Lee, Y.-H.; Park, C.-M. GH3-mediated auxin homeostasis links growth regulation with stress adaptation response in Arabidopsis. J. Biol. Chem.2007, 282, 10036–10046.

122. Ding, X.; Cao, Y.; Huang, L.; Zhao, J.; Xu, C.; Li, X.; Wang, S. Activation of the indole-3-acetic acid-amido synthetase GH3-8 suppresses expansin expression and promotes salicylate-and jasmonate-independent basal immunity in rice. Plant Cell 2008, 20, 228–240.

123. González-Lamothe, R.; El Oirdi, M.; Brisson, N.; Bouarab, K. The conjugated auxin indole-3-acetic acid-aspartic acid promotes plant disease development. Plant Cell 2012, 24, 762–777.

124. Llorente, F.; Muskett, P.; Sánchez-Vallet, A.; López, G.; Ramos, B.; Sánchez-Rodríguez, C.; Jordá, L.; Parker, J.; Molina, A. Repression of the auxin response pathway increases Arabidopsis susceptibility to necrotrophic fungi. Mol. Plant2008, 1, 496–509.

125. Ren, C.; Pan, J.; Peng, W.; Genschik, P.; Hobbie, L.; Hellmann, H.; Estelle, M.; Gao, B.; Peng, J.; Sun, C. Point mutations in Arabidopsis CULLIN1 reveal its essential role in jasmonate response. Plant J. 2005, 42, 514–524.

126. Kidd, B.N.; Kadoo, N.Y.; Dombrecht, B.; Tekeoglu, M.; Gardiner, D.M.; Thatcher, L.F.; Aitken, E.A.; Schenk, P.M.; Manners, J.M.; Kazan, K. Auxin signaling and transport promote susceptibility to the root-infecting fungal pathogen Fusarium oxysporum in Arabidopsis. Mol. Plant-Microbe Interact. 2011, 24, 733–748.

127. Rushton, P.J.; Somssich, I.E.; Ringler, P.; Shen, Q.J. WRKY transcription factors. Trends Plant Sci. 2010, 15, 247–258.

128. Dietz, K.-J.; Vogel, M.O.; Viehhauser, A. AP2/EREBP transcription factors are part of gene regulatory networks and integrate metabolic, hormonal and environmental signals in stress acclimation and retrograde signalling. Protoplasma2010, 245, 3–14.

129. Balazadeh, S.; Riaño-Pachón, D.; Mueller-Roeber, B. Transcription factors regulating leaf senescence in Arabidopsis thaliana. Plant Biol. 2008, 10, 63–75.

130. Olsen, A.N.; Ernst, H.A.; Leggio, L.L.; Skriver, K. NAC transcription factors: Structurally distinct, functionally diverse. Trends Plant Sci. 2005, 10, 79–87.

131. Dubos, C.; Stracke, R.; Grotewold, E.; Weisshaar, B.; Martin, C.; Lepiniec, L. MYB transcription factors in Arabidopsis. Trends Plant Sci. 2010, 15, 573–581.

132. Carretero-Paulet, L.; Galstyan, A.; Roig-Villanova, I.; Martínez-García, J.F.; Bilbao-Castro, J.R.; Robertson, D.L. Genome-wide classification and evolutionary analysis of the bHLH family of transcription factors in Arabidopsis, poplar, rice, moss, and algae. Plant Physiol. 2010, 153, 1398–1412.

133. Kiełbowicz-Matuk, A. Involvement of plant C2H2-type zinc finger transcription factors in stress responses. Plant Sci.2012, 185, 78–85.

134. Wang, D.; Amornsiripanitch, N.; Dong, X. A genomic approach to identify regulatory nodes in the transcriptional network of systemic acquired resistance in plants. PLoS Pathogens 2006, 2, e123.
135. Agarwal, P.; Reddy, M.; Chikara, J. WRKY: Its structure, evolutionary relationship, DNA-binding selectivity, role in stress tolerance and development of plants. Mol. Biol. Rep. 2011, 38, 3883–3896.
136. Besseau, S.; Li, J.; Palva, E.T. WRKY54 and WRKY70 co-operate as negative regulators of leaf senescence in Arabidopsis thaliana. J. Exp. Bot. 2012.
137. Robatzek, S.; Somssich, I.E. Targets of atWRKY6 regulation during plant senescence and pathogen defense. Genes Dev. 2002, 16, 1139–1149.
138. Miao, Y.; Laun, T.; Zimmermann, P.; Zentgraf, U. Targets of the WRKY53 transcription factor and its role during leaf senescence in Arabidopsis. Plant Mol. Biol. 2004, 55, 853–867.
139. Kim, K.-C.; Lai, Z.; Fan, B.; Chen, Z. Arabidopsis WRKY38 and WRKY62 transcription factors interact with histone deacetylase 19 in basal defense. Plant Cell 2008, 20, 2357–2371.
140. Xu, X.; Chen, C.; Fan, B.; Chen, Z. Physical and functional interactions between pathogen-induced Arabidopsis WRKY18, WRKY 40, and WRKY 60 transcription factors. Plant Cell 2006, 18, 1310–1326.
141. Shang, Y.; Yan, L.; Liu, Z.-Q.; Cao, Z.; Mei, C.; Xin, Q.; Wu, F.-Q.; Wang, X.-F.; Du, S.-Y.; Jiang, T. The Mg-chelatase H subunit of Arabidopsis antagonizes a group of WRKY transcription repressors to relieve ABA-responsive genes of inhibition. Plant Cell 2010, 22, 1909–1935.
142. Chen, L.; Zhang, L.; Li, D.; Wang, F.; Yu, D. WRKY8 transcription factor functions in the TMV-cg defense response by mediating both abscisic acid and ethylene signaling in Arabidopsis. Proc. Natl. Acad. Sci. USA 2013, 110, E1963–E1971.
143. Lamb, C.; Dixon, R.A. The oxidative burst in plant disease resistance. Ann. Rev. Plant Biol. 1997, 48, 251–275.
144. Han, S.; Wang, Y.; Zheng, X.; Jia, Q.; Zhao, J.; Bai, F.; Hong, Y.; Liu, Y. Cytoplastic glyceraldehyde-3-phosphate dehydrogenases interact with ATG3 to negatively regulate autophagy and immunity in Nicotiana benthamiana. Plant Cell 2015, 27, 1316–1331.
145. Van Breusegem, F.; Dat, J.F. Reactive oxygen species in plant cell death. Plant Physiol. 2006, 141, 384–390.
146. De Gara, L.; de Pinto, M.C.; Tommasi, F. The antioxidant systems vis-à-vis reactive oxygen species during plant-pathogen interaction. Plant Physiol. Biochem. 2003, 41, 863–870.
147. Gill, S.S.; Tuteja, N. Reactive oxygen species and antioxidant machinery in abiotic stress tolerance in crop plants. Plant Physiol. Biochem. 2010, 48, 909–930.
148. Baxter, A.; Mittler, R.; Suzuki, N. ROS as key players in plant stress signalling. J. Exp. Bot. 2014, 65, 1229–1240.
149. Torres, M.A.; Dangl, J.L. Functions of the respiratory burst oxidase in biotic interactions, abiotic stress and development. Curr. Opin. Plant Biol. 2005, 8, 397–403.
150. Torres, M.A.; Jones, J.D.; Dangl, J.L. Pathogen-induced, NADPH oxidase–derived reactive oxygen intermediates suppress spread of cell death in Arabidopsis thaliana. Nat. Genet. 2005, 37, 1130–1134.
151. Ye, Z.; Rodriguez, R.; Tran, A.; Hoang, H.; de los Santos, D.; Brown, S.; Vellanoweth, R.L. The developmental transition to flowering represses ascorbate peroxidase activity and induces enzymatic lipid peroxidation in leaf tissue in Arabidopsis thaliana. Plant Sci. 2000, 158, 115–127.
152. Zimmermann, P.; Zentgraf, U. The correlation between oxidative stress and leaf senescence during plant development. Cell. Mol. Biol. Letters 2005, 10, 515.
153. Navabpour, S.; Morris, K.; Allen, R.; Harrison, E.; Soheila, A.; Buchanan-Wollaston, V. Expression of senescence-enhanced genes in response to oxidative stress. J. Exp. Bot. 2003, 54, 2285–2292.

154. Wu, A.; Allu, A.D.; Garapati, P.; Siddiqui, H.; Dortay, H.; Zanor, M.-I.; Asensi-Fabado, M.A.; Munné-Bosch, S.; Antonio, C.; Tohge, T. JUNGBRUNNEN1, a reactive oxygen species–responsive NAC transcription factor, regulates longevity in Arabidopsis. Plant Cell 2012, 24, 482–506.
155. Balazadeh, S.; Wu, A.; Mueller-Roeber, B. Salt-triggered expression of the ANAC092-dependent senescence regulon in Arabidopsis thaliana. Plant Signal. Behav. 2010, 5, 733–735.
156. Balazadeh, S.; Kwasniewski, M.; Caldana, C.; Mehrnia, M.; Zanor, M.I.; Xue, G.-P.; Mueller-Roeber, B. ORS1, an H2O2-responsive NAC transcription factor, controls senescence in Arabidopsis thaliana. Mol. Plant 2011, 4, 346–360.
157. Torres, M.A. ROS in biotic interactions. Physiol. Plant. 2010, 138, 414–429.
158. Passardi, F.; Penel, C.; Dunand, C. Performing the paradoxical: How plant peroxidases modify the cell wall. Trends Plant Sci. 2004, 9, 534–540.
159. Saga, H.; Ogawa, T.; Kai, K.; Suzuki, H.; Ogata, Y.; Sakurai, N.; Shibata, D.; Ohta, D. Identification and characterization of ANAC042, a transcription factor family gene involved in the regulation of camalexin biosynthesis in Arabidopsis. Mol. Plant Microbe Interact. 2012, 25, 684–696.
160. Blume, B.; Nürnberger, T.; Nass, N.; Scheel, D. Receptor-mediated increase in cytoplasmic free calcium required for activation of pathogen defense in parsley. Plant Cell 2000, 12, 1425–1440.
161. Grant, J.J.; Yun, B.W.; Loake, G.J. Oxidative burst and cognate redox signalling reported by luciferase imaging: Identification of a signal network that functions independently of ethylene, SA and Me-JA but is dependent on MAPKK activity. Plant J. 2000, 24, 569–582.
162. Kwak, J.M.; Mori, I.C.; Pei, Z.M.; Leonhardt, N.; Torres, M.A.; Dangl, J.L.; Bloom, R.E.; Bodde, S.; Jones, J.D.; Schroeder, J.I. NADPH oxidase atRBOHD and atRBOHF genes function in ROS-dependent ABA signaling in Arabidopsis. EMBO J. 2003, 22, 2623–2633.
163. Mittler, R.; Vanderauwera, S.; Suzuki, N.; Miller, G.; Tognetti, V.B.; Vandepoele, K.; Gollery, M.; Shulaev, V.; van Breusegem, F. ROS signaling: The new wave? Trends Plant Sci. 2011, 16, 300–309.
164. Han, L.; Li, G.J.; Yang, K.Y.; Mao, G.; Wang, R.; Liu, Y.; Zhang, S. Mitogen-activated protein kinase 3 and 6 regulate Botrytis cinerea-induced ethylene production in Arabidopsis. Plant J. 2010, 64, 114–127.
165. Mou, Z.; Fan, W.; Dong, X. Inducers of plant systemic acquired resistance regulate NPR1 function through redox changes. Cell 2003, 113, 935–944.
166. Blomster, T.; Salojärvi, J.; Sipari, N.; Brosché, M.; Ahlfors, R.; Keinänen, M.; Overmyer, K.; Kangasjärvi, J. Apoplastic reactive oxygen species transiently decrease auxin signaling and cause stress-induced morphogenic response in Arabidopsis. Plant Physiol. 2011, 157, 1866–1883.
167. Desikan, R.; Reynolds, A.; Hancock, J.; Neill, S. Harpin and hydrogen peroxide both initiate programmed cell death but have differential effects on defence gene expression in Arabidopsis suspension cultures. Biochem. J. 1998, 330, 115–120.
168. Hennin, C.; Diederichsen, E.; Höfte, M. Local and systemic resistance to fungal pathogens triggered by an AVR9-mediated hypersensitive response in tomato and oilseed rape carrying the Cf-9 resistance gene. Physiol. Mol. Plant Pathol. 2001, 59, 287–295.
169. Harrach, B.D.; Fodor, J.; Pogány, M.; Preuss, J.; Barna, B. Antioxidant, ethylene and membrane leakage responses to powdery mildew infection of near-isogenic barley lines with various types of resistance. Eur. J. Plant Pathol. 2008, 121, 21–33.
170. Williams, B.; Kabbage, M.; Kim, H.-J.; Britt, R.; Dickman, M.B. Tipping the balance: Sclerotinia sclerotiorum secreted oxalic acid suppresses host defenses by manipulating the host redox environment. PLoS Pathogens 2011, 7, e1002107.
171. Avila-Ospina, L.; Moison, M.; Yoshimoto, K.; Masclaux-Daubresse, C. Autophagy, plant senescence, and nutrient recycling. J. Exp. Bot. 2014.

172. Liu, Y.; Bassham, D.C. Autophagy: Pathways for self-eating in plant cells. Ann. Rev. Plant Biol. 2012, 63, 215–237.
173. Lai, Z.; Wang, F.; Zheng, Z.; Fan, B.; Chen, Z. A critical role of autophagy in plant resistance to necrotrophic fungal pathogens. Plant J. 2011, 66, 953–968.
174. Liu, Y.; Schiff, M.; Czymmek, K.; Talloczy, Z.; Levine, B.; Dinesh-Kumar, S. Autophagy regulates programmed cell death during the plant innate immune response. Cell 2005, 121, 567–577.
175. Yoshimoto, K.; Jikumaru, Y.; Kamiya, Y.; Kusano, M.; Consonni, C.; Panstruga, R.; Ohsumi, Y.; Shirasu, K. Autophagy negatively regulates cell death by controlling NPR1-dependent salicylic acid signaling during senescence and the innate immune response in Arabidopsis. Plant Cell 2009, 21, 2914–2927.
176. Hofius, D.; Munch, D.; Bressendorff, S.; Mundy, J.; Petersen, M. Role of autophagy in disease resistance and hypersensitive response-associated cell death. Cell Death Differ. 2011, 18, 1257–1262.
177. Krattinger, S.G.; Lagudah, E.S.; Spielmeyer, W.; Singh, R.P.; Huerta-Espino, J.; McFadden, H.; Bossolini, E.; Selter, L.L.; Keller, B. A putative ABC transporter confers durable resistance to multiple fungal pathogens in wheat. Science2009, 323, 1360–1363.
178. Hulbert, S.; Bai, J.; Fellers, J.; Pacheco, M.; Bowden, R. Gene expression patterns in near isogenic lines for wheat rust resistance gene Lr34/Yr18. Phytopathology 2007, 97, 1083–1093.
179. Fu, D.; Uauy, C.; Distelfeld, A.; Blechl, A.; Epstein, L.; Chen, X.; Sela, H.; Fahima, T.; Dubcovsky, J. A kinase-start gene confers temperature-dependent resistance to wheat stripe rust. Science 2009, 323, 1357–1360.
180. Gou, J.-Y.; Li, K.; Wu, K.; Wang, X.; Lin, H.; Cantu, D.; Uauy, C.; Dobon-Alonso, A.; Midorikawa, T.; Inoue, K. Wheat stripe rust resistance protein WKS1 reduces the ability of the thylakoid-associated ascorbate peroxidase to detoxify reactive oxygen species. Plant Cell 2015.
181. Büschges, R.; Hollricher, K.; Panstruga, R.; Simons, G.; Wolter, M.; Frijters, A.; van Daelen, R.; van der Lee, T.; Diergaarde, P.; Groenendijk, J. The barley Mlo gene: A novel control element of plant pathogen resistance. Cell 1997, 88, 695–705.
182. Piffanelli, P.; Zhou, F.; Casais, C.; Orme, J.; Jarosch, B.; Schaffrath, U.; Collins, N.C.; Panstruga, R.; Schulze-Lefert, P. The barley Mlo modulator of defense and cell death is responsive to biotic and abiotic stress stimuli. Plant Physiol.2002, 129, 1076–1085.
183. Jarosch, B.; Collins, N.C.; Zellerhoff, N.; Schaffrath, U. RAR1, ROR1, and the actin cytoskeleton contribute to basal resistance to Magnaporthe grisea in barley. Mol. Plant Microbe Interact. 2005, 18, 397–404.
184. Muchero, W.; Ehlers, J.D.; Close, T.J.; Roberts, P.A. Genic SNP markers and legume synteny reveal candidate genes underlying QTL for Macrophomina phaseolina resistance and maturity in cowpea [Vigna unguiculata (L) Walp.]. BMC Genomics 2011, 12, 8.
185. Mah, K.M.; Uppalapati, S.R.; Tang, Y.; Allen, S.; Shuai, B. Gene expression profiling of Macrophomina phaseolina infected Medicago truncatula roots reveals a role for auxin in plant tolerance against the charcoal rot pathogen. Physiol. Mol. Plant Pathol. 2012, 79, 21–30.
186. Blanco-Ulate, B.; Vincenti, E.; Powell, A.L.; Cantu, D. Tomato transcriptome and mutant analyses suggest a role for plant stress hormones in the interaction between fruit and Botrytis cinerea. Front. Plant Sci. 2013, 4, 142.
187. Zhou, L.; Hu, Q.; Johansson, A.; Dixelius, C. Verticillium longisporum and V. dahliae: Infection and disease in Brassica napus. Plant Pathol. 2006, 55, 137–144.
188. Häffner, E.; Karlovsky, P.; Diederichsen, E. Genetic and environmental control of the Verticillium syndrome in Arabidopsis thaliana. BMC Plant Biol. 2010, 10, 235.
189. Veronese, P.; Narasimhan, M.L.; Stevenson, R.A.; Zhu, J.-K.; Weller, S.C.; Subbarao, K.V.; Bressan, R.A. Identification of a locus controlling Verticillium disease symptom response in Arabidopsis thaliana. Plant J. 2003, 35, 574–587.

190. Kurepa, J.; Smalle, J.; Va, M.; Montagu, N.; Inzé, D. Oxidative stress tolerance and longevity in Arabidopsis: The late-flowering mutant gigantea is tolerant to paraquat. Plant J. 1998, 14, 759–764.
191. Campbell, C.L.; Madden, L.V. Introduction to Plant Disease Epidemiology; John Wiley and Sons: New York, NY, USA, 1990.
192. Robert-Seilaniantz, A.; Navarro, L.; Bari, R.; Jones, J.D. Pathological hormone imbalances. Curr. Opin. Plant Biol. 2007, 10, 372–379.
193. Chen, Z.; Agnew, J.L.; Cohen, J.D.; He, P.; Shan, L.; Sheen, J.; Kunkel, B.N. Pseudomonas syringae type III effector avrRpt2 alters Arabidopsis thaliana auxin physiology. Proc. Natl. Acad. Sci. USA 2007, 104, 20131–20136.
194. Ludwig-Müller, J.; Prinsen, E.; Rolfe, S.A.; Scholes, J.D. Metabolism and plant hormone action during clubroot disease. J. Plant Growth Regul. 2009, 28, 229–244.
195. Dekhuijzen, H. The occurrence of free and bound cytokinins in plasmodia of Plasmodiophora brassicae isolated from tissue cultures of clubroots. Plant Cell Rep. 1981, 1, 18–20.
196. Siemens, J.; Keller, I.; Sarx, J.; Kunz, S.; Schuller, A.; Nagel, W.; Schmülling, T.; Parniske, M.; Ludwig-Müller, J. Transcriptome analysis of Arabidopsis clubroots indicate a key role for cytokinins in disease development. Mol. Plant Microbe Interact. 2006, 19, 480–494.
197. Walters, D.R.; McRoberts, N.; Fitt, B.D. Are green islands red herrings? Significance of green islands in plant interactions with pathogens and pests. Biol. Rev. 2008, 83, 79–102.
198. Thomma, B.P.; Nürnberger, T.; Joosten, M.H. Of PAMPS and effectors: The blurred PTI-ETI dichotomy. Plant Cell2011, 23, 4–15.
199. Ralhan, A.; Schöttle, S.; Thurow, C.; Iven, T.; Feussner, I.; Polle, A.; Gatz, C. The vascular pathogen Verticillium longisporum requires a jasmonic acid-independent COI1 function in roots to elicit disease symptoms in Arabidopsis shoots. Plant Physiol. 2012, 159, 1192–1203.
200. Thatcher, L.F.; Manners, J.M.; Kazan, K. Fusarium oxysporum hijacks COI1-mediated jasmonate signaling to promote disease development in Arabidopsis. Plant J. 2009, 58, 927–939.
201. El Oirdi, M.; El Rahman, T.A.; Rigano, L.; El Hadrami, A.; Rodriguez, M.C.; Daayf, F.; Vojnov, A.; Bouarab, K. Botrytis cinerea manipulates the antagonistic effects between immune pathways to promote disease development in tomato. Plant Cell 2011, 23, 2405–2421.
202. Ottmann, C.; Luberacki, B.; Küfner, I.; Koch, W.; Brunner, F.; Weyand, M.; Mattinen, L.; Pirhonen, M.; Anderluh, G.; Seitz, H.U. A common toxin fold mediates microbial attack and plant defense. Proc. Natl. Acad. Sci. USA 2009, 106, 10359–10364.
203. Barau, J.; Grandis, A.; de Andrade Carvalho, V.M.; Teixeira, G.S.; Zaparoli, G.H.A.; do Rio, M.C.S.; Rincones, J.; Buckeridge, M.S.; Pereira, G.A.G. Apoplastic and intracellular plant sugars regulate developmental transitions in witches' broom disease of cacao. J. Exp. Bot. 2014.
204. Oliver, R.P.; Solomon, P.S. New developments in pathogenicity and virulence of necrotrophs. Curr. Opin. Plant Biol.2010, 13, 415–419.
205. Liu, Z.; Faris, J.D.; Oliver, R.P.; Tan, K.-C.; Solomon, P.S.; McDonald, M.C.; McDonald, B.A.; Nunez, A.; Lu, S.; Rasmussen, J.B. SnTOX3 acts in effector triggered susceptibility to induce disease on wheat carrying the snn3 gene. PLoS pathogens 2009, 5, e1000581.
206. Hornig, H. Krankhafte Abreife—Spätschäden durch Phoma lingam und/oder Verticillium dahliae. Raps 1986, 4, 83–85.
207. Eynck, C.; Koopmann, B.; Karlovsky, P.; von Tiedemann, A. Internal resistance in winter oilseed rape inhibits systemic spread of the vascular pathogen Verticillium longisporum. Phytopathology 2009, 99, 802–811.
208. Edgar, R.; Domrachev, M.; Lash, A.E. Gene Expression Omnibus: NCBI gene expression and hybridization array data repository. Nucleic Acids Res. 2002, 30, 207–210.

209. NCBI. GEO Gene Expression Omnibus. Available online: http://www.ncbi.nlm.nih.gov/geo/query/acc.cgi?acc=GSE70021 (accessed on 8 July 2015).

210. Zheng, Z.; Qamar, S.A.; Chen, Z.; Mengiste, T. Arabidopsis WRKY33 transcription factor is required for resistance to necrotrophic fungal pathogens. Plant J. 2006, 48, 592–605.

211. Asai, T.; Tena, G.; Plotnikova, J.; Willmann, M.R.; Chiu, W.-L.; Gomez-Gomez, L.; Boller, T.; Ausubel, F.M.; Sheen, J. Map kinase signalling cascade in Arabidopsis innate immunity. Nature 2002, 415, 977–983.

212. Craigon, D.J.; James, N.; Okyere, J.; Higgins, J.; Jotham, J.; May, S. Nascarrays: A repository for microarray data generated by NASC's transcriptomics service. Nucleic Acids Res. 2004, 32, D575–D577.

213. Carvalho, B.S.; Irizarry, R.A. A framework for oligonucleotide microarray preprocessing. Bioinformatics 2010, 26, 2363–2367.

214. Smyth, G.K. Limma: Linear models for microarray data. In Bioinformatics and Computational Biology Solutions Using R and Bioconductor; Gentleman, R., Carey, V., Dudoit, S., Irizarry, R., Huber, W., Eds.; Springer: New York, NY, USA, 2005; pp. 397–420.

215. Carvalho, B. pd.aragene.1.1.st: Platform Design info for Affymetrix Aragene-1_1-st., R package version 3.8.0. Available online: http://www.bioconductor.org/packages/release/data/annotation/html/pd.aragene.1.1.st.html (accessed on 7 Jul y 2015).

216. Thimm, O.; Bläsing, O.; Gibon, Y.; Nagel, A.; Meyer, S.; Krüger, P.; Selbig, J.; Müller, L.A.; Rhee, S.Y.; Stitt, M. MAPMAN: A user-driven tool to display genomics data sets onto diagrams of metabolic pathways and other biological processes. Plant J. 2004, 37, 914–939.

217. Rodrigues-Pousada, R.A.; de Rycke, R.; Dedonder, A.; van Caeneghem, W.; Engler, G.; van Montagu, M.; van der Straeten, D. The Arabidopsis 1-aminocyclopropane-1-carboxylate synthase gene 1 is expressed during early development. Plant Cell 1993, 5, 897–911.

218. Oñate-Sánchez, L.; Anderson, J.P.; Young, J.; Singh, K.B. AtERF14, a member of the ERF family of transcription factors, plays a nonredundant role in plant defense. Plant Physiol. 2007, 143, 400–409.

219. Ren, X.; Chen, Z.; Liu, Y.; Zhang, H.; Zhang, M.; Liu, Q.; Hong, X.; Zhu, J.K.; Gong, Z. ABO3, a WRKY transcription factor, mediates plant responses to abscisic acid and drought tolerance in Arabidopsis. Plant J. 2010, 63, 417–429.

220. Sappl, P.G.; Carroll, A.J.; Clifton, R.; Lister, R.; Whelan, J.; Harvey Millar, A.; Singh, K.B. The Arabidopsis glutathione transferase gene family displays complex stress regulation and co-silencing multiple genes results in altered metabolic sensitivity to oxidative stress. Plant J. 2009, 58, 53–68.

CHAPTER 6

Evaluation of Chemical and Biological Seed Treatments to Control Charcoal Rot of Soybean

Sebastian Reznikov[1] • Gabriel R. Vellicce[1] • Victoria Gonza´lez[1] • Vicente de Lisi[1] • Atilio P. Castagnaro[1] • L. Daniel Ploper[1]

Instituto de Tecnología Agroindustrial del Noroeste Argentino (ITANOA)Estación Experimental Agroindustrial Obispo Colombres (EEAOC)-Consejo Nacional de Investigaciones Científicas y Técnicas (CONICET)Las TalitasArgentina

ABSTRACT

Charcoal rot caused by *Macrophomina phaseolina* (Mp) affects economically important crops around the world. In northwestern Argentina, the disease is present in seasons with dry and hot weather. Here, the efficacy of seed treatments with two biological products (*Trichodermaviride* or *Bacillus subtilis*) or one chemical treatment (thiophanate methyl + pyraclostrobin) were evaluated in the field to control Mp on two soybean cultivars in Tucumán, Argentina. In field tests, Mp was added to the soil at planting in the 2010/2011 and 2011/2012 growing seasons. Plant emergence (PE) was reduced for both cultivars: 6.7 and 56.8 % for NA8000 RG and 12.5 and 61.3 % for Munasqa RR in 2010/2011 and 2011/2012, respectively. On a 1–5 disease severity scale, the inoculated control of NA8000 RG scored 2.5 and Munasqa RR 1.8 during the 2010/2011 season, and 3.6 for NA8000 RG, and 2.7 for Munasqa RR during 2011/2012. Likewise, Mp CFU/g values in the inoculated control increased during the 2011/2012 season: from 2100 to 2366 in NA8000 RG and from 300 to 566 in Munasqa RR. Crop yield and 1000-seed weight were also reduced by fungal infection. Although the treatments counteracted PE reduction in both genotypes, disease severity was higher in NA8000 RG than in Munasqa RR, coincident with the highest CFU/g values. The 1000-seed weight of the two cultivars did not differ significantly among treatments. The highest yields were obtained in chemically treated plots, followed by those treated with *Trichoderma* or *Bacillus*.

Keywords: *Macrophomina phaseolina Glycine max Trichodermaviride Bacillussubtilis*

INTRODUCTION

Soybean [*Glycine max* (L.) Merr.] plants are susceptible to root and stem base rots caused by soil pathogens at all growth stages. One of these diseases is charcoal rot of soybean, caused by the polyphagous fungus *Macrophomina phaseolina* (Tassi) Goid (Mp). This pathogen infects a wide host range of nearly 500 species in more than 100 families around the world, including other important crops such as cotton (*Gossypium* spp.), chickpea (*Cicer arietinum*), corn (*Zea mays*), and common bean (*Phaseolus vulgaris*) (Srivastava et al. 2001). Morphologically, physiologically, genetically, and pathogenically, the fungus varies widely, enabling it to adapt to different environmental conditions and hence become widely distributed geographically (Su et al. 2001).

Soybean seedlings affected by Mp develop reddish brown lesions on the hypocotyl, which becomes ash-gray and then turns black (Wyllie 1989). The presence of small, black sclerotia in the cortical tissue confers the charcoal appearance that gives the disease its name (Mengistu et al. 2015). In addition to these symptoms and signs, mature plants develop chlorotic lesions on their leaves, which then die but remain attached to the stem, and finally the plants die prematurely (Ploper and Scandiani 2009). A combination of water stress and high temperatures favors disease development (Mihail 1992).

Charcoal rot of soybean is an economically important disease in North and South America, Asia, Australia, Africa and Europe (Wrather et al. 1997). Worldwide in 2006, yield losses caused by charcoal rot reached 4.2 % (Wrather et al. 2010). Yield loss in field experimental plots was estimated to be 30 % (Mengistu et al. 2011). In Argentina, hot, dry weather prevailed in the 2000/2001 growing season, which favored charcoal rot development on soybean. The provinces of Catamarca, Chaco, Córdoba, Entre Rios, Santa Fe, Salta, Santiago del Estero and Tucumán were affected with varying levels of yield losses, including total losses in some fields (Ploper et al. 2001).

Biological agents can provide an alternative to control certain plant diseases, especially when other methods such as chemicals are difficult to use. Cook and Baker (1983) successfully controlled *Sclerotium rolfsii, Rhizoctonia solani,* and *Pythium* spp. using antagonistic microorganisms such as *Trichoderma* spp., *Penicillium* spp., *Aspergillus* spp., and some bacteria. Jeyarajan et al. (1993) reported that a talc-based formulation of *Trichoderma viride*, produced on a commercial scale, was widely used by farmers to treat seeds of sesame (*Sesamum indicum*), groundnut (*Arachis hypogaea*), sunflower (*Helianthus annus*), chickpea and mung bean (*Vigna radiata*) for the biocontrol of root rot disease caused by Mp. Sankar and Jeyarajan (1996) studied *Trichoderma* spp. and *Gliocladiumvirens* as seed treatment for sesame (*Sesamum indicum*) to control Mp and concluded that treating the seeds with the formulation and storing them was more advantageous than storing the formulation and treating the seeds just before planting. Elad et al. (1986) found that *Trichoderma harzianum* inhibited the in vitro linear growth and microsclerotia production of Mp. When the biocontrol agent was applied to seeds, disease incidence was reduced in bean and melon (*Cucumis melo*) in the greenhouse and in melons and corn in the field.

To control fungal pathogens, chemical products are available for application of soil, seeds and/or foliage. Fungicide seed treatments are intended to control diseases that cause seed rot and damping-off before and after emergence. Mueller et al. (1999) used thiram, fludioxonil, and captan + pentachloronitrobenzene + thiabendazole to control *Sclerotinia sclerotiorum* in field trials and managed to reduce sclerotia formation in infected soybean seed by 98.0 %. Moreover, soybean seeds treated with thiram reduced the incidence of *Phomopsis sojae* and increased seed germination (Hepperly and Sinclair 1978). In studies on the control of *R. solani*, Martin et al. (1984) demonstrated differences in sensitivity to different fungicides among *R. solani* isolates, and Dorrance et al. (2003) assessed five fungicide seed treatments to control *R. solani* on soybean in the greenhouse. However, none of the evaluated seed treatments completely controlled the four tested isolates.

Currently, no chemicals are available to control charcoal rot in soybean. Mengistu et al. (2015) mentioned that seed treatments may be helpful if soybean seeds are infected with Mp, but there is no information on specific active ingredients effective against this pathogen.

Among fungicides tested against Mp in infected cotton seeds, carbendazim, quintozene and benomyl enhanced plant emergence and disease control (Chauhan 1986a, b, 1988; Dwivedi and Ghaube 1985). Monceren, Pencycuron, Carboxin 200, tolclofos-methyl, and Maximum AP also increased the percentage of surviving seedlings from Mp-infected seeds (Aly et al. 2001; Omar 2005). However, whether any seed treatments can efficiently control Mp of soybean in northwestern Argentina is not known. Thus, we analyzed the efficacy of one chemical and two biological products in controlling charcoal rot of soybean and their effects on plant emergence, disease severity, Mp colony forming units (CFU), crop yield and 1000-seed weight in two soybean cultivars maturity group (MG) VIII inoculated with Mp in Tucumán, Argentina.

MATERIALS AND METHODS

Macrophomina phaseolina isolation

During the 2009/2010 season, plants with typical charcoal rot symptoms were collected from commercial fields in Tucumán and taken to the Phytopathology Laboratory of the Estación Experimental Agroindustrial Obispo Colombres (EEAOC) to isolate and characterize *Macrophomina phaseolina* (Mp). Infected samples were rinsed, and 0.5 cm sections were surface sterilized with 70 % ethanol for 30 s, and with a 5 % NaClO solution for 1 min. These sections were later rinsed with sterile water and then dried with sterile air. Samples were cultivated on potato dextrose agar (PDA, Difco, Sparks, Maryland, USA) plates with 0.2 % v/v lactic acid at 28 ± 2°C for 5 days, after which cultures were maintained at 4°C. Single sclerotia were removed using a stereomicroscope and transferred to a new PDA plate acidified with 2 % lactic acid. Then, after 24–48 h, pure cultures of each isolate were obtained, which were reserved for further analysis.

Field trials

The field trials were done at the Monte Redondo substation of the EEAOC research station in Cruz Alta department, Tucumán, Argentina (26°49′S, 64°51′W). On 29 December 2010 and 28 December 2011, seeds of MG VIII soybean cultivars NA8000 RG and Munasqa RR were planted by hand in four replications of randomized blocks, and each treatment comprised four 3-m-long rows, with a planting density of 23 seeds per m. For inoculum in field trials, we used sterile rice that had been inoculated with Mp isolate number 55 and incubated for 20 days at 30 °C in darkness to promote microsclerotium development. When soybean seeds were planted, 5 g per linear meter of this rice inoculum was added by hand at planting depth. Treatments were as follows: (1) uninoculated and untreated control (neither inoculation, nor chemical or biological treatments); (2) untreated control inoculated with Mp; (3) Mp added to the soil and seed treated with a mixture of strobilurin + benzimidazole (pyraclostrobin 5 % + thiophanate methyl 45 %, 100 mL/100 kg of seed); (4) Mp added to the soil and seed treated with *Trichodermaviride*. (Trichoderma SP®, *Trichoderma viride*, 700 mL/50 kg of seed, Laboratorio San Pablo, Tucumán, Argentina); and (5) Mp added to the soil and seed treated with *Bacillussubtilis* (Bacillus subtilis SP®, *Bacillus subtilis*, 500 mL/50 kg of seed, Laboratorio San Pablo, Tucumán, Argentina).

The effects of each treatment were evaluated considering plant emergence (PE) percentage, disease severity, Mp colony forming units per gram of root (CFU/g), crop yield and 1000-seed weight (W 1000). PE was calculated as the percentage of plants that emerged 7, 14, and 21 days after planting (DAP). Charcoal rot severity was recorded at stage R7 (Fehr and Caviness 1977) using the scale of Paris et al. (2006): 1 = no discoloration and no microsclerotia visible; 2 = no discoloration of vascular tissue, with very few microsclerotia visible in the pith, vascular tissue or under the epidermis; 3 = partially discolored vascular tissue, with microsclerotia partially covering the tissue; 4 = discolored vascular tissue, with numerous microsclerotia visible in the tissue under the outer epidermis, in stem and root sections; 5 = vascular tissue with numerous microsclerotia producing a dark color inside and outside of the stem and root tissue.

Values of CFU were calculated according to the protocol proposed by Mengistu et al. (2007), using 10 plants from each plot. The samples were obtained by cutting 10 cm above and below soil level, ensuring that they included root and stem tissue, then washed and rinsed three times with tap water to remove any soil, dried at room temperature and ground. From each sample, 0.005 g of ground tissue was transferred to a test tube, disinfected with 5 % NaClO for 1 min, and rinsed with sterile distilled water for 1 min, three times. Then 5 mL of sterile PDA was added at 60 °C and poured into sterile Petri dishes, which were incubated at 28 °C for 3–5 days for CFU to be recorded.

Crop yield (kg/ha) and 1000-seed weight (g) were measured at harvest.

Data for crop yield and 1000-seed-weight were analyzed for statistical differences among treatment means using an ANOVA and LSD test at $\alpha = 0.05$ in INFOSTAT software version 2008 (Balzarini et al. 2008). Data on PE, CFU and disease severity were square-root transformed ($\sqrt{x}$) before the analysis.

To evaluate whether prevailing environmental conditions during the growing seasons corresponded to those that favor charcoal rot development, average monthly temperatures and precipitation were compared for the 2010/2011 and 2011/2012 growing seasons (Table 1).

Table 1. Mean monthly temperature and rainfall in the 2010/2011 and 2011/2012 seasons in Monte Redondo (Cruz Alta, Tucumán) compared with historical average values for this site (average temperature and precipitation in 1996–2012 and 1981–2010, respectively)

	November		December		January		February		March		April		Total
	T[a]	P[b]	T[a]	P[b]	T[a]	P[b]	T[a]	P[b]	T[a]	P[b]	T[a]	P[b]	P[b]
2010/2011	23.2	50.8	25.6	159.3	25.5	208.3	23.7	162.6	22.2	109.0	20.2	46.0	736
Differences	−0.9	−41.7	+0.1	+4.9	−0.4	+21.2	−0.8	+26.2	−0.7	−29.6	+0.5	−16.2	−35.2
2011/2012	25.5	67.3	25.4	84.1	27.1	84.6	27.3	81.3	23.8	71.4	19.9	93.0	481
Differences	+1.4	−25.2	−0.1	−70.3	+1.2	−102.5	+2.8	−55.1	+0.9	−67.2	+0.2	+30.2	−290
Mean values	24.1	92.5	25.5	154.4	25.9	187.1	24.5	136.4	22.9	138.6	19.7	62.2	771

[a] *T* monthly average temperature

[b] *P* rainfall in mm

RESULTS

Using the *Macrophomina phaseolina* (Mp) isolate number 55, we were able to reproduce the disease in the field. Inoculation reduced the number of inoculated NA8000 RG control plants that had emerged (PE), for instance, at 21 DAP in the 2010/2011 season to 86.4 %, compared with 92.6 % for the noninoculated control, although without statistically significant differences (Table 2). But in the 2011/2012 season, by 21 DAP, the inoculated control had significantly lower PE values (22.4 %), than the noninoculated control (51.9 %). Similarly, in Munasqa RR by 21 DAP in the 2010/2011 season, the PE of 79.2 % for the inoculated control differed significantly from 90.5 % PE for the noninoculated control (Table 3). Moreover, by 21 DAP in 2011/2012, the inoculated control had a PE of 25.5 %, differing significantly from 65.9 % for the noninoculated control.

Table 2. NA8000 RG emergence by 7, 14 and 21 days after planting in field trials after treatment with chemicals or biological control agents in Monte Redondo (Cruz Alta, Tucumán) in 2010/2011 and 2011/2012 seasons

Treatment	2010/2011			2011/2012		
	7	14	21	7	14	21
1-Untreated and uninoculated control	89.2 a*	92.4 a	92.6 a	30.1 a	51.9 ab	51.9 a
2-Untreated control + Mp	84.6 a	86.2 b	86.4 a	16.4 ab	30.8 c	22.4 c
3-(Pyraclostrobin + thiophanate methyl) + Mp	84.8 a	92.4 a	90.2 a	22.8 ab	62.8 a	58.9 a
4-*Trichoderma viride* + Mp	84.8 a	88.6 ab	86.4 a	19.5 ab	43.5 bc	36.3 b
5-*Bacillus subtilis* + Mp	84.1 a	88.2 ab	86.4 a	13.9 b	36.1 c	26.4 bc

* Means in each column followed by the same letter are not significantly different (LSD, $\alpha = 0.05$)

Table 3. Emergence of Munasqa RR plants by 7, 14 and 21 days after planting in field trials after treatment with chemicals or biological control agents in Monte Redondo (Cruz Alta, Tucumán), 2010/2011 and 2011/2012 seasons

Treatment	2010/2011			2011/2012		
	7	14	21	7	14	21
1-Untreated and uninoculated control	90.7 a*	90.0 a	90.5 a	52.6 ab	73.6 a	65.9 a
2-Untreated control + Mp	60.2 d	82.8 b	79.2 b	32.5 c	30.8 b	25.5 b
3-(Pyraclostrobin + thiophanate methyl) + Mp	87.1 ab	92.0 a	91.1 a	56.0 a	78.6 a	61.5 a
4-*Trichoderma viride* + Mp	75.2 c	88.1 ab	88.8 a	37.8 bc	43.3 b	34.4 b
5-*Bacillus subtilis* + Mp	78.2 bc	87.1 ab	88.1 a	32.7 c	40.6 b	32.2 b

* Means in each column followed by the same letter are not significantly different (LSD, $\alpha = 0.05$)

The effects of different treatments on PE of seeds inoculated with Mp 7, 14 and 21 DAP in NA8000 RG and Munasqa RR are also shown in Tables 2 and 3, respectively. In NA8000 RG during the 2010/2011 season, only the treatment with pyraclostrobin + thiophanate methyl (PE: 92.4 %) reduced the adverse effect of inoculation with Mp and differed significantly from the inoculated control (86.2 %) at 14 DAP. PE values at 7 and 21 DAP did not differ significantly among treatments (Table 2).

The 2011/2012 season showed environmental conditions favorable for charcoal rot (Table 1). As reflected in Table 2, significant differences were observed between the Mp-inoculated and the noninoculated control of NA8000 RG at 14 and 21 DAP; the inoculated plants had lower PE values (16.4 % by 7 DAP, 30.8 % by 14 DAP and 22.4 % by 21 DAP). At 7 DAP, none of the Mp-inoculated plants treated with either chemicals or biological agents differed significantly in PE from the untreated Mp-inoculated control (Table 2). At 14 DAP, the PE for plants in treatment 3 [(pyraclostrobin + thiophanate methyl) + Mp] was 62.8 % and differed significantly from the rest of the treatments except for the untreated and uninoculated control (51.9 %). At 21 DAP, all treatments yielded statistically significant differences in PE in relation to the inoculated control (22.4 %), except treatment 5 (*B. subtilis* + Mp) (26.4 %). Treatments 1 (untreated and uninoculated control) (51.9 %) and 3 ([pyraclostrobin + thiophanate methyl] + Mp) (58.9 %) had the highest PE values with no differences between them, whereas treatment 4 (*T. viride* + Mp) (36.3 %) and treatment 5 (*B. subtilis* + Mp) (26.4 %) gave the lowest PE values. It is worth mentioning that treatment 5 did not differ from the inoculated control (22.4 %) at this stage.

PE values for Munasqa RR after the different treatments are shown in Table 3 for three dates during the 2010/2011 and 2011/2012 seasons. In 2010/2011, the PE for the inoculated and the noninoculated control differed significantly at all dates. Seven days after planting, all treatments had PE values that contrasted significantly with those of the inoculated control (60.2 %). At 14 DAP, treatments 4 (*T. viride* + Mp) (88.1 %) and 5 (*B. subtilis* + Mp) (87.1 %) did not differ statistically from the inoculated control (82.8 %), but treatment 3 ([pyraclostrobin + thiophanate methyl] + Mp) (92.0 %) did show significant differences. At 21 DAP, all treatments differed statistically from the inoculated control (79.2 %), but not among them.

For Munasqa RR in the 2011/2012 season, the PE of plants in the untreated and uninoculated control (52.6 %) and treatment 3 [(pyraclostrobin + thiophanate methyl) + Mp] (56.0 %) differed statistically from the inoculated control (32.5 %) at 7 DAP (Table 3). On the other hand, the PE of plants in treatments 4 (*T. viride* + Mp) (37.8 %) and 5 (*B. subtilis* + Mp) (32.7 %) did not differ statistically from the inoculated control (32.5 %). At 14 DAP, the PE of inoculated control (30.8 %) plants did not differ statistically from treatments 4 (*T. viride* + Mp) (43.3 %) and 5 (*B. subtilis* + Mp) (40.6 %) but differed from treatments 1 (untreated and uninoculated control) (73.6 %) and 3 ([pyraclostrobin + thiophanate methyl] + Mp) (78.6 %). At 21 DAP, the PE for treatments 4 (*T. viride* + Mp) (34.4 %) and 5 (*B. subtilis* + Mp) (32.2 %) did not differ from the inoculated control (25.5 %), but differed significantly from the PE for treatments 1 (untreated and uninoculated control) (65.9 %) and 3 ([pyraclostrobin + thiophanate methyl] + Mp) (61.5 %).

The decrease in PE values in the 2011/2012 season compared with those in the 2010/2011 season is primarily attributable to the marked differences in environmental conditions, especially with regard to precipitation from December to January, key months for root emergence and plant establishment. According to data in Table 1, in the first season, the combined precipitation of those 2 months amounted to 367.6 mm, whereas in the second season that value dropped to 168.7 mm. This low value of precipitation was accompanied by high temperatures in January 2012 (Table 1), resulting in favorable environmental conditions for charcoal rot development in the 2011/2012 seasons.

Table 4 shows data related to crop yield, 1000-seed weight and charcoal rot severity recorded in NA8000 RG in the 2010/2011 and 2011/2012 seasons. In 2010/2011, although no significant differences were found among treatments in relation to crop yield, crop yield increased with all control treatments. For instance, treatment 3 ([pyraclostrobin + thiophanate methyl] + Mp) led to 277.7 kg/ha more soybeans than the inoculated control (4492.4 kg/ha). With respect to 1000-seed weight, no significant differences were observed among treatments. Severity value in the inoculated control was 2.5, and no significant differences were found among treatments (Table 4).

Table 4. Mean crop yield (kg/ha), 1000-seed weight (W 1000 in g) and charcoal rot severity (Sev) on a scale from 1 to 5 in NA8000 RG in field trials after treatment with chemicals or biological control agents in Monte Redondo (Cruz Alta, Tucumán), 2010/2011 and 2011/2012 seasons

Treatment	2010/2011			2011/2012		
	Yield	W 1000	Sev	Yield	W 1000	Sev
1-Untreated and uninoculated control	4515.1 a*	145.6 a	1.0 a	1445.8 a	170.0 b	1.3 c
2-Untreated control + Mp	4492.4 a	148.6 a	2.5 a	504.2 b	169.7 b	3.6 a
3-(Pyraclostrobin + thiophanate methyl) + Mp	4770.1 a	145.7 a	2.0 a	1345.8 a	179.9 ab	2.6 b
4-*Trichoderma viride* + Mp	4687.3 a	146.0 a	1.5 a	804.2 b	186.7 a	2.7 b
5-*Bacillus subtilis* + Mp	4631.2 a	144.6 a	1.0 a	741.7 b	181.9 ab	3.1 ab

* Means in each column followed by the same letter are not significantly different (LSD, $\alpha = 0.05$)

In the 2011/2012 season, treatment 3 ([pyraclostrobin + thiophanate methyl] + Mp) (1345.8 kg/ha) and treatment 1 (untreated and uninoculated control) (1445.8 kg/ha) differed statistically from treatment 2 (inoculated control) (504.2 kg/ha), treatment 4 (*T. viride* + Mp) (804.2 kg/ha) and treatment 5 (*B. subtilis* + Mp) (741.7 kg/ha) (Table 4). Treatment 3 ([pyraclostrobin + thiophanate methyl] + Mp) resulted in 841.6 kg/ha more in yield than the inoculated control. As for 1000-seed weight, treatment 4 (*T. viride* + Mp) (186.7 g) differed statistically with respect to the untreated and uninoculated (170.0 g) and the inoculated (169.7 g) controls, but not with respect to treatments 3 [(pyraclostrobin + thiophanate methyl) + Mp] (179.9 g) and 5 (*B. subtilis* + Mp) (181.9 g). The inoculated control showed high charcoal rot severity values (3.6) and differed statistically from the non-inoculated control (1.3), treatment 3 ([pyraclostrobin + thiophanate methyl] + Mp) (2.6) and treatment 4 (*T. viride* + Mp) (2.7), but not from treatment 5 (*B. subtilis* + Mp) (3.1) (Table 4). Again, the higher charcoal rot values in 2011/2012 compared with 2010/2011 can be accounted for by the favorable environment conditions that prevailed during that season (Table 1).

In Munasqa RR, yield of all treatments differed significantly from the fungus-inoculated control (3456.1 kg/ha) in the 2010/2011 season (Table 5). Treatment 3 ([pyraclostrobin + thiophanate methyl] + Mp) yielded 658.3 kg/ha more than the inoculated control, while treatments 4 (*T.viride* + Mp) and 5 (*B.subtilis* + Mp) showed increments of 635.6 kg/ha and 634.3 kg/ha, respectively. No significant differences were observed for 1000-seed weight among different treatments and the inoculated and uninoculated controls. With regard to disease severity, no significant differences were found among treatments and the inoculated control, which showed a rating of 1.8.

Table 5. Mean yield (kg/ha), 1000-seed weight (W 1000 in g) and charcoal rot severity (Sev) on a scale from 1 to 5 in Munasqa RR cultivar in field trials after treatment with chemicals or biological control agents in Monte Redondo (Cruz Alta, Tucumán), 2010/2011 and 2011/2012 seasons

Treatment	2010/2011			2011/2012		
	Yield	W 1000	Sev	Yield	W 1000	Sev
1-Untreated and uninoculated control	4166.5 a*	137.3 a	1.0 a	1866.7 a	163.2 ab	1.2 b
2-Untreated control + Mp	3456.1 b	139.4 a	1.8 a	870.8 c	169.4 a	2.7 a
3-(Pyraclostrobin + thiophanate methyl) + Mp	4114.4 a	140.8 a	1.5 a	1945.8 a	151.4 b	2.5 a
4-*Trichoderma viride* + Mp	4091.7 a	140.5 a	1.0 a	1491.7 b	160.8 ab	2.5 a
5-*Bacillus subtilis* + Mp	4090.4 a	140.0 a	1.5 a	1304.2 b	158.8 ab	2.5 a

* Means in each column followed by the same letter are not significantly different (LSD, $\alpha = 0.05$)

In the 2011/2012 season, yields for treatment 3 [(pyraclostrobin + thiophanate methyl) + Mp] (1945.8 kg/ha) and the noninoculated control (1866.7 kg/ha) differed statistically from that of the Mp-inoculated control (870.8 kg/ha) and treatments 4 (*T. viride* + Mp) (1491.7 kg/ha) and 5 (*B. subtilis* + Mp) (1304.2 kg/ha). Treatment 3 ([pyraclostrobin + thiophanate methyl] + Mp) yielded 1075.0 kg/ha more than the inoculated control and 79.1 kg/ha more than the noninoculated control. With regards to 1000-seed weight, treatment 3 ([pyraclostrobin + thiophanate methyl] + Mp) (151.4 g) differed statistically from the inoculated control (169.4 g). The Mp-inoculated control showed the highest charcoal rot severity values (2.7), but differed statistically only from the noninoculated control (1.2). The fact that higher charcoal rot severity values (2.7) were observed in the 2011/2012 season than in the 2010/2011 season (1.8) can also be accounted for by environmental conditions (Table 1).

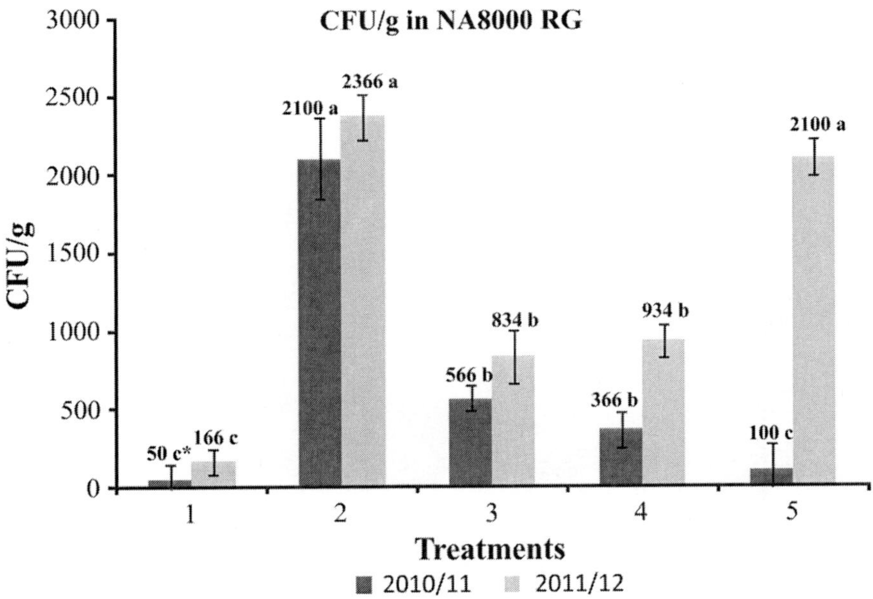

Figure 1. CFU/g of *Macrophomina phaseolina* in NA8000 RG during the 2010/2011 and 2011/2012 seasons in field trials with different seed treatments. Treatment *1* untreated and uninoculated control, *2* untreated control + Mp, *3* (pyraclostrobin + thiophanate methyl) + Mp, *4Trichodermaviride* + Mp and *5Bacillussubtilis* + Mp. * Means in each season followed by the same letter are not significantly different (LSD, $\alpha = 0.05$)

With respect to CFU/g of root, the Mp-inoculated control presented higher CFU/g values than in the other treatments for both cultivars and seasons. NA8000 RG presented 2100 CFU/g in 2010/2011 and differed significantly from all the other treatments (Figure 1). In 2011/2012, Mp-inoculated control (2366 CFU/g) differed significantly from treatments 1 (untreated and uninoculated control) (166 CFU/g), 3 ([pyraclostrobin + thiophanate methyl] + Mp) (834 CFU/g) and 4 (*T. viride* + Mp) (934 CFU/g), but not from treatment 5 (*B. subtilis* + Mp) (2100 CFU/g). On the other hand, in 2010/2011 Munasqa RR inoculated control showed 300 CFU/g and differed statistically from treatments 1 (untreated and uninoculated control) (50 CFU/g) and 4 (*T. viride* + Mp) (100 CFU/g) but did not from treatments 3 ([pyraclostrobin + thiophanate methyl] + Mp) (200 CFU/g) and 5 (*B. subtilis* + Mp) (266 CFU/g) (Figure 2). In 2011/2012 season, Munasqa RR inoculated control was 566 CFU/g and differed statistically only from the noninoculated control (166 CFU/g). All treatments had lower CFU/g values than the inoculated control in both cultivars and seasons.

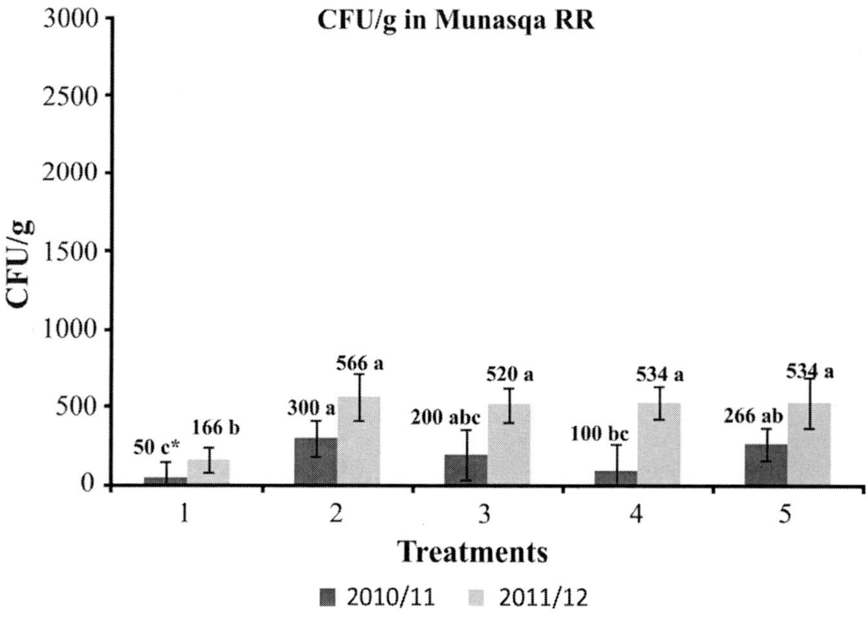

Figure 2. CFU/g of *Macrophomina phaseolina* in Munasqa RR during the 2010/2011 and 2011/2012 seasons in field trials with different seed treatments. Treatment *1* untreated and uninoculated control, *2* untreated control + Mp, *3* (pyraclostrobin + thiophanate methyl) + Mp, *4Trichodermaviride* + Mp and *5Bacillussubtilis* + Mp. * Means in each season followed by the same letter are not significantly different (LSD, $\alpha = 0.05$)

DISCUSSION

Field inoculation of soybean with *Macrophomina phaseolina* (Mp) affected all variables studied, as compared with the noninoculated control in

NA8000 RG and Munasqa RR. In the 2010/2011 season, Mp inoculation resulted in a PE reduction of 6.7 % and 12.5 % in NA8000 RG and Munasqa RR, respectively. In the 2011/2012 season, the PE reduction was even more conspicuous and amounted to 56.8 % in NA8000 RG and 61.3 % in Munasqa RR. These results are in accordance with Smith and Wyllie (1999), who reported high seedling mortality from charcoal rot in the tropics.

Disease severity values recorded at R7 in the inoculated control were also higher in the 2011/2012 season and in both cultivars: 3.6 for NA8000 RG and 2.7 for Munasqa RR, compared with 2.5 and 1.8, respectively, as recorded in 2010/2011. In addition, values of CFU/g root in the fungus-inoculated control reached 2100 in 2010/2011 and 2366 in 2011/2012 in NA8000 RG (Figure 1). In Munasqa RR, these values were 300 CFU/g in 2010/2011 and 566 CFU/g in 2011/2012 (Figure 2).

The higher PE reduction, disease severity and CFU/g values observed in the 2011/2012 season were likely the consequences of climatic conditions, with high temperatures and low rainfall levels (Table 1). Similarly, Mengistu et al. (2013) greater symptoms of charcoal rot in plants stressed by high heat and drought, especially when drought occurred during reproductive growth.

Genotypes also differ in their reaction to charcoal rot. Mengistu et al. (2013) identified sources of charcoal rot resistance in 628 accessions from soybean germoplasm by field screening in five environments. In this study, we found that Mp on Munasqa RR was less severe and developed fewer CFU/g than on NA8000 RG in both seasons.

All treatments evaluated in this research were effective in counteracting PE reduction caused by the pathogen in both seasons and cultivars. In 2011/2012 the fungicide treatment led to higher PE values than the biological treatments in both cultivars. In NA8000 RG, compared with the inoculated control, the fungicide enhanced PE by 70.3 %, *T.viride* by 26.8 % and *B.subtilis* by 7.7 %. As for Munasqa RR, the fungicide increased PE by 54.6 %, *T.viride* by 13.5 % and *B.subtilis* by 6.1 % with respect to the inoculated control. The lower effectiveness showed by the biological treatments could be due to the fact that weather conditions (Table 1) favoring Mp infection (high temperature and drought) could have affected these microorganisms negatively.

Biological and chemical seed treatments have been shown to be effective to control Mp on various crops. For cotton, several fungicides were tested as seed treatments to determine their efficacy in controlling Mp infection. Treatments with carbendazim, quintozene and benomyl enhanced plant emergence and disease control (Chauhan 1986a, b, 1988; Dwivedi and Ghaube 1985). Omar (2005) found that Monceren® 250 FS and tolclofos-methyl were the best performing fungicides in controlling Mp on cotton in the greenhouse. Pineda and Gonnella (1988) also found a reduction in the percentage of dead sesame plants affected by Mp when seeds were treated with biological antagonists such as *Trichoderma* sp. and *Aspergillus* sp.

In NA8000 RG and Munasqa RR, treatments had a similar performance trend: the highest crop yield values were obtained with the pyraclostrobin + thiophanate methyl mixture, followed by those obtained with *T.viride* and *B.subtilis.* Similar yield increases were also observed by Pineda and Avila (1988): an increment of 100 kg was obtained when applying chemical seed treatments (propineb and dicarboximide) in sesame to control Mp, as compared with yield values obtained with the untreated control.

Mp did infect Munasqa RR and NA8000 RG, but severity levels and CFU/g were higher in NA8000 RG in both seasons. The chemical and biological treatments resulted in lower disease severity and CFU/g values than in the inoculated control. These results agree with those reported by Ramezani (2008), who found lower disease incidence in eggplants (*Persea americana* L.) treated with different *Trichoderma* sp. strains.

Results from this study of chemical and biological seed treatments of soybean to control charcoal rot will be useful for farmers in Tucumán to develop more efficient management strategies for this important disease. Additional research should be extended to other regions of Argentina and neighboring soybean-producing countries in South America to further assess this plant–pathogen interaction under different environmental conditions.

REFERENCES

1. Aly AA, El-Shazly AMM, Youssef RM, Omar MR (2001) Chemical and biological control of charcoal rot of cotton caused by Macrophomina phaseolina. J Agric Sci Mansoura Univ 26:7661–7674
2. Balzarini MG, Gonzalez L, Tablada M, Casanoves F, Di Rienzo JA, Robledo CW (2008) Infostat. Manual del usuario, Editorial Brujas, Córdoba, Argentina
3. Chauhan MS (1986a) Comparative efficacy of fungicides for the control of seedling diseases of cotton due to Rhizoctonia spp. Ind Mycol Pl Pathol 16:335–337
4. Chauhan MS (1986b) Systemic and non-systemic fungicides against root rot of cotton in Haryana. Ind Mycol Pl Pathol 16:226–227
5. Chauhan MS (1988) Relative efficiency of different methods for the control of seedling disease of cotton by Rhizoctonia bataticola. Ind Mycol Pl Pathol 18:25–30
6. Cook RJ, Baker KF (1983) The nature and practice of biological control of plant pathogens. American Phytopathological Society, St. Paul

7. Dorrance AE, Kleinhenz MD, McClure SA, Tuttle NT (2003) Temperature, moisture, and seed treatment effects on Rhizoctonia solani root rot of soybean. Plant Dis 87:533–538
8. Dwivedi TS, Ghaube HS (1985) Effect of fungicides on the emergence and infection of cotton seedlings by Macrophomina phaseolina (Tassi) Goid. Ind Mycol Pl Pathol 15:295–296
9. Elad Y, Zvieli Y, Chet I (1986) Biological control of Macrophomina phaseolina (Tassi) Goid by Trichoderma harzianum. Crop Protect 5:288–292
10. Fehr WR, Caviness CE (1977) Stages of soybean development. Special report no. 80. Coop. Ext. Ser., Iowa Agric And Home Econ Exp Stn, Iowa State Univ, Ames
11. Hepperly PR, Sinclair JB (1978) Quality losses in Phomopsis-infected soybean seeds. Phytopathology 68:1684–1687
12. Jeyarajan R, Ramakrishman G, Dinakaran D, Sridhar R (1993) Development of product of Trichoderma viride and Bacillus subtilis for biocontrol of root rot disease. In: Dwivedi (ed) Biotechnology in India. Bioved Research Society, Allahabad, India
13. Martin SB, Lucas LT, Campbell CL (1984) Comparative sensitivity of Rhizoctonia solani and Rhizoctonia-like fungi to selected fungicides in vitro. Phytopathology 74:778–781
14. Mengistu A, Ray JD, Smith JR, Paris RL (2007) Charcoal rot disease assessment of soybean genotypes using a colony-forming unit index. Crop Sci 47:2453–2461
15. Mengistu A, Smith JR, Ray JD, Bellaloui N (2011) Seasonal progress of charcoal rot and its impact on soybean productivity. Plant Dis 95:1159–1166
16. Mengistu A, Bond J, Nelson R, Rupe J, Shannon G, Arelli P, Wrather A (2013) Identification of soybean accessions resistant to Macrophomina phaseolina by field screening and laboratory validation. Online. Plant Health Progress
17. Mengistu A, Wrather A, Rupe JC (2015) Charcoal rot. In: Hartman GL, Rupe JC, Sikora EF, Domier LL, Davies JA, Steffey KL (eds) Compendium of soybean diseases. APS Press, St. Paul, pp 67–69
18. Mihail JD (1992) Macrophomina. In: Rush CM (ed) Singleton LL Mihail JD. Methods for research on soilborne phytopathogenic fungi. APS Press, St. Paul, pp 134–136
19. Mueller DS, Hartman GL, Pedersen WL (1999) Development of sclerotia and apothecia of Sclerotinia sclerotiorum from infected soybean seed and its control by fungicide seed treatment. Plant Dis 83:1113–1115
20. Omar MR (2005) Pathological and biochemical studies on Macrophomina phaseolina pathogenic on cotton. Ph.D. dissertation. Suez Canal University, Ismailia, Egypt
21. Paris RL, Mengistu A, Tyler JM, Smith JR (2006) Registration of soybean germplasm line DT97-4290 with moderate resistance to charcoal rot. Crop Sci 46:2324–2325
22. Pineda JB, Avila JM (1988) Alternativas para el control de Macrophomina phaseolina y Fusarium oxysporum patógenos del ajonjolí (Sesamum indicum L.). Agron Trop 38:79–84
23. Pineda JB, Gonnella ER (1988) Evaluación del control biológico de Macrophomina phaseolina en ajonjolí (Sesamum indicum L.). Agron Trop 38:43–48
24. Ploper LD, Scandiani MM (2009) Visión general de las enfermedades radiculares de la soja en Argentina. Proceedings V Congresso Brasileiro de Soja e Mercosoja 2009 [CD-ROM]. Goiânia, Goias, pp 1–3
25. Ploper LD, González V, de Ramallo V, Gálvez R, Devani M (2001) Presencia de la podredumbre carbonosa del tallo de la soja en el centro y noroeste argentino. Avance Agroind 22:30–34
26. Ramezani H (2008) Biological control of root-rot of eggplant caused by Macrophomina phaseolina. Am Eurasian J Agric Environ Sci 4:218–220
27. Sankar P, Jeyarajan R (1996) Seed treatment formulation of Trichoderma and Gliocladium for biological control of Macrophomina phaseolina in sesamum. Indian Phytopathology 49:148–151

28. Smith GS, Wyllie TD (1999) Charcoal rot. In: Hartman GL, Sinclair JB, Rupe JC (eds) Compendium of soybean diseases, 4th edn. APS Press, St. Paul, pp 29–31
29. Srivastava AK, Singh T, Jana TK, Arora DK (2001) Microbial colonization of Macrophomina phaseolina and suppression of charcoal rot of chickpea. In: A. Sinha (ed) Microbes and plants. Vedams eBooks (P) Ltd., New Delhi, India pp 269–319
30. Su G, Suh SO, Schneider RW, Russin JS (2001) Host specialization in the charcoal rot fungus, Macrophomina phaseolina. Phytopathology 91:120–126CrossRefPubMed
31. Wrather JA, Anderson TR, Arsyad DM, Gai J, Ploper LD, Porta-Puglia A, Ram HH, Yorinori JT (1997) Soybean disease loss estimates for the top ten soybean producing countries in 1994. Plant Dis 81:107–110
32. Wrather JA, Shannon G, Balardin R, Carregal L, Escobar R, Gupta GK, Ma Z, Morel W, Ploper D, Tenuta A (2010) Effect of diseases on soybean yield in the top eight producing countries in 2006. Online. Plant Health Progress
33. Wyllie TD (1989) Charcoal rot. In: Sinclair JB, Backman PA (eds) Compendium of soybean diseases, 3rd edn. APS Press, St. Paul, pp 30–33

CHAPTER 7

TAL Effector Driven Induction of a *SWEET*gene Confers Susceptibility to Bacterial Blight of Cotton

Kevin L. Cox[1,2],*, Fanhong Meng[1,2],*, Katherine E. Wilkins[3], Fangjun Li1,[2], Ping Wang[1,2], Nicholas J. Booher[3], Sara C.D. Carpenter[3], Li-Qing Chen[4], Hui Zheng[3], Xiquan Gao[2,5,6], Yi Zheng[7], Zhangjun Fei[7], John Z. Yu[8], Thomas Isakeit[1] , Terry Wheeler[1,9], Wolf B. Frommer[10],w, Ping He[2,5], Adam J. Bogdanove[3] & Libo Shan[1,2]

[1]Department of Plant Pathology and Microbiology, Texas A&M University, College Station, Texas 77843, USA.

[2] Institute for Plant Genomics and Biotechnology, Texas A&M University, College Station, Texas 77843, USA.

[3] Plant Pathology and Plant-Microbe Biology Section, School of Integrative Plant Science, Cornell University, Ithaca, New York 14853, USA.

[4]Department of Plant Biology, School of Integrative Biology, University of Illinois at UrbanaChampaign, Champaign, Illinois 61801, USA.

[5]Department of Biochemistry and Biophysics, Texas A&M University, College Station, Texas 77843, USA.

[6] State Key Laboratory of Crop Genetics and Germplasm Enhancement, College of Agriculture, Nanjing Agricultural University, Nanjing 210095, China.

[7] Boyce Thompson Institute, Cornell University, Ithaca, New York 14853, USA.

[8] USDA-ARS, Southern Plains Agricultural Research Center, College Station, Texas 77845, USA.

[9] Texas Agricultural Experiment Station, Lubbock, Texas 79403, USA.

[10] Carnegie Science, Department of Plant Biology, 260 Panama Street, Stanford, California 94305, USA. w Present address: Institute of Molecular Physiology, Heinrich Heine University Du¨sseldorf and Max Planck Institute for Plant Breeding Research, 50829 Ko¨ln, Germany

ABSTRACT

Transcription activator-like (TAL) effectors from *Xanthomonas citri* subsp. *malvacearum* (*Xcm*) are essential for bacterial blight of cotton (BBC). Here, by combining transcriptome profiling with TAL effector-binding element (EBE) prediction, we show that *GhSWEET10*, encoding a functional sucrose transporter, is induced by Avrb6, a TAL effector determining *Xcm* pathogenicity. Activation of *GhSWEET10* by designer TAL effectors (dTALEs) restores virulence of *Xcm avrb6* deletion strains, whereas silencing of *GhSWEET10* compromises cotton susceptibility to infections. A BBC-resistant line carrying an unknown recessive *b6* gene bears the same EBE as the susceptible line, but Avrb6-mediated induction of *GhSWEET10* is reduced, suggesting a unique mechanism underlying *b6*-mediated resistance. We show via an extensive survey of *GhSWEET* transcriptional responsiveness to different *Xcm* field isolates that additional *GhSWEETs* may also be involved in BBC. These findings advance our understanding of the disease and resistance in cotton and may facilitate the development cotton with improved resistance to BBC.

INTRODUCTION

Cotton (*Gossypium* spp.) is an economically important crop and provides a significant source of fibre, feed, foodstuff, oil and biofuel products worldwide. The cotton genus is composed of at least 45 diploid and 5 tetraploid species[1]. The tetraploid (AD genome) species, including *G. hirsutum* that produces 95% of the world's cotton fibre, is likely derived from a hybridization between an A-genome-like ancestral species resembling *G. arboreum* and a D-genome-like ancestral species resembling *G. raimondii*[2]. The availability of the draft genome sequences for *G. raimondii*, *G. arboreum*, *G. hirsutum* and *G. barbadense* not only provides genetic resources to study the complex genome evolution and polyploidization process, but also lays the foundation for functional genomic approaches to dissect cotton gene functions with a goal to improve its agricultural performance in the face of biotic and abiotic stresses[3,4,5,6,7,8].

Bacterial blight of cotton (BBC), caused by *Xanthomonas citri* subsp. *malvacearum* (*Xcm*), is among the destructive diseases of cotton[9]. Following epidemics in the 1970s, the disease has occurred sporadically in the United States, but in the past several years re-emerged as a significant yield constraint. There are pressing needs to address the underlying mechanisms of susceptibility and resistance to BBC in cotton. *Xcm* injects effector proteins into plant cells via the type III secretion system to promote pathogenicity in plants. In the presence of corresponding resistance (R) proteins, some of these effectors trigger resistance and function as avirulence proteins. Interestingly, all known pathogenicity and avirulence factors (encoded by so called *'pth'* and *'avr'* genes) of *Xcm* are transcription activator-like (TAL) effectors10. TAL effectors functionally resemble eukaryotic transcription factors and upregulate host genes by directly binding to their promoters11,12. TAL effectors are highly conserved among different *Xanthomonas* spp., with an N-terminal type III translocation signal, a central repeat region (CRR) and C-terminal nuclear localization signals followed by an acidic activation domain. The proteins differ mainly in the CRR, which consists of 1.5–33.5 copies of near-perfect repeats of 33–34 amino acids. These repeats are conserved with the exception of the 12th and 13th residues of each copy, defined as the repeat variable di-residue (RVD)13,14. Each RVD targets a specific nucleotide of the promoters of host genes, creating a code, such that the sequence of RVDs defines the effector-binding element (EBE); each EBE starts, however, with a nearly invariant thymine, which is specified in a yet unclear manner by structures immediately N terminus of the CRR13,14,15,16.

Among the identified TAL effector targets include a group of pepper genes upregulated by AvrBs3 from *X. campestris* pv. *vesicatoria* (*Xcv*)17,18. These include the *R* gene *Bs3* and *UPA20*, which encodes a basic helix-loop-helix transcription factor regulating plant cell hypertrophy19. *Os8N3/Xa13 /OsSWEET11*, essential for reproductive development, is a rice susceptibility (*S*) gene targeted by TAL effector PthXo1 from *X. oryzae* pv. *oryzae* (*Xoo*)20. The *xa13* allele, which is unable to be induced by PthXo1, acts as a recessive *r* gene against *Xoo* infection21,22. Notably, four *Os8N3/ Xa13 /OsSWEET11* homologues, rice *Os11N3/OsSWEET14* and *Xa25/OsS WEET13*, pepper *UPA16* and cassava *MeSWEET10a* are also targeted by TAL effectors from *Xoo*, *Xcv* and *X. axonopodis* pv. *manihotis* (*Xam*), respectively 23,24,25. It was hypothesized that *Xanthomonas* spp. induces the expression of *SWEET* genes during infection to transport sucrose to the apoplast, thereby providing the bacteria with a carbon source20,23,25,26,27,2 8,29,30 . TAL effectors from *X. citri* sp. *citri* (*Xcc*), the causal agent for citrus bacterial canker, induce the expression of *CsSWEET1* in citrus, but the *CsSWEET1* does not contribute demonstrably to susceptibility; another TAL effector target, however, *CsLOB1*, does function as an *S* gene, promoting the characteristic pustule formation and bacterial multiplication31. Other characterized TAL effector targets include the rice bacterial blight *R* genes *Xa27* (ref. 32), *Xa23* (ref. 33) and *Xa10* (ref. 34), the bacterial leaf streak of rice *S* gene *OsSULTR3;6* (ref. 35) and the pepper bacterial spot *R* gene *Bs4C* (ref. 36).

Previous work identified at least 10 TAL effectors from the cotton pathogen *Xcm*H1005 and showed that a mutant with a deletion of seven *tal* genes was no longer able to cause observable water-soaking symptoms on Acala44 (Ac44), a BBC susceptible line of cotton. Avrb6 from *Xcm*H1005 was among the first examples of an individual TAL effector important for *Xcm* virulence. An *avrb6* mutant of *Xcm*H1005, *Xcm*H1407 (*Xcm*H1005Δ*avrb6*), showed reduced water-soaking, whereas when expressed in a relatively weak virulent strain *Xcm*N1003, Avrb6 enhances water-soaking37. Although it does not affect *in planta* bacterial multiplication, Avrb6 plays a major role in release of bacteria from the leaf interior to the leaf surface during infections37.

To identify the TAL effectors responsible for virulence and their targets, we sequenced the whole genomes of *Xcm* strains *Xcm*H1005 and *Xcm*N1003 and assembled the full repertoire of TAL effectors. Further, we deployed genome-wide gene expression profiling in cotton coupled with TAL effector DNA-binding code-assisted EBE prediction and identified the clade III sucrose transporter gene *GhSWEET10* as a target of Avrb6. We determined that *GhSWEET10* is a major *S* gene for BBC by using dTALEs to induce it independently of any other possible Avrb6 targets and by silencing it with *Agrobacterium*-mediated virus-induced gene silencing (VIGS). Significantly, other members of the clade III *GhSWEET* genes are strongly induced by different *Xcm* field strains responsible for recent re-emergence of the disease in the southern United States. Our data indicate that cotton GhSWEET10 and likely other SWEET sugar transporters in the same clade play a major role in BBC and suggest that newly evolved or horizontally transferred TAL effectors targeting these genes attribute to the re-emergence of BBC in the field. In addition, a BBC-resistant cotton line carrying *b6*, a genetically complex resistance gene that has not yet been molecularly identified, showed markedly reduced Avrb6-mediated induction of *GhSWEET10*, ostensibly due to promoter polymorphisms outside the EBE. This suggests a novel mechanism for resistance that may represent a genetic determinant of *b6.*

RESULTS

Genome and TAL effector sequences of *Xcm*H1005 and N1003

The strains *Xcm*H1005, a rifamycin-resistant derivative of *Xcm*H, and *Xcm*N1003, a spectinomycin- and rifamycin-resistant derivative of *Xcm*N, are two *Xcm* strains that cause water-soaking on Ac44 cotton38. To set the stage for identifying the role that TAL effectors in these strains play in BBC, we generated complete genome sequences using single molecule real-time sequencing (Pacific Biosciences; hereafter 'PacBio sequencing'). The H1005 genome consists of a circular chromosome of 5,212,564 bp and a plasmid, pXcmH39 of 88,283 bp. The N1003 genome consists of a circular chromosome of 5,218,607 bp and a plasmid, herein designated as pXcmN, of 59,644 bp. The general features of each genome are presented in Table 1. The overall architectures of the two chromosomes are highly similar, with a single, large inversion (Figure 1a). The plasmids are dissimilar, with only a few, relatively small regions of homology (Figure 1a). The *tal* genes of the two strains are largely distinct (Figure 1b and Supplementary Table 2). H1005 encodes 12 TAL effectors with six in the chromosome and six in pXcmH. Ten of these are Avr proteins named previously based on reactions of differential cotton lines to transformants of an otherwise-compatible *Xcm* strain carrying corresponding cosmid subclones: AvrB4, AvrB5, Avrb6, Avrb7, AvrB101, AvrB102, AvrB103, AvrB104, AvrBIn and AvrBn38·39·40. Another is PthN2, originally characterized in *Xcm*N1003 (ref. 10). The twelfth is uncharacterized, and we designated it as $Tal6_{XcmH1005}$, following a previously described naming scheme41. *Xcm*N1003 harbours nine *tal* genes, five on the chromosome and four on pXcmN. Four on the chromosome and one on the plasmid however are disrupted by a frameshift mutation or a large insertion in the repeat region or the 3′ end of the gene. One of the disrupted genes, *tal1'*, on the chromosome, carries only a last repeat and is otherwise complete until an integrase insertion further downstream. Another of the disrupted genes, *pthN'* on pXcmN, is otherwise identical to *pthN*, an *Xcm*N1003 *tal* gene shown previously to contribute to water-soaking without triggering resistance on commercial US cotton varieties10. The intact *pthN* gene is also located on pXcmN, along with *pthN2*, which also contributes to water-soaking, but less strongly10. Another intact *tal* gene, $tal7_{XcmN1003}$, also resides on the plasmid. The fourth intact gene is *avrBn* located on the chromosome. Apart from *pthN'*, *pthN*, *pthN2* and *avrBn*, no other *Xcm*N1003 *tal* genes show obvious similarity in their RVD sequences to any other *tal* genes in either strain. The divergence in *tal* gene content between the two strains, both with respect to RVD composition and genomic location of orthologs, is striking, which is in line with the fact that they were isolated from geographically distant regions38.

Table 1: Genome characteristics of *Xcm*H1005 and *Xcm*N1003.

Characteristics	*Xcm*H1005	*Xcm*N1003
Length (bp)	5,212,564 bp	5,218,607 bp
No. of protein-coding genes	4,227	4,218
GC content	64.62%	64.54%
No. of tRNA genes	54	54
No. of rRNA operons	2	2
Components	Circular chromosome (5.2 Mb); plasmid (88,283 bp)	Circular chromosome (5.2 Mb); plasmid (59,644 bp)
No. of TAL effectors (Chromosome::Plasmid)	6::6	5::4
GenBank accession no.	CP013004	CP013006

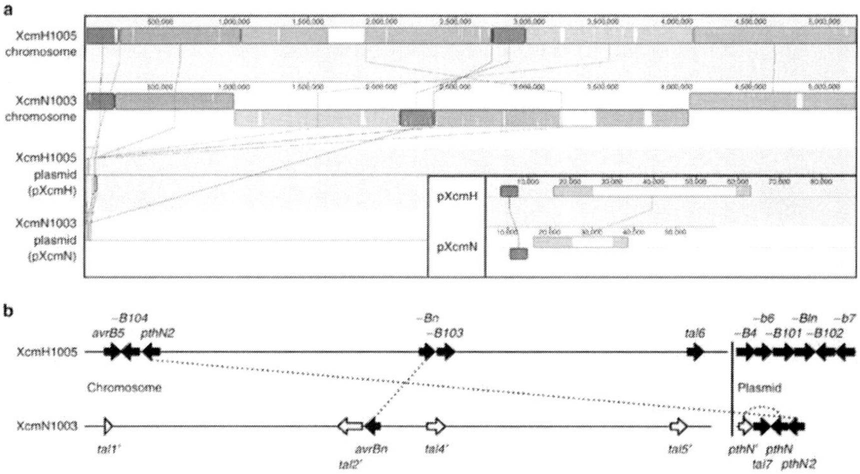

Figure 1. Comparison of whole genomes and *tal* genes of *Xcm*H1005 and *Xcm*N1003. (**a**) Alignment of the *Xcm*H1005 and *Xcm*N1003 genomes, generated using progressiveMAUVE with default parameters68. Each genome comprises a single circular chromosome and a single circular plasmid, shown linearized. Coloured, rounded rectangles represent locally collinear blocks (LCB), regions of homology without rearrangement across the aligned sequences, connected by matching coloured diagonal lines. The orientations of the LCB, forward or reverse, are indicated by their position above or below the line, respectively. The height of a column within a block reflects the average similarity relative to the other aligned sequence(s) there (see http://darlinglab.org/mauve/user-guide/viewer.html for details). Inset (bottom right) shows the alignment of the plasmids only, at a larger scale. Horizontal axes show sequence coordinates (bp). MAUVE backbone files giving the exact coordinates of all LCB are provided as Supplementary Data 1 (all molecules aligned) and Supplementary Table 1 (plasmids only). (**b**) The *tal* genes of *Xcm*H1005 and *Xcm*N1003. The genes are represented as block arrows at their relative positions in the chromosome or plasmid (horizontal lines). The arrows are magnified relative to the rest of the genome, but intergenic regions and arrow sizes relative to each other are to scale. Dashed lines connecting two arrows indicate identical encoded RVD sequences. Gene names follow the scheme of ref. 41, except for the indicated *avr* and *pth* genes, named previously10,38,39,40. For label clarity, hypens replace '*avr*'. An apostrophe following the gene name (white arrows) indicates that the coding sequence is terminated early due to a frameshift mutation or other coding sequence disruption. Details, including RVD sequences and AnnoTALE designations53, are given in Supplementary Table 2.

Candidate targets of Avrb6

Avrb6 is a TAL effector in *Xcm*H1005 that causes strong water-soaking in the cotton line Ac44 (ref. 37). In this study, we used the Ac44E-tetraploid genotype, which is a single plant selection from Brinkerhoff's original Ac44 parent42. Ac44E has similar morphology to Ac44 except it has more abundant flowers and is slightly more susceptible to BBC than Ac44 (ref. 42). To identify the cotton genes that are specifically activated by Avrb6, we used an integrative approach involving a whole-genome RNA-sequencing (RNA-Seq) analysis of protoplasts of Ac44E expressing Avrb6 and computational prediction of Avrb6 EBEs in cotton (Figure 2a). Compared to the whole plant-pathogen infection assay, the homogenous protoplast mesophyll cell system expressing the individual bacterial effector circumvents the complication of multiple pathogen effectors and elicitors that could spontaneously activate or suppress a large number of host genes. We cloned the full-length coding sequence of Avrb6 from *Xcm*H1005 into a plant expression vector with an HA-epitope tag at the C terminus under the control of the *35S* CaMV promoter. The sequence was confirmed via Sanger sequencing (Supplementary Figure 1). An immunoblot analysis with an α-HA antibody detected a major polypeptide slightly above the 100 kDa marker, which matches the predicted molecular mass of 108 kDa for Avrb6 (Figure 2b). We further subcloned the *35S::avrb6-HA* expression cassette into a binary vector and performed *Agrobacterium*-mediated transient expression assays in Ac44E cotton and *Nicotiana benthamiana*. As shown in Figure 2c, we observed a similar result as in cotton protoplasts with a major polypeptide being detected at ~108 kDa. Next, we performed RNA-Seq analysis with cotton protoplasts 12 h after transfection with the *avrb6* construct or a vector control. Approximately 39–50 million raw read pairs and 36–46 million cleaned read pairs were obtained for each sample, which corresponds to ~450 × coverage of 77,267 annotated protein-coding transcripts1. Using a twofold expression change and an adjusted *P* value <0.05 as cutoffs, and using the *G. raimondii* genome sequence as a reference (the tetraploid genome sequence was not available at the time), we identified 2,026 genes that were induced by Avrb6 (Figure 2d and Supplementary Data 2). To identify genes with potential EBEs for Avrb6, we used the TALE-NT 2.0 Target Finder tool to examine all gene promoter sequences in the *G. raimondii* genome, using the standard score ratio cutoff of 3.0 (Figure 2e). The list of genes displaying one or more putative EBEs was then intersected with the list of genes upregulated by Avrb6 to produce a list of candidate of Avrb6 targets (Table 2 and Supplementary Data 2). These candidates were ranked by the probability that the predicted EBE is functional by using a machine-learning algorithm based on the output of TALE-NT 2.0 and the genomic context of the predicted EBEs35. Interestingly, among the top candidates, Gorai.008G209000.1, a homologue of the SWEET sucrose

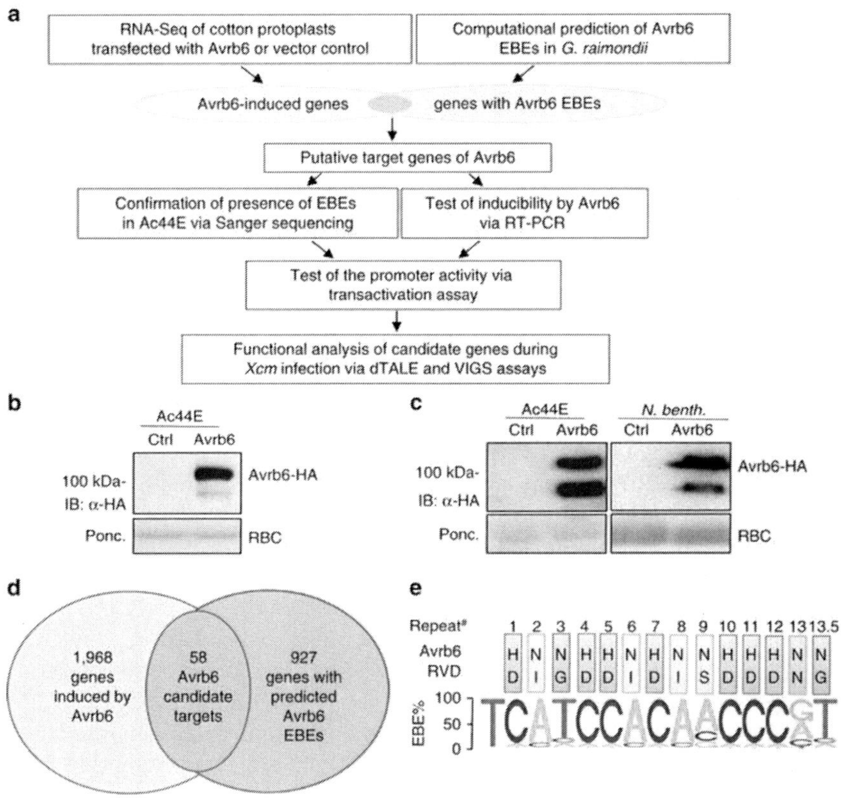

Figure 2: Transcriptome profiling coupled with EBE prediction reveals candidate target genes of Avrb6. (a) Schematic diagram of experimental design to identify Avrb6 target genes in cotton. (**b**) Avrb6 protein expression in cotton protoplasts. Cotton protoplasts isolated from Ac44E were transfected with *avrb6-HA* or an empty vector as a control (Ctrl). Samples were collected 12 h after transfection and subjected to immunoblotting with α-HA antibody (top panel). Ponceau S. staining (Ponc.) of total protein served as the protein loading control; RuBisCo (RBC) is shown (bottom panel). (**c**) Avrb6 protein expression in cotton and *N. benthamiana*. Cotyledons from 2-week-old Ac44E cotton and leaves of 4-week-old *N. benthamiana* were infiltrated with *Agrobacterium* carrying *35S::avrb6-HA* or an empty vector control (Ctrl). Inoculated tissues were collected at 48 hpi and subjected to immunoblotting. B and C were repeated three times with similar results. (**d**) Venn diagram of Avrb6-induced genes and genes with Avrb6 EBEs in *G. raimondii* genome. (**e**) The predicted EBE of Avrb6 RVDs. Coloured boxes on the top panel display the RVD of each repeat of Avrb6. Bottom panel indicates the relative frequencies of RVD associations with the four nucleotides for each repeat.

Table 2: The list of top 10 candidate target genes of Avrb6 in cotton based on their inducibility by Avrb6 and the probability of Avrb6 EBEs.

Gene ID	Gene name	Induction fold	EBE prob.	Annotation
Gorai.008G209000.1	*GhSWEET10D*	5973.1	1.0	Bidirectional sugar transporter N3
Gorai.007G067700.1	*Gh067700*	29.2	0.97	Unknown protein
Gorai.010G056300.1	*GhKBS1*	9.6	0.96	Kinase family protein
Gorai.009G327300.1	*GhHLH1*	4.0	0.89	Basic helix-loop-helix TF
Gorai.008G047400.1	*GhMDR1*	7.5	0.71	Multi-drug resistance ABC transporter
Gorai.010G246000.1	*Gh2460*	5.6	0.65	Unknown protein
Gorai.007G152300.1	*GhPPR1*	5.2	0.64	Pentatricopeptide repeat-containing protein
Gorai.008G222500.1	*GhCYP1*	471.6	0.59	Cytochrome P450
Gorai.001G131000.2	*GhZFP1*	12.3	0.51	Zinc finger family protein
Gorai.009G045100.2	*GhGLY1*	12.8	0.5	Glycine dehydrogenase

transporter genes that serve as *S* genes in rice and cassava, was induced ~6,000-fold by Avrb6 and has an EBE with a probability of 1 (Table 2). Other top candidates include Gorai.007G067700.1, Gorai.010G056300.1, Gorai.009G327300.1 and Gorai.008G047400.1. We then used the gene IDs from *G. raimondii* to identify their homologues in *G. hirsutum*. The gene IDs of the homologues identified in *G. hirsutum* are Gh_D12G1898 for Gorai.008G2 09000.1, Gh_D11G0631 for Gorai.007G067700.1, Gh_D06G0459 for Gorai.010G056300.1, Gh_D05G2954 for Gorai.009G32 7300.1 and Gh_D12G0420 for Gorai.008G047400.1. These genes were named *GhSWEET10D* (see below for explanation), *Gh067700, GhKBS1, GhHLH1* and *GhMDR1*, respectively, based on their predicted gene annotation.

Direct induction of targets by Avrb6

To independently determine the induction of the candidate genes, we performed reverse transcription (RT)–PCR assays using Ac44E cotton protoplasts transfected with Avrb6 or with an empty vector as control. The top five candidates, *GhSWEET10, Gh067700, GhKBS1, GhHLH1* and *GhMDR1*, were strongly induced by Avrb6 compared to the vector control (Figure 3a). We further confirmed induction in cotton leaves by inoculating *Xcm* strains with or without *avrb6*. HM2.2S is derived from H1005 with a deletion of at least seven *tal* genes including *avrb6*, while HM2.2S (*avrb6*) is HM2.2S carrying *avrb6* in plasmid vector pUFR127 (ref. 10). H1005Δ*avrb6* is an *avrb6* mutant derivative of *Xcm*H1005 (ref. 37). As previously reported, *avrb6* in different *Xcm* strains induced strong water-soaking on Ac44E cotton37 (Figure 3b). We observed that water-soaking was only induced by the *Xcm* strains carrying *avrb6* (HM2.2S (*avrb6*), HM2.2S (*placZ::avrb6*)—a strain that contains the plasmid pUFR135 which expresses *avrb6* driven by the *lacZ* promoter, H1005 (*placZ::avrb6*) and

H1005). As shown in Figure 3c, the induction of *GhSWEET10, Gh067700, GhKBS1* and *GhMDR1* by the strains carrying *avrb6*, HM2.2S (*avrb6*) and H1005, was stronger than that by strains lacking *avrb6*, HM2.2S and H1005Δ*avrb6*, at 12- and 24-h post inoculation (hpi). We could not amplify the *GhHLH1* gene in cotton likely due to low transcript level. To explore whether *GhSWEET10, Gh067700, GhKBS1* and *GhMDR1* are directly induced by Avrb6, we added the eukaryotic protein synthesis inhibitor cycloheximide (CHX; 50 μM final concentration) to the *Xcm* suspensions and inoculated into Ac44E cotton leaves. Each gene was still induced by strains carrying *avrb6*, H1005 and HM2.2S (*avrb6*), in the presence of CHX at 24 hpi (Figure 3d), suggesting that new protein biosynthesis is not required for induction of these genes and that they are direct targets of Avrb6. The effectiveness of CHX in suppressing protein synthesis in this context was confirmed by its suppression of Avrb6 protein expression in a separate *Agrobacterium*-mediated transient expression assay in Ac44E cotton (Supplementary Figure 2). To examine potential polymorphisms in these gene promoters between the sequenced cotton and Ac44E cotton used in this study, we amplified each promoter from Ac44E for sequencing. This revealed that each of the promoters possesses an Avrb6 EBE except *Gh067700*, which has a 1 bp deletion in the EBE in Ac44E (Supplementary Figure 3). To confirm whether Avrb6 activates the genes by targeting the EBEs, we amplified ~800 bp upstream of the translational start site of each gene, cloned each of these fragments upstream of a luciferase reporter gene (*LUC*) and performed a cotton protoplast-based transactivation assay (Figure 3e). Compared to an empty vector control, co-transfection of Avrb6 strongly activated *pGhSWEET10D::LUC* with about 110-fold induction, and also activated *pGhKBS1*:: *LUC* and *pGh MDR1::LUC* with about 15–20-fold induction (Figure 3f). Significantly, these promoters were not activated by PthN, another *Xcm* TAL effector with an RVD sequence distinct from Avrb6, indicating specificity of activation by Avrb6 (Figure 3f). Consistent with the observation that *Gh067700* in cotton Ac44E carries a nucleotide deletion in Avrb6 EBE, *pGh067700::LUC* was not activated by Avrb6. In addition, *pGhHLH1::LUC* was not significantly activated by Avrb6, though its promoter possesses a predicted Avrb6 EBE. Since the promoter of *GhSWEET10D* was strongly induced by Avrb6, we tested the role of the EBE by mutating two nucleotides to create a version of *pGhSWEET10D::LUC* with a 'CA' to 'GG' substitution in the EBE (*mEBE*; Figure 3g). Avrb6 strongly activated *pGhSWEET10D::LUC* carrying the wild-type EBE, but not the *mEBE* (Figure 3h). Taken together, these data indicate that Avrb6 directly and specifically activates *GhSWEET10D*, and likely *GhKBS1* and *GhMDR1*, and that the EBE is required for Avrb6-induced expression of *GhSWEET10D*.

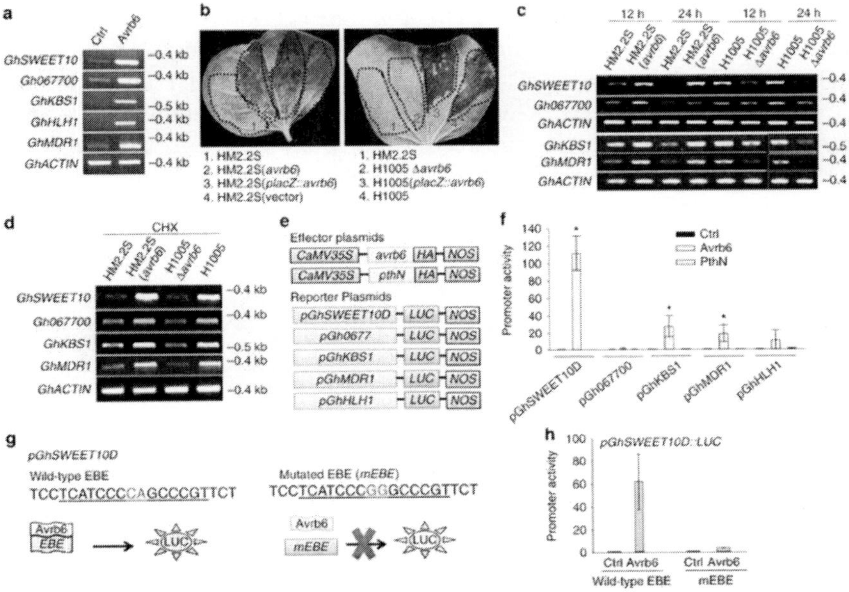

Figure 3: Avrb6 directly induces cotton gene transcription in the absence of protein synthesis. (**a**) RT–PCR analysis of Avrb6 upregulated genes in cotton protoplasts. Cotton protoplasts of Ac44E were transfected with *avrb6-HA* or a vector control (Ctrl) and incubated for 12 h before RNA isolation. *GhACTIN* was used as an internal control. (**b**) Avrb6 contributes to water-soaking development in cotton. Cotyledons from 2-week-old Ac44E cotton were syringe-inoculated with different *Xcm* strains at OD_{600}=0.1 and photographed 4 days after infiltration. Inoculation areas are indicated by dotted lines. (**c**) RT–PCR analysis of Avrb6 upregulated genes in cotton upon *Xcm* infection. Cotyledons from 2-week-old Ac44E cotton were syringe-inoculated with different *Xcm* strains. Tissues were collected at 12 and 24 hpi. (**d**) RT–PCR analysis of Avrb6 upregulated genes in the presence of CHX. Cotyledons from 2-week-old cotton were syringe-inoculated with different *Xcm* strains in 50 μM CHX. Tissues were collected at 24 hpi for RT–PCR. (**e**) Schematic diagram of the effector and reporter constructs. The reporter construct contains an expression cassette with a *LUC* reporter gene under the control of a candidate gene promoter. The effector construct contains either Avrb6 or PthN with an HA-epitope tag under the control of the CaMV *35S* promoter. (**f**) Transcriptional activity of Avrb6 in cotton protoplasts. Protoplasts were co-transfected with a reporter construct and *avrb6, pthN* or a vector control (Ctrl), and were collected 12 h after transfection. *UBQ10-GUS* was included in the transfections as an internal control. The luciferase activity was normalized with GUS activity. The data are shown as mean±s.d. (n=3) from three independent repeats. Asterisks indicate significant difference using two-

tailed *t*-test ($P<0.05$). (**g**) Schematic diagram of the transactivation assay of wild-type and mutated EBE (mEBE) in *pGhSWEET10D* in response to Avrb6. The nucleotide sequence containing the putative EBE is shown and the two nucleotides that were mutated are highlighted in blue. (**h**) Transcriptional activity of *GhSWEET10D* with wild type and mEBE in response to Avrb6. Cotton protoplasts were co-transfected with *pGhSWEET10D::LUC* carrying wild type or mEBE and Avrb6 or a vector control (Ctrl). The data are shown as mean±s.d. ($n=3$) from three independent repeats. The above experiments were repeated three times with similar results.

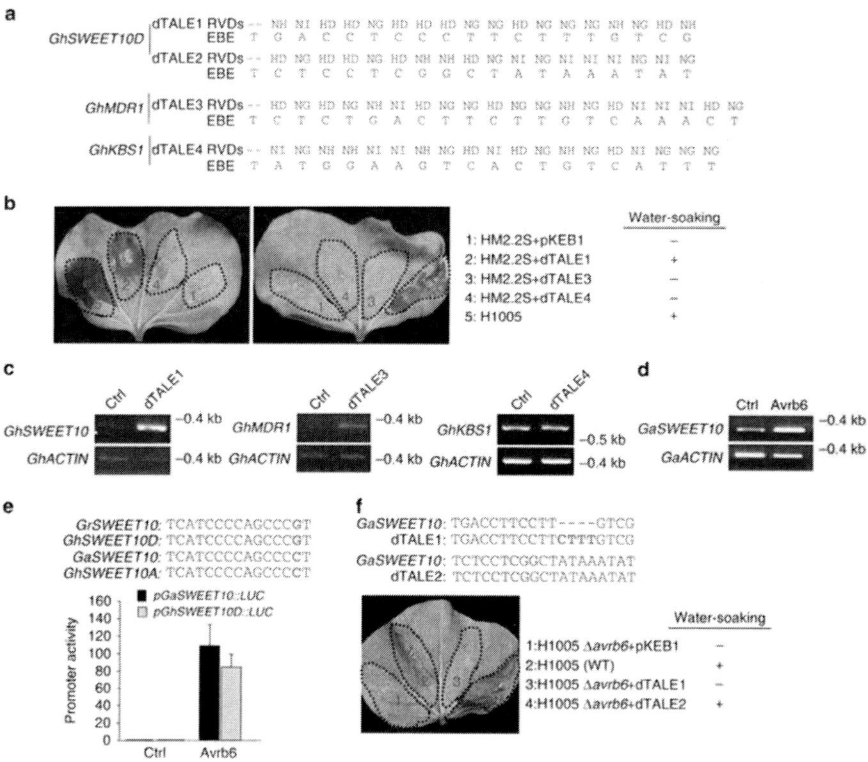

Figure 4: The dTALE-activating *GhSWEET10* induces water-soaking in cotton. (**a**) RVDs of dTALEs and the corresponding targeted EBE sequences in the *G. hirsutum* genome. (**b**) *Xcm* HM2.2S expressing a dTALE uniquely targeting *GhSWEET10D* induces water-soaking in cotton. Cotyledons from 2-week-old Ac44E plants were syringe-inoculated with different *Xcm* strains at $OD_{600}=0.1$ and photographed at 4 dpi. The table displays the presence (+) or absence (−) of water-soaking. (**c**) RT–PCR analysis of *GhSWEET10, GhKBS1* and *GhMDR1* targeted by their respective dTALEs. Two-week-old cotyledons of Ac44E were syringe-inoculated with *Xcm* strains HM2.2S ($OD_{600}=0.1$) containing different dTALEs.

Inoculated tissues were collected at 24 hpi for RT–PCR analysis. (**d**) *GaSWEET10* is induced by Avrb6 in *G. arboreum* protoplasts. Protoplasts isolated from *G. arboreum* were transfected with Avrb6 or an empty vector control (Ctrl) and samples were collected 12 h after transfection for RT–PCR analysis. (**e**) Transactivation assay of *GaSWEET10* and *GhSWEET10D* promoters in response to Avrb6 in cotton protoplasts. Alignment of Avrb6 EBEs reveals a polymorphism between *GhSWEET10D* and *GaSWEET10* promoters, which were highlighted in blue. Protoplasts isolated from commercial cotton variety FM706V were co-transfected with *pGaSWEET10::LUC* or *pGhSWEET10D::LUC* and Avrb6 or a vector control (Ctrl). The data are shown as mean±s.d. (n=3) from three independent repeats. (**f**) The dTALE matching the *GaSWEET10* promoter causes water-soaking on *G. arboreum*. The EBE sequences of two dTALEs targeted to different regions of the *GaSWEET10* promoter are shown. dTALE2 with an EBE sequence perfectly matching the *GaSWEET10* promoter but not dTALE1 with a partially matching EBE sequence restores *Xcm*-mediated water-soaking on cotton. Cotyledons from 2-week-old *G. arboreum* were syringe-inoculated with different *Xcm* strains at OD_{600}=0.1 and photographed at 4 dpi. The above experiments were repeated three times with similar results.

GhSWEET10 activation using a dTALE induces water-soaking

We generated dTALEs to specifically activate *GhSWEET10D*, *GhKBS1* or *GhMDR1* via EBEs distinct from the corresponding Avrb6 EBEs, introduced these into the *Xcm* HM2.2S strain, which lacks *avrb6*, and performed disease assays in cotton. The dTALEs were expressed from the low-copy, broad-host range vector, pKEB1, a derivative we constructed from pUFR047 (ref. 38) that carries a gateway cloning cassette (Figure 4a). Significantly, HM2.2S transformants carrying the dTALE targeting *GhSWEET10D* (dTALE1), but not those targeting *GhMDR1* (dTALE3) or *GhKBS1* (dTALE4), or the vector control, induced water-soaking in Ac44E cotton, similarly to *Xcm*H1005 (Figure 4b). To determine if the dTALEs used indeed activate the transcripts of their respective target genes, we performed RT–PCR analysis. As shown in Figure 4c, *GhSWEET10D* and *GhMDR1* were induced by the cognate dTALEs. These data indicate that activation of *GhSWEET10D* on *Xcm* infection contributes to water-soaking development, thus identifying *GhSWEET10D* as a relevant target of Avrb6 in cotton and an *S* gene. We did not detect the induction of *GhKBS1* by its dTALE, and cannot conclude whether GhKBS1 plays a role in BBC.

Avrb6 targets *GhSWEET10* in both A and D genomes of cotton

The tetraploid *G. hirsutum* has two subgenomes, A and D. The above studies of *GhSWEET10D* were mainly based on the D genome of *G. raimondii* and the corresponding D subgenome of *G. hirsutum.* We examined if Avrb6 also targets *SWEET10* in the A-genome of *G. arboreum*, which is closest to the progenitor species of the A subgenome of *G. hirsutum.* We identified a homologue of GhSWEET10D, named GaSWEET10, which bears 98% similarity to GhSWEET10D at the amino-acid level. We expressed Avrb6 in *G. arboreum* cotton protoplasts and determined whether *GaSWEET10* was induced by Avrb6. RT–PCR analysis showed that *GaSWEET10* was induced by Avrb6 (Figure 4d). To determine whether Avrb6 could specifically activate the promoter of *GaSWEET10*, we cloned the promoter, *pGaSWEET10*, upstream of the luciferase gene for a transactivation assay. As shown in Figure 4e, Avrb6 activated *pGaSWE ET10:: LUC* to a similar level as Avrb6 activation of *pGhSWEET10D: :LUC.* Notably, an alignment of the Avrb6 EBE from *pGhSWEET10D* with the one from *pGaSWEET10* revealed them to be identical except for a single nucleotide polymorphism near the 3′ end of the EBE (position 14, counting the initial thymine), where mismatches are known to be generally tolerated43. Additionally, an alignment of protein-coding sequences revealed that GrSWEET10 from *G. raimondii* is identical to D subgenome GhSWEET10D, whereas GaSWEET10 from *G. arboreum* is identical to A subgenome GhSWEET10A (Supplementary Figure 4). Similarly, the promoter sequence, including the EBE, of *pGhSWEET10A* from tetraploid Ac44E was more closely related to *pGaSWEET10* from *G. arboreum* versus *pGrSWEET10* from *G. raimondii* (Supplementary Figure 4b). We used the dTALE approach to determine if *GaSWEET10* is also a direct target of Avrb6 that causes water-soaking. As shown in Figure 4f, the dTALE2 construct, but not the dTALE1 construct specific to *pGhSWEET10D* or an empty vector introduced in H1005Δ*avrb6*-induced water-soaking in *G. arboreum.* Notably, H1005 induced relatively weak water-soaking on *G. arboreum* compared to *G. hirsutum*, likely due to *G. arboreum* carrying some strain-specific resistance as noted previously44. Taken together, these results indicate that Avrb6 can directly target SWEET10 in both the A and D genomes to contribute to the development of BBC.

GhSWEET10 silencing reduces water-soaking caused by *Xcm*

Having determined that *GhSWEET10* mediates Avrb6-induced water-soaking, we subsequently investigated whether *GhSWEET10* is genetically required for water-soaking during *Xcm* infection in cotton. We employed the *Agrobacterium*-mediated VIGS system to silence *GhSWEET10* and examined BBC development after subsequent inoculation of *Xcm*. The VIGS construct was designed to silence both *GhSWEET10A* and *GhSWEET10D* because of sequence conservation. There was no apparent difference in plant growth between the VIGS-*GhSWEET10* plants and plants inoculated with the GFP control (VIGS-Ctrl) up to 3 weeks after VIGS (Supplementary Figure 5). Three weeks after *Agrobacterium* infiltration, the silenced plants were vacuum infiltrated with *Xcm*H1005. Compared to the control, the VIGS-*GhSWEET10* plants consistently showed fewer water-soaked lesions after *Xcm*H1005 infection (Figure 5a,b). RT–PCR analysis confirmed reduced induction of *GhSWEET10* in VIGS-*GhSWEET10* plants compared to VIGS-Ctrl plants (Figure 5c). Taken together, the data indicate that silencing of *GhSWEET10* in cotton reduces the development of water-soaking following *Xcm* inoculation, providing genetic evidence that *GhSWEET10* is functionally required for normal BBC symptom development upon infection by *Xcm* strains carrying *avrb6*.

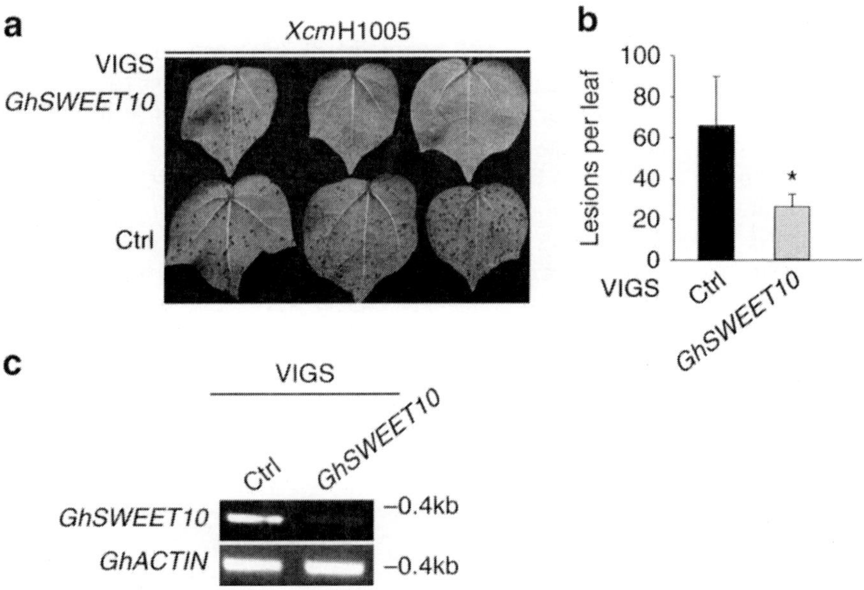

Figure 5: Silencing of *GhSWEET10* in cotton reduces water-soaking caused by *Xcm* infection. (**a**) *Xcm*-mediated water-soaking lesion development on cotton plants upon VIGS. Cotyledons from 2-week-old Ac44E plants were syringe-infiltrated with *Agrobacterium* carrying a VIGS construct to silence *GhSWEET10* (VIGS-*GhSWEET10*) or a GFP vector control (VIGS-Ctrl). Three weeks later, plants were vacuum infiltrated with *Xcm*H1005 at OD_{600}=0.01. Images were taken 2 weeks after inoculation. (**b**) Quantitative analysis of *Xcm*-mediated water-soaking lesions on cotton plants upon

VIGS. Lesions were counted on the second true leaf of each VIGS plant (at least 10 plants were inoculated for each construct in each repeat). The data are shown as mean±s.e. (n=10). The asterisk indicates a significant difference with a Student's t-test (P<0.05) when compared with the control. (**c**) RT–PCR analysis of *GhSWEET10* expression in cotton plants upon VIGS. VIGS assays were done similarly as in **a** and 3 weeks later, the second true leaf was collected for RT–PCR analysis before *Xcm* inoculation. *GhACTIN* was used as an internal control. The above experiments were repeated three times with similar results.

GhSWEET10D encodes a functional sucrose transporter

Since *GhSWEET10* belongs to the *SWEET* gene family, we examined whether GhSWEET10 functions as a sugar transporter. To measure activity, we transiently expressed the coding sequence of GhSWEET10D along with a Förster resonance energy transfer (FRET) sucrose sensor in human embryonic kidney T293 (HEK293T) cells45. Compared to the vector control, expression of GhSWEET10D enabled HEK293T cells to accumulate sucrose, which was detected as a negative ratio change, to a level comparable to the well-characterized *Arabidopsis* transporter AtSWEET11 (Figure 6a). In addition, when we transiently expressed *GhSWEET10D* under the control of the *35S* CaMV promoter in *N. benthamiana*, the sucrose concentration of the apoplastic fluids from the *35S::GhSWEET10D*-inoculated leaves was higher than that from the vector control-inoculated leaves (Supplementary Figure 6a), indicating that GhSWEET10D is able to transport sucrose *in planta*. Interestingly, *GhSWEET10D* did not restore the growth defect of yeast EBY4000, a hexose transport-deficient yeast strain, on the medium supplemented with 2% glucose or 2% fructose as the sole carbon source, indicating that GhSWEET10D does not mediate efficient fructose or glucose transport (Supplementary Figure 6b). Taken together, the data suggested that *GhSWEET10D* encodes a functional sucrose transporter.

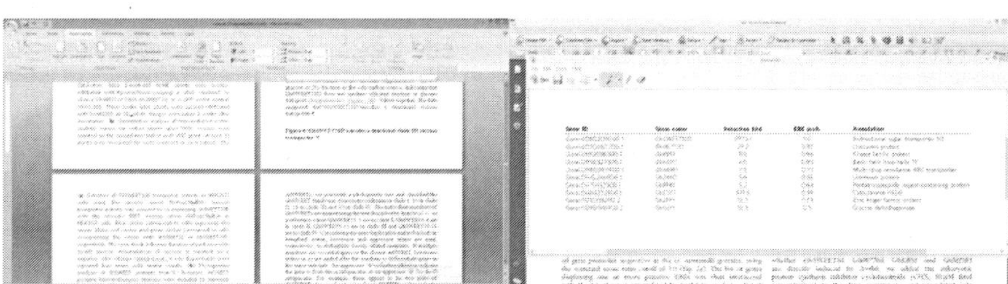

Figure 6: *GhSWEET10D* encodes a functional clade III sucrose transporter. (a) Detection of GhSWEET10D transporter activity in HEK293T cells using the sucrose sensor FLIPsuc90µδ3A. Sucrose transporter activity was assayed by co-expressing GhSWEET10D with the cytosolic FRET sucrose sensor FLIPsuc90µδ3A in HEK293T cells. Blue circles correspond to cells expressing the sensor alone; red circles and green circles correspond to cells co-expressing the sensor with AtSWEET11 or GhSWEET10D, respectively. The cyan block indicates duration of perfusion with 25 mM sucrose. Accumulation of sucrose is reported as a negative ratio change (mean−s.e.m.; n>10). Experiments were repeated four times with similar results. (**b**) Phylogenetic analysis of GhSWEET proteins from *G. hirsutum*. AtSWEET proteins from *Arabidopsis thaliana* were included to represent different clades of SWEET superfamily. The phylogenetic tree was made using the neighbour-joining method in MEGA version 6.06. The evolutionary distances were computed using the Poisson correction method and are in the units of the number of amino acid substitutions per site69. *GhSWEET10D*, which was used as the query, is highlighted with a brown box. (**c**) Phylogenetic analysis of GhSWEET10, GhSWEET14a, GhSWEET14b and the OsSWEET proteins from *Oryza sativa*. The phylogenetic tree was made as described in **b**. The clades of the SWEET family were colour coded as follows: clade I=pink, clade II=blue, clade III=black and clade IV=orange.

GhSWEET10D is a member of a large gene family in cotton

In bacterial blight of rice, several closely related *SWEET* genes can each serve as an *S* gene28. To better understand the relationship within the SWEET family in cotton, especially which SWEETs are most closely related to GhSWEET10, the full-length protein sequence of GhSWEET10D was used as the query for a BLASTp analysis against the *G. hirsutum* NBI protein database (www.cottongen.org/tools/blast). The search identified 50 additional GhSWEETs, for a total of 51 SWEETs in *G. hirsutum*, with each possessing the conserved domains found in the AtSWEET superfamily in *Arabidopsis* (Supplementary Figs 7 and 8). Based on the well-characterized phylogenetic analysis of AtSWEETs, we generated a phylogenetic tree and classified the GhSWEET family into four clades with nine in clade I, 14 in clade II, 18 in clade III and 10 in clade IV. The individual members of GhSWEETs are named using the tree described by Eom *et al.*46, as a reference where GhSWEET1-3 are in clade I, GhSWEET4-8 are in clade II, GhSWEET9-15 are in clade III and GhSWEET16-18 are in clade IV. Considering the gene duplication and polyploidy in tetraploid cotton, lowercase and uppercase letters are used, respectively, to distinguish closely related members. If multiple members are co-orthologous to the closest AtSWEET, lowercase letters (a–c) are added after the numbers to differentiate genes in the same subclade. An uppercase 'A' is further added to indicate the gene is from the A subgenome, or an uppercase 'D' for the D subgenome. For example, there appear to be two pairs of GhSWEETs that are closely related to AtSWEET1 as shown in the phylogenetic tree. Therefore, we distinguish these by naming one pair as GhSWEET1a and another pair as GhSWEET1b. Each of these pairs is further named according to their subgenome locations as GhSWEET1aA and GhSWEET1aD, and GhSWEET1bA and GhSWEET1bD, respectively (Figure 6b and Supplementary Figure 7). Like the *SWEET* genes in rice and cassava that serve as *S* genes, *GhSWEET10* is a member of the clade III family (Figure 6c).

GhSWEET10 induction by *Avrb6* is reduced in *b6* cotton

Avrb6 causes severe water-soaking in the susceptible Ac44E cotton line. However, in the near-isogenic resistant line Acb6, which harbours the apparently genetically complex BBC recessive resistance locus *b6*, Avrb6 causes the resistance-associated hypersensitive response (HR) instead (ref. 38 and Supplementary Figure 9a,b). We examined whether *b6*-mediated resistance is in part due to the inability of Avrb6 to activate *GhSWEET10* in Acb6. When we expressed Avrb6 in Ac44E and Acb6 protoplasts, respectively, the induction of *GhSWEET10* by Avrb6 was significantly lower in Acb6 than that in Ac44E (Figure 7a). We further compared *GhSWEET10* induction in Ac44E and Acb6 cotton plants inoculated with different *Xcm* strains with or without *avrb6*. Both Ac44E and Acb6 plants showed increased induction of *GhSWEET10* on inoculation with HM2.2S carrying *avrb6* compared to the plants inoculated with HM2.2S (Figure 7b). However, the induction was significantly reduced in Acb6 plants relative to Ac44E plants. A similar trend was observed when H1005 was used to inoculate both lines (Figure 7b). To examine whether the reduced induction by Avrb6 in Acb6 could be due to polymorphisms at the *GhSWEET10D* EBE, we sequenced the *GhSWEET10D* promoter region from Acb6. The EBE is perfectly conserved, but in Acb6, there is a SNP immediately adjacent to the 5′ of the EBE and a 2 bp insertion surrounded by six additional SNPs ~20 bp downstream of the EBE (Supplementary Figure 9c). We cloned the promoter of *GhSWEET10D* from Acb6 in front of the luciferase reporter for a transactivation assay in cotton protoplasts. Similar to what we observed for the endogenous *GhSWEET10D* gene, the induction of the *pGhSWEET10D*$_{Acb6}$*::LUC* reporter by Avrb6 was nearly three folds lower than that of the reporter carrying the promoter from Ac44E (*pGhSWEET10D*$_{Ac44E}$*::LUC*) (Figure 7c). The promoter of *GhSWEET10A* is almost identical in Ac44E and Acb6 (Supplementary Figure 4b). Thus, the polymorphisms outside the *GhSWEET10D* EBE may impair full induction by Avrb6. In addition, when we inoculated H1005Δ*avrb6 carrying* dTALE2, which activates *pGhSWEET10D* (Figure 4a,f), into cotyledons of Acb6, the bacterium triggered water-soaking, but not HR in Acb6 (Supplementary Figure 9d). The data suggest that *GhSWEET10D* may not mediate Avrb6-induced HR and Avrb6 targets another gene for HR in Acb6.

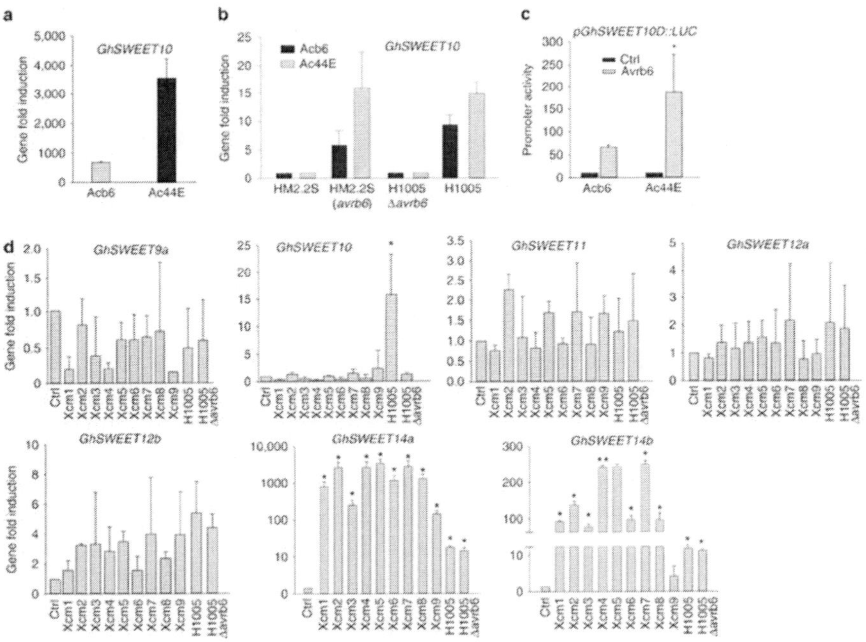

Figure 7: Induction of *GhSWEET* genes in different cotton–*Xcm*interactions. (**a**) *GhSWEET10* induction by Avrb6 in protoplasts of the resistant line Acb6 and the susceptible line Ac44E. Graph displays Avrb6-induced *GhSWEET10* reads per kilobase of transcript per million mapped reads and fold change from RNA-Seq analysis with Ac44E (compatible) and Acb6 (incompatible) cotton lines. (**b**) qRT–PCR analysis of *GhSWEET10* in Ac44E and Acb6 plants upon *Xcm* infections. Two-week-old cotyledons were syringe-inoculated with different *Xcm* strains at OD_{600}=0.1 and tissues were collected at 24 hpi. *GhUBQ1* was used as an internal control. The data are shown as mean±s.d. (*n*=3) from three independent repeats. (**c**) Transactivation assay of *pGhSWEET10D::LUC* from Acb6 and Ac44E in response to Avrb6 in cotton protoplasts. The promoter of *GhSWEET10D* was amplified from Acb6 and Ac44E, respectively, and fused with luciferase reporter. Protoplasts of cotton variety FM706V were co-transfected with the reporter construct and Avrb6 or an empty vector control (Ctrl). The data are shown as mean±s.d. (*n*=3) from three independent repeats. An asterisk indicates the significant difference using two-tailed *t*-test ($P<0.05$) between two reporter constructs after Avrb6 induction. (**d**) Induction of clade III *GhSWEET* genes by different *Xcm* isolates. Two-week-old

cotyledons were syringe-inoculated with different *Xcm* isolates (*Xcm1-Xcm9*) collected from different locations in Texas, USA at OD_{600}=0.1. Tissues were collected at 24 hpi. *GhUBQ1* was used as an internal control for qRT–PCR analysis. The data are shown as mean±s.d. (*n*=3) from three independent repeats. An asterisk indicates significant difference using two-tailed *t*-test ($P<0.01$) compared to the water control. The above experiments were repeated three times with comparable results.

Clade III *GhSWEET* responsiveness to different *Xcm* isolates

Clade III *SWEET* genes have been implicated in disease susceptibility in rice and cassava. We here identified a clade III *SWEET* gene in cotton, *GhSWEET10*, that encodes a sucrose transporter and that also plays a role in disease susceptibility. To examine how general the involvement of clade III *GhSWEET* genes in BBC is, we investigated the transcriptional changes of these genes in response to nine *Xcm* isolates collected from infected cotton fields from different locations in Texas. Notably, these isolates are suspected to be the causal agents of the recent rampant re-emergence of BBC in the United States. However, it has not been determined yet whether these isolates represent a new race. We found that the field isolates were unable to induce *GhSWEET10*. However, all of the field isolates highly induced *GhSWEET14a* (200–4,000-fold) (Figure 7d). In addition, all of the isolates, except for *Xcm9*, also significantly induced *GhSWEET14b* (20–300-fold). As expected, H1005, but not H1005Δ*avrb6*, induced *GhSWEET10* by 15-fold. Notably, *GhSWEET14a* and *GhSWEET14b* are clade III *SWEET* genes (Figure 6c). Taken together, these data suggest that in addition to *GhSWEET10*, *GhSWEET14a* and *GhSWEET14b* are potential *S* genes, and that activating clade III *GhSWEET* genes might be broadly important to the ability of different *Xcm* strains to cause BBC.

DISCUSSION

BBC caused by *Xcm* is among devastating diseases that limit cotton yields. Despite several TAL effectors being cloned from *Xcm*, no *R* or *S* genes have yet been molecularly identified, mainly because of the large and polyploidy nature of the cotton genome and limited molecular and genomic tools. The approach taken in this study successfully overcame these challenges to establish a role for SWEET-mediated susceptibility in BBC. *GhSWEET10* was strongly induced by Avrb6 during *Xcm* infection and was identified as an *S* gene for BBC. Characterization of BBC-resistant line Acb6 revealed sequence differences in the *GhSWEET10D* promoter, but not in the EBE, that impair activation by Avrb6, possibly relating to resistance mediated by the *b6* gene. Finally, surveys of different *Xcm* field isolates showed that two additional clade III *GhSWEETs* were highly induced during infection. It appears that the evolution of novel TAL effectors that induce these additional clade III *GhSWEET* members may have contributed to the recent re-emergence of BBC in the United States.

The *SWEET* family members are key *S* genes common to several diseases caused by different *Xanthomonas* species47. So far, only clade III *SWEET* genes are targeted by pathogens that use TAL effectors to facilitate disease46,48. Members of clade III *SWEET* genes are known to transport sucrose, which likely serves as a carbon source for the pathogen. Interestingly, it was recently shown that two clade III *SWEET* genes in *Arabidopsis, AtSWEET13* and *AtSWEET14*, can also transport plant hormone gibberellin to modulate growth and development49. In rice, *Xoo* can target three clade III *SWEET* genes to cause susceptibility. *Xoo* TAL effector PthXo1 targets *OsSWEET11* (ref. 20); PthXo2 targets *OsSWEET13* (ref. 27); and AvrXa7 targets *OsSWEET14* (ref. 23). In cassava, *Xam* can use TAL20$_{Xam668}$ to target *MeSWEET10a*, another clade III *SWEET* gene25. In at least one case, a host *SWEET* gene is targeted by multiple TAL effectors from different strains of the pathogen: in addition to AvrXa7, *OsSWEET14* is targeted by PthXo3, TalC and Tal5, each from a different *Xoo* strain23,28,50. The central role of SWEETs in susceptibility is further supported by our data, as we have identified *GhSWEET10* as a clade III *SWEET* gene in cotton targeted by Avrb6. In addition, we observed that several different *Xcm* field isolates strongly induce two additional clade III *GhSWEETs*, *GhSWEET14a* and *GhSWEET14b*, but not *GhSWEET10*, during infections. Since *Xcm*H1005 is a strain derived from an older race, and since these field isolates were collected within the past year, this strongly suggests that new TAL effectors have evolved in these field isolates to target new *SWEETs* and break resistance. The striking differences of the *tal* gene content and the localization of the *pthN2* gene on chromosome or plasmid in *Xcm*H1005 and *Xcm*N1003 are consistent with the notion that TAL effectors in *Xcm* may evolve rapidly. However, it is also possible that TAL effectors from these recent strains can activate another target, either another *SWEET* gene or a gene in different family, to cause susceptibility. It was revealed that TalC from *Xoo* can still cause disease on rice, despite having its corresponding EBE from *OsSWEET14* artificially mutated, suggesting that preventing TalC-mediated *SWEET14* activation alone is not sufficient to prevent susceptibility51. Additionally, even though *GhSWEET10*, *GhSWEET14a* and *GhSWEET14b* reside in clade III, it is still unclear if they function similarly. Additional biochemical analyses are needed to further understand the function of these genes in BBC.

Characterizing the spectrum of SWEET-targeting TAL effectors and the relative contributions of their targets to susceptibility, both in the older strains and the recent isolates, is an important future objective.

The CRR of TAL effectors determines their DNA-binding specificity for targeted gene activation in disease or resistance. Inventory of the repeat sequences is essential to predict important TAL effector targets. However, the repeats are generally impossible to assemble from Illumina-sequencing reads and challenge remains even with Sanger sequencing. PacBio sequencing has proven to be an effective approach however[52,53,54]. We used PacBio sequencing to reveal the whole repertoire of TAL effectors in *Xcm*H1005 and *Xcm*N1003. We identified 12 TAL effectors encoded by H1005 (six in the chromosome and six in the plasmid) and 9 encoded by N1003 (five in the chromosome and four in the plasmid). In addition to *GhSWEET* genes, another shared class of targets across TAL effectors of different *Xanthomonas* species is transcription factor genes[19,31,55]. In at least one case, the CsLOB1 transcription factor of citrus, and other members of CsLOB1 family also function as *S* genes[31,56]. It will be interesting to determine whether targets of other TAL effectors in H1005 and N1003 similarly include transcription factor genes, or yet novel types of targets.

In addition to revealing a host factor essential for disease symptom development, our study yields insight into the nature of BBC resistance mediated by *b6*. Initially characterized as recessive[57], genetic linkage mapping analysis suggests that the genetic basis of *b6*-resistance is rather complex and exhibits a quantitative inheritance. In rice, a mechanism of recessive resistance is disruption of TAL effector-dependent activation of an *OsSWEET S* gene due to polymorphism at the EBE[58]. Our data revealed that the induction of *GhSWEET10* by Avrb6 in Acb6 is compromised likely not due to the sequence polymorphism at the EBE, suggesting a novel mechanism underlying *b6*-mediated resistance. To our knowledge, this is the first report of polymorphisms outside the EBE itself that impair activation by a TAL effector. The *b6*-mediated resistance is accompanied by the HR at the site of infection, and this is triggered specifically in the presence of Avrb6 (refs 10, 38). However, a dTALE-activating *GhSWEET10* only results in water-soaking, but not HR in Acb6, suggesting that GhSWEET10 does not mediate Avb6-induced HR. This is likely that HR is due to the activation of another target by Avrb6 in Acb6, but not in Ac44E. Thus, it seems possible that *b6* comprises at least two genetic determinants, one manifesting in reduced susceptibility (lower expression of the *GhSWEET10 S* gene) and the other mediating the defence-associated HR. Perhaps both need to be present to support the full level of resistance.

Nevertheless, our findings suggest a strategy to improve BBC resistance in cotton through genome editing of TAL effector EBEs in *SWEET* genes.

Genome editing has been successfully used in rice to mutate the *OsSWEET13 S* gene and generate plants resistant to *Xoo* strains that depend on TAL effector PthXo2 as a major virulence factor27. An advantage of this strategy is that relatively minor sequence changes could be made that will destroy the EBE while being unlikely to impinge on native promoter activity and without affecting the coding region, keeping endogenous function of the *SWEET* gene intact51,59. The successful knockout of all homeologs of the *Mlo* gene in hexaploid wheat in a single genome editing experiment indicates that disrupting EBEs in *SWEET* genes in cotton is likely feasible60. As new TAL effectors important for virulence of this pathogen are characterized in the future, this system could be used to edit the EBEs in the corresponding targets, whether *GhSWEET* genes or other BBC *S* genes yet to be discovered, and to generate plants with broad-spectrum BBC resistance via pyramiding. Not least, duplicating the Acb6 genotype of *GhSWEET10* in the Ac44E background and in other susceptible genotypes would be a powerful approach to dissect the genetic determinants of *b6*-mediated resistance, potentially by uncoupling loss of susceptibility from the HR. Such characterizations would also likely lead to new strategies for improved control of BBC.

METHODS

Plants and bacterial strains and growth conditions

The *Gossypium hirsutum* lines Acala44E (Ac44E), Acalab6 (Acb6) and FM706V, and *G. arboreum* line SA-1415 were grown in 3.5-inch square pots containing Metro Mix 900 soils (Sun Gro Horticulture, Agawam, MA) and tobacco *Nicotiana benthamiana* were grown in 3.5-inch square pots containing Metro Mix 366 soils in a growth chamber at 23 °C, 30% humidity and 100 μE $m^{-2}s^{-1}$ light with a 12-h light/12-h dark photoperiod. Two-week-old cotton plants were used for *Agrobacterium*-mediated VIGS assay, *Agrobacterium*-mediated transient protein expression assay or protoplast isolation. Plants after inoculation with *Xanthomonas citri* subsp. *malvacearum* (*Xcm*) were transferred into a growth chamber at 28 °C, 50% humidity and 100 μE $m^{-2}s^{-1}$ light with a 12-h light/12-h dark photoperiod. Bacterial strains used in this study are listed in Supplementary Table 3. *E. coli* and *A. tumefaciens* strains were grown in Luria–Bertani (LB) medium with appropriate antibiotics at 37 °C or 28 °C, respectively. *Xcm* strains were grown in nutrient broth medium with appropriate antibiotics at 28 °C. The *Xcm* field strains were isolated from infected cotton fields in Texas through Koch's postulates with modifications. Water-soaked lesions were excised from infected cotton leaves using a sterile scalpel. The cut lesions were soaked and macerated in 100 µl of sterile water. The bacterial suspension was plated on nutrient agar from serial dilutions. Colonies were re-streaked and inoculated into cotton to confirm pathogenicity.

Xcm inoculations

The *Xcm* strains were inoculated into 4 ml of nutrient broth medium with appropriate antibiotics for overnight. For syringe infiltration, bacterial suspensions adjusted to OD_{600}=0.1 were inoculated into expanded cotyledons of 2-week-old plants or true leaves using a needleless syringe. Small holes were created with a 25G needle on the underside of the cotyledons to facilitate infiltration. For vacuum infiltration, a final bacterial suspension was made by adjusting the bacterial cultures to OD_{600}=0.01 mixed with 0.04% of Silwet L-77 surfactant solution. Plants that were silenced via VIGS 3 weeks earlier were submerged into the bacterial suspension inside a desiccator connected to a vacuum pump. The plants were vacuum infiltrated for 5 min at 76 mmHg.

DNA isolation for *Xcm* whole-genome sequencing

The *Xcm*H1005 and *Xcm*N1003 DNA used for PacBio sequencing was prepared using a protocol for total genomic DNA isolation with modifications54. Bacteria were cultured overnight in 30 ml glucose yeast extract media (2% glucose, 1% yeast extract) in a 250 ml flask at 28 °C on a rotary shaker at 250 rpm, collected by centrifugation at 2,580*g* for 10 min at 4 °C, then gently resuspended and washed in 20 ml NE buffer (0.15 M NaCl, 50 mM EDTA) twice to remove the extracellular polysaccharide. Cells were then gently resuspended in 2.5 ml solution containing 50 mM Tris, pH 8.0 and 50 mM EDTA, and then 0.5 ml solution containing 25 mM Tris, pH 8.0, 10 µl ReadyLyse (Epicentre) and 50 µl RNase (10 mg ml^{-1}). Suspensions were mixed gently by inversion and then incubated on ice for 45 min. Following incubation on ice, 1.0 ml STEP buffer (0.5% SDS, 50 mM Tris, pH 7.5, 40 mM EDTA, protease K at 2 mg ml^{-1}) was added, and the lysate was mixed well by inversion and incubated at 37 °C for 1 h, mixing every 10–15 min. Next, 1.8 ml of 7.5 M ammonium acetate was added and the lysates were mixed gently before being subjected to two extractions with phenol/chloroform (10 ml) and one extraction with chloroform/isoamyl alcohol (24:1, pH 8.0, 10 ml), shaking vigorously by hand to mix and separating the aqueous and organic phases by centrifugation each time at 10,300*g* for 10 min at 4 °C. Following this, the aqueous phase was transferred to a 14 ml tube and DNA was precipitated by addition of two volumes of cold 95% ethanol and gentle, repeated inversion. Once solidified, the DNA was transferred to a 2 ml micro-centrifuge tube using a Pasteur pipette with the tip previously sealed and bent into a hook over a flame. Remaining liquid was then removed by centrifugation at 825*g* for 5 min and the pellet was washed once with 70% ethanol. The remaining liquid was removed as before and the tube left open to dry in a laminar flow hood until the edges of the pellet became glossy in appearance (10–15 min). Finally, the pellet was dissolved in 100 ml TE buffer (10 mM Tris/HCl, 1 mM EDTA, pH 8.0) overnight at 4 °C and then adjusted to a concentration of 1 μg ml^{-1}.

PacBio sequencing and genome assembly

DNA library preparation and sequencing was performed according to the manufacturer's instructions by the Genomics Core Facility of the Icahn School of Medicine at Mt. Sinai (New York, NY, USA)54. Separately, DNA preparation and electrophoresis to check for small plasmids in either strain that would be lost during library preparation was using standard methods50. Sequencing was conducted to >120 × coverage, which due to improvements in sequencing technology required only two SMRT cells per strain. In all data sets, read-length distribution showed a fat tail, with 20% of coverage after adaptor removal contained in subreads greater than or equal to 15,000 bp. The whole genomes of H1005 and N1003 were assembled using HGAP3 (ref. 61) and verified by carrying out local assemblies of reads containing *tal* genes and then comparing those assemblies to the whole-genome assembly as described in Booher *et al.*54. The assemblies conclusively match published Southern blot results using a *tal* gene probe, as well as restriction enzyme mapping and sequencing or partial sequencing of cosmid clones and subclones10,38,39,40.

The initial assembly of *Xcm*H1005 showed frameshifts in *avrB104*, *avr103* and *avrb6* resulting from an extra cytosine (C) in a stretch of six Cs commonly found in the 3′ region of *tal* genes. Since *avrb6* is functional, the previously published sequence of *avrb6* shows an intact reading frame38, and residual errors in homopolymer runs sometimes persist after consensus calling by Quiver and are a known weak point of the error-correction software (https://github.com/PacificBiosciences/GenomicConsensus/blob/master/doc/FAQ.rst), we concluded that the frameshift in *avrb6* was such an error, and we corrected it in the final assembly. Likewise, since clones containing *avrB104* and *avrB103* were functional, conferring avirulence on *AcB5* and *AcBIn3*-containing cotton40, we concluded that the assembly contained the same error for each of these genes, and corrected them both. No other *tal* genes in the assembly harboured this or another frameshift within a homopolymer stretch. Note that the frameshift in *pthN′*, retained in the final assembly, is due to a deletion of a single thymine, not within a homopolymer, and *pthN'* was shown to be non-functional10.

Construction of avrb6 in the expression vectors

Avrb6 was amplified from *Xcm*H1005 plasmid DNA with primers containing restriction enzymes SpeI at the amino (N) terminus and SmaI at the carboxy (C) terminus (Supplementary Table 4) and ligated into a plant protoplast expression vector *pHBT* with a *CaMV 35S* promoter at the N terminus and a HA-epitope tag at the C terminus. To construct the *pCB302* binary vector containing *avrb6* for *Agrobacterium*-mediated transient expression assay in *N. benthamiana*, *avrb6* was released from the *pHBT* vector through digestion with SpeI and SmaI and ligated into the *pCB302* binary vector. All the clones were confirmed by sequencing.

Transient expression assays

Cotton protoplasts were isolated from 2-week-old cotyledons. Detached cotyledons were cut with a razor blade and digested in an enzyme solution (1.5% cellulose R10, 0.4% macerozyme R10, 0.4 M mannitol, 20 mM KCl, 20 mM MES pH 5.7) supplemented with 2% sucrose for 0.5 h under vacuum. Subsequently, the enzyme solution was incubated without vacuum at room temperature for 12 h. Protoplasts were released by filtering through a 30 μm-nylon mesh, washed with W5 solution (154 mM NaCl, 125 mM $CaCl_2$, 5 mM KCl, 2 mM MES pH 5.7) and diluted in MMG solution (0.4 M mannitol, 15 mM $MgCl_2$, 4 mM MES pH 5.7) to a density of 2×10^5 cells ml^{-1} (ref. 62). A measure of 100 μl of protoplasts for each sample were collected for western blotting after expressing *avrb6-HA* for 12 h. Avrb6-HA protein was detected by immunoblotting using the α-HA antibody. *Agrobacterium tumefaciens* strain GV3101 containing *pCB302-avrb6-HA* vector or an empty vector was cultured overnight in LB medium containing appropriate antibiotics at 28 °C. Bacteria were collected by centrifugation and resuspended with infiltration buffer (10 mM MES, pH 5.7, 10 mM $MgCl_2$, 200 μM acetosyringone) at OD_{600}=0.6. Leaves of 4-week-old soil-grown *N. benthamiana* or cotyledons of 2-week-old Ac44E cotton were hand infiltrated using a needleless syringe with *Agrobacterium* cultures. Leaf samples were collected 48 h after infiltration for protein sample and immunoblot analysis. Full blots are provided in Supplementary Figure 10.

Cotton RNA isolation for RT–PCR and qRT–PCR analysis

Cotton total RNA was extracted from protoplasts (500 μl protoplasts with a cell density of 2×10^5 per ml for each sample) using TRIzol reagent

(Invitrogen, USA) or leaves (one cotyledon or one true leaf for each sample) using a Spectrum plant total RNA kit (Sigma-Aldrich, USA) according to the manufacturer's protocol. RNA was then treated with DNase (Invitrogen, USA) to remove genomic DNA. Purified RNA (1 μg) was subsequently reverse transcribed using a first-strand cDNA synthesis kit (Promega, USA) with oligo(dT) as a primer. PCR amplification was performed for 30 cycles of 15 s at 98 °C, 30 s at 55 °C and 45 s at 72 °C with primers listed in Supplementary Table 4. *GhACTIN* was used as an internal control for RT–PCR. qRT–PCR analysis was performed using iTaq SYBR green Supermix (Bio-Rad, USA) with a Bio-Rad CFX384 Real-Time PCR System. Expression of each gene was normalized to the expression of *GhUBQ1*.

RNA sequencing and data analysis

A measure of 500 μl of cotton protoplasts (2×10^5 per ml) were transfected with *35S::avrb6-HA* or an empty vector control. Protoplasts were collected 12 h after transfection and RNA was extracted by the TRIzol reagent. Three independent biological repeats were performed for the RNA-Seq analysis. For each repeat, equal amount of RNA from two biological replicates was pooled for RNA-Seq library construction. RNA-Seq libraries were prepared using Illumina TruSeqTM RNA Sample Preparation Kit following the manufacturer's instruction, and sequenced on an Illumina HiSeq 2000 platform with 125-nucleotide pair-end reads at Texas AgriLife Genomics and Bioinformatics Service (College Station, TX). RNA-Seq read processing, alignment to the *G. raimondii* genome and differential gene expression analysis were performed as described in Li *et al.*63. Genes with expression fold change ≥2 and *P* value <0.05 were considered as significantly different between samples with and without Avrb6.

EBE prediction

All TAL effector EBE predictions were conducted using the TALE-NT 2.0 Target Finder tool64 to search the promoter regions defined in the *G. raimondii* genome as the 1,000 bp upstream of the transcriptional start site plus the 5′ UTR, if annotated. We identified promoter sequences for *G. raimondii* using genome assembly v2.0 and annotation v2.1 (ref. 3). Binding sites passing the standard target finder score ratio cutoff of 3.0 were ranked based on both target finder output and genomic context using the machine-learning classifier presented by Cernadas *et al.*35 and updated by Wilkins *et al.*65.

Construction of VIGS vector and *Agrobacterium*-mediated VIGS

GhSWEET10 was amplified by PCR from cDNA of *G. hirsutum* with primers described in Supplementary Table 4, and inserted into the pYL156 (pTRV-RNA2) vector with restriction enzymes EcoRI and KpnI, resulting in pYL156-*GhSWEET10*. The *Agrobacterium tumefaciens* strain GV3101 carrying the plasmids pTRV-RNA1 and pYL156 (pTRV-RNA2) or pYL156-*GhSWEET10* was collected and resuspended with infiltration buffer (10 mM MES, pH 5.7, 10 mM $MgCl_2$, 200 μM acetosyringone) at OD_{600}=0.75. The culture suspension of pTRV-RNA1 was mixed with pYL156 or pYL156-*GhSWEET10* at a 1:1 ratio and hand infiltrated using needleless syringes into two fully expanded cotyledons of 2-week-old plants66.

GhSWEET10D transporter analysis in HEK293T cells

HEK293T cells were co-transfected with FRET sucrose sensor FLIPsuc90μδ3A and *GhSWEET10D* in six-well plates and perfusion experiments were performed48. After 3.5 min buffer perfusion, cells were perfused with 25 mM sucrose for 4 min followed by 2.5 min buffer perfusion. Images were captured at the following settings: exposure time 800 ms, gain 3, binning 2, time interval 10 s. Fluorescence intensity of selected individual cells was recorded. The quantitative ratio of eCFP and Aphrodite emission at excitation of eCFP was calculated after the background was subtracted. All of ratios were normalized to the initial ratio.

Yeast mutant EBY4000 complementation growth assay

The yeast strain EBY4000 was transformed with the different plasmids following the standard protocol. Spotting assay with 10-fold serial dilutions was performed. Yeast cells were diluted to OD_{600}=0.22 in water after cells grew to OD_{600}=0.6, followed by a twofold dilution. Then, 10-fold serial dilutions were plated on synthetic medium containing either 2% maltose (as a control), 2% fructose or 2% glucose. Growth was documented by Gel-doc system after 2–4 days at 30 °C.

Transactivation assays

The reporter constructs, *pGhSWEET10D::LUC, pGh067700:: LUC, pGhKBS1:: LUC, pGhMDR1::LUC, pGhHLH1::LUC* and *pGaSWEET10::LUC,* were constructed by amplifying the promoters of candidate genes (~800 bp upstream of translational start site) from genomic DNA of Ac44E, Acb6 or *G. arboreum* and were fused to a luciferase gene at the C terminus. Protoplasts were transfected with the effector and reporter construct and expressed at room temperature for 12 h. Protoplasts were then collected and lysed and the luciferase activity was detected by Glomax Multi-Detection System (Promega, USA) with the luciferase assay substrate (Promega, USA). *UBQ10-GUS* was included in all samples as the internal transfection control and the GUS activity was analysed with a Multilabel Plate Reader (Perkin-Elmer, USA). The *pGhSWEET10D-mEBE::LUC* construct was generated by the site-directed mutagenesis PCR using plasmid DNA of *pGhSWEET10D::LUC* as the template and primers are listed in Supplementary Table 4.

Construction of dTALEs

The dTALEs were constructed using golden gate cloning using a complete plasmid kit available through AddGene67. Four types of repeats encoding the RVDs NI, NN, NG and HD that correspond to the respective nucleotide A, G, T and C were used to assemble the repeat domains of the artificial dTALEs. The dTALEs were transformed into *Xcm* stains via bacterial tri-parental mating by using a helper plasmid, pRK2073 (ref. 38). In addition, the modifier plasmid pUFR054 was used to increase the efficiency of transferring the plasmids from *E. coli* DH5α to *Xcm* strains. The bacterial mating was carried out by mixing the donor, recipient, modifier and helper strains in a 1:1:1:1 ratio. The mixed culture was plated onto nutrient agar medium with appropriate antibiotics. Colonies were confirmed via Sanger sequencing of extracted plasmid DNA.

Measurement of sucrose concentration

Leaves from 4-week-old *N. benthamiana* plants were infiltrated with *Agrobacterium* carrying *35S::GhSWEET10D-HA* or an empty vector control at OD_{600}=1.0. Leaves were detached 3 dpi and fully vacuum infiltrated with sterile Millipore water. The leaves were blot dried, cut into small pieces with a razor blade and placed in a nylon mesh strainer at the top of a 50 ml centrifuge tube. The apoplastic solution was collected through centrifugation at 3,220*g* for 20 min at 4 °C. The concentration of sucrose was determined by using a sucrose assay kit (Sigma-Aldrich).

Data availability

The *Xcm*H1005 genome assembly has been deposited in GenBank under accessions CP013004 (chromosome) and CP013005 (pXcmH) and the XcmN1003 assembly under accessions CP013006 (chromosome) and

CP013007 (pXcmN). Raw data as bas.h5 and bax.h5 files and associated metadata are available in the NCBI sequence read archive (SRA) under SRP065335 for *Xcm*H and SRP065336 for *Xcm*N. The RNA-Seq data were deposited in the NCBI SRA under accession number SRP100553. The authors declare that all other data supporting the findings of this study are included within the manuscript and its supplementary files or are available from the corresponding authors on request.

ADDITIONAL INFORMATION

How to cite this article: Cox, K. L. *et al.* TAL effector driven induction of a *SWEET* gene confers susceptibility to bacterial blight of cotton. *Nat. Commun.* **8,** 15588

Publisher's note: Springer Nature remains neutral with regard to jurisdictional claims in published maps and institutional affiliations.

REFERENCES

1. Paterson, A. H. et al. Repeated polyploidization of Gossypium genomes and the evolution of spinnable cotton fibres. Nature 492, 423–427 (2012).
2. Wendel, J. F. New world tetraploid cottons contain old world cytoplasm. Proc. Natl Acad. Sci. USA 86, 4132–4136 (1989).
3. Wang, K. et al. The draft genome of a diploid cotton Gossypium raimondii. Nat. Genet. 44, 1098–1103 (2012).
4. Li, F. et al. Genome sequence of the cultivated cotton Gossypium arboreum. Nat. Genet. 46, 567–572 (2014).
5. Li, F. et al. Genome sequence of cultivated upland cotton (Gossypium hirsutum TM-1) provides insights into genome evolution. Nat. Biotechnol. 33, 524–530 (2015).
6. Zhang, T. et al. Sequencing of allotetraploid cotton (Gossypium hirsutum L. acc. TM-1) provides a resource for fiber improvement. Nat. Biotechnol. 33, 531–537 (2015).
7. Liu, X. et al. Gossypium barbadense genome sequence provides insight into the evolution of extra-long staple fiber and specialized metabolites. Sci. Rep. 5, 14139 (2015).
8. Yuan, D. et al. The genome sequence of Sea-Island cotton (Gossypium barbadense) provides insights into the allopolyploidization and development of superior spinnable fibres. Sci. Rep. 5, 17662 (2015).
9. Delannoy, E. et al. Resistance of cotton towards Xanthomonas campestris pv. malvacearum. Annu. Rev. Phytopathol. 43, 63–82 (2005).
10. Chakrabarty, P. K., Duan, Y. P. & Gabriel, D. W. Cloning and characterization of a member of the Xanthomonas avr/pth gene family that evades all commercially utilized cotton R genes in the United States. Phytopathology 87, 1160–1167 (1997).
11. Bogdanove, A. J., Schornack, S. & Lahaye, T. TAL effectors: finding plant genes for disease and defense. Curr. Opin. Plant Biol. 13, 394–401 (2010).
12. Boch, J. & Bonas, U. Xanthomonas AvrBs3 family-type III effectors: discovery and function. Annu. Rev. Phytopathol. 48, 419–436 (2010).

13. Mak, A. N. et al. The crystal structure of TAL effector PthXo1 bound to its DNA target. Science 335, 716–719 (2012).
14. Deng, D. et al. Structural basis for sequence-specific recognition of DNA by TAL effectors. Science 335, 720–723 (2012).
15. Schreiber, T. & Bonas, U. Repeat 1 of TAL effectors affects target specificity for the base at position zero. Nucleic Acids Res. 42, 7160–7169 (2014).
16. Doyle, E. L. et al. TAL effector specificity for base 0 of the DNA target is altered in a complex, effector- and assay-dependent manner by substitutions for the tryptophan in cryptic repeat -1. PLoS ONE 8, e82120 (2013).
17. Marois, E., Van den Ackerveken, G. & Bonas, U. The Xanthomonas type III effector protein AvrBs3 modulates plant gene expression and induces cell hypertrophy in the susceptible host. Mol. Plant Microbe Interact. 15, 637–646 (2002).
18. Römer, P. et al. Plant pathogen recognition mediated by promoter activation of the pepper Bs3 resistance gene. Science 318, 645–648 (2007).
19. Kay, S. et al. A bacterial effector acts as a plant transcription factor and induces a cell size regulator. Science 318, 648–651 (2007).
20. Yang, B., Sugio, A. & White, F. F. Os8N3 is a host disease-susceptibility gene for bacterial blight of rice. Proc. Natl Acad. Sci. USA 103, 10503–10508 (2006).
21. Chu, Z. H. et al. Promoter mutations of an essential gene for pollen development result in disease resistance in rice. Genes Dev. 20, 1250–1255 (2006).
22. Yuan, M. et al. Pathogen-induced expressional loss of function is the key factor in race-specific bacterial resistance conferred by a recessive R gene xa13 in rice. Plant Cell Physiol. 50, 947–955 (2009).
23. Antony, G. et al. Rice xa13 recessive resistance to bacterial blight is defeated by induction of the disease susceptibility gene Os-11N3. Plant Cell 22, 3864–3876 (2010).
24. Kay, S. et al. Detailed analysis of the DNA recognition motifs of the Xanthomonas type III effectors AvrBs3 and AvrBs3Deltarep16. Plant J. 59, 859–871 (2009).
25. Cohn, M. et al. Xanthomonas axonopodis virulence is promoted by a transcription activator-like effector-mediated induction of a SWEET sugar transporter in cassava. Mol. Plant Microbe Interact. 27, 1186–1198 (2014).
26. Huang, S. et al. The broadly effective recessive resistance gene xa5 of rice is a virulence effector-dependent quantitative trait for bacterial blight. Plant J. 86, 186–194 (2016).
27. Zhou, J. et al. Gene targeting by the TAL effector PthXo2 reveals cryptic resistance gene for bacterial blight of rice. Plant J. 82, 632–643 (2015).
28. Streubel, J. et al. Five phylogenetically close rice SWEET genes confer TAL effector-mediated susceptibility to Xanthomonas oryzae pv. oryzae. New Phytol. 200, 808–819 (2013).
29. Romer, P. et al. Promoter elements of rice susceptibility genes are bound and activated by specific TAL effectors from the bacterial blight pathogen, Xanthomonas oryzae pv. oryzae. New Phytol. 187, 1048–1057 (2010).
30. Chen, L. Q. SWEET sugar transporters for phloem transport and pathogen nutrition. New Phytol. 201, 1150–1155 (2014).
31. Hu, Y. et al. Lateral organ boundaries 1 is a disease susceptibility gene for citrus bacterial canker disease. Proc. Natl Acad. Sci. USA 111, E521–E529 (2014).
32. Gu, K. et al. R gene expression induced by a type-III effector triggers disease resistance in rice. Nature 435, 1122–1125 (2005).
33. Wang, C. et al. XA23 is an executor R protein and confers broad-spectrum disease resistance in rice. Mol. Plant 8, 290–302 (2015).
34. Tian, D. et al. The rice TAL effector-dependent resistance protein XA10 triggers cell death and calcium depletion in the endoplasmic reticulum. Plant Cell 26, 497–515 (2014).
35. Cernadas, R. A. et al. Code-assisted discovery of TAL effector targets in bacterial leaf streak of rice reveals contrast with bacterial blight and a novel susceptibility gene. PLoS Pathog. 10, e1003972 (2014).

36. Strauss, T. et al. RNA-seq pinpoints a Xanthomonas TAL-effector activated resistance gene in a large-crop genome. Proc. Natl Acad. Sci. USA 109, 19480–19485 (2012).
37. Yang, Y. N., Defeyter, R. & Gabriel, D. W. Host-specific symptoms and increased release of Xanthomonas citri and Xanthomonas campestris pv. malvacearum from leaves are determined by the 102-bp tandem repeats of PthA and Avrb6, respectively. Mol. Plant Microbe Interact. 7, 345–355 (1994).
38. DeFeyter, R., Yang, Y. O. & Gabriel, D. W. Gene-for-genes interactions between cotton R genes and Xanthomonas campestris pv. malvacearum Avr genes. Mol. Plant Microbe Interact. 6, 225–237 (1993).
39. DeFeyter, R. & Gabriel, D. W. At least six avirulence genes are clustered on a 90-kilobase plasmid in Xanthomonas campestris pv. malvacearum. Mol. Plant Microbe Interact. 4, 423–432 (1991).
40. Yang, Y., Yuan, Q. & Gabriel, D. W. Watersoaking function(s) of XcmH1005 are redundantly encoded by members of the Xanthomonas avr/pth gene family. Am. Phytopathol. Soc. 9, 105–113 (1996).
41. Salzberg, S. L. et al. Genome sequence and rapid evolution of the rice pathogen Xanthomonas oryzae pv. oryzae PXO99A. BMC Genomics 9, 204 (2008).
42. Essenberg, M. et al. Four near-isogenic lines of cotton with different genes for bacterial blight resistance. Phytopathology 92, 1323–1328 (2002).
43. Meckler, J. F. et al. Quantitative analysis of TALE-DNA interactions suggests polarity effects. Nucleic Acids Res. 41, 4118–4128 (2013).
44. Knight, R. L. & Hutchinson, J. B. The evolution of blackarm resistance in cotton. J. Genet. 50, 36–58 (1950).
45. Hou, B. H. et al. Optical sensors for monitoring dynamic changes of intracellular metabolite levels in mammalian cells. Nat. Protoc. 6, 1818–1833 (2011).
46. Eom, J. S. et al. SWEETs, transporters for intracellular and intercellular sugar translocation. Curr. Opin. Plant Biol. 25, 53–62 (2015).
47. Talbot, N. J. Cell biology: raiding the sweet shop. Nature 468, 510–511 (2010).
48. Chen, L. Q. et al. Sucrose efflux mediated by SWEET proteins as a key step for phloem transport. Science 335, 207–211 (2012).
49. Kanno, Y. et al. AtSWEET13 and AtSWEET14 regulate gibberellin-mediated physiological processes. Nat. Commun. 7, 13245 (2016).
50. Yu, Y. et al. Colonization of rice leaf blades by an African strain of Xanthomonas oryzae pv. oryzae depends on a new TAL effector that induces the rice nodulin-3 Os11N3 gene. Mol. Plant Microbe Interact. 24, 1102–1113 (2011).
51. Blanvillain-Baufume, S. et al. Targeted promoter editing for rice resistance to Xanthomonas oryzae pv. oryzae reveals differential activities for SWEET14-inducing TAL effectors. Plant Biotechnol. J. 15, 306–317 (2017).
52. Huguet-Tapia, J. C. et al. Complete genome sequence of the African strain AXO1947 of Xanthomonas oryzae pv. oryzae. Genome Announc. 4, e01730–15 (2016).
53. Grau, J. et al. AnnoTALE: bioinformatics tools for identification, annotation, and nomenclature of TALEs from Xanthomonas genomic sequences. Sci. Rep. 6, 21077 (2016).
54. Booher, N. J. et al. Single molecule real-time sequencing of Xanthomonas oryzae genomes reveals a dynamic structure and complex TAL (transcription activator-like) effector gene relationships. Microb. Genom. 1,,
55. Sugio, A. et al. Two type III effector genes of Xanthomonas oryzae pv. oryzae control the induction of the host genes OsTFIIAgamma1 and OsTFX1 during bacterial blight of rice. Proc. Natl Acad. Sci. USA 104, 10720–10725 (2007).
56. Zhang, J. et al. Homologs of CsLOB1 in citrus function as disease susceptibility genes in citrus canker. Mol. Plant Pathol.
57. Wright, R. J. et al. D-subgenome bias of Xcm resistance genes in tetraploid Gossypium (cotton) suggests that polyploid formation has created novel avenues for evolution. Genetics 149, 1987–1996 (1998).
58. Hutin, M. et al. MorTAL Kombat: the story of defense against TAL effectors through loss-of-susceptibility. Front. Plant Sci. 6, 535 (2015).

59. Li, T. et al. High-efficiency TALEN-based gene editing produces disease-resistant rice. Nat. Biotechnol. 30, 390–392 (2012).
60. Wang, Y. et al. Simultaneous editing of three homoeoalleles in hexaploid bread wheat confers heritable resistance to powdery mildew. Nat. Biotechnol. 32, 947–951 (2014).
61. Chin, C. S. et al. Nonhybrid, finished microbial genome assemblies from long-read SMRT sequencing data. Nat. Methods 10, 563–569 (2013).
62. Gao, X. et al. Silencing GhNDR1 and GhMKK2 compromises cotton resistance to Verticillium wilt. Plant J. 66, 293–305 (2011).
63. Li, F. et al. Modulation of RNA polymerase II phosphorylation downstream of pathogen perception orchestrates plant immunity. Cell Host Microbe 16, 748–758 (2014).
64. Doyle, E. L. et al. TAL effector-nucleotide targeter (TALE-NT) 2.0: tools for TAL effector design and target prediction. Nucleic Acids Res. 40, W117–W122 (2012).
65. Wilkins, K. E. et al. TAL effectors and activation of predicted host targets distinguish Asian from African strains of the rice pathogen Xanthomonas oryzae pv. oryzicola while strict conservation suggests universal importance of five TAL effectors. Front. Plant Sci. 6, 536 (2015).
66. Gao, X. et al. Cotton GhBAK1 mediates Verticillium wilt resistance and cell death. J. Integr. Plant Biol. 55, 586–596 (2013).
67. Cermak, T. et al. Efficient design and assembly of custom TALEN and other TAL effector-based constructs for DNA targeting. Nucleic Acids Res. 39, e82 (2011).
68. Darling, A. E., Mau, B. & Perna, N. T. ProgressiveMauve: multiple genome alignment with gene gain, loss and rearrangement. PLoS ONE 5, e11147 (2010).
69. Tamura, K. et al. MEGA6: molecular evolutionary genetics analysis version 6.0. Mol. Biol. Evol. 30, 2725–2729 (2013).

CHAPTER 8

The Effect of Plant Extracts as Seed Treatments to Control Bacterial Leaf Spot of Tomato in Tanzania

Ernest R. Mbega[1,2], Carmen N. Mortensen[1], Robert B. Mabagala[2], Ednar G. Wulff[1]

[1] *Department of Agriculture and Ecology, Faculty of Life Sciences, Danish Seed Health Centre for Developing CountriesUniversity of CopenhagenCopenhagenDenmark*

[2] *Department of Crop Science and Production, African Seed Health CentreSokoine University of AgricultureMorogoroTanzania*

ABSTRACT

Bacterial leaf spot (BLS) caused by seed-borne xanthomonads is a serious disease of tomato (*Solanum lycopersicum* L.), causing significant losses in both yield and quality. To identify more effective control measures, we evaluated crude extracts from 84 plant species in in vitro and in planta assays for antibacterial activity against BLS of tomato. In the in vitro assays, 20.2 % of the tested plant extracts totally inhibited growth of bacteria when seed washings from treated seeds were plated on nutrient agar medium. In the in planta assays, 17.8 % of the tested plant extracts reduced BLS incidence by 100 % in tomato seedlings. The most effective seed treatments were obtained with extracts from *Aloe vera*, *Betula pendula*, *Coffea arabica*, *Glycyrrhiza uralensis*, *Juniperus communis*, *Ocimum basilicum*, *Quercus robur*, *Rheum palmatum*, *Rosmarinus officinalis*, *Ruta graveolens*, *Sinapis alba*, *Yucca schidigera* and *Salvia officinalis*. Seed treatment of tomato with these extracts completely inhibited *Xanthomonas perforans* in both in vitro and in planta assays. Extracts from *A. vera*, *C. arabica* and *Y. schidigera* were tested three times using tomato seeds of cultivars Tanya, Cal-J and Moneymaker in Tanzania. Treatment of tomato seeds with these extracts had a positive effect on the number of normal seeds and had no effect on seedling vigor, height and weight. These results indicate that plant extracts from *A. vera*, *C. arabica* and *Y. schidigera* are potential candidates for seed treatment against seed-borne xanthomonads of tomato in Tanzania.

Keywords: Plant extracts Antibacterial activity *Xanthomonas perforans* Tomato Seed treatment

INTRODUCTION

Bacterial leaf spot (BLS) incited by *Xanthomonas euvesicatoria*, *X. vesicatoria*, *X. perforans* and *X. gardneri* (Anonymous 2006; Jones et al. 2004) is a serious disease of tomato (*Solanum lycopersicum* L.) that occurs worldwide in regions with warm and humid climates (Jones et al. 2000; Stall et al. 1994; Tamir-Ariel et al. 2007). The BLS xanthomonads also affect pepper (*Capsicum* spp.), reducing both fruit yield and quality. Recently, *X. campestris* pv. *raphani* (Punina et al. 2009) and *X. arboricola* (Myung et al. 2010; E. R. Mbega et al., unpublished data) were also reported as pathogens of tomato. The BLS xanthomonads can survive in seeds, plant debris and volunteer plants (Kaaya et al. 2003). Seed infection by BLS pathogens of about 40.7 % has been reported in Tanzania, causing yield losses up to 45 % (Black et al. 2001; Kaaya et al. 2003).

Management of BLS is limited to foliar applications of copper-based compounds. However, presence of strains of BLS pathogens with a high degree of tolerance to copper (Carrillo-Fasio et al. 2001; Gitaitis et al. 1992; Gore and O'Garro 1999; Lee and Cho 1996; Martin et al. 2004; Scheck et al. 1996; Shenge et al. 2007) and the considerable number of *Xanthomonas* species and races causing BLS symptoms in tomato and pepper (Jones et al. 2004) have made the control of the disease difficult.

Plants synthesize a number of compounds with antibiotic and antimicrobial properties (Amadioha 2003; Kumar and Parmar 1996; Opara and Wokocha 2008; Prakash and Rao 1997). Such compounds can therefore, be exploited as an alternative approach to manage or control BLS. Some of the known advantages of using pesticides of plant origin to control plant diseases include low mammalian toxicity, minimum health hazards and environmental pollution, and low risks of development of resistance by pathogens (Amadioha 2003; Kumar and Parmar 1996; Prakash and Rao 1997). Biopesticides may also be less expensive, easily available (because of their natural occurrence) and depending on the concentration, they may have no effects on seed viability, plant growth and food quality (Opara and Wokocha 2008).

In the present study, we applied crude extracts from a large number of plant species to tomato seeds to evaluate their potential to control BLS. The effect of the most promising extracts on seed germination and tomato seedling growth was also investigated.

MATERIALS AND METHODS

Seed samples

Tomato (*S. lycopersicum* L.) seeds of cultivars Cal-J, Tanya and Moneymaker, collected from tomato growers in Tanzania, were tested for infection as described by International Seed Federation (2007). Seed samples that were free of infection by *Xanthomonas* spp. were used in the experiments. One thousand seeds per cultivar were surface-disinfested in 70 % ethanol for 1 min, then in 1 % sodium hypochlorite for 3 min and rinsed three times in sterile distilled water. The seeds were then transferred to Petri dishes containing sterile filter papers and allowed to air-dry overnight in a laminar flow chamber and stored at 4 °C until used.

Seed inoculation with bacterial suspensions

Surface-disinfested seeds of tomato cultivar Tanya were inoculated with *X. perforans* strain NCPPB 4321 based on its ability to cause severe BLS symptoms on various tomato cultivars, including those grown in Tanzania. Inoculum was prepared from 24-h-old bacterial cultures grown on nutrient agar (NA) medium (meat extract 3 g, Bacto peptone 5 g, Bacto agar 20 g, distilled water 1000 mL) at 28 °C. Bacterial cultures were flooded with 10 mL of sterile distilled water and gently scraped with a flamed Drigalski spatula. The inoculum suspension was then homogenized using a vortex mixer and suspended in sterile distilled water to obtain a ca. 10^8 CFU/mL ($OD_{600} = 0.01$) (NanoDrop, Thermo Fisher Scientific, Beverly, MA, USA). One thousand seeds of tomato were vacuum-infiltrated for 30 min with 10 mL of the bacterial suspension, and seeds were air-dried in the laminar air flow chamber at 4 °C until used.

Seed treatment with plant extracts

Plant for extractions were obtained from companies in Denmark and farmers in Tanzania as shown in Table 1. Two grams of plant material were suspended in 20 mL sterile distilled water in a 50 mL conical flask to obtain a 10 % (w/v) concentration. The conical flasks with the suspensions were briefly heated on a hot electric plate until boiling and cooled for 5 min. The suspensions were filtered through sterile cheesecloth, and the extracts were autoclaved at 115 °C for 15 min and kept at 4 °C until used. Twenty tomato seeds pre-inoculated with *X. perforans* were treated with 1 mL of the 10 % plant extract in an Eppendorf tube and placed on an agitation table at 100 rpm overnight at 25 °C. Untreated, seeds treated with copper sulphate (200 µg/mL $CuSO_4 \cdot 5H_2O$) or sterile distilled were included as controls. After overnight incubation, the treated seeds were blot-dried and allowed to air-dry for 1 h in the laminar air flow cabinet.

Table 1. Origin of plant extracts and control of bacterial leaf spot (BLS) of tomato caused by *Xanthomonas perforans* after seed treatment with extracts or control treatments

Common name	Scientific name	Family	Part tested	Origin	CFU/mL	BLS-RI (%)[a]
Aloe	*Aloe vera* L.	Aloaceae	Stem	NN, DK	0.0×10^0	100
Silver birch	*Betula pendula* Roth.	Betulaceae	Leaf	NDC, DK	0.0×10^0	100
Coffee	*Coffea arabica* L.	Rubiaceae	Seed[b]	Tanzania	0.0×10^0	100
Licorice	*Glycyrrhiza uralensis* L.	Fabaceae	Stem/leaf	NDC, DK	0.0×10^0	100
Juniper	*Juniperus communis* L.	Cupressaceae	Stem	NDC, DK	0.0×10^0	100
Basil	*Ocimum basilicum* L.	Lamiaceae	Leaf	NDC, DK	0.0×10^0	100
Oak	*Quercus robur* L.	Fagaceae	Bark	NN, DK	0.0×10^0	100
Rhubarb	*Rheum palmatum* L.	Polygonaceae	Stem	NN, DK	0.0×10^0	100
Rosemary	*Rosmarinus officinalis* L.	Lamiaceae	Stem	NN, DK	0.0×10^0	100
Rue	*Ruta graveolens* L.	Rutaceae	Stem	NDC, DK	0.0×10^0	100
White mustard	*Sinapis alba* L.	Brassicaceae	Root	NDC, DK	0.0×10^0	100
Mojave yucca	*Yucca schidigera* L.	Agavaceae	Stem	NDC, DK	0.0×10^0	100
Sage	*Salvia officinalis* L.	Lamiaceae	Stem	NDC, DK	0.0×10^0	100
Nor-grape 80	*Vitis vinifera* L.	Vitaceae	Stem/leaf	NDC, DK	0.0×10^0	94
Punica	*Punica granatum* L.	Punicaceae	Stem	NDC, DK	0.0×10^0	86.7
Olive	*Olea europaea* L.	Oleaceae	Leaf	NDC, DK	0.0×10^0	81.2
Clove	*Caryophyllus aromaticus* L.	Myrtaceae	Twig	NDC, DK	0.0×10^0	75
White willow	*Salix alba* L.	Salicaceae	Bark	NN, DK	2.0×10^0	100
Grape vine	*Vitis vinifera* L.	Vitaceae	Leaf	NN, DK	1.0×10^0	79.8
Quinoa	*Chenopodium quinoa* Wild.	Chenopodiaceae	Stem	NN, DK	1.8×10^3	75
Soapwort	*Saponaria officinalis* L.	Caryophyllaceae	Stem/leaf	NN, DK	2.2×10^1	100
Ginkgo	*Ginkgo biloba* L.	Ginkgoaceae	Leaf	NDC, DK	3.3×10^4	94
Daisy	*Bellis perennis* L.	Asteraceae	Flower	NDC, DK	4.0×10^4	90.1
Inula	*Inula helenium* L.	Asteraceae	Stem	NDC, DK	1.0×10^1	86.7
Rosemary	*Rosmarinus officinalis* L.	Lamiaceae	Leaf	NN, DK	2.5×10^3	81.6
Tea	*Camellia sinensis* L.	Theaceae	Leaf	NDC, DK	1.0×10^1	81.2
Bilberry	*Vaccinium myrtillus* L.	Vacciniaceae	Leaf	NN, DK	1.0×10^1	81.2
Celery	*Apium graveolens* L.	Apiaceae	Root	NDC, DK	1.0×10^1	80
Ginseng	*Panax ginseng* C.A. Meyer.	Araliaceae	Stem/leaf	NDC, DK	1.0×10^1	80

Burnet-saxifrage	*Pimpinella saxifraga* ssp. *nigra* (Mill)	Apiaceae	Stem	NN, DK	1.0×10^{6}	80
Thyme	*Thymus vulgaris* L.	Lamiaceae	Leaf/stem	NDC, DK	3.9×10^{3}	80
Maypop	*Passiflora incarnata* L.	Passifloraceae	Flower	NN, DK	6.0×10^{4}	79.8
Red pepper	*Capsicum frutescens* L.	Solanaceae	Fruit	NN, DK	1.8×10^{5}	73.4
Chamomile	*Matricaria chamomilla* L.	Asteraceae	Flower	NN, DK	2.0×10^{1}	73.3
Neem	*Azadirachta indica* L.	Meliaceae	Seed	Tanzania	7.0×10^{2}	68.7
Quinine	*Cinchona pubescens* Vahl.	Rubiaceae	Bark	NDC, DK	9.0×10^{1}	66.7
High mallow	*Malva sylvestris* L.	Malvaceae	Leaf	NN, DK	9.8×10^{3}	66.6
Cowslip	*Primula veris* L.	Primulaceae	Stem	NN, DK	6.0×10^{1}	60
Sisal	*Agave sisalana* Perrine.	Agavaceae	Root	Tanzania	$>1.0 \times 10^{7}$	59.9
Agapanthus	*Agapanthus* sp.	Alliaceae	Leaf	NDC, DK	1.9×10^{5}	53.2
Sisal	*Agave sisalana* Perrine.	Agavaceae	Leaf	Tanzania	2.8×10^{5}	53.2
Couch grass	*Agropyron repens* (L.) Beauv.	Poaceae	Stem	NDC, DK	1.1×10^{3}	53.1
Elm	*Ulmus campestris* L.	Ulmaceae	Bark	NN, DK	5.5×10^{2}	49.5
Sisal	*Agave sisalana* Perrine.	Agavaceae	Stem	Tanzania	1.9×10^{5}	46.8
Horsetail	*Equisetum arvense* L.	Equisateaceae	Leaf	ND, DK	1.6×10^{5}	40
Black poplar	*Populus nigra* L.	Salicaceae	Twig	NN, DK	6.0×10^{0}	40
Elderberry	*Sambucus nigra* L.	Caprifoliaceae	Leaf	ND, DK	6.0×10^{0}	40
Turmeric	*Curcuma longa* L.	Zingiberaceae	Stem	ND, DK	$>1.0 \times 10^{7}$	40
Adam's needle	*Yucca filamentosa* L.	Agavaceae	Stem	NN, DK	2.4×10^{4}	37.5
Onion	*Allium cepa* L.	Liliaceae	Bud	ND, DK	2.0×10^{6}	33.5
Eucalyptus	*Eucalyptus globulus* L.	Myrtaceae	Leaf	Tanzania	3.7×10^{3}	33.5
Avocado	*Persea americana* L.	Lauraceae	Fruit peels	Tanzania	$>1.0 \times 10^{7}$	33.5
Buckthorn	*Frangula alnus* Mill.	Rhamnaceae	Bark	ND, DK	2.0×10^{1}	29.9
Plantain	*Plantago major* L.	Plantaginaceae	Leaf	ND, DK	6.0×10^{5}	29.9
Southernwood	*Artemisia abrotanum* L.	Asteraceae	Leaves	ND, DK	5.8×10^{6}	26.8
Lemongrass	*Cymbopogon citratus* L.	Poaceae	Leaves	ND, DK	9.8×10^{5}	26.8
Eucalyptus	*Eucalyptus globulus* L.	Myrtaceae	Seeds	Tanzania	$>1.0 \times 10^{7}$	26.8
White clover	*Trifolium repens* L.	Fabaceae	Flower	NN, DK	2.6×10^{5}	26.8
Couch grass	*Agropyron repens* L.	Poaceae	Leaf	ND, DK	1.9×10^{5}	26.7
Silver birch	*Betula pendula* Roth.	Betulaceae	Bark	ND, DK	5.0×10^{3}	26.7
Scots pine	*Pinus sylvestris* L.	Pinaceae	Leaf	ND, DK	6.0×10^{3}	26.7
Red clover	*Trifolium pratense* L.	Fabaceae	Flower	NN, DK	6.8×10^{3}	26.7
Ginger	*Zingiber officinale* Rosc.	Zingiberaceae	Bud	NN, DK	7.0×10^{2}	26.7
Yucca	*Yucca* sp.	Agavaceae	Stem/leaf	NN, DK	2.6×10^{5}	25
Adam's needle	*Yucca filamentosa* L.	Agavaceae	Leaf	NN, DK	2.4×10^{5}	25
Oak	*Quercus robur* L.	Fagaceae	Leaf	NN, DK	7.0×10^{1}	18.7
Soapbark	*Quillaja saponaria* Molina.	Quillajaceae	Bark	NN, DK	2.6×10^{3}	18.7
Chicory	*Cichorium intybus* L.	Asteraceae	Stem	ND, DK	1.0×10^{5}	13.5
Water mint	*Mentha aquatica* L.	Lamiaceae	Leaf	ND, DK	4.0×10^{5}	13.5
Irish moss	*Chondrus crispus* L.	Gigartinaceae	Leaf	ND, DK	3.5×10^{5}	13.3
Tansy	*Tanacetum vulgare* L.	Asteraceae	Stem/leaf	ND, DK	1.3×10^{4}	6.2
Neem	*Azadirachta indica* L.	Meliaceae	Leaf	Tanzania	5.0×10^{5}	0
Bladderwrack	*Fucus vesiculosus* L.	Fucaceae	Flower	NN, DK	6.9×10^{4}	0
Coral plant	*Jatropha* sp.	Euphorbiaceae	Seed	Tanzania	2.2×10^{5}	0
Coral plant	*Jatropha* sp.	Euphorbiaceae	Leaf	Tanzania	1.2×10^{5}	0
Lantana	*Lantana camara* L.	Verbenaceae	Leaf	Tanzania	5.0×10^{4}	0
Spiny restharrow	*Ononis spinosa* L.	Fabaceae	Leaf	ND, DK	6.0×10^{5}	0
Silverweed	*Pontentilla anserine* L.	Rosaceae	Leaf	ND, DK	9.6×10^{3}	0
Quassia	*Quassia* sp.	Simaroubaceae	Bark	NN, DK	1.3×10^{5}	0
Nettle	*Urtica* sp.	Urticaceae	Leaf	ND, DK	1.4×10^{5}	0
Mistletoe	*Viscum album* L.	Loranthaceae	Leaf/stem	ND, DK	4.6×10^{5}	0
Cardamon	*Elletaria cardamomum* L.	Zingiberaceae		ND, DK	$>1.0 \times 10^{7}$	−10
Alfalfa	*Medicago sativa* ssp. *sativa* L.	Fabaceae	Seed	ND, DK	$>1.0 \times 10^{7}$	−10
Fenugreek	*Trigonella foenum-graecum* L.	Fabaceae	Seed	ND, DK	1.3×10^{6}	−20
Controls						
Sterile distilled water	–	–	–	–	$>1.0 \times 10^{7}$	0
Copper sulphate	–	–	–	–	0.0×10^{0}	100
Untreated seed	–	–	–	–	0	0

– not applicable, *NN, DK* Nor-Natur, Denmark, *ND, DK* Natur Drogeriet, Denmark

[a] Bacterial leaf spot reduction index (BLS-RI) was calculated as (C − T)/C × 100 %, where *C* is the incidence of BLS in tomato seedlings treated with sterile distilled water (negative control) and *T* is the incidence of BLS of tomato seedlings treated with plant extract

[b] Processed coffee (Africafe®) from Afri Tea and Coffee Blenders (1993) Ltd., Dar es Salaam, Tanzania

Evaluation of antibacterial activity of plant extracts in vitro

Washing samples (100 μL) from treated seeds were collected using sterile pipettes and serially diluted to 10^{-2} with sterile distilled water in Eppendorf tubes. An aliquot of 100 μL from each dilution was spread onto 0.6 cm thick NA medium in Petri dishes (diameter 8.5 cm × depth 1.3 cm) using a sterile glass rod. The plates were incubated at 28 °C. A pure culture of *X. perforans* NCPPB 4321 was included as a control. Yellow colonies with morphology and color similar to those of *X. perforans* were counted after 96 h. The identity of suspected colonies from each plate was confirmed by pathogenicity tests. The abaxial surfaces of four 14-day-old plants of tomato cultivar Tanya were sprayed to runoff with an inoculum suspension of 10^8 CFU/mL ($OD_{600} = 0.01$) prepared from 24-h-old bacterial cultures grown on NA at 28 °C. The inoculated seedlings were covered with polyethylene bags and kept in the growth chamber for 14 days. Seedlings sprayed with sterile saline water (containing 0.85 % of NaCl) were used as a negative control, and seedlings sprayed with suspensions prepared from the *X. perforans* culture served as positive controls. The plants were examined for symptoms 14 days after inoculation and scored as negative when no obvious symptoms were observed. Leaves with water-soaked lesions that developed into dark brown spots were scored as positive for BLS disease.

Evaluation of antibacterial activity of plant extracts in planta

To evaluate the effect of the plant extracts on the control of *X. perforans* in seedling assays, 16 tomato seeds treated with plant extracts as previously described were sown in pots (8 cm diameter) containing a 1:3 ratio of sterile sand and peat soil (Pindstrup substrate No. 2, Pindstrup Mosebrug A/S, Ryomgaard, Denmark) and kept in growth chamber at 28 °C under high relative humidity (>85 %). Twenty-one days after sowing, BLS incidence was assessed by calculating the percentage of seedlings with leaf spot symptoms in the total number of emerged seedlings. The efficacy of plant extract treatments in the control of BLS in tomato seedlings was calculated as the BLS reduction index (BLS-RI) = $(C - T)/C \times 100$ %, where C is the incidence of BLS in tomato seedlings raised from seeds treated with sterile distilled water (negative control) and T is the incidence of BLS in tomato seedlings from infected seeds treated with a given plant extract. In addition to the BLS incidence and reduction index, the height, mass and width of seedlings were also evaluated for the best-performing plant extracts. The height of the seedlings was determined by measuring the aerial part of the seedlings from the soil surface to the node of the terminal developing leaf. Fresh mass of the aerial plant part was determined using tomato seedlings cut at the base of the stem by a pair of scissors, and the seedling mass was weighed. To determine the width of the seedlings, we placed a digital caliper at a right angle to the seedlings and recorded the reading at the widest point of the stem. The measurements were repeated in three independent experiments. Plant extracts with the best ability to reduce BLS in tomato during the initial screening tests were determined by comparing the data obtained with Student–Newman–Kuels (SNK) test using SAS version 9.1 software (SAS Institute, Cary, NC, USA). The choice of plant extracts used for screenhouse evaluations in Tanzania was based on four criteria: (1) effectiveness in reducing BLS, (2) most normal seedlings in germination tests, (3) highest vigor index in seedling assays and (4) no phytotoxicity.

Effect of plant extracts on seed germination and seedling growth

The effect of selected plant extracts on seed germination, seedling vigor and mass was evaluated in *Xanthomonas*-free tomato seeds treated with plant extracts as previously described. Seed germination tests were conducted using 400 tomato seeds per treatment. The standard International Seed Testing Association top of paper method (ISTA 2005) was used. The seeds were plated uniformly (50 seeds per replicate) onto three layers of moist blotter paper in a plastic container kept at 27 ± 2 °C and RH >85 % for 14 days. Normal and abnormal seedlings and dead seeds were counted for the germination tests. The same seedlings were used to determine vigor and dry mass. The vigor test involved measurements of root and shoot lengths of seedlings and the percentage seed (normal seedlings) germination. The seedling vigor index (Vi) was calculated as Vi = (mean root length + mean shoot length) × (percentage germination) (Abdul-Baki and Anderson 1973). To determine the dry mass, we wrapped seedlings from each treatment in the germination tests in aluminium foil and dried them in an oven at 103 °C for 24 h. The dried seedlings were then allowed to cool to room temperature and weighed.

Evaluation of plant extracts for production of healthy tomato transplants in the screenhouse

Tomato seeds of cultivars Tanya, Cal-J and Moneymaker, collected from tomato growers in Tanzania and free of BLS-causing xanthomonads, were inoculated with *X. perforans* and treated with plant extracts as previously described. One hundred seeds per cultivar per treatment were sown in polyethylene plastic trays (56.5 × 26.5 × 6 cm) containing a mixture of forest soil, rice husks and farmyard manure (3:1:1). The trays were kept in the screenhouse at 25–33 °C and RH >85 %. Seven days after sowing, 40 seedlings per treatment of each cultivar (10 seedlings per replicate in four replications) were randomly selected and transferred to polyethylene sleeves (6.5 × 9.0 cm) containing the same growth substrate as previously described. The sleeves with the seedlings were placed on the screenhouse benches at the same temperature and RH. BLS incidence and severity in the tomato seedlings were assessed 21 days after sowing. Disease severity was determined based on the Horsfall and Barrett (1945) scale with minor modifications (Shenge 2006), where 1 = no disease, 2 = >0–3 % of leaves with BLS symptoms, 3 = >3–12 % of leaves with BLS symptoms, 4 = >12–25 % of leaves with BLS symptoms, 5 = >25–50 % of leaves with BLS symptoms and 6 = >50 % of leaves with BLS symptoms. In addition to disease incidence and severity, the height and mass of seedlings were also evaluated as already described. The experiment was repeated three times from March to September 2010.

Data analysis

In the in vitro assays, the average of the total number of colony forming units of *X. perforans* on NA was calculated based on three replications. Data for BLS incidence, severity, seed germination, seedling vigor, plant dry and fresh mass and plant height were analyzed using Proc GLM, and mean separation tests were calculated with the Student–Newman–Kuels (SNK) using SAS version 9.1 software (SAS Institute).

RESULTS

Results for preliminary screening of the effectiveness of plant extracts on *X. perforans* indicated that 17 of 84 tested plant extracts (20.2 %) were able to totally reduce the pathogen in in vitro assays (Table 1). The extracts were from *Aloe vera*, *Betula pendula*, processed *Coffea arabica*, *Glycyrrhiza uralensis*, *Juniperus communis*, *Ocimum basilicum*, *Quercus robur*, *Rheum palmatum*, *Rosmarinus officinalis*, *Ruta graveolens*, *Sinapsis alba*, *Yucca schidigera*, *Salvia officinalis*, *Vitis vinifera*, *Punica granatum*, *Olea europea* and *Caryophillus aromaticus*. The effect of these plant extracts on *X. perforans* was similar to that obtained when seeds were treated with copper sulphate (Table 1).

In the in planta assays, 15 of 84 plant extracts (17.8 %) completely inhibited symptoms of BLS in tomato seedlings. Such plant extracts were from *A. vera*, *B. pendula*, *C. arabica*, *G. uralensis*, *J. communis*, *O.*

basilicum, Q. robur, R. palmatum, R. officinalis, R. graveolens, S. alba, Y. schidigera and *S. officinallis* (Table 1). Only 13 of 84 extracts (corresponding to 15.5 %) of the assayed plant extracts controlled BLS in both in vitro and in planta assays. These extracts were from *A. vera, B. pendula, C. arabica, G. uralensis, J. communis, O. basilicum, Q. robur, R. palmatum, R. officinalis, R. graveolens, S. alba, Y. schidigera* and *S. officinalis* and were selected for further experiments.

The results obtained from evaluation of the best-performing plant extracts are shown in Table 2. The 13 selected plant extracts all significantly reduced ($P < 0.001$) the incidence of BLS in tomato seedlings without significantly affecting the growth of tomato seedlings. Seed treatment with plant extracts from *A. vera, C. arabica, G. uralensis* and *Y. schidigera* totally reduced ($P < 0.001$) the incidence of BLS in tomato (Table 2). The efficacy of these plant extracts to inhibit the growth of *X. perforans* was similar to the effects obtained when seeds were treated with copper sulphate (bactericide) control and untreated (disease free) seed control.

The results also showed that, treatment of tomato seeds with the best-performing plant extracts did not negatively affect the growth of tomato seedlings compared to the treatment with copper sulphate and untreated, disease-free seeds. In contrast, seed treatment with sterile distilled water (negative control) resulted in seedlings with significantly lower ($P < 0.001$) fresh mass (0.31 g) and width (1.45 mm) compared to the other seed treatments (Table 2).

The effects of tomato seed treatment with 10 % aqueous plant extracts from *A. vera, C. arabica, G. uralensis* and *Y. schidigera* on seed germination, seedling vigor and dry mass is summarised in Table 3. Treatment of tomato seeds with extracts from *A. vera, C. arabica* and *Y. schidigera* significantly increased ($P < 0.05$) the number of normal seedlings compared to seeds treated with sterile distilled water (negative control). The number of normal seedlings obtained from seeds treated with these extracts was not significantly different from those obtained from copper sulphate (positive control) and untreated tomato seeds (Table 3). The number of normal seedlings obtained with seeds treated with plant extracts of *G. uralensis* was not significantly different ($P < 0.05$) from seed treatments using plant extracts from *A. vera, C. arabica* and *Y. schidigera*, (positive and negative controls, respectively). The number of abnormal seedlings was not significantly different ($P < 0.05$) between different seed treatments (Table 3). When the number of dead tomato seeds was compared between seed treatments, there was no significant difference ($P < 0.05$) between most treatments, except for tomato seeds treated with *C. arabica* (Table 3), which significantly increased ($P < 0.05$) tomato seedling vigor. The treatment of tomato seeds with extracts from *A. vera* and *Y. schidigera* were not significantly different ($P < 0.05$) from seeds treated with *C. arabica*, positive and negative controls (Table 3).

Table 2. Effect of seed treatment with selected plant extracts on incidence of bacterial leaf spot caused by *Xanthomonas perforans* and on growth of tomato seedlings of cultivar Tanya

Treatment	Incidence (%)[a]	Height (cm)	Mass (g)	Width (mm)
Control				
Sterile distilled water	83.30a	13.50b	0.31d	1.45d
Copper sulphate	0.00e	16.13a	0.85abc	1.81abc
Untreated seed	0.00e	15.98a	0.75bc	1.78c
Plant extract				
Aloe vera	0.00e	17.28a	0.96abc	2.05ab
Betula pendula	18.80b	16.90a	0.83abc	2.00abc
Coffea arabica	0.00e	17.13a	0.98ab	2.04ab
Glycyrrhiza uralensis	0.00e	17.23a	0.96abc	2.06ab
Juniperus communis	6.30d	16.11a	0.72c	1.97abc
Ocimum basilicum	18.80b	16.84a	0.78abc	1.93abc
Quercus robur	18.80b	16.69a	0.82abc	1.97abc
Rheum palmatum	18.80b	16.50a	0.82abc	2.04ab
Rosmarinus officinalis	18.80b	16.06a	0.78abc	1.98abc
Ruta graveolens	18.80b	16.16a	0.49ab	1.82abc
Salvia officinalis	12.50c	16.56a	0.82abc	2.04ab
Sinapis alba	12.50c	16.13a	0.74c	1.90abc
Yucca schidigera	0.00e	17.24a	1.00a	2.13a
Mean	12.31	16.47	0.81	1.95
F test	***	***	***	***

*** Significant at $P = 0.01$

[a] BLS incidence = percentage of seedlings with bacterial leaf spot symptoms. Mean followed by same letters in a column are not significantly different based on the SNK test at $P = 0.05$. Each value is a mean of 48 seedlings

Table 3. The effect of treatment with selected plant extracts from *Aloe vera*, *Coffea arabica*, *Glycyrrhiza uralensis* and *Yucca schidigera* on tomato seed germination, seedling vigor and dry weight of tomato plants

Treatment	Seed germination[a]			Vigor index (%)[b]	Dry mass (g)
	NS (%)	ABS (%)	DS (%)		
Control					
Sterile distilled water	90.50b	3.00a	6.50a	702.51b	0.16a
Copper sulphate	92.80ab	2.30a	5.00ab	707.60b	0.14a
Untreated seed	92.00ab	3.50a	4.50ab	718.06b	0.16a
Plant extract					
A. vera	94.80a	1.80a	3.50ab	749.39ab	0.14a
C. arabica	95.50a	2.50a	2.00b	784.77a	0.14a
G. uralensis	92.80ab	2.80a	4.50ab	714.33b	0.16a
Y. schidigera	94.30a	2.00a	3.80ab	741.91ab	0.14a
Mean	93.20	2.50	4.30	731.22	0.14
F test	**	ns	**	**	ns

ns not significant

** Significant at $P = 0.05$

[a] Seed germination states: *NS* normal seedlings, *ABS* abnormal seedlings, *DS* dead seed. Each value is the percentage from 400 seeds test

[b] Seedling vigor index (Vi) was calculated as Vi = (mean root length + mean shoot length) × (percentage germination) (Abdul-Baki and Anderson 1973); means followed by the same letters in a column are not significantly different based on the SNK test at $P = 0.05$

Seed treatment of tomato with plant extracts from *A. vera*, *C. arabica* and *Y. schidigera* significantly reduced the incidence and severity of BLS ($P < 0.001$) in all three experiments. Such effects were similar to those obtained for seedlings grown from tomato seeds treated with copper sulphate and untreated seeds (disease-free) control (Table 4). In all three experiments, the incidence and severity of BLS disease were significantly higher ($P < 0.001$) in tomato transplants of cultivars Tanya, Cal-J and Moneymaker treated with sterile distilled water (negative control) compared to the other treatments (Table 4).

DISCUSSION

Tomato seeds were treated with aqueous extracts from 84 different plant materials to assess control of seed-borne infection of BLS of tomato caused by *X. perforans*. In the in vitro assays, 20.2 % of the tested plant extracts totally inhibited growth of *X. perforans* when seed washings from treated seeds were plated on NA. In the in planta experiments, notably 17.8 % of the tested plant extracts reduced BLS incidence by 100 % in tomato seedlings (Table 1). The most effective seed treatments, giving 100 % control in vitro and in planta, were obtained when tomato seeds were treated with plant extracts from *A. vera*, *B. pendula*, *C. arabica*, *G. uralensis*, *J. communis*, *O. basilicum*, *Q. robur*, *R. palmatum*, *R. officinalis*, *R. graveolens*, *S. alba*, *Y. schidigera* and *S. officinalis* (Table 1).

Table 4. The effect of selected plant extracts from *Aloe vera*, *Coffea arabica* and *Yucca schidigera* applied as seed treatment on incidence and severity of bacterial leaf spot (BLS) caused by *Xanthomonas perforans* on three tomato cultivars under screenhouse conditions in Morogoro, Tanzania

Cultivar	Treatment	Experiment 1		Experiment 2		Experiment 3	
		Incidence (%)[a]	Severity index[b]	Incidence (%)	Severity index	Incidence (%)	Severity index
Tanya	Control						
	Sterile distilled water	45.00a	2.15a	82.50a	2.56a	65.00	2.18a
	Copper sulphate	10.00b	1.10b	0.00b	1.00b	2.50b	1.02b
	Untreated seed	0.00b	1.00b	0.00b	1.00b	0.00b	1.00b
	Extract						
	A. vera	2.50b	1.02b	0.00b	1.00b	0.00b	1.00b
	C. arabica	0.00b	1.00b	0.00b	1.00b	0.00b	1.00b
	Y. schidigera	0.00b	1.00b	0.00b	1.00b	0.00b	1.00b
	Mean	9.58	1.21	13.75	1.26	11.25	1.20
	F test	***	***	***	***	***	***
Cal-J	Control						
	Sterile distilled water	70.00a	2.18a	80.00a	2.55a	60.00a	1.65a
	Copper sulphate	0.00b	1.00b	0.00b	1.00b	0.00b	1.00b
	Untreated seed	0.00b	1.00b	0.00b	1.00b	0.00b	1.00b
	Extract						
	A. vera	2.50b	1.02b	0.00b	1.00b	2.50b	1.02b
	C. arabica	0.00b	1.00b	0.00b	1.00b	0.00b	1.00b
	Y. schidigera	0.00b	1.00b	0.00b	1.00b	0.00b	1.00b
	Mean	12.08	1.20	13.33	1.26	10.42	1.13
	F test	***	***	***	***	***	***
Moneymaker	Control						
	Sterile distilled water	65.00a	2.28a	67.50a	2.40a	72.50a	2.00a
	Copper sulphate	0.00b	1.00b	0.00b	1.00b	0.00b	1.00b
	Untreated seed	0.00b	1.00b	0.00b	1.00b	0.00b	1.00b
	Plant extract						
	A. vera	0.00b	1.00b	0.00b	1.00b	2.50b	1.02b
	C. arabica	2.50b	1.02b	0.00b	1.00b	0.00b	1.00b
	Y. schidigera	0.00b	1.00b	2.50b	1.02b	0.00b	1.00b
	Mean	11.25	1.22	11.67	1.23	12.50	1.17
	F test	***	***	***	***	***	***

*** Significant at $P = 0.01$ b

[a] Percentage of seedlings with bacterial leaf spot symptoms

[b] Disease severity index based on Horsfall and Barrett (1945) scale of 1–6 with modifications: 1 = no disease and 6 = >50 % of leaves with BLS symptoms. Means followed by the same letters in a column are not significantly different based on SNK test at $P = 0.05$

From the in planta evaluation of the 13 best performing plant extracts (Table 2), extracts from *A. vera*, *C. arabica*, *G. uralensis* and *Y. schidigera* were the most effective and promising for control of BLS of tomato when applied as seed treatment (Table 2). Such results indicate that these plant extracts have bactericidal properties and can be used for tomato seed treatment to control xanthomonads associated with BLS. Many reports are available on the antibacterial properties of these plants. The antibacterial activity of *A. vera* against *Shigella flexneri* and *Streptococcus pyogenes* has been documented (Ferro et al. 2003). Extracts from *A. vera* also affect several Gram-positive and negative bacteria (Cock 2008). The antibacterial activity of extracts from *A. vera* was reported to be due to a direct effect on the bacterial cells caused by the presence of anthraquinones (Boateng 2000) and saponin (Reynolds and Dweck 1999; Urch 1999). Other indirect effects have been reported to be associated with the presence of polysaccharides, which stimulate plant defence responses that destroy bacteria (Lawless and Allan 2000; Pugh et al. 2001). Murthy and Manonmani (2009) reported that plant extracts from processed coffee inhibited growth of food-borne pathogens such as *Escherichia coli*, *Yersinia* and *Listeria* species. The antibacterial activity of coffee is associated with substances produced by the roasting process such as millard products, carbohydrate caramelization and thermal composition products (Daglia et al. 1994). In addition, *Glycyrrhiza* species contain α-glycyrrhetinic acid and glycyrrhizin, which inhibit DNA replication and RNA and protein synthesis of microbes (Kim et al. 2002). Other *Glycyrrhiza* species, e.g. *G. glabra*, inhibit growth of some Gram-negative bacteria such as *Salmonella* spp., *Shigella* spp. and *E. coli* (Shirazi et al. 2007). Extracts from *Y. schidigera* also have antibacterial activity, which was attributed to the presence of saponin, a compound found to inhibit microbial growth through hemolytic activity (Hassan et al. 2010). In the present study, observations of cell suspensions of *X. perforans* treated with extracts from *Y. schidigera* at 10 % concentration using confocal microscopy revealed permeabilization of bacterial cells (data not shown).

In addition to reducing BLS in tomato seedlings, plant extracts from *A. vera*, *C. arabica* and *Y. schidigera* significantly improved germination ($P < 0.05$) of tomato seedlings (Table 3). At the same time, the vigor and dry mass of tomato seedlings were not affected by the treatment of seeds with plant extracts from *A. vera*, *C. arabica*, *G. uralensis* and *Y. schidigera*, indicating that these extracts were not phytotoxic to tomato seeds and seedling development. Other seed treatments with natural compounds of plant origin to control plant pathogens have been reported to have no negative effects on plant growth, seed viability or food quality (Opara and Wokocha 2008). Based on data from the present study, extracts from *A. vera*, *C. arabica* and *Y. schidigera* were the most promising plant extracts against BLS and can therefore be used to treat tomato seed. Additionally, they fulfilled the requirements for bioactive chemicals (Hewett and Griffiths 1986) and have the advantages of low mammalian toxicity, minimal health hazards and the least environmental pollution (Amadioha 2003; Singh 1994).

Under screenhouse conditions in Tanzania, using three tomato cultivars, transplants obtained from seed treated with extracts from *A. vera*, *C. arabica* and *Y. schidigera* had significantly lower BLS incidence and severity than tomato seedlings from seed treated with sterile distilled water (Table 4). The ability of these plant extracts to control BLS in the three tomato cultivars grown in Tanzania without negatively affecting seedling growth indicated the potential of using these plant extracts as seed treatment against BLS pathogens.

We also demonstrated that aqueous extracts of 13 plant species (15.5 % of the tested extracts) had antimicrobial properties and inhibited the growth of *X. perforans* in in vitro and in *planta* assays. Plant extracts from *A. vera*, *C. arabica* and *Y. schidigera* were of particular interest as a control strategy to BLS of tomato because they consistently inhibited *X. perforans* in different experiments without negative effects on tomato seeds and seedlings. The ability of these plant extracts to control BLS xanthomonads in tomato also provides an alternative control approach against copper-resistant strains reported to be present in Tanzania (Shenge et al. 2007). These plants extracts must also be tested against BLS under farmers' conditions before they can be recommended for production of BLS free tomato transplants. More research is also needed to identify bioactive fractions from these plant extracts as well as their mechanisms of action.

ACKNOWLEDGMENTS

The authors wish to thank the Danish International Development Assistance (DANIDA) through ENRECA LIFE project No. 731 for providing funds for this work.

OPEN ACCESS

REFERENCES

1. Abdul-Baki AA, Anderson JD (1973) Vigor determination in soybean seed by multiple criteria. Crop Sci 13:630–633
2. Amadioha AC (2003) Evaluation of some plant leaf extracts against Colletotrichum lindemuthianum in cowpea. Acta Phytopathol Enthomol Hung 38:259–265

3. Anonymous (2006) List of new names and new combinations previously effectively, but not validly, published. Int J Syst Evol Microbiol 56:925–927
4. Black R, Seal S, Abubakar Z, Nono-Womdim R, Swai I (2001) Bacterial spot (Xanthomonas campestris pv. vesicatoria) of tomato and sweet pepper in Tanzania. Plant Pathol 50:810 (abstract)
5. Boateng JS (2000) Analysis of commercial samples of aloe. Ph.D. thesis, University of Strathclyde, Glasgow
6. Carrillo-Fasio JA, Garcia-Estrada RS, Allende-Molar R, Marquez-Zequera I, Millan-Ocampo S, Gaxiola-Espinoza G et al (2001) Sensitivity of Xanthomonas campestris pv. vesicatoria (Doidge) dye strains to copper. Rev Mex Fitopatol 19:72–77
7. Cock IE (2008) Antimicrobial activity of Aloe barbadensis Miller leaf gel components. Internet J Microbiol 4(2). http://www98.griffith.edu.au/dspace/bitstream/10072/21381/1/50582_1.pdf. Cited 12 April 2011
8. Daglia M, Cuzzoni MT, Dacarro C (1994) Antibacterial activity of coffee. J Agric Food Chem 42:2270–2272
9. Ferro VA, Bradbury F, Cameron P, Shakir E, Rahman SR, Stimson WH (2003) In vitro susceptibilities of Shigella flexneri and Streptococcus pyogenes to inner gel of Aloe barbadensis Miller. Antimicrob Agents Chemother 147:1137–1139
10. Gitaitis R, McCarter S, Jones J (1992) Disease control in tomato transplants produced in Georgia and Florida. Plant Dis 76:651–656
11. Gore JP, O'Garro LW (1999) Xanthomonas campestris pv. vesicatoria from bell pepper and tomato in Barbados undergoes changes in race structure, virulence and sensitivity to chemical control agents. J Phytopathol 47:397–402
12. Hassan SM, Byrd JA, Cartwright AL, Bailey CA (2010) Hemolytic and antimicrobial activities differ among saponin-rich extracts from guar, quillaja, yucca, and soybean. Appl Biochem Biotechnol 162:1008–1017
13. Hewett PD, Griffiths DC (1986) Biology of seed treatment. In: Jeffs KA (ed) Seed treatments. BCPC Publications, Thornton Health, Surrey, pp 7–12
14. Horsfall JG, Barrett RW (1945) An improved grading system for measuring plant disease. Phytopathology 35:655 (abstract)
15. International Seed Federation (ISF) (2007) Method for the detection of Xanthomonas campestris pv. vesicatoria on tomato seed. ISF Secretariat, Chemin du Reposoir, Nyon
16. International Seed Testing Association (ISTA) (2005) International rules for seed testing. Bassersdorf
17. Jones JB, Bouzar H, Stall RE, Almira EC, Roberts PD, Bowen BW, Sudberry J, Strickler PM, Chun J (2000) Systematic analysis of xanthomonads (Xanthomonas spp.) associated with pepper and tomato lesions. Int J Syst Evol Microbiol 50:1211–1219
18. Jones JB, Lacy GH, Bouzar H, Stall RE, Schaad NW (2004) Reclassification of the xanthomonads associated with bacterial spot disease of tomato and pepper. Syst Appl Microbiol 27:755–762
19. Kaaya NKF, Mortensen CN, Mabagala RB, Massomo SMS (2003) A guide on seed-borne bacterial diseases of tomato in Tanzania. Technical Bulletin, Danish Government Institute of Seed Pathology for Developing Countries (DGISP), Copenhagen, Denmark
20. Kim HK, Park Y, Kim HN, Choi BH, Jeong HG, Lee DG, Hahm KS (2002) Antimicrobial mechanism of β-glycyrrhetinic acid isolated from licorice, Glycyrrhiza glabra. Biotechnol Lett 24:1899–1902
21. Kumar J, Parmar BS (1996) Physicochemical and chemical variation in neem oils and some bioactivity leads against Spodoptera litura F. J Agric Food Chem 44:2137–2143
22. Lawless J, Allan J (2000) The chemical composition of Aloe vera. In: Aloe vera natural wonder cure. Thorsons Publishing, London, pp 161–171
23. Lee SD, Cho YS (1996) Copper resistance and race distribution of Xanthomonas campestris pv. vesicatoria on pepper in Korea. Korean J Plant Pathol 12:150–155

24. Martin HL, Hamilton VA, Kopittke RA (2004) Copper tolerance in Australian populations of Xanthomonas campestris pv. vesicatoria contributes to poor field control of bacterial spot of pepper. Plant Dis 88:921–924
25. Murthy PS, Manonmani HK (2009) Physico-chemical, antioxidant and antimicrobial properties of Indian monsooned coffee. Eur Food Res Technol 229:645–650
26. Myung IS, Jeong IH, Moon SY, Lee SW, Shim HS (2010) A new disease, arboricola leaf spot of bell pepper, caused by Xanthomonas arboricola. Plant Dis 94:271(abstract)
27. Opara EU, Wokocha RC (2008) Efficacy of some plant extracts on the in vitro and in vivo control of Xanthomonas campestris pv. vesicatoria. Agric J 3:163–170
28. Prakash A, Rao J (1997) Botanical pesticides in agriculture. Lewis Publishers, London
29. Pugh N, Ross SA, ElSohly MA, Pasco DS (2001) Characterization of aloeride, a new high-molecular-weight polysaccharide from Aloe vera with potent immunostimulatory activity. J Agric Food Chem 49:1030–1034
30. Punina NV, Ignatov AN, Pekhtereva ESH, Kornev KP, Matveeva EV, Polityko VA, Budenkov NI, Schaad NW (2009) Occurrence of Xanthomonas campestris pv. raphani on tomato plants in the Russian Federation. Acta Hortic 808:287–290
31. Reynolds T, Dweck AC (1999) Aloe vera leaf gel: a review update. J Ethnopharmacol 68:3–37
32. Scheck HJ, Pscheidt JW, Moore LW (1996) Copper and streptomycin resistance in strains of Pseudomonas syringae from Pacific Northwest nurseries. Plant Dis 80:1034–1039
33. Shenge KC (2006) Bacterial speck and spot diseases of tomato in Tanzania: pathogen characterization, epidemiology and management options. PhD thesis, Sokoine University of Agriculture, Morogoro
34. Shenge KC, Mabagala RB, Mortensen CN (2007) Identification and characterization of strains of Xanthomonas campestris pv. vesicatoria from Tanzania by biolog system and sensitivity to antibiotics. Afric J Biotech 6:015–022
35. Shirazi MH, Ranjbar R, Eshraghi S, Sadeghi G, Jonaidi N, Bazzaz N, Izadi M, Sadeghifard N (2007) An evaluation of antibacterial activity of Glycyrrhiza glabra extract on the growth of Salmonella, Shigella and ETEC E. coli. J Biol Sci 7:827–829
36. Singh DC (1994) Scope of medicinal and aromatic plants in pest management. In: Narwal SS, Tauro P (eds) International symposium, allelopathy in sustainable agriculture, forestry and environment, New Delhi, p 68
37. Stall RE, Beaulieu C, Egel D, Hodge NC, Leite RP, Minsavage GV, Bouzar H, Jones JB, Alvarez AM, Benedict AA (1994) Two genetically diverse groups of strains are included in Xanthomonas campestris pv. vesicatoria. Int J Syt Bacteriol 44:47–53
38. Tamir-Ariel D, Navon N, Burdman S (2007) Identification of genes in Xanthomonas campestris pv. vesicatoria induced during its interaction with tomato. J Bacteriol 189:6359–6371
39. Urch D (1999) Aloe vera nature's gift. Blackdown Publications, Bristol

CHAPTER 9

Suppression of Transgene RNA Silencing in Transgenic *Arabidopsis Thaliana* by a Mild Strain of *Cucumber Mosaic Virus*

Ken Nomura • Hidetoshi Uekusa • Nobuhiro Kita

Division of Agricultural Production Technology Science Kanagawa Agricultural Technology CenterHiratsukaJapan

ABSTRACT

The ability of *Cucumber mosaic virus* (CMV) subgroup II strain CMV-KT to suppress RNA silencing via posttranscriptional gene silencing (PTGS) was determined using *Turnip mosaic virus* (TuMV)-resistant transgenic *Arabidopsis thaliana* (TuR-At) carrying TuMV coat protein (CP) transgenes. CMV-KT induced only slight growth retardation in TuR-At plants, but no dwarf or mosaic symptoms, while subgroup I strain Y (CMV-Y) caused severe symptoms. Expression of the RNA-silenced TuMV-CP transgene resumed and produced TuMV-CP in either CMV-KT- or CMV-Y-infected TuR-At plants, indicating that the RNA silencing of the TuMV-CP gene was suppressed. Challenge inoculation experiments revealed that RNA silencing of the TuMV-CP transgene by CMV-KT infection is suppressed mostly in young developing leaves, but not in mature leaves, of TuR-At plants. These results demonstrated that transgene-derived virus resistance in transgenic *Arabidopsis* plants can be compromised by even a mild strain of CMV infection.

Keywords: RNA silencing Suppressor *Cucumber mosaic virus Turnip mosaic virus* Virus resistance

INTRODUCTION

Cucumber mosaic virus (CMV), the type species of genus *Cucumovirus*, consists of three genomic RNAs of different sizes (RNA1, RNA2, and RNA3) with two subgenomic RNAs (RNA4 and RNA4A) and is classified into one of three subgroups (IA, IB, or II) based on sequence homology (Palukaitis and García-Arenal. 2003; Roossinck et al. 1999). Some CMV strains carry an additional low molecular-weight satellite RNA that affects the expression of symptoms to various degrees (Devic et al. 1989; Gonsalves et al. 1982). Mutations within the viral coat protein (CP) or 2b protein gene(s) resulting in amino-acid substitutions also affect symptom development and disease severity (Shintaku et al. 1992; Suzuki et al. 1995). CMV-2b protein, which is involved in the long-distance movement and pathogenicity of the virus, functions as an RNA silencing suppressor (RSS) (Ding et al. 1994, 1995; Lewsey et al. 2009; Shi et al. 2002, 2003; Soards et al. 2002; Zhang et al. 2006). It suppresses RNA silencing via binding to Argonaute 1 and targeting small RNAs; the ability of 2b to bind small RNAs is critical for RSS activity (González et al. 2010; Goto et al. 2007). In addition to the RSS activity identified for the CMV 2b protein, RSS has been identified for proteins from various plant pathogenic viruses such as the HC-Pro from *Potyvirus*, the P25 protein from *Potexvirus*, and the P19 protein from *Tombusvirus*, all of which possess species-specific mechanisms of suppression (Anandalakshmi et al. 1998; Brigneti et al. 1998; Vargason et al. 2003; Voinnet et al. 2000).

The fact that plant viruses possess RSS implies that resistance to a specific virus via RNA silencing of the transgene would be lost when transgenic plants were co-infected with other viruses that also exhibit RSS. Actually, Savenkov and Valkonen (2001) reported that *Potato virus Y* (PVY) infection breaks down the *Potato virus A* resistance of transgenic *Nicotiana benthamiana* plants carrying the CP transgene. For PVY-resistant transgenic tobacco plants carrying the nuclear inclusion a (NIa) transgene, Mitter et al. (2003) showed that PVY resumed infectivity when the transgenic tobacco plants were pre-infected by a severe strain of CMV. The ability of a CMV severe strain to break down the transgenic resistance to a virus was also demonstrated in transgenic *Plum pox virus* (PPV)-resistant *N. benthamiana* plants containing the 5′-region of the *NIa* gene, while *Tobacco vein mottling virus* (TVMV) could not (Simón-Mateo et al. 2003). However, whether mild CMV strains can suppress the RNA silencing and whether the virulence level affects the efficiency of RNA silencing has not been determined.

Previously, we reported that transgenic *Arabidopsis thaliana* carrying the full coding region of a *Turnip mosaic virus* (TuMV)-CP gene showed complete resistance to TuMV isolates (Nomura et al. 2004). The purpose of the present study was to determine the RNA silencing suppression ability of a mild strain of CMV isolated from tomato using TuMV-resistant transgenic *A. thaliana* plants (TuR-At) carrying posttranscriptionally silenced TuMV-CP transgenes.

MATERIALS AND METHODS

Arabidopsis lines and viral strains

Three TuR-At *Arabidopsis* lines, CP112-R1, CP202-R1, and CP232-R1, carrying TuMV-CP transgenes with posttranscriptionally silenced expression, and a TuMV strain, TuMV-JO, were described previously (Nomura et al. 2004). Nontransgenic wild-type Col-0 of *Arabidopsis thaliana* was used as the control in these experiments. Mild strain CMV-KT and severe strain CMV-Y were used as counter viruses (Nomura et al. 2012; Suzuki et al. 1991).

Western blot analysis

Leaves (ca. 50 mg) were harvested and prepared for western blotting using TuMV-CP or CMV-CP rabbit polyclonal antiserum as described previously (Nomura et al. 2004).

Response of TuMV-resistant transgenic *Arabidopsis* plants to TuMV after CMV-KT pre-inoculation

Viral inoculum was prepared by grinding *N. tabacum* (tobacco) or *N. benthamiana* leaves infected either with CMV-KT or TuMV-JO in 0.1 M phosphate buffer (pH 7.0). When *A. thaliana* plants bore 10 fully expanded leaves, CMV-KT was inoculated as the first virus by rubbing the inoculum onto three fully expanded mature leaves. Plants were then kept in a growth chamber at 22 °C with 12 h light/12 h dark. 1 week after inoculation, establishment of CMV-KT infection in the inoculated plants was confirmed by western blot using rabbit polyclonal anti-CMV serum. For challenge inoculation tests, TuMV-JO was used to inoculate either three mature leaves 7 days later or three newly emerged developing leaves 14 days after the CMV-KT pre-inoculation. In both cases, establishment of TuMV infection in these challenge-inoculated transgenic plants was determined by RT-PCR 2 weeks after the challenge inoculation. Because the TuMV-inoculated transgenic plants contain TuMV-CP transgenes, RT-PCR was performed by targeting the nuclear inclusion b (NIb) gene of TuMV instead of the TuMV-CP transgene to specifically detect TuMV. A forward primer (NIbF1: 5′-GGAAACAATAGCGGACAGC-3′) and reverse primer (NIbR1: 5′-CTGGTGATAAACACAAGCTT-3′) were used with the RNA LA PCR Kit (Takara Bio, Otsu, Shiga, Japan) according to the manufacturer's instructions to amplify a 632-bp DNA fragment within the NIb gene. Amplified DNA fragments were separated by 1 % (w/v) agarose gel electrophoresis and cloned into the pT7 Blue vector (Novagen, Madison, WI, USA).

Sequencing analysis of the 2b gene of CMV-KT

A conserved region of 817 bp within the RNA2 genome of CMV-KT was cloned from total RNA extracted from CMV-KT-infected tobacco leaves by RT-PCR using the RNA LA PCR Kit (Takara Bio) with forward primer CMV.RNA2F1 (5′-TCGAGTTGAAATACAAGAAG-3′) and reverse primer CMV.RNA2R1 (5′-TGGTCTCCTTATGGAGAACC-3′) according to the manufacturer's instructions. Amplified DNA fragments were separated by 1 % (w/v) agarose gel electrophoresis and cloned into the pT7 Blue Vector (Novagen) for cycle sequencing using an automated 3130 DNA sequencer (Applied Biosystems, Foster City, CA, USA) according to the manufacturer's instructions.

RESULTS

Suppression of RNA silencing by a mild strain of CMV in transgenic *Arabidopsis* plants

In nontransgenic *Arabidopsis* plants, CMV-KT, a mild strain of CMV, induced slight growth retardation but not dwarfing or mosaic symptoms; however, CMV-Y, a severe strain, caused distinct dwarfing and mosaic, as described previously (Nomura et al. 2012). Although both CMV-Y and CMV-KT multiplied in all plants of three TuR-At-transgenic lines, CP112-R1, CP202-R1, and CP232-R1, as confirmed by western blot, the level of accumulation was higher in plants inoculated with CMV-Y than in those inoculated with CMV-KT (Figure 1a). When the same line of *Arabidopsis* plants was pre-inoculated with either CMV-Y or CMV-KT, TuMV-CP was detected by western blot in all of the TuR-At and wild-type plants that had been pre-inoculated with CMV, while no signal was detected in any of the plants that had not been inoculated with CMV (Figure 1b). Interestingly, TuMV CP accumulation levels differed depending on the tissue: TuMV CP was not detected in the CMV-pre-inoculated leaves or in mature leaves, but was detected at conspicuously high levels in the developing leaves and bolting shoots (data not shown). These results demonstrated that the RSS activity of CMV was high in developing leaves and bolting shoots and that mild strains of CMV can also suppress the RNA silencing of TuMV-CP transgenes in all of these TuR-At lines.

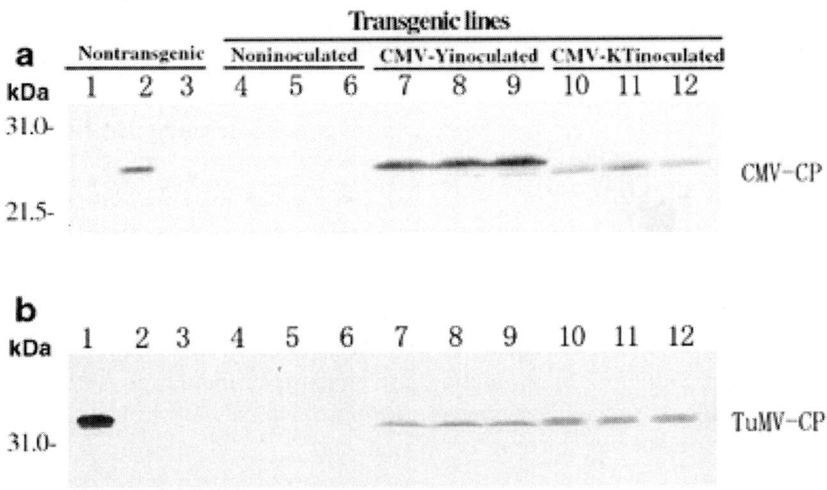

Figure 1. Response of *Turnip mosaic virus* (TuMV)-resistant transgenic *Arabidopsis thaliana* plants, carrying the coat protein (CP) gene of TuMV to inoculation with *Cucumber mosaic virus* (CMV). **A** Detection of inoculated CMV by western blot using rabbit antiserum against CMV-CP; **B** detection of TuMV-CP derived from the TuMV-CP transgene by western blot using rabbit antiserum against TuMV-CP. Nontransgenic, wild-type *A. thaliana* plants (Col-0) inoculated with either a severe CMV strain (CMV-Y) or TuMV were used as the controls. Transgenic *A. thaliana* plants containing posttranscriptionally RNA-silenced TuMV-CP gene(s). *Lane 1* Col-0 infected with TuMV, *lane 2* Col-0 infected with CMV-Y, *lane 3* noninfected Col-0, *lanes 4–6* transgenic lines (listed below) without CMV inoculation, *lanes 7–9* transgenic lines inoculated with CMV-Y, *lanes 10–12* transgenic lines inoculated with CMV-KT. Transgenic lines are CP112-R1 (in *lanes4, 7,* and *10*), CP202-R1 (in *lanes5, 8,* and *11*) and CP232-R1 (*lanes6, 9,* and *12*)

Response to TuMV infection in transgenic *Arabidopsis* plants pre-inoculated with CMV

To determine the effects of pre-inoculation with CMV-KT on the response of TuR-At to TuMV infection, the mature leaves of TuR-At plants were first inoculated with CMV-KT, and then their developing leaves were challenge-inoculated with TuMV at 14 days after the pre-inoculation. As a result, 40–60 % of the TuR-At plants lost their TuMV resistance and began to show not only growth retardation, but also typical viral symptoms comparable to those that occur in nontransgenic, wild-type plants (Figure 2a, b; Table 1). In contrast, when mature leaves were inoculated with TuMV at 7 days after CMV-KT pre-inoculation, TuMV resistance was maintained, and plants exhibited normal growth (Figure 2c). RT-PCR detection of TuMV in the TuR-At plants pre-inoculated with CMV-KT also confirmed that TuMV had multiplied in plants that had been challenge-inoculated with TuMV on developing leaves, whereas no TuMV CP was detected in plants that had been challenge-inoculated with TuMV on mature leaves (Figure 3). Therefore, these results demonstrate that suppression of RNA silencing of the TuMV-CP transgene by CMV-KT infection mostly occurs in the young developing leaves, but not in the mature leaves of TuR-At plants.

Table 1. Differential responses of *Turnip mosaic virus* (TuMV)-resistant transgenic *Arabidopsis thaliana* plants to challenge inoculation with TuMV after pre-inoculation with a mild strain of *Cucumber mosaic virus* (CMV)-KT of either developing or mature leaves

Leaf type[a] pre-inoculated	*Arabidopsis* line[b]	No. of plants with TuMV[c]			Infection rate (%)
		Inoculated	Infected	Detected	
Immature	CP112-R1	20	8	8	40
	CP202-R1	20	12	12	
	CP232-R1	20	12	12	60
	Col-0	10	10	10	100
Mature	CP112-R1	20	0	0	0
	CP202-R1	20	0	0	0
	CP232-R1	20	0	0	0
	Col-0	10	10	10	100

[a] CMV was used to rub-inoculate 3 mature leaves of plants at the 10-leaf growth stage, which were kept in a growth chamber at 22 °C with 12 h light/12 h dark until challenge inoculation with TuMV

[b] Wild type (Col-0)-derived, TuMV-resistant, transgenic *A. thaliana* lines that contain a transcriptionally RNA-silenced TuMV-coat protein (CP) transgene

[c] TuMV was used to challenge-inoculate either immature at 7 days or mature leaves at 14 days, after pre-inoculation with mild strain CMV-KT. Viral infection was confirmed by growth retardation, mosaic symptom on leaves, or both. TuMV multiplication was confirmed by RT-PCR using a specific primer pair to amplify the conserved region of the nuclear inclusion b (NIb) gene of TuMV

Figure 2. Differential response of an *Arabidopsis thaliana* wild-type plant and *Turnip mosaic virus* (TuMV)-resistant transgenic plants to TuMV infection. **A** A Col-0 plant inoculated with TuMV only; **B** CP112-R1 pre-inoculated with a mild strain of *Cucumber mosaic virus* (CMV)-KT onto mature leaves, followed by challenge inoculation with TuMV at 14 days after CMV-KT pre-inoculation; **C** CP112-R1 pre-inoculated with CMV-KT, followed by challenge inoculation with TuMV of mature leaves at 7 days after CMV-KT pre-inoculation

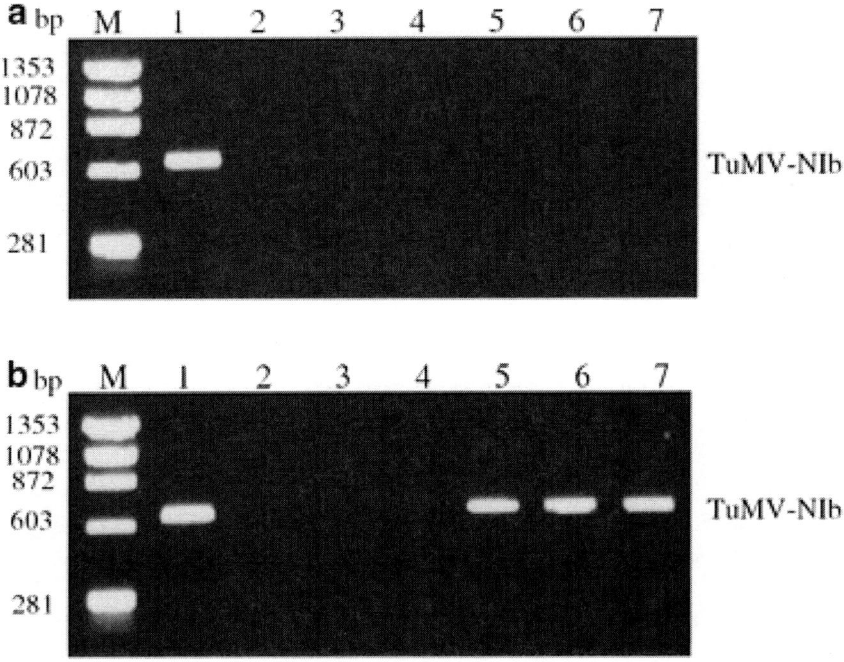

Figure 3 Detection of challenge-inoculated *Turnip mosaic virus* (TuMV) by RT-PCR in TuMV-resistant transgenic *Arabidopsis thaliana* plants carrying TuMV-coat protein transgenes after pre-inoculation with a mild strain of *Cucumber mosaic virus* (CMV)-KT of mature (**A**) or developing (**B**) leaves at 7 or 14 days after pre-inoculation, respectively. *Lane 1* TuMV inoculated nontransgenic Col-0; *lanes 2–4* noninoculated transgenic lines without CMV pre-inoculation; *lanes 5–7* TuMV-challenge-inoculated transgenic lines after pre-inoculation with CMV-KT. Transgenic lines are CP112-R1 (*lanes 2* and *5*), CP202-R1 (*lanes 3* and *6*) and CP232-R1 (*lanes 4* and *7*). RT-PCR was performed using a specific primer pair that can amplify the conserved region of nuclear inclusion b (NIb) gene of TuMV

Sequence analysis of the CMV 2b gene

After establishing that both CMV-KT and CMV-Y used in the present study could suppress RNA silencing in TuR-AT plants regardless of differences in their pathogenicity, we then determined the sequence of the CMV-KT 2b protein gene (GenBank/EMBL/DDBJ database accession: AB709965), which plays a major role in RNA silencing suppression, for comparison among various CMV strains. Homology analysis of the 2b genes demonstrated that CMV-KT shares 98.3 and 99.0 % homology with subgroup II strains CMV-Q and CMV-TN, respectively, but shares only 62.0 and 65.0 % homology with subgroup I severe strains CMV-Y and CMV-O, respectively. When the 2b sequence of CMV-KT was compared with that of CMV-Q in subgroup II, there were four single-base differences within the 2b protein gene and two of these four base changes were predicted to correspond to the amino acid substitution, L17H and I51F, while the remaining two were silent mutations. Two single-base changes in 2b protein gene were also found when comparing the 2b sequence of CMV-KT with CMV-TN, but both were silent mutations.

DISCUSSION

This study demonstrated that CMV-KT, a mild strain of CMV, can suppress the posttranscriptional RNA silencing of a transgene that confers TuMV resistance to transgenic *Arabidopsis* plants, allowing subsequent TuMV infection in the developing leaves. Under natural conditions, synergetic effects are often observed in various plant–virus interactions, including mixed infection by CMV and potyviruses. In radish, severe mosaic symptoms are induced synergistically by mixed infection of CMV and TuMV, although infection solely by CMV induces no symptoms in infected plants (Sano and Kojima 1989). CMV can also affect the host's innate metabolism and enhance disease severity after simultaneous infection of plants with potyviruses (Wang et al. 2002). In some cases, proteins that suppress RNA silencing mediate the synergistic interactions (Brigneti et al. 1998). Siddiqui et al. (2011) reported that a mild strain, CMV-Kin, enhances symptom development of tobacco plants infected by TMV. In this case, only the 2b protein of CMV is needed for this synergistic relationship. The present study suggests that pre-inoculation with a mild strain of CMV-KT allows TuMV to infect TuR-At plants by suppressing RNA silencing of the TuMV-CP transgene, probably via synergetic interaction with the CMV 2b protein.

The 2b protein of CMV was initially identified as a factor in two functions, viral movement and pathogenicity determination, and later as a factor in RSS (Brigneti et al. 1998; Ding et al. 1995; Lucy et al. 2000). In general, strains of CMV subgroup I are more virulent than those of subgroup II, and the 2b protein gene is a major determinant of viral virulence (Shi et al. 2002). The 2b protein of subgroup I strains tends to have higher RSS activity than do the 2b proteins of subgroup II strains (Ye et al. 2009). In general, the degree of symptom attenuation is positively correlated with reduced RNA silencing suppression activity. Kubota et al. (2003) reported that, in the case of a mild strain of TMV, L11, a single amino acid change within the 130-kDa protein accounts for the attenuation of its pathogenicity due to its reduced RNA silencing ability. Mutations within the HC-Pro gene of PPV, a member of the potyviruses, have been shown to be responsible for the differential ability of viral strains to suppress RNA silencing suppression and cause symptoms (González-Jara et al. 2005). Goto et al. (2007) reported that an attenuated CMV strain, CM95, had weaker RSS activity than that of its virulent revertant, and a key factor accounting for the noted difference in viral RNA silencing suppression was mapped to a single amino acid residue within the 2b protein. These results suggest that the difference in pathogenicity between CMV-KT and CMV-Y used in the present study affected the difference in suppressor activity of these strains. On the other hand, Lin et al. (2007) showed that an attenuated strain of *Zucchini yellow mosaic virus* did not affect RNA silencing suppression, implying that symptom attenuation is not always correlated with reduced levels of RNA silencing suppression. The CMV-KT used in this work causes mild mosaic symptoms in *N. tabacum* or tomato, whereas CMV-TN did not induce any symptoms in these plants (Nomura et al. 2012). The present study also indicated that differences in the pathogenicity of the mild strains CMV-KT and CMV-TN did not simply correlate with the mutations we identified within the 2b gene. As other genes and factors, such as CP genes and satellite RNAs, have been reported to also affect viral pathogenicity (Devic et al. 1989; Gonsalves et al. 1982; Shintaku et al. 1992; Suzuki et al. 1995), unknown factors other than base changes in the 2b protein gene might account for the differences observed in viral symptoms caused by CMV-Y or CMV-KT (Lewsey et al. 2009).

RNA silencing is known to be strongly suppressed in apical tissue and newly emerged leaves after CMV infection (Béclin et al. 1998; Brigneti et al. 1998). In the present study, we observed that suppression of RNA silencing was induced mostly in the new developing leaves. Breakdown of TuMV resistance by CMV-KT must have occurred in the new, developing leaves because CMV-KT cannot suppress the RNA silencing machinery in such developing tissues before viral invasion. The presence of CMV was essential for the suppression of silenced sense transgenes in new leaves (Béclin et al. 1998). These results suggest that the breakdown of TuMV resistance by CMV-KT may be due to the viral movement and distribution after infection. Further analysis is required to reveal the differential infection dynamics of various plant–virus interactions in which mixed infections with mild or weakly pathogenic viral strains affect the development and severity of virus-induced symptoms.

ACKNOWLEDGMENTS

This work was partly supported by a grant from the Regional Research Promotion Project of the Ministry of Agriculture, Forestry and Fisheries of Japan. We thank Yoko Takahashi, Kiyoko Nakata, and Mitsuko Ogasawara for their technical assistance.

REFERENCES

1. Anandalakshmi R, Pruss GJ, Ge X, Marathe R, Mallory AC, Smith TH, Vance VB (1998) A viral suppressor of gene silencing in plants. Proc Natl Acad Sci USA 95:13079–13084
2. Béclin C, Berthomé R, Palauqui JC, Tepfer M, Vaucheret H (1998) Infection of tobacco or Arabidopsis plants by CMV counteracts systemic post-transcriptional silencing of nonviral (trans)genes. Virology 252:313–317
3. Brigneti G, Voinnet O, Li WX, Ji LH, Ding SW, Baulcombe DC (1998) Viral pathogenicity determinants are suppressors of transgene silencing in Nicotiana benthamiana. EMBO J 17:6739–6746
4. Devic M, Jaegle M, Baulcombe D (1989) Symptom production on tobacco and tomato is determined by two distinct domains of the satellite RNA of cucumber mosaic virus (strain Y). J Gen Virol 70:2765–2774
5. Ding SW, Anderson BJ, Haase HR, Symons RH (1994) New overlapping gene encoded by the cucumber mosaic virus genome. Virology 198:593–601
6. Ding SW, Li WX, Symons RH (1995) A novel naturally occurring hybrid gene encoded by a plant RNA virus facilitates long distance virus movement. EMBO J 14:5762–5772
7. Gonsalves D, Provvidenti R, Edwards MC (1982) Tomato white leaf: the relation of an apparent satellite RNA and cucumber mosaic virus. Phytopathology 72:1533–1538
8. González I, Martínez L, Rakitina DV, Lewsey MG, Atencio FA, Llave C, Kalinina NO, Carr JP, Palukaitis P, Canto T (2010) Cucumber mosaic virus 2b protein subcellular targets and interactions: their significance to RNA silencing suppressor activity. Mol Plant Microbe Interact 23:294–303
9. González-Jara P, Atencio FA, Martínez-García B, Barajas D, Tenllado F, Díaz-Ruíz JR (2005) A single amino acid mutation in the Plum pox virus helper component-proteinase gene abolishes both synergistic and RNA silencing suppression activities. Phytopathology 95:894–901
10. Goto K, Kobori T, Kosaka Y, Natsuaki T, Masuta C (2007) Characterization of silencing suppressor 2b of cucumber mosaic virus based on examination of its small RNA-binding abilities. Plant Cell Physiol 48:1050–1060
11. Kubota K, Tsuda S, Tamai A, Meshi T (2003) Tomato mosaic virus replication protein suppresses virus-targeted posttranscriptional gene silencing. J Virol 77:11016–11026
12. Lewsey M, Surette M, Robertson FC, Ziebell H, Choi SH, Ryu KH, Canto T, Palukaitis P, Payne T, Walsh JA, Carr JP (2009) The role of the Cucumber mosaic virus 2b protein in viral movement and symptom induction. Mol Plant Microbe Interact 22:642–654
13. Lin SS, Wu HW, Jan FJ, Hou RF, Yeh SD (2007) Modifications of the helper component-protease of Zucchini yellow mosaic virus for generation of attenuated mutants for cross protection against severe infection. Phytopathology 97:287–296

14. Lucy AP, Guo HS, Li WX, Ding SW (2000) Suppression of post-transcriptional gene silencing by a plant viral protein localized in the nucleus. EMBO J 7:1672–1680
15. Mitter N, Sulistyowati E, Dietzgen RG (2003) Cucumber mosaic virus infection transiently breaks dsRNA-induced transgenic immunity to Potato virus Y in tobacco. Mol Plant Microbe Interact 16:936–944
16. Nomura K, Ohshima K, Anai T, Uekusa H, Kita N (2004) RNA silencing of the introduced coat protein gene of Turnip mosaic virus confers the broad-spectrum resistance in transgenic Arabidopsis. Phytopathology 94:730–736
17. Nomura K, Uekusa H, Kita N (2012) Characterization of cucumber mosaic virus isolated from tomato in Kanagawa Prefecture (In Japanese with English summary). Ann Rept Kanto Pl Prot Soc 59:43–45
18. Palukaitis P, García-Arenal F (2003) Cucumoviruses. Adv Virus Res 62:241–323
19. Roossinck MJ, Zhang L, Hellwald KH (1999) Rearrangements in the 5′ nontranslated region and phylogenetic analysis of cucumber mosaic virus RNA3 indicate radial evolution of three subgroups. J Virol 73:6752–6758
20. Sano Y, Kojima M (1989) Increase in cucumber mosaic virus concentration in Japanese radish plants co-infected with turnip mosaic virus. Ann Phytopath Soc Japan 55:296–302
21. Savenkov EI, Valkonen JPT (2001) Coat protein gene-mediated resistance to Potato virus A in transgenic plants is suppressed following infection with another potyvirus. J Gen Virol 82:2275–2278
22. Shi BJ, Palukaitis P, Symons RH (2002) Differential virulence by strains of Cucumber mosaic virus is mediated by the 2b gene. Mol Plant Microbe Interact 15:947–955
23. Shi BJ, Miller J, Symons RH, Palukaitis P (2003) The 2b protein of cucumoviruses has a role in promoting the cell-to-cell movement of pseudorecombinant viruses. Mol Plant Microbe Interact 16:261–267
24. Shintaku MH, Zhang L, Palukaitis P (1992) A single amino acid substitution in the coat protein of cucumber mosaic virus induces chlorosis in tobacco. Plant Cell 4:751–757
25. Siddiqui SA, Valkonen JPT, Rajamäki ML, Lehto K (2011) The 2b silencing suppressor of a mild strain of Cucumber mosaic virus alone is sufficient for synergistic interaction with Tobacco mosaic virus and induction of severe leaf malformation in 2b-transgenic tobacco plants. Mol Plant Microbe Interact 24:685–693
26. Simón-Mateo C, López-Moya JJ, Guo HS, González E, García JA (2003) Suppressor activity of potyviral and cucumoviral infections in potyvirus-induced transgene silencing. J Gen Virol 84:2877–2883
27. Soards AJ, Murphy AM, Palukaitis P, Carr JP (2002) Virulence and differential local and systemic spread of Cucumber mosaic virus in tobacco are affected by the CMV 2b protein. Mol Plant Microbe Interact 15:647–653
28. Suzuki M, Kuwata S, Kataoka J, Masuta C, Nitta N, Takanami Y (1991) Functional analysis of deletion mutants of cucumber mosaic virus RNA3 using an in vitro transcription system. Virology 183:106–113
29. Suzuki M, Kuwata S, Masuta C, Takanami Y (1995) Point mutations in the coat protein of cucumber mosaic virus affect symptom expression and virion accumulation in tobacco. J Gen Virol 76:1791–1799
30. Vargason JM, Szittya G, Burgyán J, Hall TMT (2003) Size selective recognition of siRNA by an RNA silencing suppressor. Cell 115:799–811
31. Voinnet O, Lederer C, Baulcombe DC (2000) A viral movement protein prevents spread of the gene silencing signal in Nicotiana benthamiana. Cell 103:157–167
32. Wang Y, Gaba V, Yang J, Palukaitis P, Gal-On A (2002) Characterization of synergy between Cucumber mosaic virus and potyviruses in cucurbit hosts. Phytopathology 92:51–58
33. Ye J, Qu J, Zhang JF, Geng YF, Fang RX (2009) A critical domain of the Cucumber mosaic virus 2b protein for RNA silencing suppressor activity. FEBS Lett 583:101–106

34. Zhang X, Yuan YR, Pei Y, Lin SS, Tuschl T, Patel DJ, Chua NH (2006) Cucumber mosaic virus-encoded 2b suppressor inhibits Arabidopsis Argonaute1 cleavage activity to counter plant defense. Genes Dev 20:3255–3268

CHAPTER 10

Effect of Methyl Salicylate (Mesa), an Elicitor on Growth, Physiology and Pathology of Resistant and Susceptible Rice Varieties

Kandaswamy Kalaivani[1], Marimuthu Maruthi Kalaiselvi[1] & Sengottayan Senthil-Nathan[2]

[1] *Post Graduate and Research Centre, Department of Zoology, Sri Parasakthi College for Women, Courtrallam-627 802, Tirunelveli, Tamil Nadu, India.*

[2] *Division of Biopesticides and Environmental Toxicology, Sri Paramakalyani Centre for Excellence in Environmental Sciences, Manonmaniam Sundaranar University, Alwarkurichi – 627 412, Tirunelveli, Tamil Nadu, India.*

ABSTRACT

Methyl salicylate (MeSA) is a volatile organic compound synthesized from salicylic acid (SA) a plant hormone that helps to fight against plant disease. Seed treatment with MeSA, is an encouraging method to the seed industry to produce more growth and yield. The aim of our study is to find out the growth, development and disease tolerance of rice seed treated with different concentrations of MeSA. Also the seed treatments were studied to determine whether they directly influenced seedling emergence and growth in rice (*Oryza sativa* L) cultivars 'IR 20, IR 50, IR 64, ASD 16, ASD 19 and ADT 46' under greenhouse condition. MeSA seed treatments at 25, 50, 75 and 100 mg/L significantly increased seedling emergence. Effects were stronger in IR 50, and IR 64 and the effects were dose dependent, although the relationship between dose and effect was not always linear. MeSA seed treated rice plant against bacterial blight were analyzed. Bacterial blight was more effectively controlled by the seed treated with 100 mg/L than others. These results suggest that seed treatment with MeSA alters plant physiology in ways that may be useful for crop production as well as protection.

INTRODUCTION

Plants are one of the basic staple food, for all forms of life. We depend on plants for food and medicine, plant derivatives, regulation of the hydrologic cycle. Plants are essential to the balance of nature[1]. For more than half of the world's population "Rice is life". Rice is an economic crop for the farmers and major food staple for the world. Rice provides the main source of food energy for at least one third of the world's population[2,3,4]. Rice is an essential part of the daily meal in Southern and Eastern parts of India and Asia. However most the rice plant are affected by the Bacterial blight caused by *Xanthomonas oryzae* pv. *oryzae (Xoo)* and it is one of the most important bacterial diseases of rice. Heavy yield losses due to *X. oryzae* was reported during last one decade[3,5,6].

Seed germination is one of the early and vital stages in the plant life cycle. Improved germination of seeds is one process which can improve the economic success of crop by ultimately enhancing the yield. The process of seed germination may be regulated by external factors, including temperature, light, soil moisture and salinity, metal or mineral composition[7,8].

Treatments of seed include applications of chemical agents, biological substances, or physical processes, to increase agricultural production. The seed treatment helps to protect seedlings from insects and disease which pose devastating consequences to crop production during critical growth stages. The application of treatments to seeds promotes stronger and healthier roots, leading to uniform seedling emergence, with improved crop performance[9,10].

Seed priming is a method of pre-sowing treatments that exposes seeds to a certain solutions that allows partial hydration until the germination process begins[11,12,13].Various seed priming techniques have been reported to enhance germination and allow for more uniform seed emergence, increasing nutrient reserves through increased physiological activities and increased yield in the crop[14].

Various common methods utilized for priming seeds includes: hydro priming, (soaking in water); osmopriming (soaking seed in osmotic solutions); thermopriming (treatment of seed with low or high temperatures); solid matrix priming (treatment of seed with solid matrices); and bio priming (hydration using biological compounds)15. Halo priming (soaking seed in salt solutions), and priming with plant growth hormones are also performed along with thermo priming, and on-farm priming16.

The mechanism of rice seed priming techniques which uses plant growth regulators may enhance rice performance. Plant hormones or bio-stimulants are also known to improve plant population and increase productivity. Presoaking seeds with optimal concentrations of plant growth hormones, such as auxin (IAA), gibberellins (GA), abscisic acid, ethylene (ET), Salicylic acid (SA) and Jasmonic acid (JA) have been shown to benefit germination performance and yield of some crop species growth under both normal and stress conditions17,18,19. The effects of seed priming by soaking seeds of sorghum, rice, or wheat in water, are reported to enhance germination rates. Some strategies for more vigorous plants report improved seedling emergence under induced stress conditions20,21.

SA is a natural phenolic compound that plays a role in the regulation of physiological and biochemical processes. SA is chemically known as 2-hydroxy benzoic acid is a member of phenolic compounds, containing of an aromatic ring bearing a hydroxyl group which is synthesized by the plant22. These hydroxyl groups are involved in lignin biosynthesis; others serve as allelopathic compounds, regulate plant responses to abiotic stress and disease resistance by inducing antimicrobial defense compounds termed phytoalexins or by signaling defense activation23,24,25,26.

Exogenous application of SA act as potential antioxidant capacity inducers, also increase the levels of hydrogen peroxide in plants. These pathways also regulate many physiological processes in plants, such as increased germination rate, cell growth, stomatal opening, photosynthesis and ion uptake27. Methyl salicylate (MeSA) is a volatile organic compound which is synthesized from SA. MeSA is associated with induced resistance plant defense which has been used in effective disease control methods28,29,30.

Seed treatment with MeSA can be used to enhance rice seed germination and seedling growth. We studied the performance of rice seed priming with different concentrations of MeSA (0, 25, 50, 75 and 100 mg/L). Variables assessed include: germination rate, growth of root and shoot heights (cm), biomass of fresh and dry weights of the shoots and roots (mg) disease incidence, mean lesion lengths, disease index and disease incidence of different rice varieties under greenhouse conditions.

RESULTS

Germination responses after seed treatment with MeSA on rice seed in greenhouse condition

The seed pretreatment of different concentration of MeSA (i.e. 0, 25, 50, 75 and 100 mg/L) had significant ($P < 0.05$) effects on the final germination percentage and germination rates (Figs 1 and 2). Treated seeds with MeSA had significantly higher percentage of germination than those in control. Treatment of 'IR 20, IR 50, IR 64, ASD 16, ASD 19 and ADT 46' rice varieties with MeSA increased the rate of seedling emergence (Figure 1A–F). Emergence of rice varieties were more pronounced in all MeSA treatment with 100 mg/L concentration being statistically different with all other treatments. For example 80% germination was observed within 8 days in IR 20 rice varieties after seed treated with 100 mg/L. All the rice varieties treated with MeSA, had greater emergence values compared with control rice plant. In 'IR 20' 20% emergence, 40% emergence, 60% emergence and 80% emergence occurred earlier in MeSA treated plants than in control plants (20%-$F_{4,30} = 3.08$; $P < 0.0031$; 40%- $F_{4,30} = 2.75$; $P < 0.046$, 60%- $F_{4,30} = 3.56$; $P < 0.08$ and 80%- $F_{4,25} = 5.10$; $P < 0.004$) (Figure 2A) there was a trend towards faster emergence with increased concentration seed treatment rates. The same developments were observed in rice variety 'IR 50' seedling emergence (20%-$F_{4,30} = 2.62$; $P < 0.055$; 40%- $F_{4,30} = 3.24$; $P < 0.025$, 60%- $F_{4,30} = 2.65$; $P < 0.053$ and 80%- $F_{4,29} = 4.24$; $P < 0.008$) (Figure 2B).

Germination percentage was influenced by MeSA seed treatment, but influence was more pronounced in 100 mg/L than other treatments. Control seeds emerged at lower than 20% at 4 days. However most of the treated rice seeds germinated at 20%, except (IR 20 and ASD 19), at 4 days, being statistically significant. The same observation was noticed in rice varieties 'IR 64' (20%-$F_{4,30} = 3.28$; $P < 0.024$; 40%- $F_{4,30} = 2.48$; $P < 0.065$, 60%- $F_{4,30} = 2.84$; $P < 0.041$ and 80%- $F_{4,26} = 2.33$; $P < 0.082$) (Figure 2C), 'ASD 16' (20%-$F_{4,30} = 3.05$; $P < 0.032$; 40%- $F_{4,30} = 2.31$; $P < 0.081$, 60%- $F_{4,28} = 5.49$; $P < 0.002$ and 80%- $F_{4,13} = 2.95$; $P < 0.061$) (Figure 2D).

The germination rate of seeds was increased after treatment with MeSA in resistant rice varieties ASD 19 (20%-$F_{4,30} = 3.06$; $P < 0.031$; 40%- $F_{4,30} = 3.80$; $P < 0.013$, 60%- $F_{4,28} = 5.61$; $P < 0.002$ and 80%- $F_{4,13} = 3.19$; $P < 0.049$) (Figure 2E) and ADT 46 (20%-$F_{4,30} = 2.88$; $P < 0.040$; 40%- $F_{4,30} = 2.78$; $P < 0.045$, 60%- $F_{4,27} = 5.46$; $P < 0.002$ and 80%- $F_{4,11} = 3.17$; $P < 0.058$) (Figure 2F). Compared with control, the seeds showed greater germination rates when treated with MeSA. The germination rates were significantly greater when treated with MeSA at 100 mg/L.

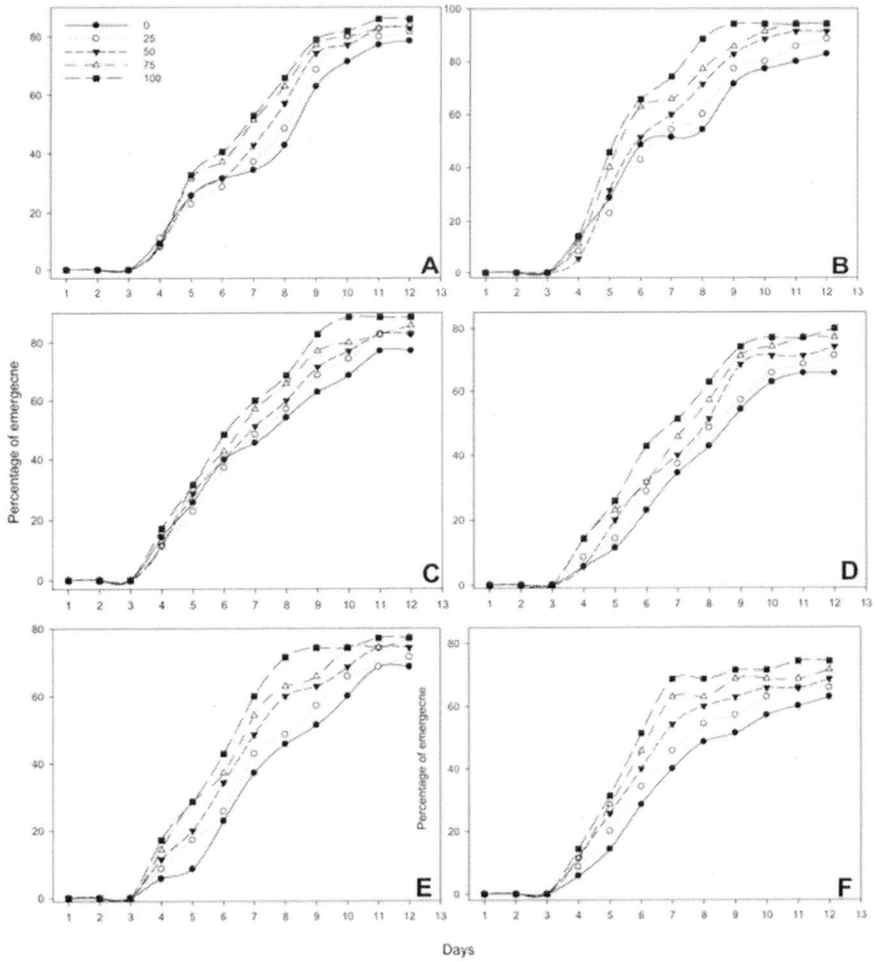

Figure 1. Percentage of emergence of rice varieties treatment with MeSA across 12 days post treatment.(A-IR 20; B-IR 50; C- IR 64; D-ASD 16; E-ASD 19; F-ADT 46).

Seed-priming treatments with MeSA significantly enhanced the seedling growth of rice compared with non-treated control. The greatest effect was observed from seed-coating treatments of 100 mg/L MeSA (Figure 3). On average, seed priming treatments significantly increased the shoot length by 40.78% ($F_{4,30} = 14.88$; $P < 0.0001$), 24.31% ($F_{4,30} = 17.05$; $P < 0.0001$),

20.90%($F_{4,30} = 10.73$; $P < 0.0001$), 35.34% ($F_{4,30} = 7.30$; $P < 0.0001$), 25.90% ($F_{4,30} = 7.88$; $P < 0.0001$) and 34.82% ($F_{4,30} = 11.10$; $P < 0.001$) for IR 20, IR 50, IR 64, ASD 16, ASD 19 and ADT 46 respectively compared with control. The same trend was also noticed in root length after treatment with 100 mg/L MeSA. For example 54.46% increased root length was observed in ASD 16 after treatment with 100 mg/L MeSA which was significantly different from control ($F_{4,\ 30} = 11.25$; $P < 0.0001$). Insignificant difference was observed between treatments of 75 and 100 mg/L MeSA.

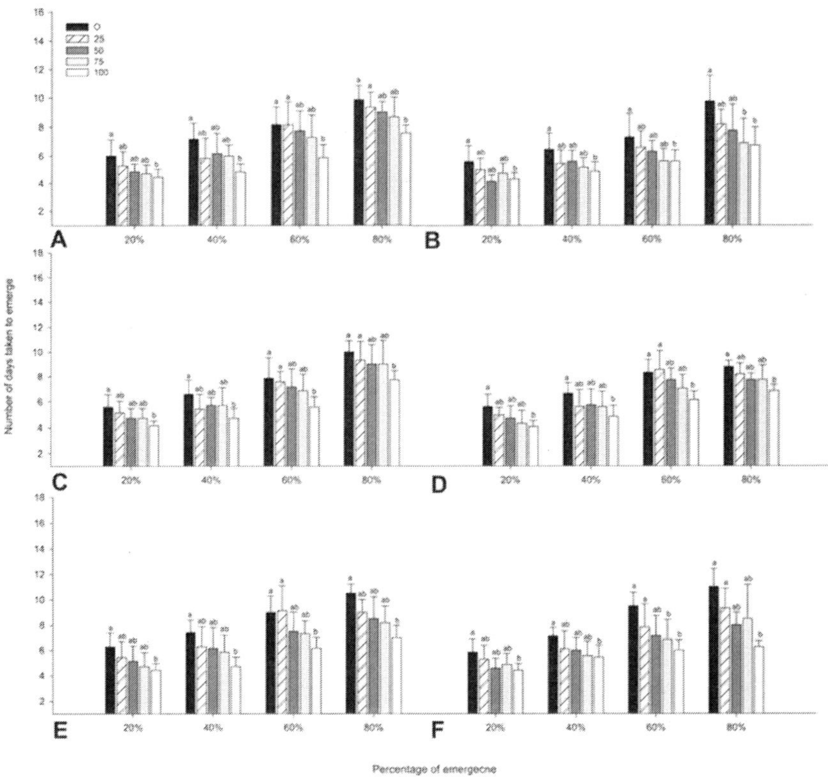

Figure 2. Days until 20%,40%, 60% and 80% emergence of rice varieties IR 20 (**A**), IR 50 (**B**), IR 64 (**C**), ASD 16 (**D**), ASD 19 (**E**), ADT 46 (**F**) under greenhouse condition after seed treatment with Me SA. Mean (±SEM) followed by the same letter in an individual experiment indicate no significant difference ($P < 0.05$) in a Tukey's test (Treatment concentration in mg/L).

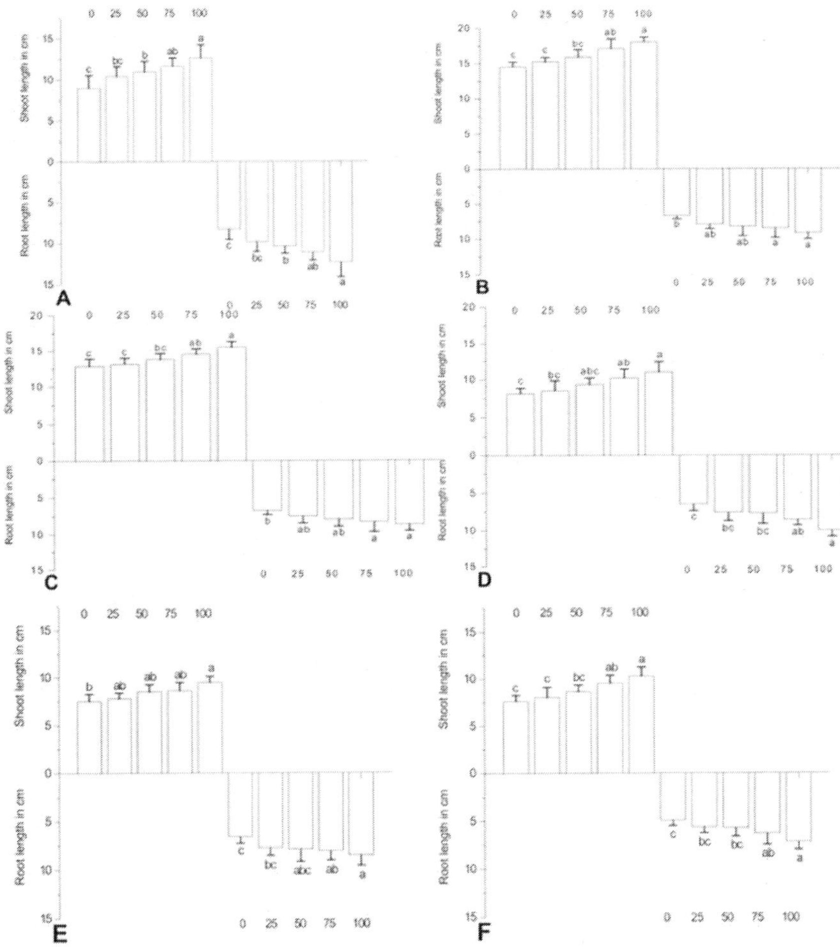

Figure 3. Root and shoot length in cm of rice varieties 22 days post treatment with MeSA (A-IR 20; B-IR 50; C- IR 64; D-ASD 16; E-ASD 19; F-ADT 46). Mean (±SEM) followed by the same letter in an individual experiment indicate no significant difference (P < 0.05) in a Tukey's test (Treatment concentration in mg/L).

Treatment of seeds with MeSA also improved root: shoot ratio of the rice seedlings under greenhouse condition. For 'ASD 16', the effect of MeSA seed treatment was significant for root: shoot ratio ($F_{4,30} = 4.44$; $P < 0.006$) (Figure 4A). Interestingly, there was a non-linear response for IR 50 and IR 64, root: shoot ratios and they were not significant between 25–100 mg/L treatments ($F_{4,\ 30} = 3.73$; $P < 1.00$ and $F_{4,\ 30} = 3.98$; $P < 1.00$ for IR 50 and IR 64 respectively) (Figure 4A). Further we have observed significant growth of root and shoot morphologically after seed were treated with MeSA (Figure 4B).

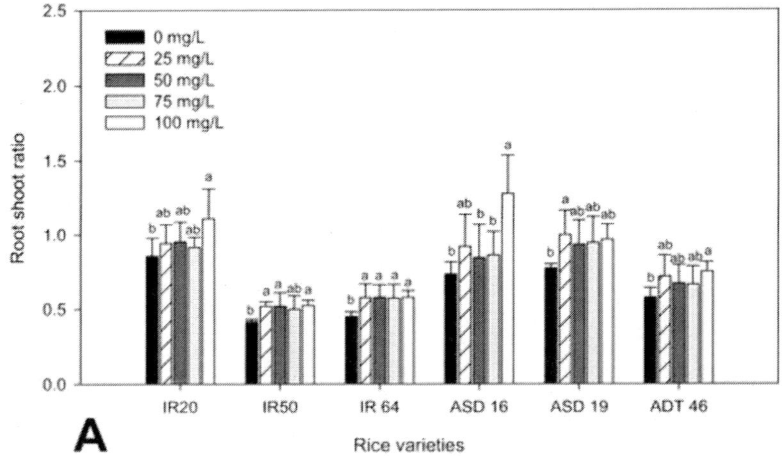

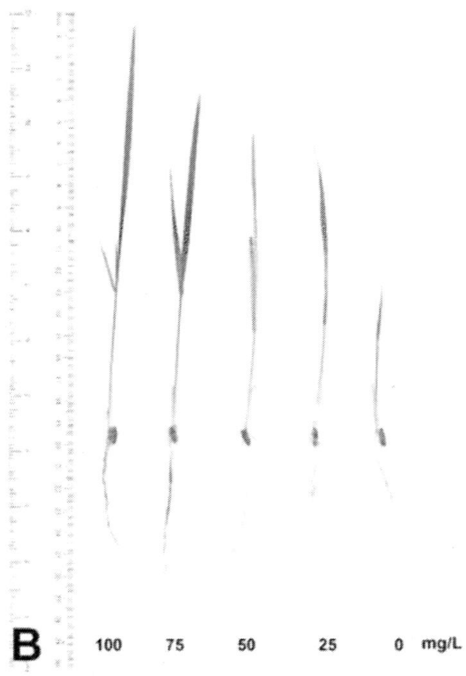

Figure 4. Root and shoot ratio of rice varieties (**A**) (A-IR 20; B-IR 50; C- IR 64; D-ASD 16; E-ASD 19; F-ADT 46) after treatment with MeSA. Mean (±SEM) followed by the same letter in an individual experiment indicate no significant difference ($P < 0.05$) in a Tukey's test (Treatment concentration in mg/L). (**B**) Root and shoot growth after treatment with MeSA of ASD 16 rice cultivar.

All the seed treatments had significant effects on fresh and dry weight of shoot and root (Figure 5). Shoot fresh weight significantly increased in seedlings raised from seeds treated with 75 and 100 mg/L MeSA. Maximum shoot fresh weight of 457.4 mg was attained in ASD 16 seeds treated with 100 mg/L MeSA after 20 days as compare to control (361.8 mg) (Figure 5). However minimum shoot fresh weight of 270.66 mg was observed in control seeds of IR 50 after 20 days. Shoot and root fresh and dry weights of 'IR 20, IR 50, IR 64, ASD 16, ASD 19 and ADT 46 treated and untreated plants are shown in Figure 5A–X. Seed treatment with 100 mg/L MeSA significantly increased shoot fresh weights ($F_{4,20} = 14.69$, $P < 0.0001$-IR 20; $F_{4,20} = 5.75$, $P < 0.003$-IR 50; $F_{4,20} = 7.69$, $P < 0.001$-IR 64; $F_{4,20} = 7.18$, $P < 0.001$-ASD 16; $F_{4,20} = 3.22$, $P < 0.035$-ASD 19; $F_{4,20} = 3.91$, $P < 0.017$-ADT 46) and dry weights ($F_{4,20} = 24.08$, $P < 0.0001$-IR 20; $F_{4,20} = 69.90$, $P < 0.0001$-IR 50; $F_{4,20} = 7.88$, $P < 0.001$-IR 64; $F_{4,20} = 29.06$, $P < 0.001$-ASD 16; $F_{4,20} = 7.60$, $P < 0.001$-ASD 19; $F_{4,20} = 13.31$, $P < 0.0001$-ADT 46) at 22 days post treatment. The same linear growth pattern was observed both in fresh and dry weight of root after seed treatment with MeSA. In general, a positive relationship was observed between seed treatment concentration rate and root and shoot weights (Figure 5).

Control of bacterial blight under greenhouse condition

The effect of MeSA seed treatment in suppressing bacterial blight disease was analyzed under greenhouse condition. Bacterial blight disease lesion length (mm) (Figure 6) and Disease index (Figure 7) was found to be significantly lower in 100 mg/L MeSA seed treated rice plant. However other lower seed treatments also found to be effective than control. The bacterial blight disease lesion length (mm) was 2.94 mm in control but the ADT 46 rice seed treated with 100 mg/L was inoculated with *X. oryzae* lesion length (mm) measured as 1.92 mm ($F_{4,20} = 4.39$, $P < 0.01$) (Figure 6). However we did not observe any significant difference between control and 25 mg/L MeSA seed treated rice ADT 46 plant ($F_{4,20} = 4.39$, $P < 0.93$).

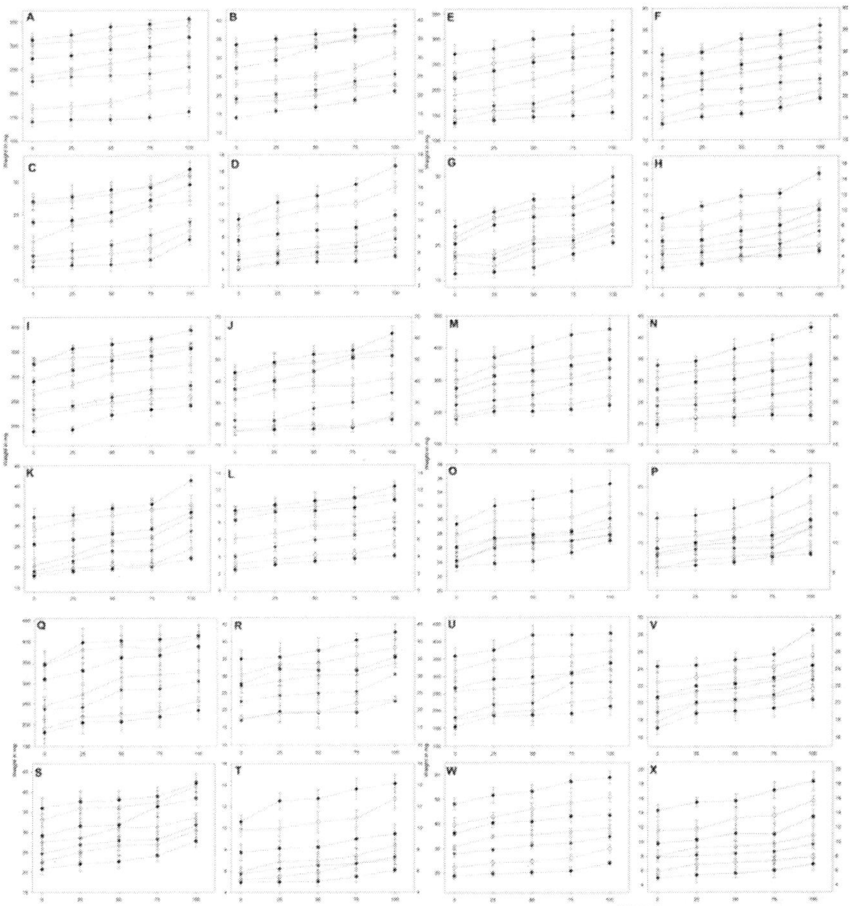

Figure 5. Fresh (**A,E,I,M,Q,U**) and dry (**B,F,J,N,R,V**) weight (mg) of shoot; fresh (**C,G,K,O,S,W**) and dry (**D,H,L,P,T,X**) weight (mg) of root of IR 20,IR 50, IR 64, ASD 16, ASD 19 and ADT 46 rice varieties after treatment with MeSA in greenhouse condition (-●-10th day -○- 12th day -▼- 14th day-△- 16th day -□- 18th day-■- 20th day -♦-22nd day) (MeSA treatment doses are mg/L).

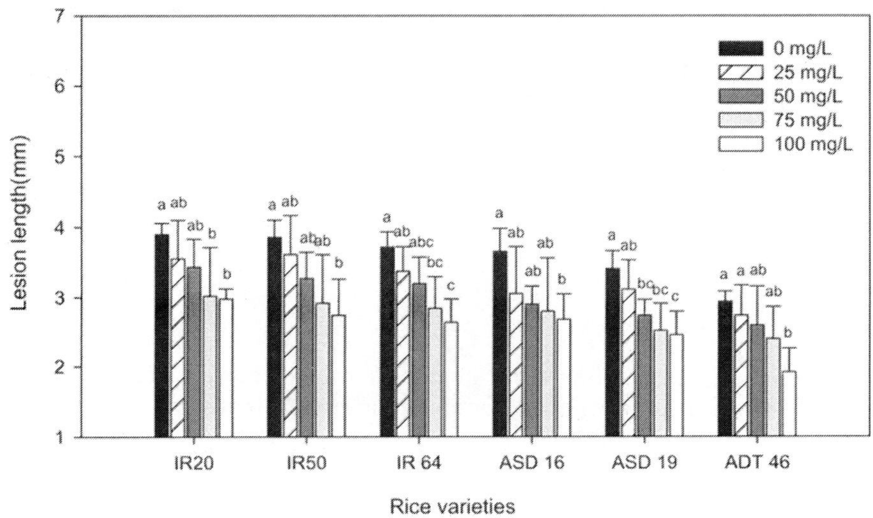

Figure 6. Mean lesion length (mm) caused by *X. oryzae* after treatment with MeSA. Means followed by the same letter were not significantly different ($P < 0.05$) according to a Tukey test) (Treatment concentration in mg/L).

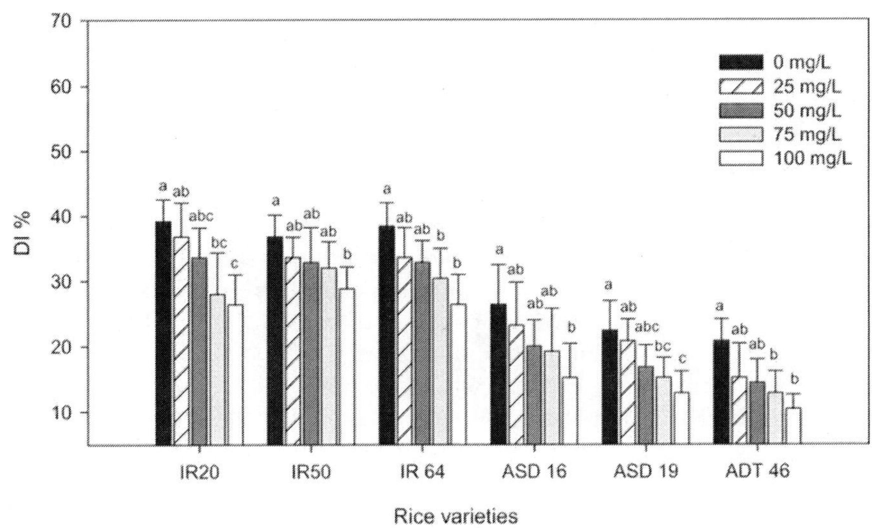

Figure 7. Disease index (DI in percentage) of rice after treatment with MeSA. Means followed by the same letter were not significantly different ($P < 0.05$) according to a Tukey test) (Treatment concentration in mg/L) Figure 1.

The same trend was observed in disease index. The disease index (DI) was higher in (39.2%) control IR 20 plant and the DI was significantly lower in IR 20 rice plant in which the seed was treated with 100 mg/L MeSA (26.40%) ($F_{4,20} = 6.33, P < 0.002$) (Figure 7). Even though three rice varieties (i.e. ASD 16, ASD 19 and ADT 46) are resistant to bacterial blight disease, seed treatment with MeSA further enhance the resistant against the bacterial blight disease and it was proved by the disease index experiment.

DISCUSSION

Treatment of seeds with MeSA increased rate of germination and rice growth. The amount of effect differed among rice varieties and effects were dose rate dependent. Also, effects appeared to be more pronounced at 75 and 100 mg/L seed treatments under normal greenhouse conditions. MeSA influences seed germination, seedling establishment, cell growth, respiration, stomatal closure, senescence-associated gene expression, basal thermotolerance, nodulation in legumes, and fruit yield25,31.

Most consistent response to MeSA seed treatment in this research was an increase in rate of emergence of seeds and biomass of the rice plants after sowing. The effect of MeSA was stronger both in 'IR 20 and IR 50' than other rice varieties, which produced higher seedling emergence rates (Kalaivani and Senthil-Nathanunpublished data).

This study shows that rice plant could be induced by exogenous application of the signaling compounds MeSA. The allelopathic potential of the root and shoot of rice thus increased after treatment with MeSA (Figs 1, 2, 3, 4, 5). The same effect was observed in the rice strains IAC165 and Huajingxian32. In the present study MeSA treatment with 100 mg/L of rice seed showed enhanced effects on emergence and biomass of rice seedlings. Other studies have also shown that natural plant products can enhance defense system of plant along with protection from pest and diseases33,34,35. Further plant hormonal and other treatments enhanced seedling vigor as measured in longer shoot and root lengths, with increased fresh and dry weights. Accelerated cell division and enlargement of cell explain the production of vigorous seedlings by pre-sowing seed treatment with plant growth regulators16.

Several studies have shown impact of MeSA seed treatment on germination and early seedling growth. Enhanced germination, emergence and growth were recorded in wheat, *Triticum aestivum* L. when the seeds were subjected to pre-sowing seed-soaking treatment in salicylic acid36. 3 Senthil-Nathan *et al.*37 reported that neem treatment on rice resulted in increased production and diseases tolerance than control.

Hayat *et al.*38 observed the grains of wheat (*Triticum aestivum* L. cv. Raj-3077) were soaked in 0, 10^{-5}, 10^{-4} and 10^{-3} M aqueous solutions of seedlings possessed significantly greater leaf numbers, fresh and dry mass. The present finding gained support from these previous reports.

Shakirova *et al.*39 conclude that pre-sowing *T. aestivum* seed treatment with SA leads to an activation of germination and seedling growth. The enhancement results in the division of root apical cells which is an important contribution to the growth stimulating effect of SA. Shakirova36 has confirmed with field experiments. In the field, wheat plants pretreated with SA were characterized with increased size of ears, grain yield, indicating prolonged effect of pre-sowing treatment of seeds, which produced stimultative effect on the productivity of wheat, at harvest.

Rate of seedling emergence was increased approximately 25% at seed treatment with 75 and 100 mg/L MeSA. Emergence of 'IR 20, IR 50 and ASD 16' seedlings was significantly faster in MeSA treated seeds, but it is not linear with treatment doses. The percentage of emergence (20, 40, 60 and 80%) at the different seed treatment rates indicated that emergence increased with seed treatment concentration rate, with faster emergence from seeds treated with 100 mg/L MeSA. The increased percentage of emergence and biomass with increasing dose is probably due to increased SA in plant cells, as reported in other crops. For example, increased percentage of emergence and growth was observed by Singh and Usha40 and Dolatabadian *et al.*41 in wheat seeds, Tavares *et al.*28 in rice, Khodary42 in maize, Jadhav and Bhamburdekar43and Szepesi *et al.*44 in tomato, Martin-Mex *et al.*45 in African violet, McCue *et al.*46 in pea seeds, Almeida *et al.*47 in wheat and barley seeds which revealed that MeSA accelerates germination greatly. Seed treatment with MeSA may help in the metabolic pentose phosphate pathway, benefiting the hydrolysis of reserves and increasing the availability of energy to the germination process and seedling emergence25.

Several reports were published in the last decade demonstrating the role of SA applied as seed soaking treatment on various physiological processes44,45,46,47. It was shown that the pre-soaking of pea seeds in SA had a beneficial effect on growth and photosynthesis48.

MeSA treatments produced measurable effects on root and shoot growth of rice seedling varieties. Gutierrez-Coronado *et al.*49 reported that foliar application of SA to the shoots of soybean plants positively affected the size of the root. Furthermore Sandoval-Yepiz50 reported similar findings with *Tagetus erecta*, using low concentrations (10^{-10} M) of SA. Increased shoot growth was observed in *Clitoria* sp. after application of salicylates51. A more in-depth study would be required to determine the specific modes of action of MeSA on non-linear growth of roots and shoots in different crops.

The increase observed in shoot and root weight when treated with MeSA, was dose dependent (75 and 100 mg/L). This response suggests that a greater portion of photosynthates was distributed to the roots than the shoots in seedlings developing from nontreated seeds as compared with seedlings developing from MeSA treated seeds[25,36,52]. The promoting effect of SA on the leaf area was attributed to its important roles on activating cell division, biosynthesis of organic foods, water uptake and stomatal resistance and tolerating water deficit better. In addition, mentioned that improving effect of SA on the ready availability of nutrients (i.e. N, P, K, Ca, and Mg) to the plant. Increased levels of mineral after treatment with SA were observed in maize[53], tomato[44] and cucumber[54]. These results are in support with many other researchers who reported an increased amount of nutrients in roots and shoot after treatments with SA[27,55,56].

According to Stout *et al.*[57] and Vazirimehr *et al.*[27] plant seeds treated with SA have higher levels of amino acids, plant growth, ion uptake, transport enzyme activity and synthesis of plant hormones, that increase the plant responses to these proteins and these events provide significant increases in production and reduce time for the establishment of the crop field, being more tolerant against abiotic stresses. The data of the present study support the above hypothesis.

For more than two decades the plant hormone salicylic acid (SA) and its derivatives (MeSA) has been studied for its use to protect the plant. However, its systemic acquired signaling act in plants, particularly in defense against plant disease causing microbes, has only become more interest during last two decades[31,57]. Our research has clearly pointed out the MeSA has a potential to control the bacterial blight disease. Also seed treated with 100 mg/L has less blight lesion and diseases index. It was evident that SA and its derivatives act as a critical defense signal has come from analyses of plants in which endogenous SA levels were triggered or altered. It was proved in series of experiments in Arabidopsis (Transgenic tobacoo) expressing the bacterial *nahG* gene, encoding the SA-metabolizing enzyme salicylate hydroxylase[31,58,59]. In addition, endogenous SA levels increase in barley resisting infection by the bacterial pathogen *P. syringae* pv. *syringae*[60]. Also in rice homolog enhances resistance to the bacterial blight-causing *X. oryzae*[61,62,63].

MeSA seed treatments (75 and 100 mg/L) significantly increased the emergence and early growth of rice plants. But the effect was non-linear across treatment concentrations and rice varieties. Further we conclude that use of MeSA in seed treatment has great potential for increasing uniform emergence, root and shoot length, root and shoot biomass. Thus, treatment of seeds with SA may produce favorable effects on rice growth. Also MeSA seed treatment has better protection against blight diseases.

METHODS

Plant material and seed treatments

A commercial formulation of methyl salicylate (Methyl salicylate, Sigma-Aldrich) was used for all experiments. The rice varieties 'IR 20, IR 50, IR 64, ASD 16, ASD 19 and ADT 46' were used for this study. 'IR 20, IR 50, IR 64' are susceptible, conventional, semi-dwarf, long-grain aromatic variety with relatively high seedling vigor, whereas 'ASD 16, ASD 19 and ADT 46' are moderate resistance, conventional, semi-dwarf long grain variety with low seedling vigor. Rice were surface disinfected by immersing in 2% sodium hypochlorite for about two minutes and washed several times with sterilized water, then dried in sterilized filter paper.

Salicylate formulations and treatments consisted of increasing levels of 0, 25, 50, 75 and 100, mg/L of MeSA. A stock solution of MeSA 100 mg/L was used for successive dilutions in distilled water. The seeds were treated by applying 2 ml/kg of seed in each respective MeSA concentration according to methodology described by Tavares *et al.*28. The solution of MeSA with the different doses was directly placed at the bottom of the zip lock plastic bag. Then 250 g of seeds were added into each bag. They were agitated for 10 minutes. The seeds were spread out on clean paper and left to dry at room temperature for 24 hours. Treated seeds were used for further experiments.

Seed treatments under greenhouse conditions

'IR 20, IR 50, IR 64, ASD 16, ASD 19 and ADT 46' rice plants were grown by sowing five seeds of a single treatment rate (0, 25, 50, 75 and 100 mg/L) in 0.5 L pots in a greenhouse. There were 7 pots (replicates) per treatment. The potting soil was composed of two parts of autoclaved silt loam soil with one part each of sand and peat moss. The experiment was conducted in early January and, as temperature regulation in the greenhouse was poor, temperatures fluctuated from 22° to 32 °C. Lighting was ambient, and pots were watered as needed.

Pots were observed daily for emergence of hypocotyls above the surface of the soil after sowing. The experiment was terminated after 12 days. For statistical analysis, days until 20%, 40% and 60% seedling emergence were calculated for each pot. At 12th day after sowing, all seedlings were removed from pots, counted, and cleaned with water. Maximum root length (cm/plant) shoot length (cm/plant), weight (both wet and dry weight in mg) for each plant were determined by using a ruler and monopan balance.

Germination was scored daily by counting emerging plant shoots. The experiment was terminated after 22 days. Percent emergence was calculated as in the greenhouse experiment. Roots and shoots were weighed after cleaning and patting dried with a paper towel. After taking fresh weights, plants were dried in an oven at 40 °C for two days and weighed.

Effect of MeSA seed treatment on the bacterial blight disease of rice under greenhouse conditions

Rice seed treated with MeSA were sown in earthen pots. The potting soil was composed of two parts of autoclaved silt loam soil with one part each of sand and peat moss and maintained under greenhouse conditions. Five replicates of five pots, each containing 5 plants, were maintained. *X. oryzae* inoculum was sub-cultured and grown individually in nutrient broth for 48 h and then centrifuged for 8000 × *g* for 15 m and then resuspended in sterile distilled water at 2×10^7 CFU/ml. Inoculation was done on twenty-day old plant leaves by scissors-dip method30. In each pot, the number of plants showing symptoms was counted and lesion length was measured at 14 days post-inoculation. The values are represented as percent disease incidence (DI) (Formula 1) and mean lesion length (MLL) (Formula 2).

$$\mathrm{DI}(\%) = \frac{\text{Number of leaves showing symptoms}}{\text{Total number of leaves inoculated}} \times 100 \qquad (1)$$

mean variance of disease index and disease incidence of five replications containing five plants in each pot were determined by Tukey-family error test ($P < 0.05$) by using Minitab®17

$$\text{Mean lesion lengths (MLL)} = \frac{n_1 + n_2 + n_3 + n_4 + n_5}{5} \times 100 \qquad (2)$$

Statistical analysis

Treatment effects on the percentage of emergence, root and shoot weight, and root: shoot ratio, in rice plants were analyzed by analysis of variance (one-way ANOVA), and the treatment means were compared by Tukey-family error test ($P < 0.05$) by using Minitab®17 software package. The data on above said experiments were arcsine transformed (except percentage of emergence) before undergoing statistical analysis.

ADDITIONAL INFORMATION

How to cite this article: Kalaivani, K. *et al.* Effect of methyl salicylate (MeSA), an elicitor on growth, physiology and pathology of resistant and susceptible rice varieties. *Sci. Rep.* **6**, 34498;

REFERENCES

1. Larcher, W. Physiological plant ecology: ecophysiology and stress physiology of functional groups (Springer Science & Business Media, 2003).
2. Fresco, L. Rice is life. Journal of Food Composition and Analysis 18, 249–253 (2005).
3. Senthil-Nathan, S. Biology, Behavioral and Population Dynamics of the Rice Leaffolder Complex: Dynamics of Insect Behavior (Scientific Publishers, Jodphur, pp. 156–167, 2011).
4. Senthil-Nathan, S., Kalaivani, K. & Murugan, K. Behavioural responses and changes in biology of rice leaffolder following treatment with a combination of bacterial toxin and botanical pesticides. Chemosphere 64, 1650–1658 (2006).
5. Niño-Liu, D. O., Ronald, P. C. & Bogdanove, A. J. Xanthomonas oryzae pathovars: model pathogens of a model crop. Mol. Plant Pathol. 7, 303–324 (2006).
6. Nisha, S. et al. Effect of plant compounds on induced activities of defense-related enzymes and pathogenesis related protein in bacterial blight disease susceptible rice plant. Physiol. Mol. Plant Pathol. 80, 1–9 (2012).
7. Kuo, Y.-H., Rozan, P., Lambein, F., Frias, J. & Vidal-Valverde, C. Effects of different germination conditions on the contents of free protein and non-protein amino acids of commercial legumes. Food Chem. 86, 537–545 (2004).
8. Ghaderi-Far, F., Gherekhloo, J. & Alimagham, M. Influence of environmental factors on seed germination and seedling emergence of yellow sweet clover (Melilotus officinalis). Planta Daninha 28, 463–469 (2010).
9. Ling, L. et al. Effects of cold plasma treatment on seed germination and seedling growth of soybean. Sci. Rep. 4 (2014).
10. Bewley, J. D. & Black, M. Physiology and Biochemistry of Seeds in Relation to Germination: Volume 2: Viability, Dormancy, and Environmental Control (Springer Science & Business Media, 2012).
11. Maguire, J. D. Speed of germination-aid in selection and evaluation for seedling emergence and vigor. Crop Sci. 2, 176–177 (1962).
12. Heydecker, W., Higgins, J. & Gulliver, R. L. Accelerated germination by osmotic seed treatment. Nature 246, 42–44 (1973).
13. Harris, D. et al. On-farm seed priming in semi-arid agriculture: development and evaluation in maize, rice and chickpea in India using participatory methods. Exp. Agri. 35, 15–29 (1999).

14. Tian, Y. et al. Responses of seed germination, seedling growth, and seed yield traits to seed pretreatment in maize (Zea mays L.). Sci. World J. 1–8 (2014).
15. Jisha, K. C., Vijayakumari, K. & Puthur, J. T. Seed priming for abiotic stress tolerance: an overview. Acta Physiol. Planta 35, 381–396 (2013).
16. Ghassemi-Golezani, K., Aliloo, A. A., Valizadeh, M. & Moghaddam, M. Effects of different priming techniques on seed invigoration and seedling establishment of lentil (Lens culinaris Medik). J. Food Agri. Env. 6, 222 (2008).
17. Reddy, M., Hynes, R. & Lazarovits, G. Relationship between in vitro growth inhibition of pathogens and suppression of preemergence damping-off and postemergence root rot of white bean seedlings in the greenhouse by bacteria. Can. J. Microbiol. 40, 113–119 (1994).
18. Senthil-Nathan, S., Kalaivani, K., Choi, M.-Y. & Paik, C.-H. Effects of jasmonic acid-induced resistance in rice on the plant brownhopper, Nilaparvata lugens Stål (Homoptera: Delphacidae). Pestic. Biochem. Physiol. 95, 77–84 (2009).
19. Jia, X., Meng, Q., Zeng, H., Wang, W. & Yin, H. Chitosan oligosaccharide induces resistance to Tobacco mosaic virus in Arabidopsis via the salicylic acid-mediated signalling pathway. Sci. Rep. 6 (2016).
20. Harris, D. et al. On-farm seed priming: using participatory methods to revive and refine a key technology. Agri. Sys. 69, 151–164 (2001).
21. Moghanibashi, M., Karimmojeni, H., Nikneshan, P. & Behrozi, D. Effect of hydropriming on seed germination indices of sunflower (Helianthus annuus L.) under salt and drought conditions. Plant Knowl. J. 1, 10 (2012).
22. Santner, A. & Estelle, M. Recent advances and emerging trends in plant hormone signalling. Nature 459, 1071–1078 (2009).
23. Raskin, I. Role of salicylic acid in plants. Ann. Rev. Plant Biol. 43, 439–463 (1992).
24. Humphreys, J. M. & Chapple, C. Rewriting the lignin roadmap. Curr. Opi. Plant Biol. 5, 224–229 (2002).
25. Yusuf, M., Hayat, S., Alyemeni, M. N., Fariduddin, Q. & Ahmad, A. In Salicylic Acid 15–30 (Springer, 2013).
26. Wiesel, L. et al. A transcriptional reference map of defence hormone responses in potato. Sci, Rep. 5 (2015).
27. Vazirimehr, M., Rigi, K. & Branch, Z. Effect of salicylic acid in agriculture. Int. J. Plant Anim. Environ. Sci. 4, 291–296 (2014)
28. Tavares, L. C., Rufino, C. A., Oliveira, S. d., Brunes, A. P. & Villela, F. A. Treatment of rice seeds with salicylic acid: seed physiological quality and yield. J. Seed Sci. 36, 352–356 (2014).
29. Tang, F., Fu, Y.-Y. & Ye, J.-R. The effect of methyl salicylate on the induction of direct and indirect plant defense mechanisms in poplar (Populus× euramericana 'Nanlin 895'). J. Plant Inter. 10, 93–100 (2015).
30. Kauffman, H., Reddy, A., Hsieh, S. & Merca, S. Improved technique for evaluating resistance of rice varieties to Xanthomonas oryzae. Plant Dis. Rep. (1973).
31. Vlot, A. C., Dempsey, D. M. A. & Klessig, D. F. Salicylic acid, a multifaceted hormone to combat disease. Ann. Rev. Phytopathol. 47, 177–206 (2009).
32. Bi, H. H., Zeng, R. S., Su, L. M., An, M. & Luo, S. M. Rice allelopathy induced by methyl jasmonate and methyl salicylate. J. Chem. Ecol. 33, 1089–1103 (2007).
33. Senthil-Nathan, S. A review of biopesticides and their mode of action against insect pest's. In Environmental Sustainability- Role of Green Technologies (Springer-Verlag, pp. 49–63, 2015).
34. Senthil-Nathan, S. Physiological and biochemical effect of Neem and other Meliaceae plants secondary metabolites against Lepidopteran insects. Front. Physiol. 4, 359 (2013).
35. Senthil-Nathan, S. Effects of Melia azedarach on nutritional physiology and enzyme activities of the rice leaffolder Cnaphalocrocis medinalis (Guenée) (Lepidoptera: Pyralidae). Pestic. Biochem. Physiol. 84, 98–108 (2006).
36. Shakirova, F. In Salicylic acid: a plant hormone 69–89 (Springer, 2007).

37. Senthil-Nathan, S. et al. Toxicity and physiological effects of neem pesticides applied to rice on the Nilaparvata lugens Stal, the brown planthopper. Ecotoxicol. Environ. Saf. 72, 1707–1713 (2009).
38. Hayat, S., Fariduddin, Q., Ali, B. & Ahmad, A. Effect of salicylic acid on growth and enzyme activities of wheat seedlings. Acta Agro. Hung. 53, 433–437 (2005).
39. Shakirova, F. M., Sakhabutdinova, A. R., Bezrukova, M. V., Fatkhutdinova, R. A. & Fatkhutdinova, D. R. Changes in the hormonal status of wheat seedlings induced by salicylic acid and salinity. Plant Sci. 164, 317–322 (2003).
40. Singh, B. & Usha, K. Salicylic acid induced physiological and biochemical changes in wheat seedlings under water stress. Plant Growth Regul. 39, 137–141 (2003).
41. Dolatabadian, A., Modarres Sanavy, S. A. M. & Sharifi, M. Effect of salicylic acid and salt on wheat seed germination. Acta Agri. Sca. Sec. B-Soil Plant Sci. 59, 456–464 (2009).
42. Khodary, S. E. Effect of salicylic acid on the growth, photosynthesis and carbohydrate metabolism in salt-stressed maize plants. Int. J. Agric. Biol. 6, 5–8 (2004).
43. Jadhav, S. & Bhamburdekar, S. Effect of salicylic acid on germination performance in groundnut. Int. J. Appl. Biol. Pharm. Technol. 2, 224–227 (2011).
44. Szepesi, A. et al. Role of salicylic acid pre-treatment on the acclimation of tomato plants to salt-and osmotic stress. Acta Biol. Szeg. 49, 123–125 (2005).
45. Martin-Mex, R., Villanueva-Couoh, E., Herrera-Campos, T. & Larqué-Saavedra, A. Positive effect of salicylates on the flowering of African violet. Sci. Hort. 103, 499–502 (2005).
46. McCue, P. et al. A model for enhanced pea seedling vigour following low pH and salicylic acid treatments. Proc. Biochem. 35, 603–613 (2000).
47. Almeida, A. S. et al. Physiological performance of wheat and barley seeds treated with bioactivator. Am. J. Exp. Agri 2, 90 (2012).
48. Wang, W. et al. Pre-sowing seed treatments in direct-seeded early rice: consequences for emergence, seedling growth and associated metabolic events under chilling stress. Sci. Rep. 6 (2016).
49. Gutiérrez-Coronado, M. A., Trejo-López, C. & Larqué-Saavedra, A. Effects of salicylic acid on the growth of roots and shoots in soybean. Plant Physiol. Biochem. 36, 563–565 (1998).
50. Sandoval-Yapiz, M. Reguladores de crecimiento XXIII: efecto del acido salicilico en la biomasa del cempazuchitl (Tagetes erecta), Tesis de Licenciatura Instituto Tecnologico Agropecuario, Conkal, Yucatan, Mexico (2004).
51. Martin-Mex, R. & Larqué-Saavedra, A. In 28th Annual Meeting Plant Growth Regulation Society of America, Miami Beach Florida, USA pp. 97–99 (2001).
52. Larqué-Saavedra, A. & Martin-Mex, R. In Salicylic Acid: A Plant Hormone. 15–23 (Springer, 2007).
53. Gunes, A. et al. Salicylic acid induced changes on some physiological parameters symptomatic for oxidative stress and mineral nutrition in maize (Zea mays L.) grown under salinity. J. Plant Physiol. 164, 728–736 (2007).
54. Yildirim, E., Turan, M. & Guvenc, I. Effect of foliar salicylic acid applications on growth, chlorophyll, and mineral content of cucumber grown under salt stress. J. Plant Nutr. 31, 593–612 (2008).
55. Shi, Q. & Zhu, Z. Effects of exogenous salicylic acid on manganese toxicity, element contents and antioxidative system in cucumber. Environ. Exp. Botany 63, 317–326 (2008).
56. Karlidag, H., Yildirim, E. & Turan, M. Salicylic acid ameliorates the adverse effect of salt stress on strawberry. Sci. Agri. 66, 180–187 (2009).
57. Stout, M. J., Thaler, J. S. & Thomma, B. P. Plant-mediated interactions between pathogenic microorganisms and herbivorous arthropods. Annu. Rev. Entomol. 51, 663–689 (2006).
58. Gaffney, T. et al. Requirement of salicylic acid for the induction of systemic acquired resistance. Science 261, 754–754 (1993).

59. Delaney, T. P., Uknes, S., Vernooij, B. & Friedrich, L. A central role of salicylic acid in plant disease resistance. Science 266, 1247 (1994).
60. Vallélian-Bindschedler, L., Métraux, J.-P. & Schweizer, P. Salicylic acid accumulation in barley is pathogen specific but not required for defense-gene activation. Mol. Plant-Microbe Inter. 11, 702–705 (1998).
61. Chern, M., Fitzgerald, H. A., Canlas, P. E., Navarre, D. A. & Ronald, P. C. Overexpression of a rice NPR1 homolog leads to constitutive activation of defense response and hypersensitivity to light. Mol. Plant-Microbe Inter. 18, 511–520 (2005).
62. Chern, M. S. et al. Evidence for a disease-resistance pathway in rice similar to the NPR1-mediated signaling pathway in Arabidopsis. Plant J. 27, 101–113 (2001).
63. Yuan, Y. et al. Functional analysis of rice NPR1-like genes reveals that OsNPR1/NH1 is the rice orthologue conferring disease resistance with enhanced herbivore susceptibility. Plant Biotechnol. J. 5, 313–324 (2007).

CHAPTER 11

Legitimacy and Implications of Reducing *Colletotrichum kahawae* to Subspecies in Plant Pathology

Dora Batista [1,2,3*], DiogoN.Silva [1,3], AnaVieira [1,3], AnaCabral [1,2], AnaS.Pires [1], Andreia Loureiro [1,2], LeonorGuerra-Guimarães[1,2], AnaP.Pereira [1], HelenaAzinheira [1,2], Pedro Talhinhas [1,2], MariadoCéuSilva [1,2] and VitorVárzea [1,2]

[1] *CentrodeInvestigação das Ferrugensdo Cafeeiro, Instituto Superiorde Agronomia, Universidadede Lisboa, Oeiras, Portugal,*

[2] *Linking Landscape, Environment, Agricultureand Food, Instituto Superiorde Agronomia, Universidadede Lisboa, Lisboa, Portugal,*

[3] *Computational Biologyand Population Genomics Group, Centre for Ecology, Evolutionand Environmental Changes, Faculdadede Ciências, Universidadede Lisboa, Lisboa, Portugal*

Colletotrichum kahawae Waller and Bridge is a highly aggressive and specialized fungal pathogen of coffee, causing the devastating Coffee Berry Disease (CBD), particularly at high altitudes. The disease arises from the unique ability of the pathogen to infect green developing coffee berries. This pathogen is currently confined to the African continent in all countries that grow Arabica coffee (*Coffea arabica* L.), leading to up to 80% yield losses, if no control measures are applied (Silva et al., 2006; Vossen and Walyaro, 2009; Hindorf and Omondi, 2011). For such huge economic impact, it is ranked as a quarantine pathogen and even as a biological weapon (Australia Group, 2014). Consequently, the pathogen's potential dispersal to other Arabica coffee cultivation regions is greatly feared, particularly to those at high altitude also found in Latin America and Asia. Recently, this recognized species was brought down to a subspecific level (*C. kahawae* subsp. *kahawae*) based on molecular data (Weir et al., 2012), clustering together with a generalist and cosmopolitan group of *Colletotrichum* isolates unable to cause CBD (*C. kahawae* subsp. *ciggaro*). Since then a growing number of studies have reported the identification of *C. kahawae* in various hosts and regions of the world (Liu et al., 2013; Afanador-Kafuri et al., 2014; Mosca et al., 2014; Schena et al., 2014; Ismail et al., 2015; Garibaldi et al., 2016a,b; Perrone et al., 2016). Although these reports are referring to *C. kahawae* subsp. *ciggaro*, some of them could not distinguish the pathogen at the subspecific level, and this is leading to a wave of confusion of whether the long accepted species *C. kahawae*, the CBD pathogen, has escaped from Africa and extended its host range. Given the extreme impact that this situation may trigger and the subsequent biosecurity implications, there is a practical need to completely distinguish these pathogens taxonomically as to avoid the risk of misidentification, and caution should be taken on assigning/reassigning taxonomic ranking and nomenclature. Here we consider the evidences sustaining and contradicting the classification proposed by Weir et al. (2012), and discuss the risks and practical implications of changing the CBD pathogen's species status in a plant pathology context.

After a long period of confusion in nomenclature, *Colletotrichum kahawae* (epithet derived from the Swahili word for coffee) was proposed by Waller et al. (1993) as a formal name to refer specifically to *Colletotrichum* isolates causing CBD and taxonomically distinguishing them from other *Colletotrichum* strains found on coffee. The CBD pathogen was previously known as either *C. coffeanum* var. *virulans* (Rayner, 1952), a form of *C. coffeanum* (Nutman and Roberts, 1960), CBD-strain (Gibbs, 1969), or *C. coffeanum* Noack sensu Hindorf (Hindorf, 1970, 1974). Interestingly, the name *C. coffeanum* was not based on type material associated with CBD, but on samples from Brazil where the disease does not exist, and it clearly referred to *C. gloeosporioides* (Waller et al., 1993), thus in view of its distinctive characteristics the applicability of these names was questionable. Waller et al. (1993) argued that, although closely related with *C. gloeosporioides*, the CBD pathogen is sufficiently distinct in pathogenicity and related ecology, morphological, cultural, and biochemical characteristics to warrant recognition as a distinct species. This was further confirmed based on multi-locus phylogenies (Prihastuti et al., 2009; Silva et al., 2012a).

Recently, an alteration of the species taxonomic status of *C. kahawae* was proposed by Weir et al. (2012), placing the CBD pathogen again in an elusive position. Using a multi-locus sequencing approach, both Silva et al. (2012b) and Weir et al. (2012) showed that the CBD pathogen share a remarkable genetic similarity with a group of isolates found on a wide range of hosts and geographical origins [referred as Undescribed Group 1 (UG1) by Silva et al., 2012b]. Indeed, for a set of barcoding gene sequences

(*ACT, CAL, CHS-1, GAPDH, TUB2, SOD2,* and *ITS*) the two taxa are indistinguishable. In their revision of the *C. gloeosporioides* species complex, Weir et al. (2012) accepted species according to the genealogical concordance phylogenetic species recognition (Taylor et al., 2000) criteria alone, and as such both groups should indeed be regarded as a single species. Thus, standing solely from a phylogenetic point of view, Weir et al. (2012) recognized them as two subspecies: the CBD pathogen (*C. kahawa*e subsp. *kahawae*) and the genetic similar *Colletotrichum* isolates (*C. kahawae* subsp. *ciggaro*; as a side note to the nomenclature chosen by the authors, the Portuguese word for cigarette is *cigarro*). However, as further demonstrated by Silva et al. (2012b) through population genetic, evolutionary, and pathological data, these clearly represent ecologically distinct and reproductively isolated biological entities, which have separated quite recently. According to the authors, UG1 (*C. kahawae* subsp. *ciggaro sensu* Weir et al., 2012) seems to be a sibling lineage from which *C. kahawae* may have undergone a very recent ecological speciation through host jump into *C. arabica*, and thus still share a considerable genetic background, although have become quite differentiated, as shown by the significant and elevated differentiation indexes across all studied loci and a complete segregation of polymorphic sites (Silva et al., 2012b). Its unparalleled adaptation of infecting green coffee berries enables the CBD pathogen to occupy a unique ecological niche, which separates it on a functional level from all other *Colletotrichum* species (Figure (Figure1).1). Quite a list of distinctive features can follow. In addition to its biology and slow growth form, the CBD pathogen can be distinguished metabolically by its inability to use either citrate or tartrate as a sole carbon source (Waller et al., 1993 and references herein; Varzea et al., 2002; Weir et al., 2012). At the genetic level, single genes or a combination of genes, specifically glutamine synthetase (*GS*) (Weir et al., 2012), mating type 1-2-1 (*MAT1-2-1*), and a fragment of DNA lyase *Apn*2 (*Apn*25L) (Silva et al., 2012b), reliably distinguish the CBD pathogen, as well as the genome size (Pires et al., 2016).

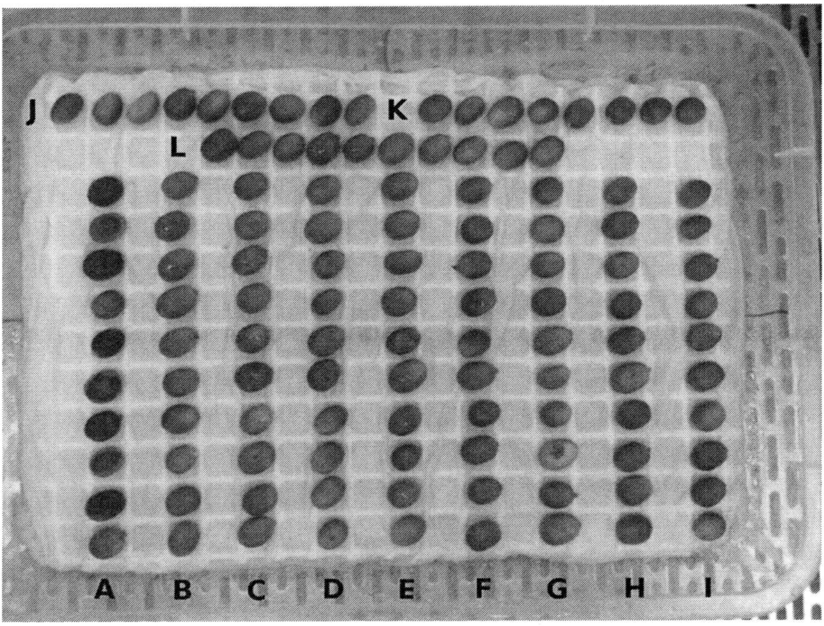

Figure 1. Pathogenicity tests on coffee green berries showing that only C. kahawaesubsp. kahawae isolates produced necrotic and sunken lesions characteristic of coffee berry disease symptoms (isolate Cam1, A),when compared to isolates from C. Kahawae subsp. ...

In the light of all these evidences reinforcing the differentiation of the CBD pathogen as a distinct and quite separate taxon in the coffee mycobiota, the taxonomic treatment proposed by Weir et al. (2012) is debatable and should be carefully interpreted. The authors acknowledge its distinctive features and biosecurity importance and to reflect this particularity retain a distinct taxonomic label for the CBD pathogen, while choosing to accept it only as biologically specialized population within a monophyletic supported species. It is well known that the species concepts are still in a state of flux in *Colletotrichum* taxonomy (Crouch et al., 2014), and particularly the low levels of genetic divergence across the *C. gloeosporioides* complex provide a technical challenge for species resolution (Silva et al., 2012a). Even more so, Weir et al. (2012) recognize the continuing uncertainty about sensible species limits within the *C. gloeosporioides* complex and that more powerful genes than the limited set sampled in their study, may provide finer resolution within the species-level clades that they accept. Precisely for this reason, extreme caution should be taken when assigning taxonomic ranking within members of the *C. gloesporioides* complex and additional data should be consider to complement genetic diversity and phylogenetic information, especially when biosecurity issues are at stake. In this case, the name *Colletotrichum kahawae* has a strong practical meaning and an indissoluble connection with the coffee disease. Bringing a volatile group of fungi able to infect a wide range of organs and hosts worldwide under the same specific epithet (which means coffee), is likely to disrupt the accuracy required in the nomenclature for pathogen identification and cause a great deal of confusion, even if distinction at the subspecific level is provided.

As a predictable consequence, several disease notes have reported the identification of *C. kahawae* in various hosts and regions out of Africa, including *Rubus glaucus* in Colombia (Afanador-Kafuri et al., 2014), olive, mango, and mandarin in Italy (Mosca et al., 2014; Schena et al., 2014; Ismail et al., 2015; Perrone et al., 2016), *Banksia* sp. in Portugal (Madeira) and Spain, and *Leucospermum* sp. in the USA (Hawaii) (Liu et al., 2013). While these studies assigned their isolates to *C. kahawae* subsp. *ciggaro*, others did not applied the genetic markers able to distinguish the two subspecies, and dangerously reported the presence of *C. kahawae* (no subspecies given) on rocket (*Eruca sativa*) (Garibaldi et al., 2016a) and American sweetgum (*Liquidambar stryraciflua*) (Garibaldi et al., 2016b) in Italy, suggesting the adoption of measures to contain the spread of the pathogen. The current classification promotes the chance of misidentification and uncertainty of the taxon addressed, leading to inaccurate records, distress within the coffee research and technical community, and severe biosecurity implications. On one hand, this might provoke a situation of assuming that *C. kahawae* has become able of infecting other hosts, and as such, of causing a devastating disease similar to that in coffee. On the other, *C. kahawae* is listed under strict quarantine regulations in many coffee growing countries, such as Australia and China (Jayawardena et al., 2016), and upon the information of the pathogen's potential detection, a state of emergency could be declared, particularly in major producing countries which have their economy strongly based on coffee production. Accurate naming of *Colletotrichum* to species level is crucial with regard to species identification and diversity, plant pathology and quarantine, agriculture, bio-control, plant breeding, whole genome sequencing, developing, and maintaining knowledge data bases, commercial applications and understanding evolutionary history (Jayawardena et al., 2016). If cryptic species are mistaken for a single species, then the species' integrity and understanding is compromised. Whether the *C. kahawae* subspecies' recognized by Weir et al. (2012) are after all genetically good species, will most likely be only suitably resolved with genomic data, an endeavor that is close to be accomplished (Vieira et al., 2016). For now, it should be considered that, as shown by Silva et al. (2012b) and Pires et al. (2016), notwithstanding the remarkable genetic similarity, there are enough evidences to accept them as distinct species, and in this light they would be better described and viewed as cryptic species.

AUTHOR CONTRIBUTIONS

DB, MCS, and VV conceived the idea. DB, DNS, AV, AC, ASP, AL, LGG, APP, HA, PT, MCS, and VV contributed to develop the concept and structure of the manuscript. DB wrote the manuscript. DNS prepared and edited the figure. DNS, AV, AL, LGG, HA, PT, MCS, and VV provided revisions to the manuscript. All authors read and approved the work for publication. The views and opinion presented in this article are those of the authors.

CONFLICT OF INTEREST STATEMENT

The authors declare that the research was conducted in the absence of any commercial or financial relationships that could be construed as a potential conflict of interest.

ACKNOWLEDGMENTS

Fundação para a Ciência e a Tecnologia (FCT, Portugal) is acknowledged for grants SFRH/BD/893972012, SFRH/BD/86736/2012, SFRH/BPD/84508 /2012 and SFRH/BPD/104629/2014, attributed to AV, DS, AC, and DB, respectively, and Unit funding UID/AGR/04129/2013.

REFERENCES

1. Afanador-Kafuri L., González A., Ga-án L., Mejía J. F., Cardona N., Alvarez E. (2014). Characterization of the *Colletotrichum* species causing anthracnose in andean blackberry in Colombia. Plant Dis. 98, 1503–1513.
2. Australia Group (2014). Australia Group Common Control List Handbook – Volume II: Biological Weapons-Related Common Control Lists. Available online at: http://www.australiagroup.net
3. Crouch J., O'Connell R., Gan P., Buiate E., Torres M. F., Beirn L., et al. (2014). The genomics of *Colletotrichum*, in Genomics of Plant-Associated Fungi: Monocot Pathogens, eds Dean R. A., Lichens-Park A., Kole C., editors. (Berlin: Springer-Verlag;), 69–102.
4. Garibaldi A., Gilardi G., Franco-Ortega S., Gullino M. L. (2016b). First report of leaf spot caused by *Colletotrichum kahawae* on american sweetgum (*Liquidambar styraciflua*) in Italy. J. Plant Pathol. 98:36

5. Garibaldi A., Gilardi G., Puglisi I., Cacciola S. O., Gullino M. L. (2016a). First report of leaf spot caused by *Colletotrichum kahawae* on cultivated rocket (*Eruca sativa*) in Italy. Plant Dis. 100:1240.
6. Gibbs J. N. (1969). Inoculum sources for coffee berry disease. Ann. Appl. Biol.64, 515–522.
7. Hindorf H. (1970). *Colletotrichum* spp. isolated from *Coffea arabica* L. in Kenya. Z. Pflanzenkrankh. Pflanzenschutz 77, 328–331.
8. Hindorf H. (1974). *Colletotrichum*-Arten aus dem Kaffeeanbaugebiet von Kiambu in Kenia/ *Colletotrichum* spp. collected in coffee growing areas of the Klambu-district of Kenya. Z. Pflanzenkrankh. Pflanzenschutz 81, 108–113.
9. Hindorf H., Omondi C. O. (2011). A review of three major fungal diseases of *Coffea arabica* L. in the rainforests of Ethiopia and progress in breeding for resistance in Kenya. J. Adv. Res. 2, 109–120.
10. Ismail A. M., Cirvilleri G., Yaseen T., Epifani F., Perrone G., Polizzi G. (2015). Characterisation of *Colletotrichum* species causing anthracnose disease of mango in Italy. J. Plant Pathol. 97, 167–171.
11. Jayawardena R. S., Hyde K. D., Jeewon R., Li X. H., Liu M., Yan J. Y. (2016). Mycosphere essay 6: why is it important to correctly name *Colletotrichum* species? Mycosphere 7, 1076–1092.
12. Liu F., Damm U., Cai L., Crous P. W. (2013). Species of the *Colletotrichum gloeosporioides* complex associated with anthracnose diseases of *Proteaceae*. Fungal Divers. 61, 89–105.
13. Mosca S., Li Destri Nicosia M. G., Cacciola S. O., Schena L. (2014). Molecular analysis of *Colletotrichum* species in the carposphere and phyllosphere of olive. PLoS ONE 9:e114031.
14. Nutman F. J., Roberts F. M. (1960). Investigations on a disease of *Coffea arabica* caused by a form of *Colletotrichum coffeanum* Noack I. Some factors affecting infection by the pathogen. Trans. Br. Mycol. Soc. 43, 489–505.
15. Perrone G., Magistà D., Ismail A. M. (2016). First report of *Colletotrichum kahawae* subsp. *ciggaro* on mandarin in Italy. J. Plant Pathol. 98:12
16. Pires A. S., Azinheira H. G., Cabral A., Tavares S., Tavares D., Castro M., et al. (2016). Cytogenomic characterization of *Colletotrichum kahawae*, the causal agent of coffee berry disease, reveals diversity in minichromosome profiles and genome size expansion. Plant Pathol. 65, 968–977.
17. Prihastuti H., Cai L., Chen H., McKenzie E. H. C., Hyde K. D. (2009). Characterisation of *Colletotrichum species* associated with coffee berries in northern Thailand. Fungal Divers. 39, 89–109.
18. Rayner R. W. (1952). Coffee berry disease. a survey of investigations carried out up to 1950. E. Afr. Agric. J. 17, 130–158.
19. Schena L., Mosca S., Cacciola S. O., Faedda R., Sanzani S. M., Agosteo G. E., et al. (2014). Species of the *Colletotrichum gloeosporioides* and C. *boninense* complexes associated with olive anthracnose. Plant Pathol. 63, 437–446.
20. Silva D. N., Talhinhas P., Cai L., Manuel L., Gichuru E. K., Loureiro A., et al. . (2012b). Host-jump drives rapid and recent ecological speciation of the emergent fungal pathogen *Colletotrichum kahawae*. Mol. Ecol. 21, 2655–2670.
21. Silva D. N., Talhinhas P., Várzea V., Cai L., Paulo O. S., Batista D. (2012a). Application of the Apn1/MAT locus to improve the systematics of the *Colletotrichum gloeosporioides* complex: an example from coffee (*Coffea* spp.) hosts. Mycologia 104, 396–409.
22. Silva M. C., Várzea V., Guerra-Guimarães L., Azinheira H. G., Fernandez D., Petitot A. S., et al. (2006). Coffee resistance to the main diseases: leaf rust and coffee berry disease. Braz. J. Plant Physiol. 18, 119–147.
23. Taylor J. W., Jacobson D. J., Kroken S., Kasuga T., Geiser D. M., Hibbett D. S., et al. . (2000). Phylogenetic species recognition and species concepts in fungi. Fungal Genet. Biol. 31, 21–32.

24. Vossen H. A. M., Walyaro D. J. (2009). Additional evidence for oligogenic inheritance of durable resistance to coffee berry disease (*Colletotrichum kahawae*) in arabica coffee (*Coffea arabica* L.). Euphytica 165, 105–111.
25. Varzea V. M. P., Rodrigues C. J., Jr., Lewis B. G. (2002). Distinguishing characteristics and vegetative compatibility of *Colletotrichum kahawae* in comparison with other related species from coffee. Plant Pathol. 51, 202–207.
26. Vieira A., Silva D. N., Várzea V., Silva M. C., Paulo O. S., Batista D. (2016). Adaptation of *Colletotrichum kahawae* to *Coffea arabica*: a genomic approach, in 2nd cE3c Annual Meeting: Frontiers in E3, Book of Abstracts, ed FCUL (Lisboa), editor. 60–61.
27. Waller J. M., Bridge P. D., Black R., Hakiza G. (1993). Characterization of the coffee berry disease pathogen, *Colletotrichum kahawae* sp. nov. Mycol. Res. 97, 989–994.
28. Weir B. S., Johnston P. R., Damm U. (2012). The *Colletotrichum gloeosporioides* species complex. Stud. Mycol. 73, 115–180.

CHAPTER 12

Phytoplasmas–The "Crouching Tiger" Threat of Australian Plant Pathology

Jian Liu[1,2,3], David Gopurenko[3,4], Murray J. Fletcher[3], Anne C. Johnson[3] and Geoff M. Gurr[1,2,3*]

[1]*State Key Laboratory of Ecological Pest Control for Fujian and Taiwan Crops, Fujian Agriculture and Forestry University, Fuzhou, China*

[2]*Institute of Applied Ecology, Fujian Agriculture & Forestry University, Fuzhou, China*

[3]*Graham Centre for Agricultural Innovation (Charles Sturt University & NSW Department of Primary Industries), Orange, NSW, Australia*

[4]*NSW Department of Primary Industries, Wagga Wagga Agricultural Institute, Wagga Wagga, NSW, Australia*

ABSTRACT

Phytoplasmas are insect-vectored bacteria that cause disease in a wide range of plant species. The increasing availability of molecular DNA analyses, expertise and additional methods in recent years has led to a proliferation of discoveries of phytoplasma-plant host associations and in the numbers of taxonomic groupings for phytoplasmas. The widespread use of common names based on the diseases with which they are associated, as well as separate phenetic and taxonomic systems for classifying phytoplasmas based on variation at the 16S rRNA-encoding gene, complicates interpretation of the literature. We explore this issue and related trends through a focus on Australian pathosystems, providing the first comprehensive compilation of information for this continent, covering the phytoplasmas, host plants, vectors and diseases. Of the 33 16Sr groups reported internationally, only groups I, II, III, X, XI and XII have been recorded in Australia and this highlights the need for ongoing biosecurity measures to prevent the introduction of additional pathogen groups. Many of the phytoplasmas reported in Australia have not been sufficiently well studied to assign them to 16Sr groups so it is likely that unrecognized groups and sub-groups are present. Wide host plant ranges are apparent among well studied phytoplasmas, with multiple crop and non-crop species infected by some. Disease management is further complicated by the fact that putative vectors have been identified for few phytoplasmas, especially in Australia. Despite rapid progress in recent years using molecular approaches, phytoplasmas remain the least well studied group of plant pathogens, making them a "crouching tiger" disease threat.

Keywords: "*Candidatus* Phytoplasma", 16S rRNA, biosecurity, taxonomy, biodiversity, vector, seed transmission, host range

INTRODUCTION

Phytoplasmas are insect-vectored bacteria that cause disease in a wide range of plant species (Lee et al., 2000; IRPCM, 2004; Bertaccini et al., 2014; Marcone, 2014). They contrast with other phloem-limited bacteria (Gram-negative proteobacteria such as liberibacters and phlomobacters Bove and Garnier, 2003) which are vectored by the same types of insects, in lacking a cell wall and in having a much reduced genome size (0.53–1.2 kb; Streten and Gibb, 2006). Spiroplasmas, another group of insect vectored plant pathogenic microbes, share the absence of a cell wall but differ from phytoplasmas in that some are culturable *in vitro*. In this mini review we seek to raise awareness of the importance of this group of plant pathogens, summarizing five key issues that mean the threat they pose to agricultural, ornamental and natural vegetation is not fully appreciated, making them worthy of the "crouching tiger" description. We focus principally on the phytoplasmas of Australia, as a sub-set of the large and rapidly-expanding global literature. We consider the phytoplasmas, associated plant diseases, plant hosts and the putative insect vectors (Table 1). Work from locations other than Australia is mentioned where appropriate to provide context and show where the issues we identify are generic rather than Australian-specific as well as to point out opportunities that are open for future study of Australian phytoplasma pathosystems. The only other reviews of phytoplasmas in Australia are at least a decade old and confined to tropical (Wilson et al., 2001) and sub-tropical regions (Streten and Gibb, 2006) whereas the scope of the present review includes the agriculturally important temperate zone. Reviews of varying levels of comprehensiveness are available for other geographical areas: New Zealand (Veerakone et al., 2015), Latin America (Pérez-López et al., 2016), the Pacific region (Davis and Ruabete, 2010) and Southern Italy (Marcone, 2011). Reflecting the large body of literature now available on phytoplasmas, the only reviews with an international scope are confined to particular crops such as coconut palm (Gurr et al., 2016), date palm (Gurr et al., 2015), fruit trees (Adams et al., 2001) and sugar cane (Smith et al., 2001).

Table 1. Taxonomic and biological information on phytoplasmas in australia (empty cells denote the absence of available information).

16Sr group	"Candidatus Phytoplasma" name	Phytoplasma trivial name	Host plant species	Potential vectors	Location*	References*
I	asteris	Buckland Valley grapevine yellows	*Vitis vinifera*		Victoria	Constable et al., 2003; Streten and Gibb, 2006
I-J	australiense[a]	Australian grapevine yellows	*Vitis vinifera*		South Australia	Davis et al., 1997a
II	aurantifolia	Australian lucerne yellows	*Medicago sativa*, *Carica papaya*	*Orosius argentatus*, *Austroagallia torrida*, *Orosius* spp., *Batracomorphus* sp.	South Australia, New South Wales, Northern Territory	Padovan and Gibb, 2001*; Pilkington et al., 2003*; Yang et al., 2013
II	aurantifolia	Bonamia pannosa little leaf	*Bonamia pannosa*		Western Australia, Northern Territory	Schneider et al., 1999; Padovan and Gibb, 2001*
II	aurantifolia	Cactus witches' broom	*Carica papaya*		Northern Territory	Padovan and Gibb, 2001
II	aurantifolia	Cocky apple witches' broom	*Planchonia careya*		Queensland	Davis et al., 2001
II	aurantifolia	Pigeon pea little leaf	*Arachis* spp., *Crotalaria* spp., *Indigofera* sp., *Macroptilium bracteatum*, *Pterocaulon* sp., *Stylosanthes scabra*, *Vigna radiata*		Queensland, Northern Territory	Schneider et al., 1999; Padovan and Gibb, 2001; Wilson et al., 2001; Davis et al., 2003
II	aurantifolia	Waltheria little leaf	*Mitracarpus hirtus*, *Saccharum* sp., *Spermacoce* sp., *Waltheria indica*, *Carica papaya*		Northern Territory	Schneider et al., 1999; Tran-Nguyen et al., 2000; Padovan and Gibb, 2001; Wilson et al., 2001
II-D	aurantifolia	Sweet potato little leaf	*Alysicarpus vaginalis*, *Aphyllodium* sp., *Arachis* spp., *Cajanus marmoratus*, *Carica papaya*, *Catharanthus roseus*, *Centrosema pascuorum*, *Citrus* sp., *Cleome viscosa*, *Crotalaria* spp., *Cucurbita maxima*, *Cyanthillium* spp., *Desmodium* spp., *Emilia sonchifolia*, *Indigofera* spp., *Ipomoea batatas*, *Macroptilium gracile*, *Medicago sativa*, *Mitracarpus hirtus*, *Nicotiana tabacum*, *Pachyrhizus erosus*, *Physalis minima*, *Rhynchosia minima*, *Senna obtusifolia*, *Sesamum indicum*, *Stylosanthes* spp., *Tridax procumbens*, *Vigna* spp.	*Austroagallia torrida*, *Orosius* spp., *Batracomorphus* sp.	Torres Strait, Northern Territory, Western Australia, New South Wales	Gibb et al., 1995*; Liu et al., 1996; Davis et al., 1997b; Schneider and Gibb, 1997; De La Rue et al., 1999, 2001; Padovan and Gibb, 2001*; Wilson et al., 2001; Davis et al., 2003; Streten and Gibb, 2006; Saqib et al., 2006; Tran-Nguyen et al., 2012;
II	aurantifolia	Tomato big bud	*Achyranthes aspera*, *Aeschynomene* spp., *Alysicarpus rugosus*, *Amaranthus* sp., *Apium graveolens*, *Arachis* spp., *Boerhavia* sp., *Brugmansia x candida*, *Cajanus cajan*, *Capsicum annuum*, *Carica papaya*, *Catharanthus roseus*, *Cenchrus ciliaris*, *Cichorium intybus*, *Citrus paradisi*, *Crotalaria* spp., *Cleome viscosa*, *Cucurbita maxima*, *Cynodon dactylon*, *Daucus carota*, *Emilia sonchifolia*, *Eragrostis falcata*, *Eriachne obtusa*, *Euphorbia milii*, *Evolvulus* sp., *Gerbera* sp., *Goodenia* sp., *Guizotia abyssinica*, *Ipomoea* spp., *Lactuca sativa*, *Lycopersicon esculentum*, *Macroptilium* spp., *Medicago sativa*, *Mucuna pruriens*, *Passiflora* sp., *Phlox* sp., *Physalis minima*, *Ptilotus distans*, *Rhynchosia minima*, *Saccharum* sp., *Sarcochilus hartmanii* × *S. falcatus*, *Sesamum indicum*, *Sida cordifolia*, *Solanum melongena*, *Stylosanthes scabra*, *Trifolium repens*, *Vigna* spp., *Vitis vinifera*, *Zinnia elegans*.	*Austroagallia torrida*	Northern Territory, New South Wales, Queensland, Western Australia, Victoria	Gibb et al., 1995; Davis et al., 1997b; Gowanlock et al., 1998; De La Rue et al., 1999; Tran-Nguyen et al., 2000, 2003; Wilson et al., 2001; Pilkington et al., 2004*; Streten and Gibb, 2006

II	aurantifolia	Chickpea little leaf	*Cicer arietinum*		Western Australia	Saqib et al., 2006
II	australasia[b]	Papaya yellow crinkle	*Carica papaya*		Queensland	Gibb et al., 1996, White et al., 1998
II	australasia[b]	Papaya mosaic	*Carica papaya*		Queensland	Gibb et al., 1996; White et al., 1998
II		Tree medic witches' broom	*Medicago arborea*		South Australia	Yang et al., 2013
II		Pigeonpea phyllody	*Cajanus cajan*		South Australia	Yang et al., 2013
II		Pigeon pea little leaf	*Arachis hypogaea, Catharanthus roseus, Crotalaria* spp., *Desmodium triflorum, Pterocaulon* sp., *Sesuvium portulacastrum, Stylosanthes* spp., *Vigna radiata*		Northern Territory, Queensland, Torres Strait	De La Rue et al., 2001; Wilson et al., 2001; Davis et al., 2003; Streten and Gibb, 2006
II-D	australasia[b]	Pale purple coneflower witches' broom	*Echinacea pallida*		Tasmania	Pearce et al., 2011
III	pruni	Poinsettia branching	*Euphorbia pulcherrima*			Schneider et al., 1999
III	pruni	Weeping tea tree witches' broom	*Melaleuca* spp.		Queensland	Davis et al., 2003
X		Allocasuarina yellows	*Allocasuarina muelleriana*		South Australia	Gibb et al., 2003
XI-B	australiense	Cynodon white leaf	*Cynodon dactylon, Dactyloctenium aegyptium*		Northern Territory, Western Australia	Schneider et al., 1999; Tran-Nguyen et al., 2000; Blanche et al., 2003
XI-B	australiense	Sorghum grassy shoot	*Chloris inflata, Dactyloctenium* spp., *Sorghum stipoideum, Whiteochloa cymbiformis, Whiteochloa* spp.		Western Australia, Northern Territory	Tran-Nguyen et al., 2000; Blanche et al., 2003
XII-B	australiense	Pumpkin yellow leaf curl	*Cucurbita maxima, C. moschata*		Queensland, Western Australia, Northern Territory	Streten et al., 2005
XII-B	australiense	Cenchrus bunchy shoot	*Cenchrus setiger*		Western Australia	Tran-Nguyen et al., 2000
XII-B	australiense	Strawberry green petal disease	*Fragaria x ananassa*		Queensland	Padovan et al., 2000
XII-B	australiense	Strawberry lethal yellows	*Fragaria x ananassa*		Queensland	Padovan et al., 2000
XII-B	australiense[a]	Australian grapevine yellows	*Vitis vinifera, Carica papaya*		South Australia, Queensland	Davis et al., 1997a,b; Davis and Sinclair, 1998; Davis et al., 2003
XII	australiense	Australian lucerne yellows	*Medicago sativa*		New South Wales	Getachew et al., 2007
XII	australiense	Paulownia witches' broom	*Paulownia* sp.		Western Australia	Bayliss et al., 2005
XII	australiense	Papaya dieback	*Carica papaya*		Queensland	Gibb et al., 1996; White et al., 1998
		Galactia little leaf	*Galactia tenuiflora*		Northern Territory	Schneider et al., 1999; Padovan and Gibb, 2001
		Sorghum bunchy shoot	*Sorghum stipoideum*			Tran-Nguyen et al., 2000
		Stylosanthes little leaf	*Arachis pintoi, Carica papaya, Saccharum* sp., *Sesuvium portulacastrum, Stylosanthes scabra*	*Austroagallia torrida, Orosius* spp., *Batracomorphus* sp.	Northern Territory, Torres Straight, Queensland, New South Wales	Schneider et al., 1999; Tran-Nguyen et al., 2000; De La Rue et al., 2001; Padovan and Gibb, 2001; Davis et al., 2003; Gopurenko et al., 2016
		Sugarcane white leaf	*Saccharum* spp.		Western Australia, Queensland	Tran-Nguyen et al., 2000
		Vigna little leaf	*Carica papaya, Tridax procumbens, Vigna lanceolata*	*Austroagallia torrida, Batracomorphus* sp.	Northern Australia	Schneider et al., 1999; De La Rue et al., 2001; Padovan and Gibb, 2001
		Mundulla yellows disease[c]	*Eucalyptus camaldulensis, E. baxteri, E. leucoxylon*		South Australia	Hanold et al., 2006

[illegible] Denotes reference for vector data.

* Location data are from the listed references but not every plant species was diseased in every location.

[a] Davis and Sinclair (1998) moved the AGY phytoplasma from the 16SrI group into the stolbur group (16SrXII) and designated it subgroup B.

[b] A new taxon, Ca. Phytoplasma australasia proposed (White et al., 1998) to include the phytoplasma associated with papaya yellow crinkle and papaya mosaic (as well as tomato big bud) but revised to 'Ca. australasiae' to include the papaya-associated phytoplasmas but not TBB (Bertaccini et al., 2006).

[c] Tentative data only for a phytoplasma etiology.

Issue 1: Phytoplasmology is a Young Science

Difficulties in studying phytoplasmas greatly limited early progress because phytoplasmas cannot be grown in axenic culture. Researchers were reliant initially on symptomology and transmission experiments, sometimes using grafting or the parasitic plant dodder (*Cuscuta* spp.) as in Australian work by Gibb et al. (1995), to study symptoms and host ranges but were unable to determine the nature of the pathogen or differentiate phytoplasmas from plant pathogenic viruses. Electron microscopy allowed phytoplasma bodies to be visualized in plant and insect vector tissue and differentiation of phytoplasmas from viruses. Bertaccini and Duduk (2010) provide a useful summary of the development of methods in phytoplasmology. Enzyme-linked immunosorbent assay (ELISA)-based methods began to emerge in the 1980s allowing more rapid detection and identification. The development of polymerase chain reaction (PCR) and restriction fragment length polymorphism (RFLP) methods for the detection and identification of phytoplasmas since the early 1990s allowed major advances, particularly in diagnostics and development of a genetic system for phenetic group classifications of phytoplasmas. By 1998, the international total of 34 representative phytoplasma strains were differentiated into 14 groups and 32 sub-groups based on similarity coefficients derived from RFLP analyses (Lee et al., 1998; Duduk and Bertaccini, 2011). More recent work has extended these counts to 33 groups and at least 100 sub-groups (Dickinson and Hodgetts, 2013; Davis et al., 2015; Zhao and Davis, 2016).

The increasingly widespread availability of molecular methods, equipment and expertise in recent decades has led to a proliferation of discoveries of phytoplasma-plant host associations and in taxonomic groupings for phytoplasmas. Many articles on phytoplasma pathosystems published this century are "first report" or "first record" articles for phytoplasmas in geographical regions or report known phytoplasmas from new host plant species. More fundamentally, new taxa of phytoplasma are being reported on a frequent basis. Progress has accelerated with the development of new approaches for detection and study of phytoplasmas. A prominent example is loop-mediated isothermal amplification (LAMP; Obura et al., 2011; Dickinson, 2015). This approach offers the advantages of low cost and high sensitivity, low risk of cross-contamination because reaction vessels do not need to be opened and, most especially, use in a kit form that requires very little training and can be used in the field (Hodgetts et al., 2011; Dickinson, 2015; Kogovšek et al., 2015). Though LAMP assays have yet to be used in studies of Australian phytoplasma pathosystems they have recently proven useful in studies of Bogia coconut syndrome in nearby Papua New Guinea (Lu et al., 2016).

In parallel to the use of simple LAMP diagnostics, next-generation sequencing (NGS) technologies have provided genomic characterization of phytoplasmas, and this has profoundly advanced the understanding of phytoplasma evolution and pathogenicity over recent years (Marcone, 2014). To date, complete genomes of six phytoplasma strains (Oshima et al., 2004; Bai et al., 2006; Kube et al., 2008; Tran-Nguyen et al., 2008; Andersen et al., 2013; Orlovskis et al., 2017) and an additional 16 draft or incomplete genomes have been reported (refer Genomes OnLine database [GOLD]; https://gold.jgi.doe.gov).

Comparative genomic analyses of phytoplasma and other mollicutes have provided direct evidence of reduced genomic complexity in phytoplasmas, characterized by encoding for fewer metabolic functions and pathways (Oshima et al., 2004). This highlights the obligate adaptation to, and reliance on, varied metabolic substrates available in hosts and vectors by phytoplasmas (Kube et al., 2013). Genomic studies have also detailed genes putatively encoding for phytoplasma virulence factors, including "effector" proteins which phytoplasmas produce and secrete to alter host cell activities, thereby modifying host development and reducing host defences against herbivorous arthropod vectors (Hogenhout and Loria, 2008; Hogenhout et al., 2008, 2009). Interestingly, comparative genomic analysis of five diverse phytoplasma groups failed to detect a consistent shared set of predicted secreted effector encoding genes (Wang et al., 2014). This suggests virulence encoding genes are likely to be diverse among phytoplasma strains, and may explain the wide range of pathogenicity in different 16Sr groups. Other work (Chung et al., 2013) indicates that horizontal gene flow of mobile genetic elements among some divergent phytoplasma strains may have facilitated horizontal transfer of effector genes linked to the mobile elements, adding a novel pathway for increasing the adaptive potential of phytoplasmas with regards to their hosts. NGS methods have been applied also for population metagenomics, allowing valuable insight into the ecology and dynamics of phytoplasmas and their relationships to hosts (Nicolaisen et al., 2011). Finally, phylogenetic analyses of phytoplasmas using NGS data are likely to improve understanding of systematic and taxonomic relationships in the genus, traditionally reliant on analyses of the 16S rRNA gene region (see Issue 3) and (in some cases) use of additional informative loci for comparison of very recently diverged strains (Al-Abadi et al., 2016).

Anticipated widespread use of affordable NGS services will result in a proliferation of the availability of published phytoplasma genomes and meta-genomic analyses, and this will ultimately lead to more advanced understanding of evolutionary relationships of species in this genus, and their interactive pathways with hosts and vectors including for Australian pathosystems.

Issue 2: Complex Taxonomic Nomenclature

A major impediment to comprehension of the phytoplasma literature and comparisons between studies, particularly for the non-specialist, is the taxonomic nomenclature with three systems currently in use. First, reflecting the history of phytoplasmology described above, the early literature uses disease common names based on symptoms (e.g., little leaf, yellows, witches' broom) often coupled with the host plant's name. These disease common names have been applied also to the phytoplasmas and continue to be used frequently in recent literature. Examples from the Australian literature include Buckland Valley grapevine yellows phytoplasma, tomato big bud phytoplasma and Cockey apple witches' broom phytoplasma (Table 1). This allows great scope for confusion because a given phytoplasma can be found in multiple plant species and can cause different disease symptoms in different hosts.

Second, as molecular methods became available, workers were able to group and phenetically classify phytoplasmas using restricted fragment length polymorphism (RFLP) analysis of a PCR amplified portion of the 16S rRNA gene with a defined set of restriction enzymes (Lee et al., 1998). The RFLP profiles generated for different phytoplasmas are generally consistent with sequence-based phylogenetic analyses of the 16S rRNA gene, particularly in the co-identification and grouping of related strains. The 33 16Sr groups currently defined each have a similarity of <85% compared with any representative phytoplasma from within an established 16Sr group (Zhao and Davis, 2016). Table 1 summarizes available information on the 16Sr groups reported in Australian studies. Of the 33 16Sr groups reported internationally, only groups I, II, III, X, XI and XII have been recorded in Australia and this highlights the need for ongoing biosecurity measures to prevent the introduction of additional pathogen groups.

Third, phytoplasmas are classified in the provisional genus "*Candidatus* Phytoplasma" (IRPCM, 2004). To date, there are 42 formally described species and ten potentially novel phytoplasma species (Davis et al., 2015). This number exceeds the current number of 16s rRNA groups because some of these groups contain several "*Candidatus* Phytoplasma" species. At least 100 subgroups are known (Dickinson and Hodgetts, 2013). According to Phytoplasma/Spiroplasma Working Team-Phytoplasma Taxonomy Group, a novel "*Ca*. Phytoplasma" species description should refer to a single, unique 16S rRNA gene sequence (>1,200 bp), and a strain can be recognized as a novel "*Ca*. Phytoplasma" species if its 16S rRNA gene sequence has <97.5% similarity to that of any previously described "*Ca*. Phytoplasma" species (Duduk and Bertaccini, 2011). Additional biological characters such as antibody specificity, host range and vector transmission specificity as well as genetic markers can also be used in an integrative taxonomy approach for species differentiation. Of the 42 recognized "*Ca*. Phytoplasma" species, only *Ca*. Phytoplasma asteris, *Ca*. Phytoplasma aurantifolia, *Ca*. Phytoplasma australasiae, *Ca*. Phytoplasma australiense and *Ca*. Phytoplasma pruni are reported in Australia (Table 1) but uncertainty exists because many papers appear without *Ca*. Phytoplasma names which are used consistently only in the case of the GenBank database.

The general literature uses a mix of *Ca*. Phytoplasma names, 16Sr group and sub-group names, and common names that often reflect the host plant or symptom (Table 1). Though formally named *Ca*. Phytoplasma species each align with a group or sub group in the 16Sr system, many groups and sub-groups do not currently have a "*Candidatus* Phytoplasma" species name.

Issue 3: Large and Poorly Understood Biodiversity among Pathogens and Wide Host Ranges

Whilst it is important to note that a given type of phytoplasma can attack more than one plant species, and that a given plant species can be attacked by multiple types of phytoplasma, the list of phytoplasma disease common names from Australia (Table 1) is instructive in indicating the wide range of plant species that are affected, even for specific phytoplasmas. Examples extend over forage, broadacre and horticultural crops of a perennial and annual nature, as well as ornamental and uncultivated (i.e., natural vegetation) species.

There is an especially poor knowledge of phytoplasma pathosystems in temperate Australia, largely because of the tropical and sub-tropical zone focus of most of the key Australian workers in the last 20 years. Many of the phytoplasmas reported in Australia have not been sufficiently well studied to assign them to 16Sr groups so it is likely that unrecognized groups—as well as sub-groups—are present, particularly in native vegetation.

Great efforts are currently being made to define the extent and diversity of phytoplasmas in Australia (most recently the discovery of Stylosanthes little leaf 16Sr XII phytoplasma from lucerne in New South Wales Gopurenko et al., 2016) and elsewhere but this remains a challenge because of the apparent high levels of biological diversity. As we approach a definitive list of 16Sr groups and sub groups, "*Ca*. phytoplasma" species names need to be assigned that avoid scope for confusion over the host plant, perhaps by wider use of names that reflect the country in which the phytoplasma was first discovered (as is the case for *Ca*. Phytoplasma australiense, though noting the potential confusion with *Ca*. Phytoplasma australasiae and the redundant *Ca*. Phytoplasma australasia Table 1) or a more neutral name based, perhaps, on the discoverer rather than the use of host plant taxon names such as pini, palmicola, pruni and oryzae, all of which are reported internationally though not from Australia.

Reflecting and compounding the three, related issues outlined above, there is no universally recognized taxonomic resource for phytoplasmas. For other higher taxa, there is a name-bearing type specimen lodged in an accessible scientific collection that can be checked by subsequent researchers, allowing comparison with unknown or new taxa. With phytoplasmas, however, the extreme difficulties associated with axenic culture mean that the "type" is a DNA sequence in GenBank®, the National Institutes of Health's annotated collection of all publicly available DNA sequences (Benson et al., 2013), and not a biological specimen. Further, there is no common platform for registering phytoplasma groups and sub-groups (Zhao and Davis, 2016). The *International Journal of Systematics and*

Evolution (formerly the *International Journal of Systematic Bacteriology*) is the official journal of record for novel prokaryotic taxa since it is the official publication of the International Committee on Systematics of Prokaryotes and the Bacteriology and Applied Microbiology Division of the International Union of Microbiological Societies. Accordingly, new "*Ca.* Phytoplasma" species are published in this journal. Importantly, however, new 16Sr groups can be (and are) published in other journals provided that the group contains a previously described "*Ca.* Phytoplasma" species. This lack of a common platform for group names means that a given 16Sr group number can inadvertently be applied by different authors to different "*Ca.* Phytoplasma" taxa. Zhao and Davis (2016) provide the example that "*Ca.* Phytoplasma" allocasuarinae was assigned as 16SrXXXIII (Bertaccini et al., 2014) but the same 16Sr group name was also subsequently used for another new (Chilean grapevine phytoplasma) group (Pérez-López et al., 2016). This led to the recent suggestion (Zhao and Davis, 2016) that 16Sr groups and sub-groups should be registered via a web-based portal linked to iPhyClassifier, itself a web-based resource that allows identification and classification of phytoplasmas by simulating restriction enzyme digestions and electrophoresis to produce "virtual" RFLP patterns for submitted query sequences.

Issue 4: Challenges Associated with Vector Studies

The insect vectors of phytoplasmas are yet to be investigated thoroughly in Australia and more generally and this is a major constraint on the development of disease management approaches that focus on vector control. The Australian literature provides putative vector names only for Australian lucerne yellows, sweetpotato little leaf, tomato big bud, Stylosanthes little leaf and Vigna little leaf phytoplasmas (Table 1).

Since phytoplasmas are phloem-limited pathogens, they are transmitted by phloem-feeding hemipteran insects, especially leafhoppers and planthoppers (Weintraub and Beanland, 2006) though psyllids are responsible in some non-Australian pathosystems (Carraro et al., 1998). The identification of potential vector species involves field surveys to determine Hemiptera species that are spatially and temporally associated with plant symptoms followed by the use of PCR to detect phytoplasma DNA in the insects. However, the detection of phytoplasma DNA in an insect does not establish vector status because DNA may be confined to the gut as a result of feeding on an infected host plant rather than the pathogen having colonized the salivary glands making it a competent vector (Vega et al., 1993; Danielli et al., 1996). Definitive proof of vector status can be obtained from transmission tests confining putative vectors on phytoplasma-free host plants in insect-proof cages. Whilst this has been done in a preliminary manner for Australian lucerne yellows (e.g., Pilkington et al., 2004), vector transmission testing is logistically demanding, especially if host plants are large perennial species, if the pathogen is vectored inefficiently or if symptoms develop slowly (Gurr et al., 2016). A relatively new method that is intermediate in ease of use and the level of proof involves holding individual insects in vessels from which they can feed on a sucrose solution through a parafilm barrier with subsequent PCR-based detection of phytoplasma DNA in the medium (Tanne et al., 2001). Though this approach involves many samples (i.e., one per insect rather than one per plant as in the case of a cage transmission test) a recent study has employed LAMP to readily handle the large numbers of samples associated with screening multiple putative vector species of Bogia coconut syndrome (Lu et al., 2016).

Issue 5: Possibility of Seed Transmission

Among the crop species of great economic importance in Australia, tomato, canola and maize have all been experimentally shown to exhibit seed transmission of phytoplasmas, a phenomenon previously considered unlikely and this adds to the importance of ongoing biosecurity and research efforts. In that study, PCR was used to detect phytoplasma DNA in laboratory grown seedlings arising from seed of phytoplasma infected plants of oilseed rape, tomato and corn (Calari et al., 2011). Caution is required in extrapolating from those tests with herbaceous annuals, however, because earlier reports of phytoplasma DNA in embryos of various woody perennial plant species including coconut have not been followed-up with conclusive evidence of seed transmissibility (Nipah et al., 2007) but see Oropeza et al. (2017).

CONCLUSION AND PERSPECTIVE

The ubiquity of PCR capacity and the advent of LAMP kits for simple diagnostics and NGS for advanced genomic and meta-genomic analyses will spur the discovery of many new phytoplasma diseases and allow rapid advances in understanding of phytoplasma biodiversity and biology in Australia and internationally. This has great practical relevance to address issues such as vector identity and the field-relevance of seed transmission. Because Australia is an island nation with well-established biosecurity measures, improvements in knowledge of phytoplasma biodiversity, host range and detection will be particularly valuable in the prevention of and appropriate responses to phytoplasma incursions. Further sustained research effort by specialists is important but this needs to be complemented by efforts to make this group of pathogens better known and easier to understand, especially taxonomically.

AUTHOR CONTRIBUTIONS

All authors listed, have made substantial, direct and intellectual contribution to the work, and approved it for publication.

FUNDING

This work was supported by a Graham Centre New Initiative Grant to the authors. GG is supported by a Chinese Government Thousand Talents Program fellowship (KX1452102).

CONFLICT OF INTEREST STATEMENT

The authors declare that the research was conducted in the absence of any commercial or financial relationships that could be construed as a potential conflict of interest.

REFERENCES

1. Adams, A. N., Davies, D. L., and Kirby, M. J. (2001). Virus and phytoplasma detection in fruit trees. *Outlook Agric.* 30, 45–54.
2. Al-Abadi, S. Y., Al-Sadi, A. M., Dickinson, M., Al-Hammadi, M. S., Al-Shariqi, R., Al-Yahyai, A., et al. (2016). Population genetic analysis reveals a low level of genetic diversity of "*Candidatus* Phytoplasma aurantifolia" causing witches' broom disease in lime. *BMC Microbiol.* 5:1701.
3. Andersen, M. T., Liefting, L. W., Havukkala, I., and Beever, R. E. (2013). Comparison of the complete genome sequence of two closely related isolates of "*Candidatus* Phytoplasma australiense" reveals genome plasticity. *BMC Genomics* 14:529.
4. Bai, X., Zhang, J., Ewing, E., Miller, S. A., Radek, A., Schevchenko, D., et al. (2006). Living with genome instability: the adaptation of phytoplasmas to diverse environments of their insect and plant hosts. *J. Bacteriol.* 188, 3682–3696.
5. Bayliss, K. L., Saqib, M., Dell, B., Jones, M. G. K., and Hardy, G. E. S. J. (2005). First record of "*Candidatus* Phytoplasma australiense" in *Paulownia* trees. *Australas. Plant Pathol.* 34, 123–124.
6. Benson, D. A., Cavanaugh, M., Clark, K., Karsch-Mizrachi, I., Lipman, D. J., Ostell, J., et al. (2013). GenBank. *Nucleic Acids Res.* 41, D36–D42.

7. Bertaccini, A., and Duduk, B. (2010). Phytoplasma and phytoplasma diseases: a review of recent research. *Phytopathol. Mediterr.* 48, 355–378.
8. Bertaccini, A., Duduk, B., Paltrinieri, S., and Contaldo, N. (2014). Phytoplasmas and phytoplasma diseases: a severe threat to agriculture. *Am. J. Plant Sci.* 5, 1763–1788.
9. Blanche, K. R., Tran-Nguyen, L. T. T., and Gibb, K. S. (2003). Detection, identification and significance of phytoplasmas in grasses in northern Australia. *Plant Pathol.* 52, 505–512.
10. Bove, J., and Garnier, M. (2003). Phloem-and xylem-restricted plant pathogenic bacteria. *Plant Sci.* 164, 423–438.
11. Calari, A., Paltrinieri, S., Contaldo, N., Sakalieva, D., Mori, N., Duduk, B., et al. (2011). Molecular evidence of phytoplasmas in winter oilseed rape, tomato and corn seedlings. *Bull. Insectol.* 64, S157–S158.
12. Carraro, L., Osler, R., Loi, N., Ermacora, P., and Refatti, E. (1998). Transmission of European stone fruit yellows phytoplasma by *Cacopsylla pruni*. *J. Plant Pathol.* 80, 233–239.
13. Chung, W.-C., Chen, L.-L., Lo, W.-S., Lin, C.-P., and Kuo, C.-H. (2013). Comparative analysis of the peanut witches'-broom phytoplasma genome reveals horizontal transfer of potential mobile units and effectors. Robinson-Rechavi, M., ed. *PLoS ONE* 8:e62770.
14. Constable, F. E., Whiting, J. R., Jones, J., Gibb, K. S., and Symons, R. H. (2003). The distribution of grapevine yellows disease associated with the buckland valley grapevine yellows phytoplasma. *J. Phytopathol.* 151, 65–73.
15. Danielli, A., Bertaccini, A., Vibio, M., Mori, N., Murari, E., Posenato, G., et al. (1996). Detection and molecular characterization of phytoplasmas in the planthopper *Metcalfa pruinosa* (Say) (Homoptera: Flatidae). *Phytopathol. Mediterr.* 35, 62–65.
16. Davis, R., Dally, E. L., Gundersen, D. E., Lee, I. M., and Habili, N. (1997a). "*Candidatus* Phytoplasma australiense," a New Phytoplasma Taxon Associated with Australian Grapevine Yellows. *Int. J. Syst. Evol. Microbiol.* 47, 262–269.
17. Davis, R. E., and Sinclair, W. A. (1998). Phytoplasma identity and disease etiology. *Phytopathology* 88, 1372–1376.
18. Davis, R. I., Henderson, J., Jones, L. M., McTaggart, A. R., O'Dwyer, C., Tsatsia, F., et al. (2015). First record of a wilt disease of banana plants associated with phytoplasmas in Solomon Islands. *Australas. Plant Dis. Notes* 10:14.
19. Davis, R. I., Jacobson, S. C., Rue, S. J., Tran-Nguyen, L., Gunua, T. G., and Rahamma, S. (2003). Phytoplasma disease surveys in the extreme north of Queensland, Australia, and the island of New Guinea. *Australas. Plant Pathol.* 32, 269–277.
20. Davis, R. I., Jacobson, S. C., Waldeck, G. J., De La Rue, S. J., and Gibb, K. S. (2001). A witches' broom of cocky apple (*Planchonia careya*) in north Queensland. *Australas. Plant Pathol.* 30, 179–179.
21. Davis, R. I., and Ruabete, T. K. (2010). Records of plant pathogenic viruses and virus-like agents from 22 Pacific island countries and territories: a review and an update. *Australas. Plant Pathol.* 39, 265–291.
22. Davis, R., Schneider, B., and Gibb, K. (1997b). Detection and differentiation of phytoplasmas in Australia. *Aust. J. Agric. Res.* 48, 535–544.
23. De La Rue, S. J., Schneider, B., and Gibb, K. S. (1999). Genetic variability in phytoplasmas associated with papaya yellow crinkle and papaya mosaic diseases in Queensland and the Northern Territory. *Australas. Plant Pathol.* 28, 108–114.
24. De La Rue, S., Padovan, A., and Gibb, K. (2001). *Stylosanthes* is a Host for Several Phytoplasmas, One of which Shows Unique 16S-23S Intergenic Spacer Region Heterogeneity. *J. Phytopathol.* 149, 613–619.
25. Dickinson, M. (2015). "Loop-mediated isothermal amplification (LAMP) for detection of phytoplasmas in the field," in *Plant Pathology*, ed C. Lacomme (New York, NY: Springer), 99–111
26. Dickinson, M., and Hodgetts, J. (eds) (2013). *Phytoplasma: Methods and Protocols, Dordrecht.* Netherlands: Science+Business Media.

27. Duduk, B., and Bertaccini, A. (2011). Phytoplasma classification: taxonomy based on 16S ribosomal gene, is it enough? *Phytopathogenic Mollicutes* 1, 3–13.
28. Firrao, G., Gibb, K., and Streten, C. (2005). Short taxonomic guide to the genus "*Candidatus* Phytoplasma." *J. Plant Pathol.* 87, 249–263.
29. Getachew, M. A., Mitchell, A., Gurr, G. M., Fletcher, M. J., Pilkington, L. J., and Nikandrow, A. (2007). First report of a "*Candidatus* phytoplasma australiense"-related strain in lucerne (*Medicago sativa*) in Australia. *Plant Dis.* 91, 111–111.
30. Gibb, K., Padovan, A., and Mogen, B. (1995). Studies on sweet potato little-leaf phytoplasma detected in sweet potato and other plant species growing in Northern Australia. *Phytopathology* 85, 169–174.
31. Gibb, K., Persley, D., Schneider, B., and Thomas, J. (1996). Phytoplasmas associated with papaya diseases in Australia. *Plant Dis.* 80, 174–178.
32. Gibb, K. S., Tran-Nguyen, L. T. T., and Randles, J. W. (2003). A new phytoplasma detected in the South Australian native perennial shrub, *Allocasuarina muelleriana*. *Ann. Appl. Biol.* 142, 357–364.
33. Gopurenko, D., Fletcher, M. J., Liu, J., and Gurr, G. M. (2016). Expanding and exploring the diversity of phytoplasmas from lucerne (*Medicago sativa*). *Sci. Rep.* 6:37746.
34. Gowanlock, D. H., Ogle, H. J., and Gibb, K. S. (1998). Phytoplasma associated with virescence in an epiphytic orchid in Australia. *Australas. Plant Pathol.* 27, 265–268.
35. Gurr, G. M., Bertaccini, A., Gopurenko, D., Krueger, R. R., Alhudaib, K. A., Liu, J., et al. (2015). "Phytoplasmas and their insect vectors: implications for date palm," in *Sustainable Pest Management in Date Palm: Current Status and Emerging Challenges*, eds W. Wakil, J. Romeno Faleiro and A. T. Miller (Cham: Springer International Publishing), 287–314.
36. Gurr, G. M., Johnson, A., Ash, G., Wilson, B., Ero, M., Pilotti, C., et al. (2016). Coconut lethal yellowing diseases: a phytoplasma threat to palms of global economic and social significance. *Front. Plant Sci.* 7:5121.
37. Hanold, D., Gowanlock, D., Stukely, M. J. C., Habili, N., and Randles, J. W. (2006). Mundulla Yellows disease of eucalypts: descriptors and preliminary studies on distribution and etiology. *Australas. Plant Pathol.* 35, 199–215.
38. Hodgetts, J., Tomlinson, J., Boonham, N., González-Martín, I., Nikolić, P., Swarbrick, P., et al. (2011). Development of rapid in-field loop-mediated isothermal amplification (LAMP) assays for phytoplasmas. *Bull. Insectol.* 64, S41–S42.
39. Hogenhout, S. A., and Loria, R. (2008). Virulence mechanisms of Gram-positive plant pathogenic bacteria. *Curr. Opin. Plant Biol.* 11, 449–456.
40. Hogenhout, S. A., Oshima, K., Ammar, E.-D., Kakizawa, S., Kingdom, H. N., and Namba, S. (2008). Phytoplasmas: Bacteria that manipulate plants and insects. *Mol. Plant Pathol.* 9, 403–423.
41. Hogenhout, S. A., Van der Hoorn, R. A. L., Terauchi, R., and Kamoun, S. (2009). Emerging concepts in effector biology of plant-associated organisms. *Mol. Plant Microbe Interact.* 22, 115–22.
42. IRPCM (2004). "*Candidatus* Phytoplasma," a taxon for the wall-less, non-helical prokaryotes that colonize plant phloem and insects. *Int. J. Syst. Evol. Microbiol.* 54, 1243–1255.
43. Kogovšek, P., Hodgetts, J., Hall, J., Prezelj, N., Nikolić, P., Mehle, N., et al. (2015). LAMP assay and rapid sample preparation method for on-site detection of flavescence dorée phytoplasma in grapevine. *Plant Pathol.* 64, 286–296.
44. Kube, M., Schneider, B., Kuhl, H., Dandekar, T., Heitmann, K., Migdoll, A. M., et al. (2008). The linear chromosome of the plant-pathogenic mycoplasma '*Candidatus* Phytoplasma mali'. *BMC Genomics* 9:306.
45. Kube, M., Siewert, C., Migdoll, A. M., Duduk, B., Holz, S., Rabus, R., et al. (2013). Analysis of the complete genomes of *Acholeplasma brassicae, A. palmae* and *A. laidlawii* and their comparison to the obligate parasites from "*Candidatus* Phytoplasma." *J. Mol. Microbiol. Biotechnol.* 24, 19–36.
46. Lee, I.-M., Davis, R. E., and Gundersen-Rindal, D. E. (2000). Phytoplasma: phytopathogenic mollicutes. *Annu. Rev. Microbiol.* 54, 221–255.

47. Lee, I.-M., Gundersen-Rindal, D. E., Davis, R. E., and Bartoszyk, I. M. (1998). Revised classification scheme of phytoplasmas based on RFLP Analyses of 16S rRNA and ribosomal protein gene sequences. *Int. J. Syst. Bacteriol.* 48, 1153–1169.
48. Liu, B., White, D., Walsh, K., and Scott, P. (1996). Detection of phytoplasmas in dieback, yellow crinkle, and mosaic diseases of papaya using polymerase chain reaction techniques. *Aust. J. Agric. Res.* 47, 387–394.
49. Lu, H., Wilson, B. A. L., Ash, G. J., Woruba, S. B., Fletcher, M. J., You, M., et al. (2016). Determining putative vectors of the Bogia Coconut Syndrome phytoplasma using loop-mediated isothermal amplification of single-insect feeding media. *Sci. Rep.* 6:35801.
50. Marcone, C. (2011). Current status of phytoplasma diseases of medicinal and nutraceutical plants in Southern Italy. *Bull. Insectol.* 64, S233–S234.
51. Marcone, C. (2014). Molecular biology and pathogenicity of phytoplasmas. *Annals of Applied Biology*, 165, 199–221.
52. Nicolaisen, M., Contaldo, N., Makarova, O., Paltrinieri, S., and Bertaccini, A. (2011). Deep amplicon sequencing reveals mixed phytoplasma infection within single grapevine plants. *Bull. Insectol.* 64, S35–S36.
53. Nipah, J. O., Jones, P., Hodgetts, J., and Dickinson, M. (2007). Detection of phytoplasma DNA in embryos from coconut palms in Ghana, and kernels from maize in Peru. *Bull. Insectol.* 60, 385–386.
54. Obura, E., Masiga, D., Wachira, F., Gurja, B., and Khan, Z. R. (2011). Detection of phytoplasma by loop-mediated isothermal amplification of DNA (LAMP). *J. Microbiol. Methods* 84, 312–316.
55. Orlovskis, Z., Canale, M. C., Haryono, M., Lopes, J. R. S., Kuo, C.-H., and Hogenhout, S. A. (2017). A few sequence polymorphisms among isolates of Maize bushy stunt phytoplasma associate with organ proliferation symptoms of infected maize plants. *Ann. Bot.* 119, 869–884.
56. Oropeza, C., Cordova, I., Puch-Hau, C., Castillo, R., Chan, J. L., and Sáenz, L. (2017). Detection of lethal yellowing phytoplasma in coconut plantlets obtained through *in vitro* germination of zygotic embryos from the seeds of infected palms. *Ann. Appl. Biol.*
57. Oshima, K., Kakizawa, S., Nishigawa, H., Jung, H. Y., Wei, W., Suzuki, S., et al. (2004). Reductive evolution suggested from the complete genome sequence of a plant-pathogenic phytoplasma. *Nature Genet.* 36, 27–29.
58. Padovan, A. C., and Gibb, K. S. (2001). Epidemiology of phytoplasma diseases in papaya in Northern Australia. *J. Phytopathol.* 149, 649–658.
59. Padovan, A., Gibb, K., and Persley, D. (2000). Association of "*Candidatus* Phytoplasma australiense" with green petal and lethal yellows diseases in strawberry. *Plant Pathol.* 49, 362–369.
60. Pearce, T. L., Scott, J. B., and Pethybridge, S. J. (2011). First report of a 16SrII-D subgroup phytoplasma associated with pale purple coneflower Witches'-Broom disease in Australia. *Plant Dis.* 95, 773–773.
61. Pérez-López, E., Luna-Rodríguez, M., Olivier, C. Y., and Dumonceaux, T. J. (2016). The underestimated diversity of phytoplasmas in Latin America. *Int. J. Syst. Evol. Microbiol.* 66, 492–513.
62. Pilkington, L. J., Gibb, K. S., Gurr, G. M., Fletcher, M. J., Nikandrow, A., Elliott, E., et al. (2003). Detection and identification of a phytoplasma from lucerne with Australian lucerne yellows disease. *Plant Pathol.* 52, 754–762.
63. Pilkington, L. J., Gurr, G. M., Fletcher, M. J., Nikandrow, A., and Elliott, E. (2004). Vector status of three leafhopper species for Australian lucerne yellows phytoplasma. *Aust. J. Entomol.* 43, 366–373.
64. Saqib, M., Bayliss, K. L., Dell, B., Hardf, G. E. S., and Jones, M. G. K. (2005). First record of a phytoplasma-associated disease of chickpea (*Cicer arietinum*) in Australia. *Australas. Plant Pathol.* 34, 425–426.
65. Schneider, B., and Gibb, K. S. (1997). Detection of phytoplasmas in declining pears in Southern Australia. *Plant Dis.* 81, 254–258.

66. Schneider, B., Padovan, A., Rue, S., Eichner, R., Davis, R., Bernuetz, A., et al. (1999). Detection and differentiation of phytoplasmas in Australia: an update. *Aust. J. Agric. Res.* 50, 333–342.
67. Smith, G. R., Braithwaite, K. S., and Cronjé, C. P. R. (2001). "The viral and phytoplasma forms of yellow leaf syndrome of sugarcane," in *Proceedings International Society Sugar Cane Technologists*, ed D. M. Hogarth (Brisbane, QLD: International Society Of Sugar Cane Technologists), 614–617.
68. Streten, C., Conde, B., Herrington, M., Moulder, J., and Gibb, K. (2005). *Candidatus* Phytoplasma australiense is associated with pumpkin yellow leaf curl disease in Queensland, Western Australia and the Northern Territory. *Australasian Plant Pathol.* 34, 103–105.
69. Streten, C., and Gibb, K. S. (2006). Phytoplasma diseases in sub-tropical and tropical Australia. *Australas. Plant Pathol.* 35, 129–146.
70. Tairo, F., Jones, R. A. C., and Valkonen, J. P. T. (2006). Phytoplasma from little leaf disease affected sweetpotato in Western Australia: detection and phylogeny. *Ann. Appl. Biol.* 149, 9–14.
71. Tanne, E., Boudon-Padieu, E., Clair, D., Davidovich, M., Melamed, S., and Klein, M. (2001). Detection of phytoplasma by polymerase chain reaction of insect feeding medium and its use in determining vectoring ability. *Phytopathology* 91, 741–746.
72. Tran-Nguyen, L., Blanche, K. R., Egan, B., and Gibb, K. S. (2000). Diversity of phytoplasmas in northern Australian sugarcane and other grasses. *Plant Pathol.* 49, 666–679.
73. Tran-Nguyen, L. T. T., Kube, M., Schneider, B., Reinhardt, R., and Gibb, K. S. (2008). Comparative genome analysis of "*Candidatus* Phytoplasma australiense" (subgroup *tuf*-Australia I; *rp-A*) and "*Ca.* Phytoplasma asteris" strains OY-M and AY-WB. *J. Bacteriol.* 190, 3979–3991.
74. Tran-Nguyen, L. T. T., Persley, D. M., and Gibb, K. S. (2003). First report of phytoplasma disease in capsicum, celery and chicory in Queensland, Australia. *Australas. Plant Pathol.* 32, 559–560.
75. Tran-Nguyen, L. T. T., Smith, S. H., and Liberato, J. R. (2012). Sweet potato little leaf strain V4 phytoplasma associated with snake bean in the Northern Territory, Australia. *Australas. Plant Dis. Notes* 7, 147–150.
76. Veerakone, S., Tang, J. Z., Ward, L. I., Liefting, L. W., Perez-Egusquiza, Z., Lebas, B. S. M., et al. (2015). A review of the plant virus, viroid, liberibacter and phytoplasma records for New Zealand. *Australas. Plant Pathol.* 44, 463–514.
77. Vega, F. E., Davis, R. E., Barbosa, P., Dally, E. L., Purcell, A. H., and Lee, I. (1993). Detection of a plant pathogen in a nonvector insect species by the polymerase chain reaction. *Phytopathology* 83, 621–624.
78. Wang, Q., Guo, Y., Wang, N., Li, Y., Chen, W., Chen, W., et al. (2014). Identification of a conserved core genome with group-specific genes from comparative genomics of ten different *Candidatus* Phytoplasma strains. *J. Phytopathol.* 162, 650–659.
79. Weintraub, P. G., and Beanland, L. (2006). Insect vectors of phytoplasmas. *Annu. Rev. Entomol.* 51, 91–111.
80. White, D. T., Blackall, L. L., Scott, P. T., and Walsh, K. B. (1998). Phylogenetic positions of phytoplasmas associated with dieback, yellow crinkle and mosaic diseases of papaya, and their proposed inclusion in "*Candidatus* Phytoplasma australiense" and a new taxon, "*Candidatus* Phytoplasma australasia." *Int. J. Syst. Evol. Microbiol.* 48, 941–951.
81. Wilson, D., Blanche, K. R., and Gibb, K. S. (2001). Phytoplasmas and disease symptoms of crops and weeds in the semi-arid tropics of the Northern Territory, Australia. *Australas. Plant Pathol.* 30, 159–163.
82. Yang, S., Habili, N., Aoda, A., Dundas, I., Paull, J., and Randles, J. (2013). Three group 16SrII phytoplasma variants detected in co-located pigeonpea, lucerne and tree medic in South Australia. *Australas. Plant Dis. Notes* 8, 125–129.

83. Zhao, Y., and Davis, R. E. (2016). Criteria for phytoplasma 16Sr group/subgroup delineation and the need of a platform for proper registration of new groups and subgroups. *Int. J. Syst. Evol. Microbiol.* 66, 2121–2123.

CHAPTER 13

Assessment of Early Blight (*Alternaria Solani*) Resistance in Tomato using a Droplet Inoculation Method

Reni Chaerani[1], Remmelt Groenwold[1], Piet Stam[2], Roeland E. Voorrips[1]

[1]Plant Research International Wageningen The Netherlands

[2]Laboratory of Plant Breeding, Department of Plant Sciences Wageningen University Wageningen The Netherlands

[3]Indonesian Center for Agricultural Biotechnology and Genetic Resources Research and Development BogorIndonesia

ABSTRACT

A droplet inoculation method was used for evaluation of tomato resistance to early blight, a destructive foliar disease of tomato caused by Alternaria solani (Ellis and Martin) Sorauer. In this test method, leaflets are inoculated with small droplets of a spore suspension in either water or a 0.1% agar solution. Early blight resistance was evaluated based on lesion size. The droplet method better discriminated the level of resistance ($P < 0.001$) for a range of spore densities in comparison with the more commonly used spray inoculation method. Lesions generated by droplet inoculation at 7 days after inoculation ranged from small flecks to almost complete blight with an exponential-like distribution of lesion sizes. Significant correlations ($r = 0.52, 0.58$, and 0.63, $P < 0.001$) were observed across three glasshouse tests of 54 accessions including wild species using the droplet method. The most resistant accessions included wild species: one accession of Solanum arcanum, three accessions of Solanum peruvianum, one accession of Solanum neorickii, and one of Solanum chilense. Solanum pennellii and Solanum pimpinellifolium accessions were susceptible, whereas Solanum habrochaites and Solanum lycopersicum accessions ranged from susceptible to moderately resistant. The droplet test method is simple to apply, offers a fine discrimination of early blight resistance levels, and allows objective evaluation.

Keywords: Early blight · Alternaria solani · Tomato · Resistance · Screening method

INTRODUCTION

Early blight of tomato, caused by Alternaria solani (Ellis and Martin) Sorauer, is a serious disease in warm and humid regions (Sherf and MacNab 1986) and in semiarid areas where frequent and prolonged night dew occurs (Rotem and Reichert 1964). Early blight (EB) reduces the photosynthetic area, and, in severe cases, can defoliate plants.

Cultivars highly resistant to EB are not known for cultivated tomato [Solanum lycopersicum (Peralta et al. 2005; formerly known as Lycopersicon esculentum)]. All breeding lines and released cultivars range in susceptibility from susceptible to moderately resistant (Vakalounakis 1983; Gardner 1988; Poysa and Tu 1996; Banerjee et al. 1998; Vloutoglou 1999; Gardner and Shoemaker 1999). Several wild species [Solanum habrochaites (syn. Lycopersicon hirsutum), Solanum pimpinellifolium (syn. Lycopersicon pimpinellifolium), Solanum peruvianum (syn. Lycopersicon peruvianum), and Solanum chilense (syn. Lycopersicon chilense)] have been identified as potential sources of resistance (Nash and Gardner 1988; Kalloo and Banerjee 1993; Poysa and Tu 1996; Foolad et al. 2000; Thirthamalappa and Lohithaswa 2000). Some of these, primarily S. habrochaites accession PI 126445, have been used to develop moderately resistant breeding lines (Gardner 1988; Gardner and Shoemaker 1999). Identification of additional sources of resistance could facilitate the development of resistant cultivars.

Field evaluations can identify sources of resistance, but the major drawbacks are the lengthy duration of the tests, uncontrollable environmental conditions necessary for infection, and the presence of other foliar pathogens (Locke 1948; Foolad et al. 2000; Pandey et al. 2003).

Glasshouse tests using spray inoculation of a spore suspension on seedlings are widely used since the establishment of efficient screening and spore inoculum production techniques by Barksdale (1969). The EB lesions resulting from spray inoculation are scattered on the leaves so that the observer must estimate the combined area of all lesions on all leaflets as a percentage of the total leaf area. Disease estimates are rapid but rather subjective. Another disadvantage of the spray inoculation method is that the inoculum may not be uniformly distributed on the leaves. Furthermore, the method is not sensitive enough to discriminate moderately resistant plants from those that are susceptible (Gardner 1990).

An alternative method to obtain more precise and reliable disease readings is offered by placing individual droplets of a fungal inoculum suspension on the leaflets. This method was first introduced by Locke (1948) to find sources of resistance to EB. Detached leaflets were inoculated with a mycelial suspension in a laboratory assay, and the disease reaction was evaluated using a diagram of a graded series of lesions with known diameters (Locke 1948, 1949). Henning and Alexander (1959) used the droplet method to investigate the existence of A. solani races by inoculating leaflets still attached to plants. Nash and Gardner (1988) applied the method, which they called point inoculation, on a whole plant assay and measured the EB lesion diameter. EB resistance of two parents and the F1s were tested in a glasshouse. Their results correlated well with field tests, but were based only on a few genotypes.

Large numbers of accessions have never been screened in the glasshouse using the droplet inoculation method. We describe here some improvements on the method and its application to identify potential EB resistance sources in a collection of tomato accessions.

MATERIALS AND METHODS

Plant material and culture conditions

Tomato seeds were germinated on moistened filter paper in 90-mm-diameter petri dishes for 5–7 days in darkness at 19°C. Germinating seeds were planted in peat soil in boxes or plastic pots. Plants were grown in a glasshouse in Wageningen, The Netherlands, at day/night temperatures of 22°/20°C. Tomato accessions used in the screening experiments are listed in Table 3. They were propagated one generation before use; where possible, inbred lines were obtained by selfing, but in the case of S. peruvianum, halfsib families were harvested after intercrossing five plants per accession. In cases with clear morphological differences between the five plants of the original accession, two lines or two half-sib families were included in the screening experiments.

Fungal culture and inoculum preparation

An Alternaria solani isolate obtained from infected tomato leaves in Sukabumi, West Java, Indonesia, was propagated on V8 juice agar in 90-mm-diameter petri dishes. The dishes were incubated at 21°–22°C with a 12-h diurnal period of fluorescent light for 10–17 days. The cultures were induced to sporulate as described by Barksdale (1969). The number of spores in the suspension was counted in five 10-μl samples. The yield per plate was about 0.7–13.0 × 105 spores.

Conditions during infection

For the first 40 h immediately after inoculation, plants were incubated on a glasshouse bench lined with a wet mat and covered with a transparent plastic tunnel. Periodic misting to maintain high humidity was supplied from a humidifier. After the initial incubation, each side of the tunnel was opened, and the humidifier was turned off for 8 h during the day to allow the plant surface to dry. Minimum light intensity in the plastic tunnel was approximately 14 μmol m−2 s−1 ; when necessary, daylight was supplemented with light from high-pressure sodium vapor lamps (17 μmol m−2 s−1) for 16 h day−1 . The temperature and relative humidity were recorded with a thermohygrograph.

Effect of spore concentration on early blight severity with two inoculation methods

One moderately resistant (FT94-978; 99-213) and one susceptible (HRC90.145) S. lycopersicum line were grown for 3 weeks on peat soil in boxes of 34 × 29.5 × 4 cm. Two inoculation methods, the droplet and the standard spray inoculation methods, were compared. Spore density was varied (0, 1, 2, 4, 10, and 20 × 103 ml−1 water) to find the most discriminating level.

Plants inoculated using the droplet method were raised in boxes of 12 plants, with two rows of three plants of each genotype. A single drop (10 μl) of a spore suspension was placed on an interveinal space of the upper surface of three apical leaflets. The two first expanded leaves were used.

The six spore concentrations were randomized over the six plants of each genotype in each box. The experiment was replicated over three boxes.

In the spray inoculation treatment, boxes contained four rows of four plants, with the two genotypes in alternating rows. Each pair of rows was sprayed with one spore concentration until runoff. The experiment was replicated four times (24 boxes).

The boxes were covered with a transparent lid and placed in the tunnel with intermittent misting for 15 min at 45-min intervals. After the first 24 h, the lids were removed. The temperature during the day was 20°–27°C and during the night was 16°–24°C. The relative humidity ranged from 40% to 72% during the day and from 85% to 100% during the night. Symptom evaluations were done 7 days after inoculation. Length and width of lesions after droplet inoculation were measured. EB severity on each leaf of the sprayed plants was recorded on a scale of 0 to 5, where 0 = no visible lesions on leaf; 1 = up to 10% leaf area affected; 2 = 11%– 25%; 3 = 26%–50%; 4 = 51%–75%; and 5 = more than 75% leaf area affected or leaf abscised (Vakalounakis 1983). Leaves that were not completely unfurled during the inoculation were not assessed. The disease scales were converted nto percentage of EB index (PEBI) for each plant using the following formula (Pandey et al. 2003):

$$\text{PEBI} = \frac{\text{sum of all ratings}}{\text{no. of leaves sampled} \times \text{maximum disease scale}} \times 100$$

Resistance reaction of selected accessions with two inoculation methods in the glasshouse

The repeatability of the droplet inoculation method in determining early blight resistance in a wider range of accessions was compared with the spray inoculation method. Nine accessions including wild species that were found to differ in mean EB lesion size in preliminary experiments were planted in pots and inoculated at 6 weeks after germination.

For spray inoculation, plants were sprayed with spores in water until runoff. For droplet inoculations, a single drop (10 µl) of 0.1% agar solution with 104 spores ml−1 was placed on an interveinal area of three apical leaflets of the four topmost expanded leaves. With agar, the droplets were more likely to adhere to the leaves. The spore concentration of 104 spores ml−1 was selected based on the most optimal inoculum level found for the spray inoculation (see Results), was used for both inoculation methods. Two plants of each accession were tested in three replications for the droplet inoculation and four replications for the spray inoculation.

Noninoculated plants were used as controls for both inoculation methods. The plants were placed in a humidified tunnel directly after inoculation and received periodic misting for 45 s at 8-min intervals. Daytime temperatures ranged from 25° to 27°C and nighttime temperatures from 20° to 22°C. The relative humidity ranged from 59% to 69% during the day and was 98% during the night. Disease reactions were recorded at 7 days after inoculation using the procedure for the respective inoculation methods as previously described.

Glasshouse screening of tomato accessions

Glasshouse screening 2001 (autumn)

Forty-one accessions including wild species were tested (Table 3; GH I and GH II). For 11 accessions, two or three lines or half-sib families were tested because the original accession was not morphologically uniform. The plants were raised in boxes of 34 × 29.5 × 4 cm. Each box contained 12 plots of two plants of 12 different accessions, which were randomized in the boxes. The plants were inoculated using the droplet method at 3 weeks after sowing, when most of them had two fully expanded leaves. Boxes were closed with transparent lids for 24 h and placed in the tunnel. The misting period was 15 min h−1 . The length and the perpendicular width of lesions were recorded 7 days after inoculation. The experiment was replicated five times at weekly intervals; each replicate was treated as a block in the statistical analysis.

In the first two replicates, the three apical leaflets of the two basal leaves of two plants of each accession were inoculated with 2 × 104 spores ml−1 water. However, the basal leaves of some wild species, both with and without lesions were lost earlier (3 days after inoculation) than those of the cultivated tomato, possibly due to faster development and senescence. Early senescence and defoliation were accelerated by inoculation with the pathogen. On some susceptible accessions, lesions expanded rapidly and caused early development of blight. Because of these problems, the droplet inoculation procedures were modified in the subsequent replicates. These first two replicates were treated as a separate experiment, designated as "glasshouse test I" (GH I).

In the subsequent three replicates (GH II) the three apical leaflets of the four topmost leaves were inoculated to achieve a more uniform physiological age of leaves, and a lower inoculum density (4 × 103 spores ml−1 water) was used to prevent blight symptoms from developing too rapidly. Three replicates in time were performed. In the first replicate of GH II, some accessions were represented by less than two plants due to poor germination

Temperatures ranged from 20° to 23°C during the day and from 17° to 19°C during the night. Relative humidity ranged from 43% to 64% during the day and from 97% to 100% during the night.

Glasshouse screening 2002 (summer)

The same 41 accessions were retested together with 13 additional accessions in five replicated tests, performed at weekly intervals (Table 3; GH III). Plants were grown in 12-cm-diameter pots (one seedling per pot) to facilitate inoculation and evaluation. Four weeks after sowing, plants were inoculated at the three apical leaflets of the four topmost expanded leaves with a single drop (10 µl) of 0.1% agar solution carrying 104 spores ml−1 . Each replicate included one plant of each accession. Control plants, one plant of each species, which were inoculated with agar solution without spores, were also included in the tests. Plants were placed in a tunnel and exposed to a fine mist for 45 s to 1 min at intervals of 6–8 min. Five replicate tests were performed during the season at 1-week intervals. The length and the perpendicular width of the lesions were measured at 7 days after inoculation. Day temperatures ranged from 22° to 27°C, night temperatures from 20° to 22°C. Relative humidity ranged from 40% to 66% during the day and from 91% to 93% during the night.

Experimental design and statistical analyses

The elementary data consisted of lesion size (length × width) for the droplet inoculation, and PEBI per plant for the spray inoculation. Heterogeneity in the variances was observed in the data from both inoculation methods. Logarithmic and arcsine-square root transformation was applied before statistical analysis to the lesion size and PEBI data, respectively, to stabilize the variances.

Table 1. Means of disease parameters after droplet inoculation (lesion size) and spray inoculation (percentage of early blight index, PEBI)

Spore density ($\times 10^3$ ml^{-1})	Lesion size mean (mm^2)[a]		t value	PEBI mean (%)[b]		t value
	HRC90.145	FT94-978; 99-213		HRC90.145	FT94-978; 99-213	
1	29.92	10.50	2.16*	18.12	14.68	1.63
2	86.50	16.60	4.38****	28.93	19.23	2.61*
4	108.64	37.93	5.24****	44.88	27.64	2.53*
10	135.83	53.09	4.16****	74.91	53.31	4.16****
20	213.80	80.91	6.18****	61.62	75.21	2.98***

* $P < 0.05$; ** $P < 0.01$; *** $P < 0.005$; **** $P < 0.001$

[a] Average of three replicates of 18 leaflets each (6 leaflets × 3 plants); values are back transformations of log (x)

[b] Average of four replicates of four plants each; values are back transformations of arcsine [$\sqrt{(x/100)}$]

Table 2. Mean early blight lesion size and percentage of early blight index (PEBI) of nine selected accessions

Species	Accessions	Lesion size (mm^2)[a]	PEBI[b]
S. peruvianum	PE44	1.19 a	67.43 abc
S. peruvianum	PE33	6.67 b	59.03 a
S. lycopersicum	HRC90.158	7.73 bc	66.28 abc
S. pimpinellifolium	G1.1554	12.59 cd	86.69 cd
S. lycopersicum	NC EBR-4	15.07 d	62.43 ab
S. habrochaites	G1.1561	15.21 d	84.98 bcd
S. habrochaites	864086-2; PI272745	29.72 e	97.71 d
S. habrochaites	PE36	40.18 e	60.40 ab
S. lycopersicum	FT97-515; 99-214	45.92 e	90.86 d

[a]Length × width measured 7 days after inoculation. Each value is an average of three replicates of 24 leaflets (3 leaflets × 4 leaves × 2 plants); data are back transformations of log(x). Values within a column followed by the same letters are not significantly different at $P = 0.05$
[b]Based on individual leaf scores: 0 = no visible lesions on leaves; 1 = up to 10% leaf area affected; 2 = 11%–25%; 3 = 26%–50%; 4 = 51%–75%; and 5 = more than 75% leaf area affected or leaf shed. Each value is an average of four replicates of two plants; data are back transformation of arcsine [$\sqrt{(x/100)}$]

Student's t-test was performed for the data from the spore concentration experiment to compare means. All other experimental data were analyzed by analysis of variance (ANOVA) as a randomized complete block design. Mean separations were done by means of least significant difference (LSD) tests (P ≤ 0.05). GH II was analyzed using the unbalanced treatment structure procedure of ANOVA because of the unequal number of plants per block and per accession. All analyses were done using the GenStat 6 statistical package (Payne et al. 2002).

RESULTS

The effect of spore concentration on EB severity under two inoculation methods

EB lesions resulting from both the droplet and spray inoculations appeared within 3 days after inoculation. Some droplet inoculations failed to yield substantial lesions; either small spots (≤1 mm in diameter) formed, or no symptoms developed. This was observed for both accessions. Symptomless inoculations were scored as missing values.

With both inoculation methods, FT94-978; 99-213 (moderately resistant) developed significantly smaller lesions or lower PEBI than HRC90.145

(susceptible) at all spore concentrations, except for PEBI at 20 × 103 ml−1 (Table 1). At all concentrations, the droplet method better discriminated the resistance level than the spray method, as indicated by higher and more significant t values ($P < 0.001$). The most significant difference in PEBI between FT94-978; 99-213 and HRC90.145 was observed at a spore concentration of 104 ml−1 , while the differences were highly significant at all concentrations above 103 ml−1 for the droplet inoculation method.

Resistance reaction of selected accessions with two inoculation methods in glasshouse tests

The comparison of the droplet and spray inoculations was expanded to a set of nine accessions representing four tomato species, which from preliminary experiments were known to represent large differences in EB resistance. The wild accessions occasionally developed spontaneous blisters or necroses under glasshouse conditions. When EB developed on S. habrochaites leaves with necroses, the lesions would expand rapidly and result in severe blight symptom, complicating the measurement of lesion size. Leaflets with blight symptoms so severe that lesion size could not be measured were scored as missing values.

The droplet method allowed a better separation of accessions than the spray inoculation method (Table 2) in accordance with the results of the spore concentration experiment. Accession reactions with the two inoculation methods were not significantly correlated ($r = 0.44$, $P > 0.2$). S. habrochaites PE36 had inconsistent results between the two inoculation methods: it was ranked as susceptible with the droplet method but resistant with the spray method. Occasional spontaneous necroses on this accession inoculated with the droplet method exacerbated EB lesions. Excluding this accession from the analysis increased the correlation considerably ($r = 0.66$).

EB lesions on petioles were observed on sprayed plants. Large, sunken lesions on petioles often caused loss of the leaf and thus raised the PEBI of accessions that had small leaf lesions when inoculated by the droplet method. The petiole lesions were generated randomly because the spray inoculation was not purposely directed to petioles.

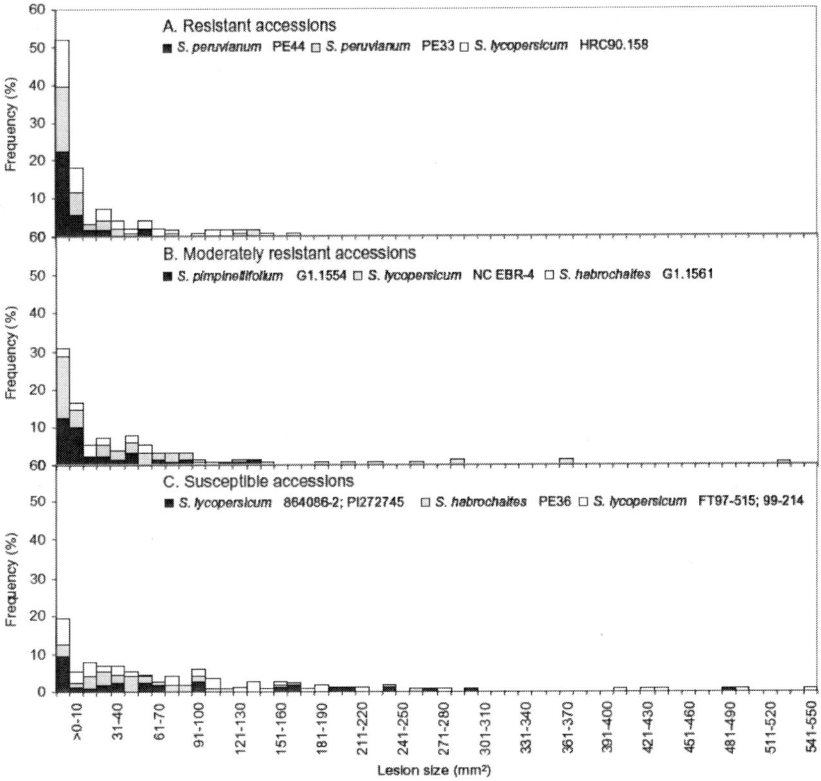

Figure 1A–C. Frequency distribution of lesion sizes on droplet-inoculated plants in three groups of three accessions: A resistant, B moderately resistant, and C susceptible accessions

Lesion size distribution

Inoculations using the droplet method did not always develop into a noticeable lesion, irrespective of the level of resistance of a plant. We first assumed that the spores dried out before successfully penetrating the host tissue during the initial hours of incubation in the tunnel. However, incorporation of 0.1% agar solution into the spore suspension, which immobilized the droplets and slowed evaporation, did not influence the probability of lesion formation.

After grouping the nine accessions into three categories (resistant, moderately resistant, and susceptible), an exponential-like distribution of lesion sizes was observed in each group (Figure 1). The more resistant accessions had a higher frequency of small lesions, a lower frequency of larger lesions, and a lower mean lesion size. Symptomless leaflets were observed with an average of 0.9%, 7%, and 18.2% for susceptible, intermediate, and resistant accessions, respectively; for flecks (≤1 mm2), these frequencies were 2%, 9%, and 17%. In a microscopic evaluation of leaflets with flecks after staining with lactophenol–cotton blue, infection was evident by the presence of germ tube penetration, but mycelial proliferation was absent. An exponential-like distribution of lesion size was also observed for plants inoculated by spraying (data not shown).

Glasshouse screenings

Lesions near leaf veins of the susceptible S. lycopersicum accessions were often accompanied by smaller lesions of angular shape without concentric rings along the vein and at the distal ends of the vein. These smaller lesions rapidly expanded and eventually merged with the primary lesions, resulting in almost completely blighted leaves before lesion measurement at 7 days after inoculation. Flecks, which did not further expand, and symptomless inoculations were again observed on all tested accessions. Leaflets with severe blight and those without appreciable lesions were scored as missing values.

The means of the lesion sizes from each glasshouse test were weighted with the reciprocals of the variances to obtain overall accession mean values. This adjustment was necessary since variances of the means among the three tests were unequal. After calculating the weighted average, we observed a continuous range of resistance levels, from highly susceptible (S. pennellii LA716, average lesion size 107.65 mm2) to highly resistant (S. arcanum LA2157, 1.40 mm2) (Table 3). However, no complete resistance was found. Among the glasshouse tests, significant correlations were observed ($P < 0.001$), with correlation coefficients of 0.58 (between GH I and II), 0.52 (GH I and III), and 0.63 (GH II and III). Most of the resistant accessions belonged to wild species (S. arcanum, S. peruvianum, S. neorickii, and S. chilense). However, both susceptible and moderately resistant S. habrochaites accessions were found, and other wild accessions belonging to S. pimpinellifolium and especially S. pennellii were susceptible. Among different lines of the previously reported moderately resistant accession HRC91.341 (Poysa and Tu 1996), there were significant differences in EB lesion sizes. One line of this accession was susceptible in GH II but resistant or moderately resistant in GH I and GH III, while the other line was resistant in all three tests. Lines derived from NC EBR-2 and NC EBR-3 also had a different reaction in GH II compared with GH I and GH III.

Single lesions resulting from the droplet inoculation permitted detailed observation of lesion phenotypes. Necrotic lesions on S. neorickii (syn. L. parviflorum), some accessions of S. habrochaites and S. peruvianum, and S. lycopersicum NC EBR-6 were surrounded by narrow chlorotic halos, whereas those on S. pennellii, S. chilense, and other accessions of S. habrochaites and S. peruvianum were not accompanied by chlorotic halos. S. pimpinellifolium and other S. lycopersicum genotypes had a range of intermediate to wide halos. The extent of the halos did not seem to correlate with the size of the necrotic lesions (data not shown).

DISCUSSION

The droplet inoculation method offers better discrimination among accessions than the spray method. This was observed in the spore concentration experiment and in the nine-accession experiment. Variance of lesion size increased with increasing means in both experiments. Distribution of lesion sizes seemed to be exponential, with many small lesions and few large ones. This trend was observed at all levels of resistance.

The discrepancy between the results of the droplet and spray methods can partly be explained by two factors. First, some wild species developed severe necroses in the glasshouse experiment even without inoculation. After spray inoculation, these necroses were often indistinguishable from EB lesions, whereas after droplet inoculation they were simply recognized and treated as missing values. Sec ond, spray inoculation may lead to random leaf shedding due to petiole lesions and therefore erratic, high symptom scores. The droplet method offers a possibility to test the effect of petiole lesions in a controlled way.

Across the three glasshouse screenings, significant correlations were observed. GH II and GH III yielded a better separation than GH I. Also the correlation between GH II and GH III was higher than that between GH I and the other two tests. This may be due to: (1) the lower number of observations in GH I, (2) the higher inoculum density in GH I, and (3) the difference in the selection of leaves between GH I and the other two tests. We used a lower inoculum concentration (40 to 200 spores per droplet) in the glasshouse screenings but observed overall larger lesion sizes than Nash and Gardner (1988) who applied more inoculum (500 to 750 spores per droplet) on basal leaves. This indicates that our tests were performed in near-optimal conditions for infection and disease development. Another difference between our work and the study of Nash and Gardner (1988) was that they selected one lesion per plant to be measured, whereas we measured all inoculations without selection. As a result, we observed a

Table 3. Early blight lesion sizes of Lycopersicon accessions in glasshouse tests

Species	Accessions	Source[a]	GH I[b,c]	GH II[d]	GH III[e]	Weighted average[f]
Solanum arcanum	LA2157	1	NT	NT	1.40 a	1.40
S. peruvianum	PE44	4	NT	NT	1.46 ab	1.46
S. peruvianum	PI 390665	4	NT	NT	4.07 c–h	4.07
S. peruvianum	PE33	4	11.38 a	1.54 a	3.30 c–f	6.23
S. neorickii	G1.1601	2	NT	NT	6.68 g–m	6.68
S. chilense	G1.1556	2	NT	NT	6.68 g–m	6.68
S. lycopersicum	NC EBR-6	3	16.90 ab	2.59 a–d	5.02 d–k	9.56
S. lycopersicum	NC EBR-6	3	17.10 ab	4.56 a–i	2.61 bc	10.49
S. lycopersicum	HRC86.320	4	24.27 a–d	1.57 a	4.32 c–l	10.61
S. lycopersicum	cv. Santacruz	6	NT	NT	11.67 m–p	11.67
S. habrochaites	LA2650	4	18.58 ab	3.60 a–g	16.79 p–u	12.24
S. lycopersicum	HRC90.159	4	25.82 a–d	2.92 a–e	3.16 c–f	12.56
S. lycopersicum	HRC86.320	4	32.36 a–e	2.06 ab	2.93 cd	13.29
S. habrochaites	PE36	4	31.48 a–e	2.34 a–c	10.45 m–p	15.00
S. peruvianum	PI390665	4	16.94 ab	11.00 f–k	16.52 p–u	15.19
S. lycopersicum	HRC90.159	4	30.55 a–e	3.90 a–i	4.84 d–j	15.65
S. lycopersicum	HRC91.341	4	30.97 a–e	4.62 a–j	4.56 c–l	16.40
S. lycopersicum	NC EBR-1	3	28.91 a–e	4.73 a–j	10.69 m–p	16.94
S. lycopersicum	NC EBR-3	3	22.49 a–d	9.59 e–k	7.59 i–n	16.97
S. lycopersicum	HRC90.190	4	32.58 a–e	4.98 a–j	4.82 d–j	17.36
S. peruvianum	PI270435	4	NT	NT	17.50 p–u	17.50
S. lycopersicum	cv. Sufan n.1	6	NT	NT	17.54 p–u	17.54
S. habrochaites	LA1777	2	NT	NT	17.54 p–u	17.54
S. lycopersicum	NC EBR-2	3	21.43 a–c	15.10 jk	10.40 l–p	18.46
S. lycopersicum	NC EBR-5	3	42.46 b–e	3.10 a–f	5.73 f–l	18.67
S. lycopersicum	HRC86.329	4	36.39 a–e	3.71 a–h	12.94 n–r	18.88
S. lycopersicum	HRC86.329	4	32.66 a–e	5.16 a–j	14.16 o–s	19.23
S. lycopersicum	HRC90.157	4	30.76 a–e	8.24 c–k	3.83 c–g	19.28
S. lycopersicum	HRC86.321	4	41.88 b–e	4.00 a–i	5.33 e–k	19.60
S. lycopersicum	HRC90.159	4	39.08 a–e	5.93 b–k	3.12 c–e	19.72
S. lycopersicum	NC EBR-3	3	33.50 a–e	6.04 b–k	NT	20.34
S. lycopersicum	HRC90.157	4	41.40 b–e	4.33 a–i	8.36 j–o	20.60
S. lycopersicum	FT94-978; 99-213	5	39.72 a–e	4.60 a–j	9.73 l–p	20.62
S. lycopersicum	NC EBR-4	3	36.90 a–e	6.50 b–k	15.63 p–t	22.50
S. lycopersicum	FT94-968; 99-212	5	33.11 a–e	10.80 f–k	11.35 m–p	22.86
S. habrochaites	G1.1561	2	42.95 b–e	6.67 b–k	7.62 i–n	23.33
S. lycopersicum	HRC91.341	4	34.28 a–e	12.80 g–k	7.76 i–n	23.63
S. lycopersicum	FT97-515; 99-214	5	40.18 b–e	6.38 b–k	17.50 p–u	23.69
S. lycopersicum	FT94-978; 99-213	5	42.07 b–e	7.80 c–k	9.04 k–p	24.27
S. lycopersicum	cv. Allround	1	NT	NT	24.32 s–u	24.32
S. pimpinellifolium	G1.1554	1	47.53 b–e	8.59 d–k	6.75 g–m	24.70
S. lycopersicum	HRC90.158	4	49.66 b–e	7.24 b–k	4.60 c–j	25.13
S. lycopersicum	NC EBR-2	3	39.36 a–e	13.20 h–k	10.21 l–p	26.73
S. lycopersicum	PI79532	6	NT	NT	26.98 t–v	26.98
S. lycopersicum	FT94-968; 99-212	4	48.19 b–e	9.57 e–k	9.77 l–p	28.15
S. lycopersicum	cv. Vogliotti	6	NT	NT	29.17 uv	29.17
S. lycopersicum	HRC89.302	4	44.26 b–e	12.20 g–k	21.04 q–u	30.15
S. lycopersicum	864084-2; PI 273048	5	68.39 c–e	12.10 g–k	7.05 h–m	36.63
S. lycopersicum	HRC89.302	4	53.33 b–e	16.40 jk	21.09 r–u	36.74
S. lycopersicum	cv. Moneymaker	1	75.34 de	12.40 g–k	17.62 p–u	42.52
S. lycopersicum	HRC86.327	4	72.11 c–e	13.70 i–k	45.08 v	45.88
S. lycopersicum	HRC90.145	4	71.45 c–e	20.50 k	28.84 uv	48.49
S. lycopersicum	864086-2; PI 272745	5	94.84 e	18.80 k	45.08 v	59.49
S. pennellii	LA716	1	NT	NT	107.65 w	107.65

Lesion sizes are given in units of mm^2 (length × width) measured 7 days after inoculation. Data are back transformations of log(x)
NT, Not tested
[a] 1, Plant Research International, Wageningen, The Netherlands; 2, Dr. P. Lindhout, Laboratory of Plant Breeding, Wageningen University, The Netherlands; 3, Dr. R.G. Gardner, North Carolina Agricultural Research Institute, North Carolina State University, Raleigh, North Carolina, USA; 4, Dr. V. Poysa, Agriculture and Agri-Food Canada, Harrow Research Center, Harrow, Ontario, Canada; 5, Prof. M. Mutschler, Department of Plant Breeding, Cornell University, Ithaca, New York, USA; 6, Nunhems Zaden BV, Haelen, the Netherlands
[b] Values followed by the same letters within a column are not significantly different at $P = 0.05$
[c] Inoculated with 2×10^4 spores ml^{-1} water. Each value is the average of two replicates of 12 leaflets (3 leaflets × 2 leaves × 2 plants)
[d] Inoculated with 4×10^3 spores ml^{-1} water. Each value is the average of three replicates of 12 or 24 leaflets (3 leaflets × 4 leaves × 1 plants or 3 leaflets × 4 leaves × 2 plants)
[e] Inoculated with 10^4 spores ml^{-1} in 0.1% agar. Each value is an average of five replicates of 12 leaflets (3 leaflets × 4 leaves × 1 plant)
[f] $\sum(Y_i/s_i^2)/\sum(1/s_i^2)$

large range of lesion types, from small lesions (≤1 mm in diameter) to almost blighted leaflets and also symptomless leaflets within the same accession. Small lesions (≤1 mm in diameter) occurred on all genotypes, but their frequency corresponded with the resistance level. This was indicated by a high correlation between lesion size and the percentage of small lesions, most notably in GH II and GH III where a lower spore concentration was applied: $r = -0.73$ (GH II, $P < 0.001$) and $r = -0.77$ (GH III, $P < 0.001$).

Variability in pathogenicity of A. solani isolates has been widely described (Bonde 1929; Henning and Alexander 1959; Rotem 1966), but no evidence has been presented for the existence of pathological races. The results of our study, in which we used a single highly pathogenic Indonesian isolate, can therefore be expected to be representative for other Indonesian isolates as well.

The droplet inoculation method is simple to apply and allows an objective evaluation of EB severity. The method has been used to evaluate EB resistance components (O'Leary and Shoemaker 1983). Single lesions created by the droplet method allow detailed observation of lesion phenotype such as differential formation of halos among genotypes. The importance of the chlorotic halo as an indicator of resistance has not been studied so far. Improvements of the method have been made by incorporating agar into the spore suspension and the use of upper leaves as opposed to the bottom leaves.

The considerable amount of time required to measure lesions may make this method less attractive for largescale screenings, but this problem can be circumvented by determining the percentage of small lesions. The described advantages of the droplet inoculation method make this the method of choice where a fine discrimination of resistance level and accurate quantitative data are required, for example, in quantitative trait locus studies of resistance or in assessing breeding material during advanced backcross programs.

ACKNOWLEDGMENTS

We thank Dr. Suhardi (Indonesian Ornamental Plants Research Institute, IOPRI) and Dr. E. Sofiari (Indonesian Vegetables Research Institute, IVEGRI) for providing the Alternaria solani isolate, and D. Geurtsen and A. Hermsen for their assistance in the glasshouse. Financial support was provided by the Royal Netherlands Academy of Arts and Sciences in the framework of the Scientific Programme Indonesia–Netherlands.

REFERENCES

1. Banerjee, MK, Chhabra, ML, Saini, PS 1998Responses of tomato cultivars to Alternaria blightTest Agrochem Cult195051
2. Barksdale, TH 1969Resistance of tomato seedlings to early blight Phytopathology 59443446
3. Bonde, R 1929Physiological strains of Alternaria solaniPhytopathology19533548
4. Foolad, MR, Ntahimpera, N, Christ, BJ, Lin, GY 2000Comparison of field, greenhouse, and detached-leaflet evaluations of tomato germ plasm for early blight resistancePlant Dis84967972
5. Gardner, RG 1988NC EBR-1 and NC EBR-2 early blight resistant tomato breeding linesHortScience23779781
6. Gardner, RG 1990Greenhouse disease screen facilitates breeding resistance to tomato early blightHortScience25222223
7. Gardner, RG, Shoemaker, PB 1999"Mountain Supreme" early blight-resistant hybrid tomato and its parents, NC EBR-3 and NC EBR-4HortScience34745746
8. Henning, RG, Alexander, LJ 1959Evidence of existence of physiologic races of Alternaria solaniPlant Dis Rep43298308
9. Kalloo, G, Banerjee, MK 1993Early blight resistance in Lycopersicum esculentum Mill. transferred from L. pimpinellifolium (L.) Mill. and L. hirsutum f. glabratum MullGartenbauwissenschaft58238239
10. Locke, SB 1948A method for measuring resistance to defoliation diseases in tomato and other Lycopersicon speciesPhytopathology38937942
11. Locke, SB 1949Resistance to early blight and Septoria leaf spot in the genus LycopersiconPhytopathology39829836
12. Nash, AF, Gardner, RG 1988Tomato early blight resistance in a breeding line derived from Lycopersicon hirsutum PI 126445Plant Dis72206209
13. O'Leary, DJ, Shoemaker, PB 1983Components of resistance for tomato early blightPhytopathology73803
14. Pandey, KK, Pandey, PK, Kallo, G, Banerjee, MK 2003Resistance to early blight of tomato with respect to various parameters of disease epidemicsJ Gen Plant Pathol69364371
15. Payne, RW, Harding, SA, Murray, DA, Soutar, DM, Baird, DB, Welham, SJ, Kane, AF, Gilmour, AR, Thompson, R, Webster, R, Wilson, GT 2002GenStat for Windows6th ednVSNOxford
16. Peralta, IE, Knapp, S, Spooner, DM 2005New species of wild tomatoes (Solanum Section Lycopersicon: Solanaceae) from northern PeruSyst Bot30424434
17. Poysa, V, Tu, JC 1996Response of cultivars and breeding lines of Lycopersicon spp. to Alternaria solaniCan Plant Dis Surv7658
18. Rotem, J 1966Variability in Alternaria porrii f. sp. solaniIsrael J Bot154857
19. Rotem, J, Reichert, I 1964Dew – a principal moisture factor enabling early blight epidemics in a semiarid region of IsraelPlant Dis Rep48211215
20. Sherf, AF, MacNab, AA 1986Vegetable diseases and their controlWileyNew York
21. Thirthamalappa, , Lohithaswa, HC 2000Genetics of resistance to early blight (Alternaria solani Sorauer) in tomato (Lycopersicum esculentum L.)Euphytica113187193
22. Vakalounakis, DJ 1983Evaluation of tomato cultivars for resistance to Alternaria blightAnn Appl Biol102138139
23. Vloutoglou, I 1999Evaluation of tomato cultivars and hybrids for resistance to Alternaria solani infectionTest Agrochem Cult204849

CHAPTER 14

Synthesis and Functions of Jasmonates in Maize

Eli J. Borrego and Michael V. Kolomiets *

Department of Plant Pathology and Microbiology, Texas A&M University, College Station, TX 77843, USA

ABSTRACT

Of the over 600 oxylipins present in all plants, the phytohormone jasmonic acid (JA) remains the best understood in terms of its biosynthesis, function and signaling. Much like their eicosanoid analogues in mammalian system, evidence is growing for the role of the other oxylipins in diverse physiological processes. JA serves as the model plant oxylipin species and regulates defense and development. For several decades, the biology of JA has been characterized in a few dicot species, yet the function of JA in monocots has only recently begun to be elucidated. In this work, the synthesis and function of JA in maize is presented from the perspective of oxylipin biology. The maize genes responsible for catalyzing the reactions in the JA biosynthesis are clarified and described. Recent studies into the function of JA in maize defense against insect herbivory, pathogens and its role in growth and development are highlighted. Additionally, a list of JA-responsive genes is presented for use as biological markers for improving future investigations into JA signaling in maize.

Keywords: jasmonic acid; maize; lipoxygenase; oxylipins; plant-insect interactions; plant-microbe interactions

1. IMPORTANCE OF MAIZE AS A CROP AND A GENETIC MODEL

Despite contributing over 50% of the annual calories for humans [1] and 34% of the production for animal feed [2], little is known about fundamental hormone biology in monocot plants compared to greater advances with dicot plants, primarily *Arabidopsis*. Maize is a diploid large grain cereal possessing unique reproductive organs where the male and female sexual organs are spatially separated. It also has a rich cultural heritage. Of all of the top most cultivated monocot crops, maize (*Zea mays* L.) is the only one with New World origins. Several characteristics of maize facilitate its use as a model organism to explore fundamental processes: the maize genome was recently sequenced for the B73 [3] and Palomero Toluqueño [4] lines; its separated sexual organs allow for effortless controlled pollination; it possesses tremendous genetic diversity; and it is backed by an intensely collaborative scientific community [5].

Maize and its closest crop relative, *Sorghum bicolor*, belong to the Panicoideae subfamily of the grasses [6] and utilize C4 carbon metabolism, an adaptation to high light intensities, temperatures and low water availability [7]. In contrast, most other economically-important cereals, such as wheat, rice and barley, belong to the Pooideae subfamily and utilize the more common C3 carbon metabolism. In recent years, scientific communities have sought to establish *Brachypodium distachyon* [8] and *Setaria viridis* [9] as monocot models for the Pooideae and Panicoideae, respectively [10]. Modern maize arose from an ancient hybridization event between two closely related diploid species that resulted in doubling of the genome (i.e., allopolyploidization) between 8.5 and 13 million years ago [11]. Eventually, the genome of this ancestor was reduced to the diploid state, but with uneven gene loss [3]. This resulted in many maize genes, including jasmonic acid (JA) biosynthesis genes, occurring as pairs throughout the genome.

2. JASMONATES BELONG TO OXYLIPINS, A GROUP OF SIGNALS BETTER UNDERSTOOD IN MAMMALS

JAs belong to the immense group of oxygenated fatty acid products known collectively as oxylipins. Oxylipins are derived from either enzymatic or autoxidation of free or membrane-esterified fatty acids. Since all metabolic processes in cells are enclosed within fatty acid-containing membranes, oxylipins are universally produced across all kingdoms of life. Undeniably, oxylipins are best understood in mammals and in these organisms are termed eicosanoids. Eicosanoids are further categorized into subgroups based on their chemical structure and the enzymes (e.g., lipoxygenases, cyclooxygenases and epoxygenases) that catalyze the fatty acid oxygenation. The major subgroup of enzymatically-derived mammalian oxylipins are leukotrienes, prostaglandins, prostacyclins, thromboxanes, lipoxins, eoxins, hydroxyeicosatetraenoic acids (HETEs) and epoxyeicosatrienoic acids (EETs), while chemically-produced eicosanoids are known as isoprostanes [12]. In mammals, oxylipins regulate diverse physiological processes, such as vasoconstriction, vasodilation, pain response and fever generation. A central characteristic of eicosanoid activity is their cell-type-dependent responses, which occur from the receptor-ligand binding of specific eicosanoid with its particular receptor [13]. In addition to receptor-ligand signaling, oxylipins also possess direct antimicrobial activity [14] and are capable of altering cellular redox status [15].

Eicosanoids were first identified in 1935 with the discovery of prostaglandins in seminal fluid [16,17]; however, significant strides in eicosanoid function did not occur until after the development of synthetic eicosanoid standards and improved analytical techniques. Figure 1 represents the number of articles published each year from 1949 through 2015 as indexed by ISI Web of Science using eicosanoids (terms used: "prostaglandin", "thromboxane" and "leukotrienes"), jasmonates (terms used: "jasmonate", "jasmonic", "jasmonyl" and "jasmone" [18]) and plant oxylipins (term used: "oxylipins") and related oxylipin terms. The first dramatic increase in the number of articles published with the terms associated with eicosanoids in the early 1970s corresponds to the development of methods of the synthesis of prostaglandins [19]. Another substantial increase in the early 1990s coincides with the development of electrospray ionization, an essential component for the detection of biomolecules through mass spectrometry [20].

Together, these techniques have allowed for the elucidation of the mechanisms behind the activity of oxylipins on numerous human physiological processes, which has resulted in the understanding of oxylipins and their biosynthetic enzymes as important therapeutic targets in modern clinical medicine. The pharmacological manipulation of oxylipin biosynthesis has been used to relieve fever and pain for several hundred years [21]. The first of these therapies arose from plant extract enriched in salicylic acid and later with synthetic salicylates (e.g., aspirin) and nonsteroidal anti-inflammatory drugs (NSAIDs; e.g., ibuprofen), all of which inhibit cyclooxygenase activity [22]. Selective inhibitors of oxylipin biosynthetic enzymes [23] or substrate availability (e.g., glucocorticoids) [24] have been developed and are routinely prescribed or used over-the-counter. Cyclooxygenase inhibitors make up one third of the prescriptions and 60% of drug costs [25]. Impressively, nearly 80% of the U.S. Food and Drug Administration-approved drugs [26] target pathways under oxylipin regulation.

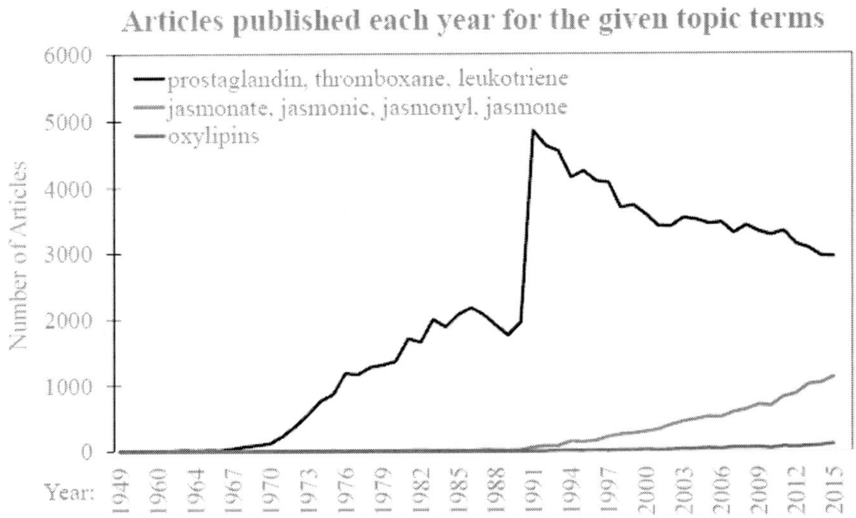

Figure 1. Articles published each year for the given topic terms related to mammalian and plant oxylipins. The x-axis represents year from 1949 to 2015 and y-axis represents the number of articles published each year for the selected topic terms. The line representing the terms related to mammalian oxylipins (terms "prostaglandin", "thromboxane", and "leukotriene"), known as eicosanoids, and is colored black. The line representing jasmonates (terms "jasmonate", "jasmonic", "jasmonyl", and "jasmone") is colored red. Plant oxylipins (term "oxylipins") is colored purple. Terms were searched in the Web of Science™ citation indexing service and organized per year.

3. PLANT OXYLIPINS

In plants, oxylipins are generated primarily from the enzymatic oxygenation of polyunsaturated fatty acids, linoleic (C18:2) or linolenic acid (C18:3) via lipoxygenase (LOX) or alpha-dioxygenase [27,28].

Chemical oxygenation of fatty acids produces a subset of oxylipins, known as phytoprostanes and phytofurans [29,30]. The structural similarities of mammalian prostaglandins and JA provided plant biologists a springboard to explore the role of oxylipins in the plants and to explore the commonality of oxylipin biology in both kingdoms. Of the more than 600 estimated oxylipins [31] produced from plants as a whole, JA has been in the limelight for nearly all investigations into plant oxylipins. Despite this focus, in sharp contrast to human oxylipins, the number of articles published about JA lags behind (Figure 1). Nonetheless, JA remains the best-understood plant-derived oxylipin and serves as the model oxylipin species in plant biology. Its biosynthesis, signaling and roles in defense, development, tolerance, growth and symbiotic interactions were identified through work with *Arabidopsis*, tobacco and tomato [32,33].

It is important to point out that JA biosynthesis has been reported to occur in fungi from several genera, including pathogenic and symbiotic species [34,35,36]. It appears that at least for some fungi, JA biosynthesis evolved independently, but developed in a manner similar to plant JA production [37]. The capability of fungi to produce JA suggests that fungi modify plants into more suitable hosts through JA-mediated signaling. This phenomenon is similar to how coronatine, a JA structural mimic, is utilized by pathogenic bacteria to facilitate pathogenesis. In that system, coronatine facilitates stomata opening for bacterial entry into the plant [38] and suppresses plant inducible defenses [39]. In turn, JA produced by fungi may be used as molecular cues for defense against diverse pathogens [40].

4. JASMONATE BIOSYNTHESIS IN MAIZE

Less than a handful of JA biosynthesis enzymes have been characterized in maize thus far. This review describes the current literature regarding the genes identified via genetic or recombinant protein evidence, but also proposes additional uncharacterized gene candidates with a strong likelihood of their involvement in JA biosynthesis. The majority of studies in JA production arose from investigations utilizing the B73 yellow dent inbred line [41], the first maize genome that has been sequenced. As such, this section regarding JA biosynthesis is B73-centric, and it is reasonable to expect that diverse inbred lines and landraces possess varying copy numbers for specific JA biosynthesis gene family members. Another gene in the octadecanoid pathway, *LIPOXYGENASE5* (*ZmLOX5*), was shown to be present in zero, one and two copy numbers in the inbred lines tested [42]. A diagram of JA biosynthesis with maize-specific enzymes is presented (Figure 2), but for a more detailed description of JA biosynthesis in plants, please refer to the comprehensive historical and current reviews ([32,33,43]).

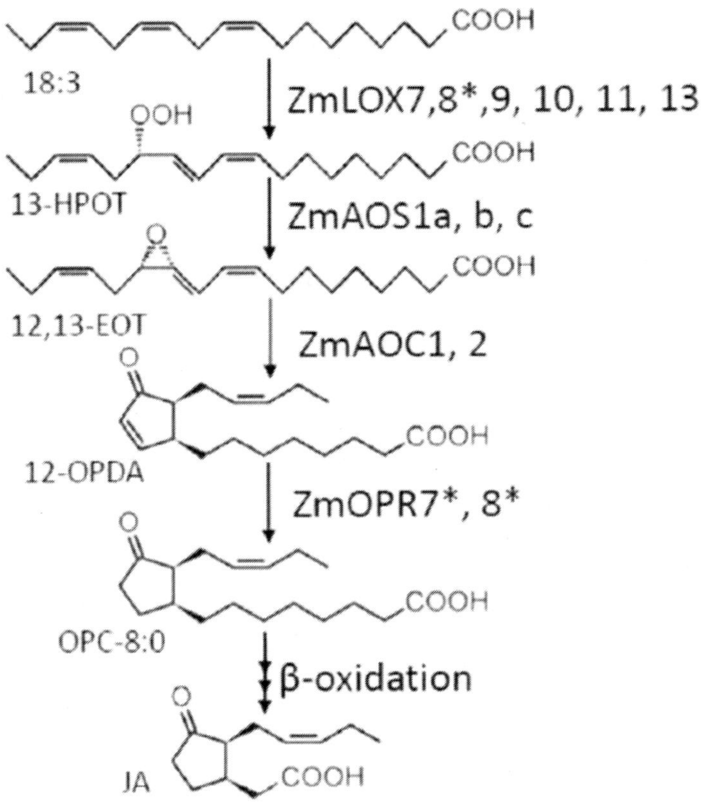

Figure 2. Overview of maize JA biosynthetic pathway. Genes marked with * have been functionally characterized [44,45] and other genes are predicted from phylogenic analysis.

5. 13-Lipoxygenases

These non-heme iron containing oxygenases are the namesake for an immense pathway of fatty acid metabolism, known as the lipoxygenase (LOX) pathway, consisting of at least seven enzyme branches [46]. Depending on plant species and tissue specificity [47], this pathway produces diverse subsets of alcohols, aldehydes, divinyl ethers, esters, epoxides, hydroxides, hydroperoxides, ketols, ketones and triols [46,48] (Figure 3). These metabolites are implicated in nearly all plant biological processes that include defense, development and trans-kingdom communication [40,49].

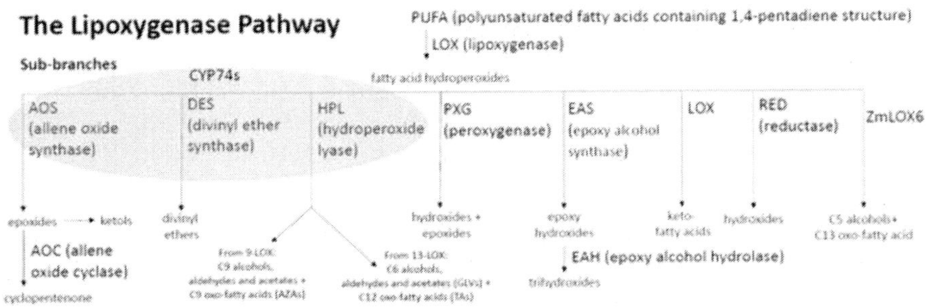

Figure 3. Major metabolite classes produced by the eight sub-branches in the Lipoxygenase (LOX) pathway. LOXs catalyze the dioxygenation of polyunsaturated fatty acids (PUFA), namely linoleic (C18:2) or linolenic (C18:3) fatty acids at the 9- or 13-carbon position. The subsequent hydroperoxide fatty acid can be fluxed into seven sub-branches [31,46]. Allene oxide synthase (AOS), divinyl ether synthase (DES), and hydroperoxide lyase (HPL) are cytochrome P450 monooxygenases (CYP) and part of the CYP74 clade (green oval). The AOS sub-branch is responsible for production of jasmonates (JAs). Divinyl ether synthases (DES) have been identified in some dicots, but not in monocots, including maize. The hydroperoxide lyase (HPL) sub-branch mediates the cleavage of 13-hydroperoxide octadecadi(tri)enoic (HPOD/T) into C6- alcohols, aldehydes, and acetates collectively known as green leaf volatiles (GLVs), and as well as the oxo-fatty acid, traumatin which oxidizes into the dicarboxylic acid, traumatic acid (TA). 9-HPL converts 9-HPOD/T to produce C-9 alcohols, aldehydes, acetates, and C9-oxo-fatty acid that can be oxidized into the dicarboxylic acid, azelaic acid (AZA). Peroxygenase (PXG) catalyzes the formation of hydroxides and epoxides. LOX activity on hydroperoxides yields keto fatty acids. Epoxide alcohol synthase (EAS) synthesizes epoxy-hydroxides which are hydrolyzed to trihydroxy fatty acids. Hydroxides can be converted nonenzymatically from hydroperoxy fatty acids or through action of reductases. ZmLOX6 is a unique monocot specific sub-branch which produces a C13-oxo-fatty acid and C5 alcohols from 13-HPOT specifically [48].

In the maize B73 genome, this enzyme family is comprised of 13 members categorized into either the 9- or 13-LOX groups depending on the position the oxygen is incorporated into the fatty acid carbon backbone [50,51]. Only 13-LOXs are capable of catalyzing 13(*S*)-hydroperoxide octadecatrienoic acid (13-HPOT, (9Z,11E,13S,15Z)-13-

hydroperoxyoctadeca-9,11,15-trienoic acid), from α-linolenic acid (C18:3) with appropriate regio- and stereo-specificity required for JA biosynthesis. The initial steps of JA biosynthesis (lipases, LOX, AOS and AOC) are localized in plastids, and therefore, most 13-LOXs possess plastid transit peptides (CTP) for close proximity to linolenate.

Six 13-LOXs are present in the maize genome (Figure 4) and cluster into two groups. ZmLOX7 (Zm00001d025524; GRMZM2G070092), 8 (Zm00001d003533; GRMZM2G104843) and 9 (Zm00001d027893; GRMZM2G017616) cluster together and ZmLOX10 (Zm00001d053675; GRMZM2G015419), and 11 (Zm00001d015852; GRMZM2G009479) and 13 (Zm00001d031449; GRMZM5G822593) form a distinct group [51,52,53,54]. The two groups share relatively low amino acid sequence identity (Figure 5) of around 45%, yet the members of the same group share up to 90% sequence identity. While any of these 13-LOXs in maize may produce substrate for JA biosynthesis, thus far, only ZmLOX8 has been shown to be required for basal JA biosynthesis in tassel [44]. Christensen et al. (2013) have shown that ZmLOX8 is the major 13-LOX isoform in wound-induced JA biosynthesis in leaves. ZmLOX8 localization to the chloroplasts was determined by analysis of the LOX8:mCherry fluorescent protein [44]. In addition to wound-inducible JA, ZmLOX8 is essential for normal JA production during tassel primordia development, discovered in a forward genetics approach characterizing the *tasselseed1* mutant of maize [44]. ZmLOX8 shares 94% sequence identity with ZmLOX7, suggesting that ZmLOX7 may also be involved in JA biosynthesis. Another functionally-characterized maize 13-LOX, ZmLOX10, is solely responsible for green leaf volatile (GLV) biosynthesis and wound-induced JA; however, whether it is directly involved in JA production or signaling JA biosynthesis via GLVs needs to be further elucidated [51]. ZmLOX10-YFP (yellow fluorescent protein) was previously shown to not localize to chloroplasts and instead be found in small microbodies [51]; however, recent analysis with improved microscopy techniques positively confirm ZmLOX10-YFP presence in chloroplasts (M. K., unpublished data). Thus, it cannot be ruled out that ZmLOX10 provides substrate towards JA biosynthesis during stress response in a tissue- or stimulus-specific manner. A cautionary note, careful examination of the subcellular localization of ZmLOX9-YFP, did not find it to be localized to chloroplasts [55].

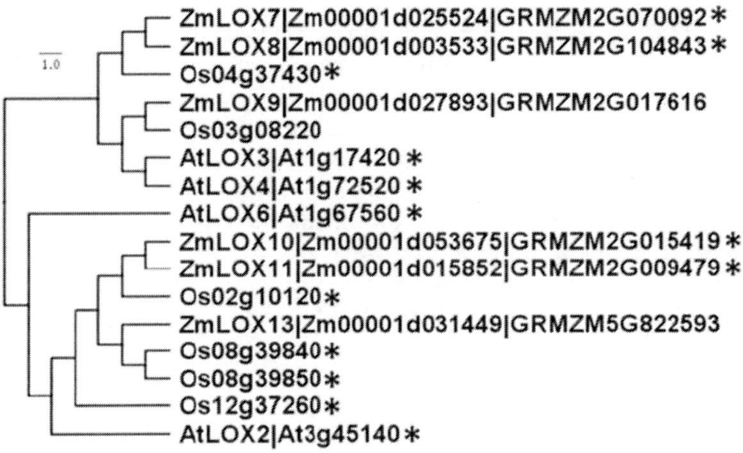

Figure 4. Amino acid sequences of 13-LOX family members from *Arabidopsis*, maize, and rice compared with ClustalX and visualized with Figtree. Gene labels follow established nomenclature from *Arabidopsis* [53], maize [51], and rice [52]. * denotes predicted CTP by ChloroP.

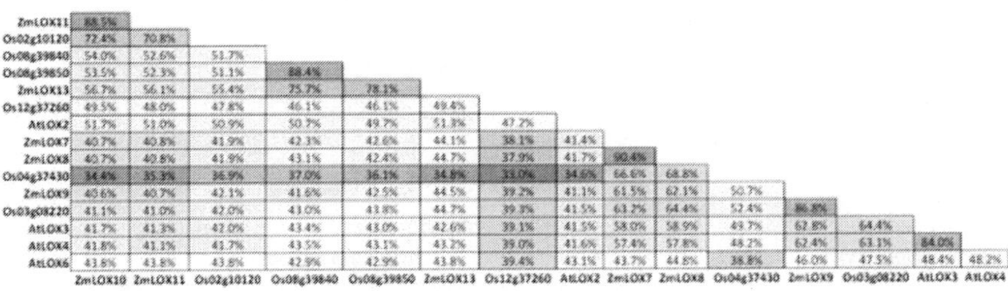

	ZmLOX10	ZmLOX11	Os02g10120	Os08g39840	Os08g39850	ZmLOX13	Os12g37260	AtLOX2	ZmLOX7	ZmLOX8	Os04g37430	ZmLOX9	Os03g08220	AtLOX3	AtLOX4
ZmLOX11	88.5%														
Os02g10120	72.4%	70.8%													
Os08g39840	54.0%	52.6%	51.7%												
Os08g39850	53.5%	52.3%	51.1%	88.4%											
ZmLOX13	56.7%	56.1%	55.4%	75.7%	78.1%										
Os12g37260	49.5%	48.0%	47.8%	46.1%	46.1%	49.4%									
AtLOX2	51.7%	51.0%	50.9%	50.7%	49.7%	51.3%	47.2%								
ZmLOX7	40.7%	40.8%	41.9%	42.3%	42.6%	44.1%	38.1%	41.4%							
ZmLOX8	40.7%	40.8%	41.9%	43.1%	42.4%	44.7%	37.9%	41.7%	90.4%						
Os04g37430	34.4%	35.3%	36.9%	37.0%	36.1%	34.8%	33.0%	34.6%	66.6%	68.8%					
ZmLOX9	40.6%	40.7%	42.1%	41.6%	42.5%	44.5%	39.2%	41.1%	61.5%	62.1%	50.7%				
Os03g08220	41.1%	41.0%	42.0%	41.0%	41.8%	44.7%	39.3%	41.5%	63.2%	64.4%	52.4%	86.8%			
AtLOX3	41.7%	41.3%	42.0%	43.4%	43.0%	42.6%	39.1%	41.5%	58.0%	58.9%	49.7%	62.8%	64.4%		
AtLOX4	41.8%	41.1%	41.7%	43.5%	43.1%	43.2%	39.0%	41.6%	57.4%	57.8%	48.2%	62.4%	63.1%	84.0%	
AtLOX6	43.8%	43.8%	43.8%	42.9%	42.9%	43.8%	39.4%	43.1%	43.7%	44.8%	38.8%	46.0%	47.5%	48.4%	48.2%

Figure 5. Protein sequence identity comparison of the 13-LOX family members from *Arabidopsis*, maize, and rice. Analyses was performed with Sequence Identity and Similarity software (SIAS; http://imed.med.ucm.es/Tools/sias.html). The color code (red to white to blue) scale represents the percent sequence identity from greatest to least.

Another class of LOXs, termed dual positional LOXs, has dioxygenase activity on both the 9- and 13- carbon positions [56,57,58,59]. In maize, two dual positional LOXs exist, ZmLOX1 (Zm00001d042541; GRMZM2G156861; DQ335760) [60] and ZmLOX2 (Zm00001d042540; GRMZM2G156861 [sic]; DQ335761). Both lack a predicted plastid transit peptide sequence as determined by ChloroP [61]. Interestingly, when expressed transgenically in rice, ZmLOX1 was able to associate with chloroplast membranes in the presence of calcium [62]. A similar dual positional LOX from rice was determined to localize into chloroplasts and when silenced, failed to produce increased JA following brown plant hopper (*Nilaparvata lugens*) infestation [59]. Therefore, genetic evidence is required to definitively assign a role to the dual positional maize LOXs, ZmLOX1 and 2. Unfortunately, these two genes share 87% amino acid identity and are tandemly duplicated, suggesting functional redundancy. To completely characterize the role of maize dual positional LOXs, the creation of double mutants is necessary, but problematic. It is likely that next generation genome editing tools will be required for this endeavor [63]. In summary, the most likely candidate isoforms for JA biosynthesis are ZmLOX7, 8, 9, 10, 11 and 13.

6. ALLENE OXIDE SYNTHASE

The 13-LOX product 13-HPOT is catalyzed into an epoxide, 12,13(*S*)-epoxy-octadecatrienoic acid, by 13-AOS. 13-AOS, along with the closely related 9-AOS, hydroperoxide lyase (HPL) and divinyl ether synthase (DES), belong to the cytochromeP450 monooxygenase (CYP) enzyme family and part of the CYP74 group (Figure 3) [64]. The CYP74s are involved in the metabolism of the immediate LOX products, hydroperoxy fatty acids. All HPLs, DESs and AOSs utilize the same substrate, but form distinct products, and the enzymes cluster into different clades. The CYP74A clade contains all 9-, 9/13- and 13-AOSs. 13-AOS serves as the first committed enzymatic step for the production of JA [65]. Interestingly, these enzymes are extremely similar, and even a single amino acid substitution repurposed the AOS from *Arabidopsis* into a functional HPL [66].

Previous research reported three maize AOS isoforms [67], but after analysis of the maize B73 genome, five AOSs were identified (Figure 6). These five isoforms cluster into two clades, termed here AOS1 and 2. AOS1 clade contains ZmAOS1a (Zm00001d034186; GRMZM2G376661), b (Zm00001d034184; GRMZM2G07265) and c (Zm00001d013185; GRMZM2G033098), and AOS2 contains ZmAOS2a (Zm00001d028282; GRMZM2G002178) and b (Zm00001d048021; GRMZM2G067225). Currently, no specific 13-AOS has been characterized in maize. Nearly all JA-producing 13-AOSs are localized in plastids and, similarly to 13-LOXs, encode CTP sequences with few exceptions, a notable example being the AOS from barley (*Hordeum vulgare*) [68]. The three maize AOS1s contain predicted CTPs [61] that are expected to target them to plastids. The ZmAOS1 clade members also cluster with OsAOS1 at ~86% sequence identity. OsAOS1 was shown to contain a CTP and identified through a preliminary mapping of a mutant in phytochrome-mediated inhibition of *coleoptile photomorphogenesis growth* (*cpm1*) [69] that was later determined to be NPH3 (*NONPHOTOTROPIC HYPOCOTYL3*) [70]. Two previously predicted rice AOSs were later determined to encode for HPLs and have been appropriately renamed [71] and are not included in this analysis.

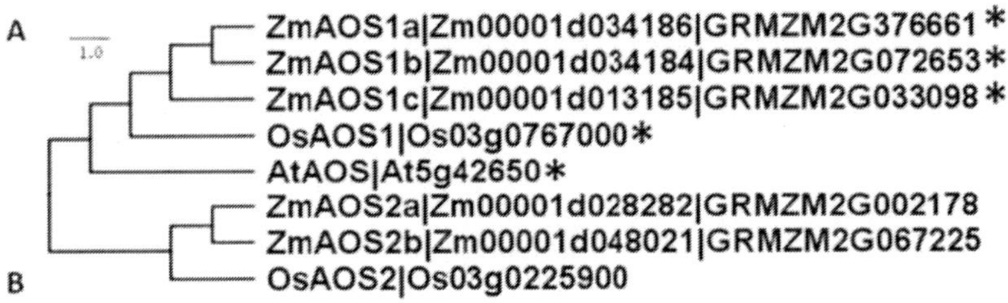

B

	ZmAOS1a	ZmAOS1b	ZmAOS1c	OsAOS1	AtAOS	ZmAOS2a	ZmAOS2b
ZmAOS1b	98.8%						
ZmAOS1c	94.3%	93.9%					
OsAOS1	86.1%	85.7%	87.5%				
AtAOS	55.2%	55.2%	56.4%	56.1%			
ZmAOS2a	54.1%	53.7%	53.7%	54.6%	55.0%		
ZmAOS2b	54.1%	53.7%	53.9%	54.6%	54.4%	91.5%	
OsAOS2	52.7%	52.3%	52.5%	53.3%	52.7%	75.1%	74.1%

Figure 6. Amino acid sequences of CYP74A family members from *Arabidopsis*, maize, and rice compared with ClustalX and visualized with Figtree. (**A**) Gene labels follow nomenclature for maize (this chapter) and rice [72]. * denotes predicted CTP by ChloroP. (**B**) Identity analyses was performed with SIAS. The color code (red to white to blue) scale represents the percent sequence identity from greatest to least.

Similar to the dual positional specific LOXs of maize [62] and rice [59], the barley AOS [68], rice AOS2 (Os03g0225900) [73] and recently maize (ZmAOS2b) [74] were reported to possess dual substrate specificity capabilities of catalyzing either 9- or 13-HPOD/T into the corresponding allene oxides. None of the AOS2 clade members (Figure 6) encode predicted CTPs. Recombinant ZmAOS2b was determined to catalyze 9-HPOD into 9,10-epoxy-10,12-octadecadienoic acid (9,10-EOD). In that study, maize root extract possessed a cyclase activity to convert 9,10-EOD into (9*S*,13*S*)-10-oxo-phytoenoic acid (10-OPEA). In rice, while attempting to produce male-sterile plants with an RNAi approach, silencing OsAOS2 was determined to be more effective than silencing OsAOS1 [75]. This observation suggests that either members of the AOS2 clade produce JA directly due to their dual substrate specificity or function through the production of a yet to be determined oxylipin signal that promotes JA-mediated male fertility. In agreement with this hypothesis, a 9-LOX mutant of maize had reduced JA content [76].

Very few studies have explored expressional analysis of the AOS1 clade [67]. It is important to bring to attention that most published reports analyze the expression of the maize AOS2 clade [45,77,78,79,80,81]. These data should be treated with caution until genetic evidence reveals which of the AOS isoforms are directly involved in JA production in maize.

7. ALLENE OXIDE CYCLASE

Cyclization of 12,13(*S*)-EOT into stereospecific (9*S*,13*S*)-12-oxo-phytodienoic acid (12-OPDA) occurs via allene oxide cyclase (AOC) and represents the first enzymatic step to yield a member of the jasmonate family of oxylipins, 12-OPDA. It is predicated that instead of directly catalyzing the reaction, AOC constrains 12,13(*S*)-EOT into the appropriate configuration for self-cyclization in a stereospecific manner [82,83].

The maize genome encodes two AOC isoforms, which are termed here ZmAOC1 (Zm00001d029594; GRMZM2G077 316) and ZmAOC2 (Zm00001d047340; GRMZM2G415793) and cluster closer with the sole rice AOC (Figure 7) [72] (~78% sequence identity) than with the four AOCs from *Arabidopsis* [84]. The availability of an additional AOC in maize compared to rice may convey functional redundancy, but also provide a tissue- and stress-specific regulation, as was observed for the tissue- and promoter-activity of the four AOC isoforms in *Arabidopsis* [84]. A maize AOC isolated from seed was shown to utilize only 18:3-derived, but not 18:2-derived 12,13-epoxy fatty acids; however, the identity of this AOC has not yet been determined [85]. While in maize no functional analysis of AOC have been reported, in rice, the JA-deficient *hebiba* (Japanese for "snake leaf") mutation [86] and *coleoptile photomorphogenesis2* (*cpm2*) [72] mutants lack the sole AOC. Recently, it was determined that in addition to missing AOC, *hebiba* is also deleted in DWARF14-like [87], which imparts defects root interactions with arbuscular mycorrhize through the karrikin (the so-called, "smoke hormone") receptor [88] and 25 other genes [89]. Because of the complicated nature of the *hebiba* mutation, the function of JA in rice will need to be revisited by using JA-deficient mutants disrupted in single genes. Interestingly, *cpm2* and *hebiba* displayed increased tolerance to salt [90], but it remains to be elucidated whether JA-deficiency in maize impacts salt tolerance. Given that the two maize

AOCs share ~90% sequence identity, it is reasonable to expect functional redundancy will require double knockout mutants to be used to establish their role in JA synthesis. Fortunately, ZmAOC1 and 2 are located on chromosomes 1 and 9, respectively, so the double mutant can be generated through conventional breeding.

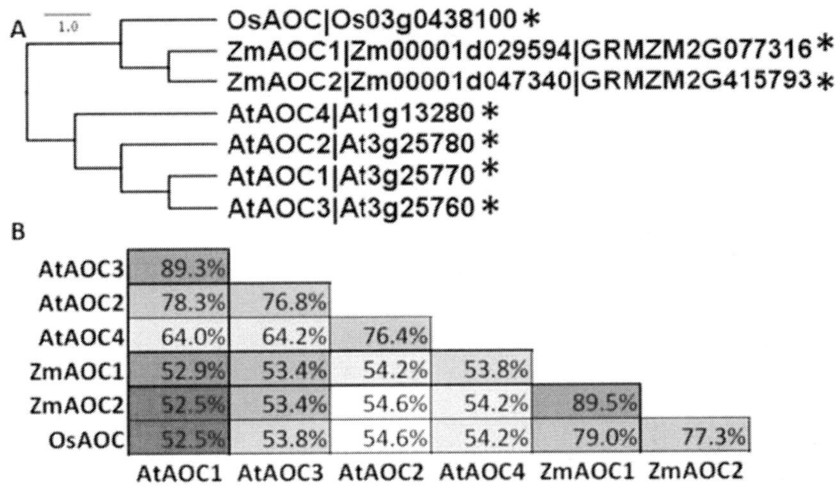

Figure 7. Amino acid sequences of AOC family members from *Arabidopsis*, maize, and rice compared with ClustalX and visualized with Figtree. (**A**) Gene labels follow nomenclature from *Arabidopsis* [84], maize (this chapter), and rice [72]. * denotes predicted CTP by ChloroP. (**B**) Identity analyses was performed with SIAS. The color code (red to white to blue) scale represents the percent sequence identity from greatest to least.

8. OXO-PHYTODIENOIC ACID REDUCTASE AND BETA-OXIDATION

Reduction of 12-OPDA occurs in the peroxisome via Type II 12-OPDA reductases (OPRs). Type II OPRs reduce the cyclopentenone *cis*-(+)-12-OPDA into the cyclopentanone, OPC-8:0 (8-[3-oxo-2-*cis*-[(*Z*)-2-pentenylcyclopentyl]octanoic acid).

The maize genome encodes eight OPR isoforms, yet only two are Type II OPRs [91], ZmOPR7 (Zm00001d032049; GRMZM2G148281) and ZmOPR8 (Zm00001d050107; GRMZM2G082087). ZmOPR7 and 8 share

~87% and ~70% sequence identity, respectively, with rice [92] and *Arabidopsis* (Figure 8) [93,94] JA-producing OPRs. In the B73 inbred line, *ZmOPR7* and/or *ZmOPR8* are induced by JA, ethylene and abscisic acid in leaves [91]. OPR7 and OPR8 share ~95% sequence identity, and recent functional characterization showed that ZmOPR7 and 8 are functionally redundant and both are capable of reducing 12-OPDA [45]. Supporting the case of redundancy for the two genes in JA biosynthesis, only *opr7opr8* double mutants, but not single mutants displayed reduced JA [45]. As expected, 12-OPDA levels of *opr7opr8* double mutants were similar to WT. To date, the *opr7opr8* double mutant provides the best genetic evidence for the role of JA in maize. The phenotypes of the double mutant are described in detail below.

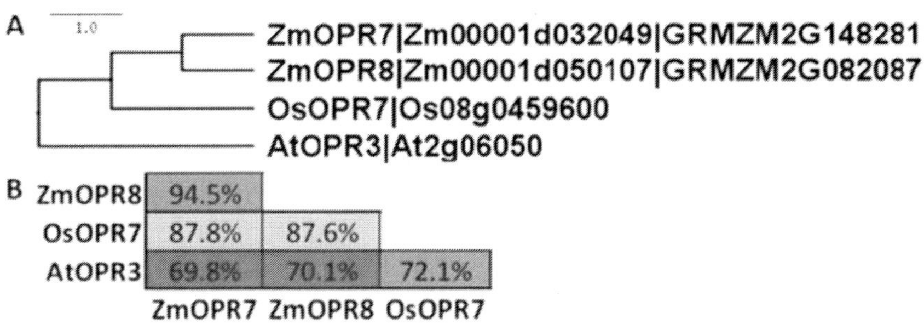

Figure 8. Amino acid sequences of Type II OPR family members from *Arabidopsis*, maize, and rice compared with ClustalX and visualized with Figtree. (**A**) Gene labels follow nomenclature from maize [91] and rice [92]. (**B**) Identity analyses was performed with SIAS. The color code (red to white to blue) scale represents the percent sequence identity from greatest to least.

Three rounds of beta-oxidation are required to shorten the octanoic acid side change of OPC-8:0 to ethanoic acid. The three rounds of beta-oxidation are performed by three distinct enzymes, ACX (acylcoenzyme A (CoA) oxidases), MFP (multifunctional proteins), and KAT (3-ketoacyl-CoA thiolases) [32]. The product of the three reactions is JA.

9. JA-AMINO ACID CONJUGATION

Many chemically-diverse jasmonic acid derivatives were reported from different plant species [33]. In summary, JA can be converted to one of its derivatives by conjugation with a number of amino acids, decarboxylation, glycosylation, hydroxylation of a number of carbon positions, autocyclization of hydroxy-JAs into lactones, methylation, sulfonation or by more than one modification. The conjugation of JA with the ethylene precursor, ACC (1-aminocyclopropane-1-carboxylic acid), has also been reported [95]. JAs were also found with an unsaturated cyclopentenone (didehydro-JAs), with a saturated pentanyl moiety (dihydro-JAs) or with its ethanoic acid chain elongated into three (homo-JA) or four carbons (dihomo-JA) [96]. In addition, JA is capable of undergoing stereo-rearrangement of its ethanonic acid and pentenyl chains and can be found in the 3*R*, 7*R* or 3*R*, 7*S* (7-*iso*-JA) configurations. These different jasmonates possess different effects on JA-inducible gene expression, and while most have shown no discernible activity, several were able to induce or suppress JA-responsive genes [97]. Recently, (+)-7-*iso*-jasmonoyl-L-isoleucine (JA-Ile) was determined to be the most effective ligand for the SCF(COI1) receptor [98], currently the only JA receptor identified in plants.

The conjugation of JA with Ile occurs through the JAR1 (JASMONATE RESISTANT1) enzyme [99]. The JAR1 enzyme belongs to the large firefly luciferase gene family. In *Arabidopsis*, only one protein, AtJAR1 was shown to utilize JA as a substrate [99]. In rice, two proteins were shown to produce JA-Ile in vitro, OsJAR1 and OsJAR2 [100]. The maize genome encodes five JAR1-like isoforms that group into two clusters (Figure 9). The JAR1 cluster contains ZmJAR1a (Zm00001d011377; GRMZM2G09127) and ZmJAR1b (Zm00001d039346; GRMZM2G162413), which share ~60% sequence identity with AtJAR1. The JAR2 cluster contains ZmJAR2a (Zm00001d008957; GRMZM2G001421), ZmJAR2b (Zm00001d039346|GRMZM2G060991) and ZmJAR2c (Zm00001d039345; GRMZM2G061005). These isoforms only share ~56% sequence identity with AtJAR1, but 68 to 78% sequence identity with OsJAR2.

To date, no maize JAR1-like isoform has been characterized. In rice, JAR1 was determined to be involved in photomorphogenesis [101], floret development [102] and phytoalexin production [103]. The presence of multiple JAR1-like isoforms in maize will require utilization of double mutants for functional characterization.

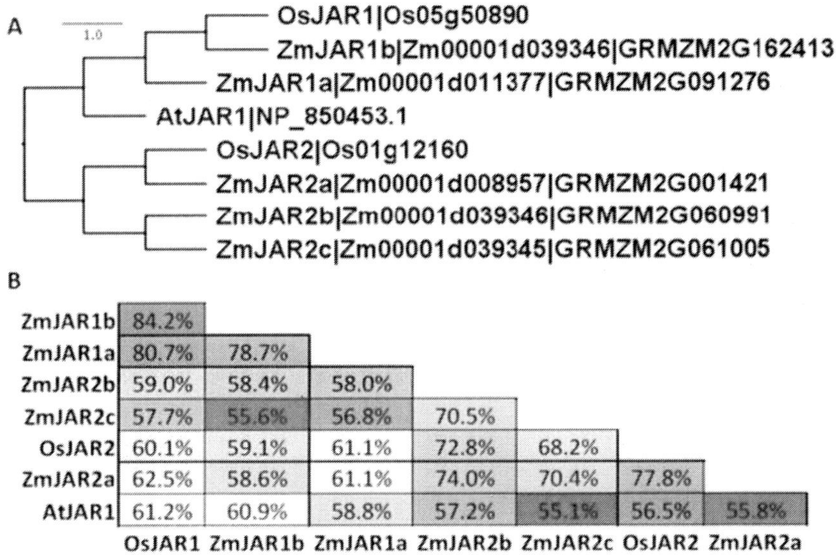

Figure 9. Amino acid sequences of JAR1-like family members from *Arabidopsis*, maize, and rice compared with ClustalX and visualized with Figtree. (**A**) Gene labels follow nomenclature from maize (this chapter) and rice [100]. (**B**) Identity analyses was performed with SIAS. The color code (red to white to blue) scale represents the percent sequence identity from greatest to least.

10. 9-OXYLIPIN JASMONATE ANALOGS (10-OPEA, 10-OPDA AND DERIVATIVES)

In addition to the 13-LOX-derived jasmonates, plants possess a parallel pathway involving 9-LOX reactions that also produce cyclopenta(e)nones, analogous to JAs [67]. These 9-LOX products, 10-oxo-11-phytoenoic acid (10-OPEA) and 10-oxo-11,15-phytodienoic acid (10-OPDA), are structurally related to 12-OPDA and first described in potato [104]. The 9-LOX isoforms responsible for their biosynthesis have not been identified yet in maize or any other plant species. In maize, following 9-LOX activity on C18:2, the 9-hydroperoxy fatty acid (9-HPOD) is catalyzed into 9,10-epoxy octadecadienoic acid (9,10-EOD) through the action of ZmAOS2b [74]. The 9,10-EOD is cyclized into 10-OPEA by an, as yet uncharacterized, AOC-like enzyme [74].

Recently, these cyclopentanones were described to possess potent signaling, insecticidal and antimicrobial activities in maize [67,105]. Transcriptional analysis uncovered that JA, 12-OPDA or 10-OPEA regulate unique subsets of genes, suggesting distinct signaling activities [67]. In wounded silk and root tissue, 10-OPEA accumulates as the predominant cyclopentanone species followed by 10-OPDA [105]. Low concentrations of 10-OPEA added to artificial diet media reduce the growth of *Helicoverpa zea* (corn earworm) by half compared to insects grown on control media. 10-OPEA and its derivatives accumulated in local tissue following inoculation by *Cochliobolus heterostrophus*, the causal agent of southern corn leaf blight, to greater levels than either 12-OPDA or JA. While not having direct antimicrobial effects against *C. heterostrophus*, concentrations as low as 170 μM reduced growth of *Fusarium verticillioides* and *Aspergillus flavus*. Direct application of 10 μL of 2 mM of 10-OPEA caused dramatic phytotoxicity due to programmed cell death (PCD). Most surprisingly, 10-OPEA promoted the greatest necrosis and ion leakage compared to all other PCD-promoting compounds tested, including fumonisin B1 [67].

11. REGULATION OF JA PRODUCTION BY THE NON-JA PRODUCING BRANCHES OF THE LOX PATHWAY

11.1. The Role of GLV-Mediated Signaling on JA Production

In maize, ZmLOX10 was determined to serve as the sole 13-LOX isoform responsible for basal, herbivore- and wound-induced GLV production in aboveground tissue [51]. Following ZmLOX10 activity, the C18:2- or C18:3-derived 13-hydroperoxy fatty acids are cleaved into C6 aldehydes and C12-oxo-fatty acids, via the action of hydroperoxide lyase (HPL) [106]. The C6 aldehydes can be subsequently converted into C6 alcohols, acetylated or hydroxylated [107]. Transposon-disrupted *lox10* knockout mutant plants are completely devoid of GLVs and were utilized as a genetic tool to understand the role of GLVs in regulation of JA production. Compared to wild-type (WT) plants, *lox10* mutants accumulated significantly lower levels of 12-OPDA and JA in locally-damaged tissue after wounding treatment compared to WT. The *lox10* mutants also showed reduced transcript levels of the JA biosynthesis genes, *ZmLOX8* and *ZmOPR7/8*, suggesting that GLVs are required for normal wound-induced JA via transcriptional regulation.

11.2. The Role of 9-LOX-Mediated Signaling on JA Production

Recent evidence suggests that 9-oxylipins regulate maize JA production, in both a positive or negative manner. ZmLOX3 is a 9-LOX and has been shown to serve as a negative regulator of JA biosynthesis in roots and seed. The *lox3* knockout mutant roots produced elevated basal levels of JA and displayed an increased expression of JA biosynthesis genes [77], suggesting that LOX3-produced 9-oxylipins suppress the production of JA. Interestingly, transcript levels for GLV-biosynthesis genes, Zm*LOX10* and *ZmHPL*, were also elevated in untreated *lox3* roots. In seeds infected by *A. flavus*, *lox3* mutant accumulated increased levels of linolenic acid and JA compared to WT. These studies suggest that ZmLOX3 suppresses JA biosynthesis in diverse organs. In contrast to ZmLOX3, another predominant 9-LOX, ZmLOX12 appears to function as a positive regulator for JA production [76]. In response to *F. verticillioides* infection, *lox12* knockout mutants were unable to induce accumulation of 12-OPDA, JA and JA-Ile. The decrease in JAs was accompanied with decreased transcript accumulation of JA biosynthetic genes and increased susceptibility of *lox12* mutants to *F. verticillioides*. Taken together, these reports suggest that 9-LOX-derived oxylipins help fine-tune JA biosynthesis in organ- and stress-inducible fashion.

12. JA IN MAIZE DEFENSE AGAINST INSECTS

12.1. Plant Defenses, Genetic Evidence and Insect Elicitors

Plant defenses against insects are classified according to nature (physical or chemical), mode of action (direct or indirect) and timing (constitutive or inducible) [108]. Physical defenses are anatomical structures that provide increased durability to herbivory. Chemical defenses utilize toxins or other molecules to deter feeding or nutrition acquisition. Direct defenses explicitly target the insect herbivore physiology and survival, while indirect defenses rely on the attraction of predators of the herbivores. Constitutive defenses are preformed and persist through herbivory, while inducible defenses are activated only during the perception of an herbivore or herbivory-associated tissue damage. While the role of JA in dicot insect defense has been well studied [109], especially in tobacco and tomato, there is limited information on the role of JA in maize insect defenses.

As in other studied plants, JA in maize is critical to successfully thwarting insect attack. The strongest evidence for the role of JA in maize insect defense stems from studies using the JA-deficient *opr7opr8* double mutants' response to herbivory by beet armyworm (*Spodoptera exigua*) [45]. Double mutants were consumed by beet armyworm more readily than WT, which was accompanied with an increased larval weight. The significance of wound-induced JA in insect defenses was also confirmed by the analysis of the *lox10* mutant that was unable to induce accumulation of JA in response to wounding [51]. Similar to *opr7opr8*, mutant seedlings of *lox10* lost more mass due to insect feeding and supported greater larval weight gain compared to WT.

As in other plants, JA serves as a major regulator for diverse insect resistance mechanisms [110], but its precise mode of action in maize is still under investigation. Here, we present recently published studies highlighting the impact of JA on diverse lines of maize defenses against herbivory. In maize, JA biosynthesis in response to insect feeding has been studied extensively using insect elicitors. Of the insect-derived elicitors described so far, volicitin, known chemically as *N*-(17-hydroxylinolenoyl)-L-glutamine (17-OH-C18:3-Gln) [111] and *N*-linolenoyl-glutamine (C18:3-Gln) [112], possesses the greatest activity on maize JA activation [113]. An application of either elicitor caused JA accumulation to spike after treatment. Volicitin, crude regurgitant and mechanical wounding increased local JA concentration when applied to leaf tissue, but only the treatments by crude regurgitant or volicitin induced the accumulation of JA in the distal portion of the leaf [78]. In terms of basal tissue, treatment with volicitin was unable to induce JA accumulation after application [78]. However, treatment with C18:3-Gln induced expression of *ZmMYC7E*, a putative orthologue of the *Arabidopsis* transcription factor AtMYC2, in maize (Zm00001d030028; GRMZM2G001930), after application in both basal tissue and in untreated systemic leaves [80].

12.2. Direct Defenses

One type of direct physical defense is trichomes. Trichomes obstruct insects and are capable of storing toxic defensive metabolites that are released upon damage. JA positively regulates trichome formation. Exogenous application of JA on maize leaves induced the development of trichomes [114]. Another physical defense is lignification, which strengthens cell walls to increase resistance against insect mastication. Recently, treatment with MeJA resulted in degradation of repressors of the lignin biosynthetic pathway [115]. Direct chemical defense responses include the production of metabolites that intoxicate or repel herbivores or suppress nutrient acquisition (e.g., protease inhibitors). Benzoxazinoids are indole derived hydroxamic acids (e.g., DIMBOA, 2,4-dihydroxy-7-methoxy-1,4-benzoxazin-3-one) that can have direct insecticidal activity [116]. ZmIGL (INDOLE-3-GLYCEROL PHOSPHATE LYASE) is capable of generating free indole for benzoxazinoid biosynthesis and is inducible by MeJA application [117]. Similarly, application of JA induced accumulation of 2-(2-hydroxy-4,7-dimethoxy-1,4-benzoxazin-3-one)-β-D-glucopyranose (HDMBOA-Glc) in levels similar to insect herbivory [118].

Several recent studies have identified potential mechanisms of JA-mediated herbivory defense in maize. For example, a new class of diterpenoids derived from *ent*-kaurane [119] (precursor for gibberellic acid) was increased upon insect feeding when leaves were pre-treated with a combination of JA and ethylene [120]. Kauralexin A3 or B3 treatment on maize stems reduced feeding by European corn borer (*Ostrinia nubilalis*) [120]; however, kauralexins may not have direct insecticidal activity, as artificial diet enriched with high concentrations failed to affect *O. nubilalis* growth [119].

Cysteine proteases are defensive enzymes produced by plants, which when consumed by insects damage their gut [121]. In maize, the best-characterized defensive cysteine protease is ZmMIR1 (MAIZE INSECT RESISTANCE1, Zm00001d036542; GRMZM2G150276) [122,123]. In the place of JA, another phytohormone, ethylene (ET) appears to be the major regulator of ZmMIR1. Treatment with 2-chloroethylenphosphonic acid (CEP), an ET-releasing chemical, was able to induce *ZmMIR1* transcript accumulation more dramatically than MeJA treatment [124]. Application of an ET inhibitor reduced aphid-induced *ZmMIR1* expression and improved aphid performance [125]. Plant protease inhibitors inactivate insect digestive enzymes [126] and slows their growth and development [127]. In several dicot species, including tomato, tobacco and *Arabidopsis*, JA is the major hormone mediating the induction of protease inhibitor production [128,129]. Similarity, *ZmMPI* (MAIZE PROTEASE INHIBITOR, Zm00001d011080, GRMZM2G028393) is induced more strongly by MeJA treatment than by CEP [124] and is upregulated in *lox3* knockout mutant roots, overproducers of JA [77]. *ZmMPI1* is clearly JA-dependent, as induction of this gene by wounding is significantly reduced in the JA-deficient *opr7opr8* double mutants. In addition to protease inhibitors that inactivate native insect digestive proteins, plants are capable of producing non-protein amino acids that are incorrectly incorporated into protein synthesis. MeJA transiently induced the production of 5-hydroxynorvaline, and aphids fed diets supplemented with 5-

hydroxynorvaline at physiological concentrations suppressed aphid reproduction [130].

In addition to signaling for resistance, JAs can display direct toxicity on insects in a manner similar to what was described for their 9-LOX analogues [67]. The α,β-unsaturated carbonyl structure makes a subset of jasmonate (e.g., 12-OPDA) reactive electrophilic species (RES) and subsequently toxic metabolites. To circumvent this toxicity, ear- and army-worms possess isomerases that can rearrange the double bond in the cyclopentenone ring to disrupt the α,β-unsaturated carbonyl structure, which detoxifies 12-OPDA [131]. It was shown that silencing of an isomerase gene involved in the 12-OPDA detoxification reduces the insect performance [132].

12.3. Indirect Defenses

Indirect defense is another strategy of plants to resist insect herbivory, and it relies on the emission of volatile organic compounds (VOCs) to recruit beneficial predators of insect herbivores [133]. Following insect damage, JA accumulation was positively associated with increased indole and sesquiterpene emissions [134] and was correlated with the site of terpene release [135]. While mechanical damage and volicitin induced similar levels of JA in intact or excised leaves, treatment with either JA or volicitin results in a greater release of sesquiterpenes from excised leaves compared to intact leaves [136]. Additionally, ethylene enhanced volatile emission from volicitin and JA treatment, but not mechanical damage. Treatment with JA increased the attractiveness of maize plants to the parasitoid wasp, *Cotesia kariyai* [137]; however, JA did not interfere with the ability of *C. kariyai* to discriminate between armyworm (*Mythimna separata*) infested and uninfested maize plants. Interestingly, application of both JA and a derivative of a JA precursor, methyl-C18:3, produced a comparable increase of *C. kariyai* attraction to maize plants [138]. However, the volatile profiles emitted from JA or methyl-C18:3 were substantially different. Methanol and the GLV, Z-3-hexenyl acetate, were found to be the most attractive volatiles, respectively, from the methyl-C18:3 or JA treatment. It remains to be seen if these observations can be repeated under natural conditions [139].

Another indirect defensive role of VOCs is their ability to induce intra- and inter-plant defenses in a phenomenon known as priming. The priming of plants is the ability to activate defenses earlier and to a greater level after exposure to a stimulus [140]. One such stimulus is the volatile blend released from an injured neighbor or by exogenous volatile treatment. The effect of the priming response is determined by subjecting the exposed plants to stress such as wounding or herbivory after 24 h of priming treatment and monitoring JA accumulation or VOC emissions. The priming state persists for at least five days following initial treatment [141] and is likely reliant on chromatin modification to prepare the transcriptional machinery to activate defense genes similar to what is observed with systemic acquired resistance (SAR) against pathogens [142]. JA accumulation is responsive to the volatile-mediated priming. Maize plants pretreated overnight with GLVs facilitated enhanced JA induction following regurgitant application compared with non-pretreated plants [143]. Similarly, a subset of JA-responsive defensive genes also displays a priming pattern in response to airborne volatiles [144].

While GLVs [143], indole [145] and other VOCs [144] are better understood in their role in priming, much less is known concerning the role of jasmonates during priming activation. JA itself can be converted into several volatiles that serve roles in the priming response. One example of a JA-derived volatile capable of inducing a priming response is the decarboxylated JA derivative, *cis*-jasmone. When cis-jasmone is pre-applied to maize plants and incubated for 24 h, treated plants produced greater VOC emissions rapidly after leafhopper (*Cicadulina storeyi*) infestation [146]. Treatment with another volatile JA derivative, MeJA, induces *ZmIGL* [117], thus potentially indirectly signaling for indole-mediated priming [145]. Curiously, multiple tests in several laboratories, including the groups of Engelberth and Kolomiets, have failed to detect any measurable amounts of MeJA from maize, suggesting that the inbred lines that were tested do not produce MeJA. It remains to be seen if other maize lines or teosintes produce this volatile jasmonate.

13. JA IN MAIZE PATHOGEN AND NEMATODE DEFENSE

JA-mediated defenses are the principle countermeasures against pathogens with a necrotrophic lifestyle [147]. In maize, JA-deficient *opr7opr8* mutants are not capable of surviving under field conditions due to "damping-off" disease [45]. In that study, the major pathogen responsible for "damping-off" in the diseased roots of *opr7opr8* was identified to be *Pythium aristosporum*. Supporting the genetic evidence that maize requires JA for immunity against *P. aristosporum*, exogenous application of JA as a soil drench could rescue normal survival of the plants. Any study with *opr7opr8* double mutants requires the use of sterile soil to exclude the plants from soil-borne pathogens. JA is also required for immunity against another common opportunistic soil-borne fungus, *F. verticillioides*, as the *opr7opr8* mutants succumb to mesocotyl inoculation as rapidly as 8 h post-infection by this pathogen [76]. In another study, JA accumulation and JA biosynthetic gene induction correlated with increased resistance against *F. verticillioides* stalk rot in a CO_2-dependent manner [148]. A recent genome-wide association study of maize infected by *A. flavus* identified the JA pathway as the most impacted between resistant and susceptible genotypes [149]. To date, however, no genetic evidence was reported that clearly determines the role of JA in either defense or potentially even benefiting the pathogenicity of this important mycotoxin-producing fungus. In this regard, it is worthy to note that the *lox3* knockout mutant was more susceptible to *A. flavus* and possessed increased JA accumulation in response to infection by this pathogen [150]. To unambiguously establish the role of JA in maize interactions with *A. flavus*, the utilization of JA-deficient *opr7opr8* [45] will be required. The male infertility defect of this double mutant will make this analysis technically challenging, as this pathogen is a seed resident.

A potential mechanism for JA-mediated signaling in response to fungal pathogens is through the peptide signal, ZmPEP1 [151]. In a pattern reminiscent of the interplay between the peptide hormone, systemin and JA in tomato [152], exogenous ZmPEP1 and JA application induce each other's accumulation. ZmPEP1 treatment also induces HDMBOA-Glc accumulation and benzoxazinoid biosynthetic genes. Pretreatment with ZmPEP1 increased resistance to fungal pathogens, *C. heterostrophus* and *Colletotrichum graminicola*.

In contrast to the role of JA in defense, JA may also facilitate increased pathogenesis. Elevated endogenous JA correlates with increased susceptibility to root-knot nematodes (*Meloidogyne incognita*) infestation [77]. The *lox3* mutant maize line, disrupted in the predominantly root-expressed 9-LOX, possessed more JA content in roots compared with WT maize seedlings [77]. Furthermore, *lox3* mutant seedlings supported more root-knot nematode eggs compared to WT seedlings. Additionally, *lox3* mutants were more attractive to *M. incognita* juveniles than WT as determined by a six-arm root olfactometer choice assay The elevated JA content corresponded to increased expression of the JA biosynthetic genes, as well as the 10-OPEA producing *ZmAOS2b*, suggesting the suppression of the JA pathway at the transcriptional level via ZmLOX3-mediated signaling. It will be of interest to determine the role of 9-LOX-derived cyclopenta(e)nones during maize-nematode interactions. A definitive role of JA in susceptibility to nematodes will require the creation of triple mutant *lox3opr7opr8*, but because of male sterility and increased susceptibility of JA-deficient *opr7opr8* to *Pythium*, it will be difficult to maintain.

14. JA IN MAIZE SYMBIOSIS

Colonization of roots by beneficial microorganisms can instill beneficial traits upon the plant host [153]. One of the most intriguing traits is induced systemic resistance (ISR) that primes the plant for increased resistance against pathogens. This phenomenon is believed to rely on JA and ET-mediated signaling in studied dicot species, but relatively little is known about the role of JA in maize symbiosis compared to other plants. A JA role in the positive regulation of ISR in maize came from a recent report showing that the JA-overproducing *lox3* mutant display constitutively activated ISR effective against multiple foliar and stem pathogens due to over production of as-yet unidentified long distance signals found in xylem-enriched exudates [154]. Because *lox3* mutant roots constitutively overexpress all of the JA biosynthesis genes and produce elevated levels of JA in roots [77], it is likely that JA and/or other jasmonates are responsible for the constitutively-active ISR displayed by *lox3* mutant. Additional evidence for JA's role in ISR signaling was obtained from the study of maize ISR responses conferred by root colonization with bacterial symbiote *Pseudomonas putida*. Maize roots inoculated with this beneficial bacterium were characterized by increased expression of jasmonic acid biosynthesis genes and increased resistance against foliar pathogen, *C. graminicola* [155]. Interestingly, priming by *P. putida* produced greater emissions of JA-induced aromatic volatiles and terpenes [156].

15. JA IN MAIZE GROWTH, DEVELOPMENT AND SENESCENCE

The strongest evidence for the role of JA in maize growth, development and senescence came from the functional analyses of JA-deficient *opr7opr8* double mutants. *opr7opr8* mutants are defective of total JA accumulation in young tissue [45]. Fitting with the role of JA in negatively regulating plant vegetative growth [157], seedlings of *opr7opr8* double mutants possess long coleoptiles, first leaf sheathes and first and second leaf blades that displayed increased growth compared with WT seedling tissues [158]. In addition to the increase growth in leaf-associated tissue seen in *opr7opr8* mutants, an increase in root lateral root length and density was also detected [158].

The most striking developmental effect of JA in maize is on the formation of the male reproductive organ called the tassel [44,45]. Lack of JA in developing tassels of *lox8* mutants (*tasselseed1*, *ts-1*) [44] or *opr7opr8* mutants failed to abort pistils in the developing florets [159], resulting in the so-called *tasselseed* mutant phenotype. While these mutants are capable of setting viable seeds when cross pollinated, they are completely male sterile. Exogenous JA application was able to rescue the defective masculinization of the male florets in both *lox8* single [44], *opr7opr8* double mutants [45] and another related gene, *tasselseed2* (*ts2*) [44]. TS2 was determined to encode a short-chain alcohol dehydrogenase required for normal tassel masculinization, as well [160]. Subsequent reports list TS2 as a 3β/17β-hydroxysteroid dehydrogenase [161], with implications in linking JA-mediated signaling with brassinosteroid-dependent tassel masculinization [162]. Unlike the mutation of the LOX8 gene, which impacts JA production in the tassel, but not in other organs, likely due to redundancy in JA producing LOXs (i.e., LOX7 and 9), *opr7opr8* mutants are completely devoid of JA in many other organs and possess additional reproductive developmental defects relating to the female sexual organ [45], known as the ear. While WT maize plants only develop one to two ears upon sexual maturity, *opr7opr8* double mutants initiate ear axillary buds at nearly every internode along the maize stem. Furthermore, developing ears of *opr7opr8* possess extremely elongated ear shoots three- to four-times the length of WT. Similar to the chemical complementation rescue of normal tassel, treatment with JA successfully restored the aberrant female sexual organ phenotype to normal.

Normal JA levels accelerate maize leaf senescence. The first and second leaf of double mutant *opr7opr8* seedlings had an increased lifespan compared to WT leaves. The senescence delay correlated with decreased ethylene emissions, *cis*-zeatin 9-riboside production and abscisic acid accumulation in *opr7opr8* double mutants compared with WT [45].

16. JA IN MAIZE ANTHOCYANIN PIGMENTATION AND PHOTOMORPHOGENESIS

JA signaling is necessary for anthocyanin pigmentation of maize brace roots and leaf auricle [45]. While brace roots and the leaf auricle of WT, *opr7* and *opr8* single mutants displayed a normal deep red pigmentation due to prominent anthocyanin accumulation, brace roots of *opr7opr8* did not display any anthocyanin pigmentation and remained a yellow-green color [45]. The depigmented phenotype of *opr7opr8* was restored by exogenous JA application. The lack of anthocyanin correlated with a decrease in expression of two anthocyanin biosynthetic genes, *FLAVANONE 3-HYDROXYLASE* (*F3H*) and *DIHYDROFLAVONOL REDUCTASE* (*A1*). Both genes had nearly an eight-fold decrease in expression in *opr7opr8* double mutants compared with WT levels [45]. JA is required for maize seedlings to experience normal light-induced development. When germinated and grown in darkness for seven days, *opr7opr8* double mutants grew longer mesocotyls and coleoptiles compared with WT, indicative of exaggerated skotomorphogenesis [158]. However, when grown under continuous red light (an external signal for photomorphogenesis) and compared to WT, *opr7opr8* double mutant continued to display longer mesocotyl and coleoptiles.

17. MAIZE GENES REGULATED BY JA

The best-characterized JA-dependent wound-responsive maize genes were determined from a macroarray analysis comparing gene expression of *opr7opr8* double mutant seedling leaves to WT following mechanical damage [45]. Table 1 shows the genes that show clear dependence on intact JA biosynthesis. We recommended the use of these genes as molecular markers for JA responsive transcription. Primer sequences are available for probe construction or for semi-quantitative reverse transcription polymerase chain reaction (semi-q RT-PCR) [45]. Primer sequences for qPCR analysis are available for the lipoxygenase genes [67].

Table 1. JA-dependent wound-responsive genes, adapted from [45].

Gene	Gene annotation	Gramene ID †	GRMZM ID †	Genbank † or TIGR ID †
		α,β-Unsaturated carbonyl detoxification		
ZmOPR3	12-oxo-phytodienoic acid reductase 3	Zm00001d037182	GRMZM2G156712	AY921640
ZmOPR4	12-oxo-phytodienoic acid reductase 4	Zm00001d013493	N/A in RefGen_v3	AY921641
ZmOPR5	12-oxo-phytodienoic acid reductase 5	Zm00001d003584	GRMZM2G087192	AY921642
		Direct defense		
ZmMIR1	maize insect resistance 1 (mir1) cysteine proteinase	Zm00001d036542	GRMZM2G150276	NM_001112101
ZmMPI	proteinase inhibitor	Zm00001d011080	GRMZM2G028393	X78988
		Ethylene associated		
ZmACS2	1-aminocyclopropane-1-carboxylate synthase 2	Zm00001d002592	GRMZM2G164405	AY359569
ZmERF2	homologue to AtERF1 and 2	Zm00001d002762	GRMZM2G055180	NM_001158578
ZmERF6	homologue to AtERF1 and 2	Zm00001d034920	GRMZM2G381441	NM_001176924
		Flavonoid metabolism		
ZmF3H	flavanone 3-hydroxylase	Zm00001d001960	GRMZM2G062396	NM_001112225
		Lipase		
ZmPLC	phospholipase C	Zm00001d014903	GRMZM5G889467	NM_001111784
		Oxylipin biosynthesis		

ZmLOX3	lipoxygenase 3	Zm00001d033623	GRMZM2G109130	AF329371
ZmLOX4	lipoxygenase 4	Zm00001d033624	GRMZM2G109056	DQ335762
ZmLOX5	lipoxygenase 5	Zm00001d013493	GRMZM2G102760	DQ335763
ZmLOX6	lipoxygenase 6	Zm00001d002000	GRMZM2G040095	DQ335764
ZmLOX9	lipoxygenase 9	Zm00001d027893	GRMZM2G017616	DQ335767
ZmLOX10	lipoxygenase 10	Zm00001d053675	GRMZM2G015419	DQ335768
ZmLOX11	lipoxygenase 11	Zm00001d015852	GRMZM2G009479	DQ335769
ZmAOS2a	allene oxide synthase 2a	Zm00001d028282	GRMZM2G002178	
ZmAOS2b	allene oxide synthase 2b	Zm00001d048021	GRMZM2G067225	AY488135
		Transcription factors		
ZmMYC2a	homologue to AtMYC2	Zm00001d007536	GRMZM2G303463	
ZmWRKY14	WRKY transcription factor 14	Zm00001d043569	GRMZM2G040298	EU973705
ZmWRKY46	WRKY transcription factor 46	Zm00001d052357	GRMZM2G063216	EU956406
		Repressors of JA signaling		
ZmJAZ1	ZIM domain containing protein	Zm00001d026477	GRMZM2G143402	NM_001156069
ZmJAZ3	ZIM domain containing protein	Zm00001d048268	GRMZM2G036288	NM_001157436
ZmJAZ4	ZIM domain containing protein	Zm00001d027899	GRMZM2G343157	NM_001157673
ZmJAZ5	ZIM domain containing protein	Zm00001d027901	GRMZM2G445634	NM_001156053
ZmJAZ6	ZIM domain containing protein	Zm00001d048263	GRMZM2G036351	NM_001157328
ZmJAZ7	ZIM domain containing protein	Zm00001d014253	GRMZM2G173596	NM_001159100
ZmJAZ8	ZIM domain containing protein	Zm00001d033050	GRMZM2G145412	EU970040
ZmJAZ10	ZIM domain containing protein	Zm00001d020614	GRMZM2G116614	AZM5_3798
ZmJAZ11	ZIM domain containing protein	Zm00001d006860	GRMZM2G101769	AZM5_1941
ZmJAZ12	ZIM domain containing protein	Zm00001d027900	GRMZM5G838098	NM_001154831
ZmJAZ19	ZIM domain containing protein	Zm00001d036494	GRMZM2G080509	AZM5_87469

† Database information: Gramene (http://www.gramene.org/), GRMZM (Maize Genetics and Genomics Database, https://www.maizegdb.org/), Genbank (https://www.ncbi.nlm.nih.gov/genbank/), TIGR (The TIGR Maize Database, http://maize.jcvi.org/).

18. CONCLUSIONS

Given that monocots contribute, either directly or indirectly, to a substantial proportion of calories for humans and the growing interests in utilizing monocots as a renewable fuel alternative, it is crucial to improve monocot yield through environmentally-conscientious practices. Fine-tuning of phytohormone signaling can provide an avenue for this improvement, but many more studies are required to provide a comprehensive understanding of hormone biology in monocots. Generating a level of understanding comparable to dicot models for the biosynthesis, signaling and function of jasmonates in maize (and other monocots) will require a coalition of multiple research teams with diverse expertise. While not covered in this review, recent studies with *Arabidopsis* suggests that various jasmonates and their derivatives have their own independent roles in multiple physiological processes [15,157,158]. Additionally, many of these JA derivatizations are under enzymatic control, and orthologous for the genes involved are found in the maize genome. Together, this suggests that plant responses to JA are under a tight regulation via conversion of one jasmonate species to another. Another point that requires more investigation is the cross-talk between 9-oxylipins and JA production. This complexity that is emerging in JA biology resembles the intricacies found in mammalian systems. It is reasonable to expect that in the near future, other plant oxylipins will take their place alongside JA as recognized phytohormones.

ACKNOWLEDGMENTS

This work was supported by grants from the United States National Science Foundation (IOS-0925561 and IOS-0951272) and United States Department of Agriculture National Institute of Food and Agriculture (AFRI-1008528, AFRI-1005224 and AFRI-1008478).

AUTHOR CONTRIBUTIONS

Eli J. Borrego and Michael V. Kolomiets wrote the paper.

CONFLICTS OF INTEREST

The authors declare no conflict of interest. The founding sponsors had no role in the design of the study; in the collection, analyses or interpretation of data; in the writing of the manuscript; nor in the decision to publish the results.

REFERENCES

1. Awika, J.M. Major cereal grains production and use around the world. Adv. Cereal Sci. Implic. Food Proc. Health Promot.2011, 1089, 1–13.
2. Food and Agriculture Organization of the United Nations. Food Outlook: Biannual Report of Global Food Markets; Food and Agriculture Organization of the United Nations: Rome, Italy, 2013.
3. Schnable, P.S.; Ware, D.; Fulton, R.S.; Stein, J.C.; Wei, F.; Pasternak, S.; Liang, C.; Zhang, J.; Fulton, L.; Graves, T.A.; et al. The B73 maize genome: Complexity, diversity, and dynamics. Science 2009, 326, 1112–1115.
4. Vielle-Calzada, J.P.; de la Vega, O.M.; Hernandez-Guzman, G.; Ibarra-Laclette, E.; Alvarez-Mejia, C.; Vega-Arreguin, J.C.; Jimenez-Moraila, B.; Fernandez-Cortes, A.; Corona-Armenta, G.; Herrera-Estrella, L.; et al. The palomero genome suggests metal effects on domestication. Science 2009.
5. Andorf, C.M.; Cannon, E.K.; Portwood, J.L., 2nd; Gardiner, J.M.; Harper, L.C.; Schaeffer, M.L.; Braun, B.L.; Campbell, D.A.; Vinnakota, A.G.; Sribalusu, V.V.; et al. Maizegdb update: New tools, data and interface for the maize model organism database. Nucleic Acids Res. 2016, 44, D1195–D1201.
6. Kellogg, E.A. Evolutionary history of the grasses. Plant Physiol. 2001, 125, 1198–1205.

7. Gowik, U.; Westhoff, P. The path from C3 to C4 photosynthesis. Plant Physiol. 2011, 155, 56–63.
8. Vogel, J.P.; Garvin, D.F.; Mockler, T.C.; Schmutz, J.; Rokhsar, D.; Bevan, M.W.; Barry, K.; Lucas, S.; Harmon-Smith, M.; Lail, K.; et al. Genome sequencing and analysis of the model grass brachypodium distachyon. Nature 2010, 463, 763–768.
9. Bennetzen, J.L.; Schmutz, J.; Wang, H.; Percifield, R.; Hawkins, J.; Pontaroli, A.C.; Estep, M.; Feng, L.; Vaughn, J.N.; Grimwood, J.; et al. Reference genome sequence of the model plant Setaria. Nat. Biotechnol. 2012, 30, 555–561.
10. Brutnell, T.P.; Bennetzen, J.L.; Vogel, J.P. Brachypodium distachyon and Setaria viridis: Model genetic systems for the grasses. Annu. Rev. Plant Biol. 2015, 66, 465–485.
11. Estep, M.C.; McKain, M.R.; Vela Diaz, D.; Zhong, J.; Hodge, J.G.; Hodkinson, T.R.; Layton, D.J.; Malcomber, S.T.; Pasquet, R.; Kellogg, E.A. Allopolyploidy, diversification, and the miocene grassland expansion. Proc. Natl. Acad. Sci. USA 2014, 111, 15149–15154.
12. Buczynski, M.W.; Dumlao, D.S.; Dennis, E.A. An integrated omics analysis of eicosanoid biology. J. Lipid Res. 2009, 50, 1505.
13. Funk, C.D. Prostaglandins and leukotrienes: Advances in eicosanoid biology. Science 2001, 294, 1871–1875.
14. Prost, I.; Dhondt, S.; Rothe, G.; Vicente, J.; Rodriguez, M.J.; Kift, N.; Carbonne, F.; Griffiths, G.; Esquerre-Tugaye, M.T.; Rosahl, S.; et al. Evaluation of the antimicrobial activities of plant oxylipins supports their involvement in defense against pathogens. Plant Physiol. 2005, 139, 1902–1913.
15. Park, S.W.; Li, W.; Viehhauser, A.; He, B.; Kim, S.; Nilsson, A.K.; Andersson, M.X.; Kittle, J.D.; Ambavaram, M.M.; Luan, S.; et al. Cyclophilin 20–3 relays a 12-oxo-phytodienoic acid signal during stress responsive regulation of cellular redox homeostasis. Proc. Natl. Acad. Sci. USA 2013, 110, 9559–9564.
16. v. Euler, U.S. Über die spezifische blutdrucksenkende substanz des menschlichen prostata-und samenblasensekretes. J. Mol. Med. 1935, 14, 1182–1183.
17. Goldblatt, M.W. Properties of human seminal plasma. J. Physiol. 1935, 84, 208–218.
18. Wasternack, C. How jasmonates earned their laurels: Past and present. Plant Growth Regul. 2015, 34, 761–794.
19. Corey, E.J.; Weinshenker, N.M.; Schaaf, T.K.; Huber, W. Stereo-controlled synthesis of DL-prostaglandins F2α and E2. J. Am. Chem. Soc 1969, 91, 5675–5677.
20. Fenn, J.B.; Mann, M.; Meng, C.K.; Wong, S.F.; Whitehouse, C.M. Electrospray ionization for mass spectrometry of large biomolecules. Science 1989, 246, 64–71.
21. Pountos, I.; Georgouli, T.; Bird, H.; Giannoudis, P.V. Nonsteroidal anti-inflammatory drugs: Prostaglandins, indications, and side effects. Int. J. Infereron Cytokine Mediator Res. 2011, 3, 19–27.
22. Vane, J.R. Inhibition of prostaglandin synthesis as a mechanism of action for aspirin-like drugs. Nat.-New Biol. 1971, 231, 232–235.
23. Meirer, K.; Steinhilber, D.; Proschak, E. Inhibitors of the arachidonic acid cascade: Interfering with multiple pathways. Basic Clin. Pharmacol. Toxicol. 2014, 114, 83–91.
24. Goppelt-Struebe, M.; Wolter, D.; Resch, K. Glucocorticoids inhibit prostaglandin synthesis not only at the level of phospholipase A2 but also at the level of cyclo-oxygenase/pge isomerase. Br. J. Pharmacol. 1989, 98, 1287–1295.
25. Laine, L. Approaches to nonsteroidal anti-inflammatory drug use in the high-risk patient. Gastroenterology 2001, 120, 594–606.
26. U.S. Food and Drug Administration. National Drug Code Directory. Available online: http://www.fda.gov/Drugs/InformationOnDrugs/ucm142438.htm (accessed on13 October 2016).
27. Andreou, A.; Brodhun, F.; Feussner, I. Biosynthesis of oxylipins in non-mammals. Prog. Lipid Res. 2009, 48, 148–170.
28. Hamberg, M.; Ponce de Leon, I.; Sanz, A.; Castresana, C. Fatty acid alpha-dioxygenases. Prostaglandins Other Lipid Mediat. 2002, 68–69, 363–374.

29. Mueller, M.J. Archetype signals in plants: The phytoprostanes. Curr. Opin. Plant Biol. 2004, 7, 441–448.
30. Cuyamendous, C.; Leung, K.S.; Durand, T.; Lee, J.C.; Oger, C.; Galano, J.M. Synthesis and discovery of phytofurans: Metabolites of alpha-linolenic acid peroxidation. Chem. Commun. (Camb.) 2015, 51, 15696–15699.
31. Schomburg, A. Oxylipin Profiling Database. Available online: http://www.oxylipins.uni-goettingen.de/path.php?path=Pathway&caption=Oxylipin%20Pathway (accessed on 15 November 2016).
32. Wasternack, C. Jasmonates: An update on biosynthesis, signal transduction and action in plant stress response, growth and development. Ann. Bot. 2007, 100, 681–697.
33. Wasternack, C.; Hause, B. Jasmonates: Biosynthesis, perception, signal transduction and action in plant stress response, growth and development. An update to the 2007 review in annals of botany. Ann. Bot. 2013, 111, 1021–1058.
34. Miersch, O.; Bohlmann, H.; Wasternack, C. Jasmonates and related compounds from Fusarium oxysporum. Phytochemistry 1999, 50, 517–523.
35. Miersch, O.; Bruckner, B.; Schmidt, J.; Sembdner, G. Cyclopentane fatty-acids from Gibberella fujikuroi. Phytochemistry1992, 31, 3835–3837.
36. Miersch, O.; Günther, T.; Fritsche, W.; Sembdner, G. Jasmonates from different fungal species. Nat. Prod. Lett. 1993, 2, 293–299.
37. Tsukada, K.; Takahashi, K.; Nabeta, K. Biosynthesis of jasmonic acid in a plant pathogenic fungus, Lasiodiplodia theobromae. Phytochemistry 2010, 71, 2019–2023.
38. Melotto, M.; Underwood, W.; Koczan, J.; Nomura, K.; He, S.Y. Plant stomata function in innate immunity against bacterial invasion. Cell 2006, 126, 969–980.
39. Brooks, D.M.; Bender, C.L.; Kunkel, B.N. The Pseudomonas syringae phytotoxin coronatine promotes virulence by overcoming salicylic acid-dependent defences in Arabidopsis thaliana. Mol. Plant Pathol. 2005, 6, 629–639.
40. Borrego, E.J.; Kolomiets, M.V. Lipid-mediated signaling between fungi and plants. In Biocommunication of Fungi; Springer: New York, NY, USA, 2012; pp. 249–260.
41. Russell, W. Registration of B70 and B73 parental lines of maize (reg. Nos. Pl16 and pl17). Crop Sci. 1972, 12, 721.
42. De La Fuente, G.N.; Murray, S.C.; Isakeit, T.; Park, Y.S.; Yan, Y.; Warburton, M.L.; Kolomiets, M.V. Characterization of genetic diversity and linkage disequilibrium of ZmLOX4 and ZmLOX5 loci in maize. PLoS ONE 2013, 8, e53973.
43. Yan, Y.; Borrego, E.; Kolomiets, M.V. Jasmonate Biosynthesis, Perception and Function in Plant Development and Stress Responses; InTech Open Access Publisher: Rijeka, Croatia, 2013.
44. Acosta, I.F.; Laparra, H.; Romero, S.P.; Schmelz, E.; Hamberg, M.; Mottinger, J.P.; Moreno, M.A.; Dellaporta, S.L. Tasselseed1 is a lipoxygenase affecting jasmonic acid signaling in sex determination of maize. Science 2009, 323, 262–265.
45. Yan, Y.; Christensen, S.; Isakeit, T.; Engelberth, J.; Meeley, R.; Hayward, A.; Emery, R.J.; Kolomiets, M.V. Disruption of OPR7 and OPR8 reveals the versatile functions of jasmonic acid in maize development and defense. Plant Cell 2012, 24, 1420–1436.
46. Feussner, I.; Wasternack, C. The lipoxygenase pathway. Annu. Rev. Plant Biol. 2002, 53, 275–297.
47. Ghanem, M.E.; Ghars, M.A.; Frettinger, P.; Perez-Alfocea, F.; Lutts, S.; Wathelet, J.P.; du Jardin, P.; Fauconnier, M.L. Organ-dependent oxylipin signature in leaves and roots of salinized tomato plants (Solanum lycopersicum). J. Plant Physiol. 2012, 169, 1090–1101.
48. Gao, X.Q.; Stumpe, M.; Feussner, I.; Kolomiets, M. A novel plastidial lipoxygenase of maize (Zea mays) ZmLOX6 encodes for a fatty acid hydroperoxide lyase and is uniquely regulated by phytohormones and pathogen infection. Planta 2008, 227, 491–503.

49. Christensen, S.A.; Kolomiets, M.V. The lipid language of plant-fungal interactions. Fungal Genet. Biol. 2011, 48, 4–14.
50. Park, Y.S.; Kunze, S.; Ni, X.; Feussner, I.; Kolomiets, M.V. Comparative molecular and biochemical characterization of segmentally duplicated 9-lipoxygenase genes ZmLOX4 and ZmLOX5 of maize. Planta 2010, 231, 1425–1437.
51. Christensen, S.A.; Nemchenko, A.; Borrego, E.; Murray, I.; Sobhy, I.S.; Bosak, L.; DeBlasio, S.; Erb, M.; Robert, C.A.; Vaughn, K.A.; et al. The maize lipoxygenase, ZmLOX10, mediates green leaf volatile, jasmonate and herbivore-induced plant volatile production for defense against insect attack. Plant J. 2013, 74, 59–73.
52. Umate, P. Genome-wide analysis of lipoxygenase gene family in Arabidopsis and rice. Plant Signal. Behav. 2011, 6, 335–338.
53. Chauvin, A.; Caldelari, D.; Wolfender, J.L.; Farmer, E.E. Four 13-lipoxygenases contribute to rapid jasmonate synthesis in wounded Arabidopsis thaliana leaves: A role for lipoxygenase 6 in responses to long-distance wound signals. New Phytol. 2013, 197, 566–575.
54. Nemchenko, A.; Kunze, S.; Feussner, I.; Kolomiets, M. Duplicate maize 13-lipoxygenase genes are differentially regulated by circadian rhythm, cold stress, wounding, pathogen infection, and hormonal treatments. J. Exp. Bot.2006, 57, 3767–3779.
55. Wu, Q.; Luo, A.; Zadrozny, T.; Sylvester, A.; Jackson, D. Fluorescent protein marker lines in maize: Generation and applications. Int. J. Dev. Biol. 2013, 57, 535–543.
56. Schiller, D.; Contreras, C.; Vogt, J.; Dunemann, F.; Defilippi, B.G.; Beaudry, R.; Schwab, W. A dual positional specific lipoxygenase functions in the generation of flavor compounds during climacteric ripening of apple. Hortic. Res. 2015, 2, Article 15003.
57. Hughes, R.K.; West, S.I.; Hornostaj, A.R.; Lawson, D.M.; Fairhurst, S.A.; Sanchez, R.O.; Hough, P.; Robinson, B.H.; Casey, R. Probing a novel potato lipoxygenase with dual positional specificity reveals primary determinants of substrate binding and requirements for a surface hydrophobic loop and has implications for the role of lipoxygenases in tubers. Biochem. J. 2001, 353, 345–355.
58. Palmieri-Thiers, C.; Canaan, S.; Brunini, V.; Lorenzi, V.; Tomi, F.; Desseyn, J.L.; Garscha, U.; Oliw, E.H.; Berti, L.; Maury, J. A lipoxygenase with dual positional specificity is expressed in olives (Olea europaea L.) during ripening. Biochim. Biophys. Acta 2009, 1791, 339–346.
59. Wang, R.; Shen, W.; Liu, L.; Jiang, L.; Liu, Y.; Su, N.; Wan, J. A novel lipoxygenase gene from developing rice seeds confers dual position specificity and responds to wounding and insect attack. Plant Mol. Biol. 2008, 66, 401–414.
60. Kim, E.S.; Choi, E.; Kim, Y.; Cho, K.; Lee, A.; Shim, J.; Rakwal, R.; Agrawal, G.K.; Han, O. Dual positional specificity and expression of non-traditional lipoxygenase induced by wounding and methyl jasmonate in maize seedlings. Plant Mol. Biol. 2003, 52, 1203–1213.
61. Emanuelsson, O.; Nielsen, H.; Von Heijne, G. Chlorop, a neural network-based method for predicting chloroplast transit peptides and their cleavage sites. Protein Sci. 1999, 8, 978–984.
62. Cho, K.; Han, Y.; Woo, J.C.; Baudisch, B.; Klosgen, R.B.; Oh, S.; Han, J.; Han, O. Cellular localization of dual positional specific maize lipoxygenase-1 in transgenic rice and calcium-mediated membrane association. Plant Sci. 2011, 181, 242–248.
63. Svitashev, S.; Young, J.K.; Schwartz, C.; Gao, H.; Falco, S.C.; Cigan, A.M. Targeted mutagenesis, precise gene editing, and site-specific gene insertion in maize using CAS9 and guide RNA. Plant Physiol. 2015, 169, 931–945.
64. Itoh, A.; Schilmiller, A.L.; McCaig, B.C.; Howe, G.A. Identification of a jasmonate-regulated allene oxide synthase that metabolizes 9-hydroperoxides of linoleic and linolenic acids. J. Biol. Chem. 2002, 277, 46051–46058.
65. Park, J.H.; Halitschke, R.; Kim, H.B.; Baldwin, I.T.; Feldmann, K.A.; Feyereisen, R. A knock-out mutation in allene oxide synthase results in male sterility and defective wound signal transduction in Arabidopsis due to a block in jasmonic acid biosynthesis. Plant J. 2002, 31, 1–12.

66. Lee, D.S.; Nioche, P.; Hamberg, M.; Raman, C.S. Structural insights into the evolutionary paths of oxylipin biosynthetic enzymes. Nature 2008, 455, 363–368.
67. Christensen, S.A.; Huffaker, A.; Kaplan, F.; Sims, J.; Ziemann, S.; Doehlemann, G.; Ji, L.; Schmitz, R.J.; Kolomiets, M.V.; Alborn, H.T. Maize death acids, 9-lipoxygenase-derived cyclopente (a) nones, display activity as cytotoxic phytoalexins and transcriptional mediators. Proc. Natl. Acad. Sci. USA 2015, 112, 11407–11412.
68. Maucher, H.; Hause, B.; Feussner, I.; Ziegler, J.; Wasternack, C. Allene oxide synthases of barley (Hordeum vulgare cv. Salome): Tissue specific regulation in seedling development. Plant J. 2000, 21, 199–213.
69. Haga, K.; Iino, M. Phytochrome-mediated transcriptional up-regulation of allene oxide synthase in rice seedlings. Plant Cell Physiol. 2004, 45, 119–128.
70. Haga, K.; Takano, M.; Neumann, R.; Iino, M. The rice COLEOPTILE PHOTOTROPISM1 gene encoding an ortholog of Arabidopsis NPH3 is required for phototropism of coleoptiles and lateral translocation of auxin. Plant Cell 2005, 17, 103–115.
71. Chehab, E.W.; Raman, G.; Walley, J.W.; Perea, J.V.; Banu, G.; Theg, S.; Dehesh, K. Rice hydroperoxide lyases with unique expression patterns generate distinct aldehyde signatures in Arabidopsis. Plant Physiol. 2006, 141, 121–134.
72. Riemann, M.; Haga, K.; Shimizu, T.; Okada, K.; Ando, S.; Mochizuki, S.; Nishizawa, Y.; Yamanouchi, U.; Nick, P.; Yano, M.; et al. Identification of rice allene oxide cyclase mutants and the function of jasmonate for defence against Magnaporthe oryzae. Plant J. 2013, 74, 226–238.
73. Yoeun, S.; Rakwal, R.; Han, O. Dual positional substrate specificity of rice allene oxide synthase-1: Insight into mechanism of inhibition by type Π ligand imidazole. BMB Rep. 2013, 46, 151–156.
74. Ogorodnikova, A.V.; Gorina, S.S.; Mukhtarova, L.S.; Mukhitova, F.K.; Toporkova, Y.Y.; Hamberg, M.; Grechkin, A.N. Stereospecific biosynthesis of (9S,13S)-10-oxophytoenoic acid in young maize roots. Biochim. Biophys. Acta Mol. Cell Biol. Lipids 2015, 185, 1262–1270.
75. Bae, H.K.; Kang, H.G.; Kim, G.J.; Eu, H.J.; Oh, S.A.; Song, J.T.; Chung, I.K.; Eun, M.Y.; Park, S.K. Transgenic rice plants carrying RNA interference constructs of AOS (allene oxide synthase) genes show severe male sterility. Plant Breed. 2010, 129, 647–651.
76. Christensen, S.A.; Nemchenko, A.; Park, Y.S.; Borrego, E.; Huang, P.C.; Schmelz, E.A.; Kunze, S.; Feussner, I.; Yalpani, N.; Meeley, R.; et al. The novel monocot-specific 9-lipoxygenase ZmLOX12 is required to mount an effective jasmonate-mediated defense against Fusarium verticillioides in maize. Mol. Plant Microbe Interact. 2014, 27, 1263–1276.
77. Gao, X.; Starr, J.; Gobel, C.; Engelberth, J.; Feussner, I.; Tumlinson, J.; Kolomiets, M. Maize 9-lipoxygenase ZMLOX3 controls development, root-specific expression of defense genes, and resistance to root-knot nematodes. Mol. Plant Microbe Interact. 2008, 21, 98–109.
78. Engelberth, J.; Seidl-Adams, I.; Schultz, J.C.; Tumlinson, J.H. Insect elicitors and exposure to green leafy volatiles differentially upregulate major octadecanoids and transcripts of 12-oxo phytodienoic acid reductases in Zea mays. Mol. Plant Microbe Interact. 2007, 20, 707–716.
79. Dafoe, N.J.; Thomas, J.D.; Shirk, P.D.; Legaspi, M.E.; Vaughan, M.M.; Huffaker, A.; Teal, P.E.; Schmelz, E.A. European corn borer (Ostrinia nubilalis) induced responses enhance susceptibility in maize. PLoS ONE 2013, 8, e73394.
80. Engelberth, J.; Contreras, C.F.; Viswanathan, S. Transcriptional analysis of distant signaling induced by insect elicitors and mechanical wounding in Zea mays. PLoS ONE 2012, 7, e34855.
81. Farag, M.A.; Fokar, M.; Abd, H.; Zhang, H.; Allen, R.D.; Pare, P.W. (Z)-3-hexenol induces defense genes and downstream metabolites in maize. Planta 2005, 220, 900–909.
82. Hofmann, E.; Zerbe, P.; Schaller, F. The crystal structure of Arabidopsis thaliana allene oxide cyclase: Insights into the oxylipin cyclization reaction. Plant Cell 2006, 18, 3201–3217.

83. Wasternack, C.; Kombrink, E. Jasmonates: Structural requirements for lipid-derived signals active in plant stress responses and development. ACS Chem. Biol. 2010, 5, 63–77.
84. Stenzel, I.; Otto, M.; Delker, C.; Kirmse, N.; Schmidt, D.; Miersch, O.; Hause, B.; Wasternack, C. Allene oxide cyclase (AOC) gene family members of Arabidopsis thaliana: Tissue- and organ-specific promoter activities and in vivo heteromerization. J. Exp. Bot. 2012, 63, 6125–6138.
85. Ziegler, J.; Wasternack, C.; Hamberg, M. On the specificity of allene oxide cyclase. Lipids 1999, 34, 1005–1015.
86. Riemann, M.; Muller, A.; Korte, A.; Furuya, M.; Weiler, E.W.; Nick, P. Impaired induction of the jasmonate pathway in the rice mutant hebiba. Plant Physiol. 2003, 133, 1820–1830.
87. Gutjahr, C.; Gobbato, E.; Choi, J.; Riemann, M.; Johnston, M.G.; Summers, W.; Carbonnel, S.; Mansfield, C.; Yang, S.Y.; Nadal, M.; et al. Rice perception of symbiotic arbuscular mycorrhizal fungi requires the karrikin receptor complex. Science 2015, 350, 1521–1524.
88. Flematti, G.R.; Ghisalberti, E.L.; Dixon, K.W.; Trengove, R.D. A compound from smoke that promotes seed germination. Science 2004, 305, 977.
89. Nordstrom, K.J.; Albani, M.C.; James, G.V.; Gutjahr, C.; Hartwig, B.; Turck, F.; Paszkowski, U.; Coupland, G.; Schneeberger, K. Mutation identification by direct comparison of whole-genome sequencing data from mutant and wild-type individuals using k-mers. Nat. Biotechnol. 2013, 31, 325–330.
90. Hazman, M.; Hause, B.; Eiche, E.; Nick, P.; Riemann, M. Increased tolerance to salt stress in OPDA-deficient rice allene oxide cyclase mutants is linked to an increased ROS-scavenging activity. J. Exp. Bot. 2015, 66, 3339–3352.
91. Zhang, J.; Simmons, C.; Yalpani, N.; Crane, V.; Wilkinson, H.; Kolomiets, M. Genomic analysis of the 12-oxo-phytodienoic acid reductase gene family of Zea mays. Plant Mol. Biol. 2005, 59, 323–343.
92. Tani, T.; Sobajima, H.; Okada, K.; Chujo, T.; Arimura, S.; Tsutsumi, N.; Nishimura, M.; Seto, H.; Nojiri, H.; Yamane, H. Identification of the OsOPR7 gene encoding 12-oxophytodienoate reductase involved in the biosynthesis of jasmonic acid in rice. Planta 2008, 227, 517–526.
93. Schaller, F.; Biesgen, C.; Mussig, C.; Altmann, T.; Weiler, E.W. 12-oxophytodienoate reductase 3 (OPR3) is the isoenzyme involved in jasmonate biosynthesis. Planta 2000, 210, 979–984.
94. Stintzi, A.; Browse, J. The Arabidopsis male-sterile mutant, opr3, lacks the 12-oxophytodienoic acid reductase required for jasmonate synthesis. Proc. Natl. Acad. Sci. USA 2000, 97, 10625–10630.
95. Staswick, P.E.; Tiryaki, I. The oxylipin signal jasmonic acid is activated by an enzyme that conjugates it to isoleucine in Arabidopsis. Plant Cell 2004, 16, 2117–2127.
96. Hamberg, M.; Gardner, H.W. Oxylipin pathway to jasmonates: Biochemistry and biological significance. Biochim. Biophys. Acta 1992, 1165, 1–18.
97. Miersch, O.; Kramell, R.; Parthier, B.; Wasternack, C. Structure-activity relations of substituted, deleted or stereospecifically altered jasmonic acid in gene expression of barley leaves. Phytochemistry 1999, 50, 353–361.
98. Fonseca, S.; Chini, A.; Hamberg, M.; Adie, B.; Porzel, A.; Kramell, R.; Miersch, O.; Wasternack, C.; Solano, R. (+)-7-iso-jasmonoyl-l-isoleucine is the endogenous bioactive jasmonate. Nat. Chem. Biol. 2009, 5, 344–350.
99. Staswick, P.E.; Tiryaki, I.; Rowe, M.L. Jasmonate response locus JAR1 and several related Arabidopsis genes encode enzymes of the firefly luciferase superfamily that show activity on jasmonic, salicylic, and indole-3-acetic acids in an assay for adenylation. Plant Cell 2002, 14, 1405–1415.
100. Wakuta, S.; Suzuki, E.; Saburi, W.; Matsuura, H.; Nabeta, K.; Imai, R.; Matsui, H. OsJAR1 and OsJAR2 are jasmonyl-l-isoleucine synthases involved in wound- and pathogen-induced jasmonic acid signalling. Biochem. Biophys. Res. Commun. 2011, 409, 634–639.

101. Riemann, M.; Riemann, M.; Takano, M. Rice JASMONATE RESISTANT 1 is involved in phytochrome and jasmonate signalling. Plant Cell Environ. 2008, 31, 783–792.
102. Xiao, Y.; Chen, Y.; Charnikhova, T.; Mulder, P.P.; Heijmans, J.; Hoogenboom, A.; Agalou, A.; Michel, C.; Morel, J.B.; Dreni, L.; et al. OsJAR1 is required for JA-regulated floret opening and anther dehiscence in rice. Plant Mol. Biol.2014, 86, 19–33.
103. Shimizu, T.; Miyamoto, K.; Miyamoto, K.; Minami, E.; Nishizawa, Y.; Iino, M.; Nojiri, H.; Yamane, H.; Okada, K. OsJAR1 contributes mainly to biosynthesis of the stress-induced jasmonoyl-isoleucine involved in defense responses in rice. Biosci. Biotechnol. Biochem. 2013, 77, 1556–1564.
104. Hamberg, M. New cyclopentenone fatty acids formed from linoleic and linolenic acids in potato. Lipids 2000, 35, 353–363.
105. Christensen, S.A.; Huffaker, A.; Hunter, C.T.; Alborn, H.T.; Schmelz, E.A. A maize death acid, 10-oxo-11-phytoenoic acid, is the predominant cyclopentenone signal present during multiple stress and developmental conditions. Plant Signal. Behav. 2015, 12, e1120395.
106. Matsui, K. Green leaf volatiles: Hydroperoxide lyase pathway of oxylipin metabolism. Curr. Opin. Plant Biol. 2006, 9, 274–280.
107. Matsui, K.; Sugimoto, K.; Mano, J.; Ozawa, R.; Takabayashi, J. Differential metabolisms of green leaf volatiles in injured and intact parts of a wounded leaf meet distinct ecophysiological requirements. PLoS ONE 2012, 7, e36433.
108. Howe, G.A.; Jander, G. Plant immunity to insect herbivores. Annu. Rev. Plant Biol. 2008, 59, 41–66.
109. Schuman, M.C.; Baldwin, I.T. The layers of plant responses to insect herbivores. Annu. Rev. Entomol. 2016, 61, 373–394.
110. War, A.R.; Paulraj, M.G.; Ahmad, T.; Buhroo, A.A.; Hussain, B.; Ignacimuthu, S.; Sharma, H.C. Mechanisms of plant defense against insect herbivores. Plant Signal. Behav. 2012, 7, 1306–1320.
111. Turlings, T.C.J.; Alborn, H.T.; Loughrin, J.H.; Tumlinson, J.H. Volicitin, an elicitor of maize volatiles in oral secretion of Spodoptera exigua: Isolation and bioactivity. J. Chem. Ecol. 2000, 26, 189–202.
112. Lait, C.G.; Alborn, H.T.; Teal, P.E.; Tumlinson, J.H., 3rd. Rapid biosynthesis of N-linolenoyl-L-glutamine, an elicitor of plant volatiles, by membrane-associated enzyme(s) in Manduca sexta. Proc. Natl. Acad. Sci. USA 2003, 100, 7027–7032.
113. Schmelz, E.A.; Engelberth, J.; Alborn, H.T.; Tumlinson, J.H., 3rd; Teal, P.E. Phytohormone-based activity mapping of insect herbivore-produced elicitors. Proc. Natl. Acad. Sci. USA 2009, 106, 653–657.
114. Maes, L.; Goossens, A. Hormone-mediated promotion of trichome initiation in plants is conserved but utilizes species and trichome-specific regulatory mechanisms. Plant Signal. Behav. 2010, 5, 205–207.
115. Velez-Bermudez, I.C.; Salazar-Henao, J.E.; Fornale, S.; Lopez-Vidriero, I.; Franco-Zorrilla, J.M.; Grotewold, E.; Gray, J.; Solano, R.; Schmidt, W.; Pages, M.; et al. A MYB/ZML complex regulates wound-induced lignin genes in maize. Plant Cell 2015, 27, 3245–3259.
116. Niemeyer, H.M. Hydroxamic acids derived from 2-hydroxy-2 h-1, 4-benzoxazin-3 (4 h)-one: Key defense chemicals of cereals. Agric. Food Chem. 2009, 57, 1677–1696.
117. Frey, M.; Spiteller, D.; Boland, W.; Gierl, A. Transcriptional activation of IGL, the gene for indole formation in Zea mays: A structure-activity study with elicitor-active n-acyl glutamines from insects. Phytochemistry 2004, 65, 1047–1055.
118. Dafoe, N.J.; Huffaker, A.; Vaughan, M.M.; Duehl, A.J.; Teal, P.E.; Schmelz, E.A. Rapidly induced chemical defenses in maize stems and their effects on short-term growth of Ostrinia nubilalis. J. Chem. Ecol. 2011, 37, 984–991.
119. Yang, Y.-L.; Chang, F.-R.; Wu, C.-C.; Wang, W.-Y.; Wu, Y.-C. New ent-kaurane diterpenoids with anti-platelet aggregation activity from Annona quamosa. J. Nat. Prod. 2002, 65, 1462–1467.

120. Schmelz, E.A.; Kaplan, F.; Huffaker, A.; Dafoe, N.J.; Vaughan, M.M.; Ni, X.; Rocca, J.R.; Alborn, H.T.; Teal, P.E. Identity, regulation, and activity of inducible diterpenoid phytoalexins in maize. Proc. Natl. Acad. Sci. USA 2011, 108, 5455–5460.
121. Mohan, S.; Ma, P.W.; Pechan, T.; Bassford, E.R.; Williams, W.P.; Luthe, D.S. Degradation of the S. frugiperda peritrophic matrix by an inducible maize cysteine protease. J. Insect Physiol. 2006, 52, 21–28.
122. Lopez, L.; Camas, A.; Shivaji, R.; Ankala, A.; Williams, P.; Luthe, D. MIR1-CP, a novel defense cysteine protease accumulates in maize vascular tissues in response to herbivory. Planta 2007, 226, 517–527.
123. Pechan, T.; Ye, L.; Chang, Y.; Mitra, A.; Lin, L.; Davis, F.M.; Williams, W.P.; Luthe, D.S. A unique 33-kd cysteine proteinase accumulates in response to larval feeding in maize genotypes resistant to fall armyworm and other Lepidoptera. Plant Cell 2000, 12, 1031–1040.
124. Ankala, A.; Luthe, D.S.; Williams, W.P.; Wilkinson, J.R. Integration of ethylene and jasmonic acid signaling pathways in the expression of maize defense protein MIR1-CP. Mol. Plant Microbe Interact. 2009, 22, 1555–1564.
125. Louis, J.; Basu, S.; Varsani, S.; Castano-Duque, L.; Jiang, V.; Williams, W.P.; Felton, G.W.; Luthe, D.S. Ethylene contributes to MAIZE INSECT RESISTANCE1-mediated maize defense against the phloem sap-sucking corn leaf aphid. Plant Physiol. 2015, 169, 313–324.
126. Tamayo, M.C.; Rufat, M.; Bravo, J.M.; San Segundo, B. Accumulation of a maize proteinase inhibitor in response to wounding and insect feeding, and characterization of its activity toward digestive proteinases of Spodoptera littoralis larvae. Planta 2000, 211, 62–71.
127. Zhu-Salzman, K.; Zeng, R. Insect response to plant defensive protease inhibitors. Annu. Rev. Entomol. 2015, 60, 233–252.
128. Farmer, E.E.; Johnson, R.R.; Ryan, C.A. Regulation of expression of proteinase inhibitor genes by methyl jasmonate and jasmonic acid. Plant Physiol. 1992, 98, 995–1002.
129. Farmer, E.E.; Ryan, C.A. Octadecanoid precursors of jasmonic acid activate the synthesis of wound-inducible proteinase inhibitors. Plant Cell 1992, 4, 129–134.
130. Yan, J.; Lipka, A.E.; Schmelz, E.A.; Buckler, E.S.; Jander, G. Accumulation of 5-hydroxynorvaline in maize (Zea mays) leaves is induced by insect feeding and abiotic stress. J. Exp. Bot. 2014, eru385.
131. Dabrowska, P.; Freitak, D.; Vogel, H.; Heckel, D.G.; Boland, W. The phytohormone precursor OPDA is isomerized in the insect gut by a single, specific glutathione transferase. Proc. Natl. Acad. Sci. USA 2009, 106, 16304–16309.
132. Shabab, M.; Khan, S.A.; Vogel, H.; Heckel, D.G.; Boland, W. Opda isomerase GST16 is involved in phytohormone detoxification and insect development. FEBS J. 2014, 281, 2769–2783.
133. Heil, M. Indirect defence via tritrophic interactions. New Phytol. 2008, 178, 41–61.
134. Schmelz, E.A.; Alborn, H.T.; Banchio, E.; Tumlinson, J.H. Quantitative relationships between induced jasmonic acid levels and volatile emission in Zea mays during Spodoptera exigua herbivory. Planta 2003, 216, 665–673. [PubMed]
135. Kollner, T.G.; Lenk, C.; Schnee, C.; Kopke, S.; Lindemann, P.; Gershenzon, J.; Degenhardt, J. Localization of sesquiterpene formation and emission in maize leaves after herbivore damage. BMC Plant Biol. 2013, 13, 15.
136. Schmelz, E.A.; Alborn, H.T.; Tumlinson, J.H. Synergistic interactions between volicitin, jasmonic acid and ethylene mediate insect-induced volatile emission in Zea mays. Physiol. Plant. 2003, 117, 403–412.
137. Ozawa, R.; Shiojiri, K.; Sabelis, M.W.; Arimura, G.; Nishioka, T.; Takabayashi, J. Corn plants treated with jasmonic acid attract more specialist parasitoids, thereby increasing parasitization of the common armyworm. J. Chem. Ecol.2004, 30, 1797–1808.
138. Ozawa, R.; Shiojiri, K.; Sabelis, M.W.; Takabayashi, J. Maize plants sprayed with either jasmonic acid or its precursor, methyl linolenate, attract armyworm

parasitoids, but the composition of attractants differs. Entomol. Exp. Appl. 2008, 129, 189–199.
139. Von Mérey, G.E.; Veyrat, N.; de Lange, E.; Degen, T.; Mahuku, G.; Valdez, R.L.; Turlings, T.C.; D'Alessandro, M. Minor effects of two elicitors of insect and pathogen resistance on volatile emissions and parasitism of Spodoptera frugiperda in mexican maize fields. Biol. Control. 2012, 60, 7–15.
140. Kim, J.; Felton, G.W. Priming of antiherbivore defensive responses in plants. Insect Sci. 2013, 20, 273–285.
141. Ali, M.; Sugimoto, K.; Ramadan, A.; Arimura, G. Memory of plant communications for priming anti-herbivore responses. Sci. Rep. 2013, 3.
142. Jaskiewicz, M.; Conrath, U.; Peterhansel, C. Chromatin modification acts as a memory for systemic acquired resistance in the plant stress response. EMBO Rep. 2011, 12, 50–55.
143. Engelberth, J.; Alborn, H.T.; Schmelz, E.A.; Tumlinson, J.H. Airborne signals prime plants against insect herbivore attack. Proc. Natl. Acad. Sci. USA 2004, 101, 1781–1785.
144. Ton, J.; D'Alessandro, M.; Jourdie, V.; Jakab, G.; Karlen, D.; Held, M.; Mauch-Mani, B.; Turlings, T.C. Priming by airborne signals boosts direct and indirect resistance in maize. Plant J. 2007, 49, 16–26.
145. Erb, M.; Veyrat, N.; Robert, C.A.; Xu, H.; Frey, M.; Ton, J.; Turlings, T.C. Indole is an essential herbivore-induced volatile priming signal in maize. Nat. Commun. 2015, 6.
146. Oluwafemi, S.; Dewhirst, S.Y.; Veyrat, N.; Powers, S.; Bruce, T.J.; Caulfield, J.C.; Pickett, J.A.; Birkett, M.A. Priming of production in maize of volatile organic defence compounds by the natural plant activator cis-jasmone. PLoS ONE2013, 8, e62299.
147. Glazebrook, J. Contrasting mechanisms of defense against biotrophic and necrotrophic pathogens. Annu. Rev. Phytopathol. 2005, 43, 205–227.
148. Vaughan, M.M.; Huffaker, A.; Schmelz, E.A.; Dafoe, N.J.; Christensen, S.; Sims, J.; Martins, V.F.; Swerbilow, J.; Romero, M.; Alborn, H.T.; et al. Effects of elevated [CO2] on maize defence against mycotoxigenic Fusarium verticillioides. Plant Cell Environ. 2014, 37, 2691–2706.
149. Tang, J.D.; Perkins, A.; Williams, W.P.; Warburton, M.L. Using genome-wide associations to identify metabolic pathways involved in maize aflatoxin accumulation resistance. BMC Genom. 2015, 16.
150. Gao, X.; Brodhagen, M.; Isakeit, T.; Brown, S.H.; Gobel, C.; Betran, J.; Feussner, I.; Keller, N.P.; Kolomiets, M.V. Inactivation of the lipoxygenase ZmLOX3 increases susceptibility of maize to Aspergillus spp. Mol. Plant Microbe Interact. 2009, 22, 222–231.
151. Huffaker, A.; Dafoe, N.J.; Schmelz, E.A. Zmpep1, an ortholog of Arabidopsis elicitor peptide 1, regulates maize innate immunity and enhances disease resistance. Plant Physiol. 2011, 155, 1325–1338.
152. Stenzel, I.; Hause, B.; Maucher, H.; Pitzschke, A.; Miersch, O.; Ziegler, J.; Ryan, C.A.; Wasternack, C. Allene oxide cyclase dependence of the wound response and vascular bundle-specific generation of jasmonates in tomato–amplification in wound signalling. Plant J. 2003, 33, 577–589.
153. Pieterse, C.M.; Leon-Reyes, A.; Van der Ent, S.; Van Wees, S.C. Networking by small-molecule hormones in plant immunity. Nat. Chem. Biol. 2009, 5, 308–316.
154. Constantino, N.N.; Mastouri, F.; Damarwinasis, R.; Borrego, E.J.; Moran-Diez, M.E.; Kenerley, C.M.; Gao, X.; Kolomiets, M.V. Root-expressed maize LIPOXYGENASE 3 negatively regulates induced systemic resistance to Colletotrichum graminicola in shoots. Front. Plant Sci. 2013, 4.
155. Planchamp, C.; Glauser, G.; Mauch-Mani, B. Root inoculation with Pseudomonas putida KT2440 induces transcriptional and metabolic changes and systemic resistance in maize plants. Front. Plant Sci. 2014, 5.
156. Neal, A.L.; Ton, J. Systemic defense priming by Pseudomonas putida KT2440 in maize depends on benzoxazinoid exudation from the roots. Plant Signal. Behav. 2013, 8, e22655.

157. Yamane, H.; Sugawara, J.; Suzuki, Y.; Shimamura, E.; Takahashi, N. Syntheses of jasmonic acid related-compounds and their structure-activity-relationships on the growth of rice seedings. Agric. Biol. Chem. 1980, 44, 2857–2864.
158. Yan, Y.; Huang, P.-C.; Borrego, E.; Kolomiets, M. New perspectives into jasmonate roles in maize. Plant Signal. Behav.2014, 9, e970442.
159. Calderon-Urrea, A.; Dellaporta, S.L. Cell death and cell protection genes determine the fate of pistils in maize. Development 1999, 126, 435–441. [PubMed]
160. DeLong, A.; Calderon-Urrea, A.; Dellaporta, S.L. Sex determination gene TASSELSEED 2 of maize encodes a short-chain alcohol dehydrogenase required for stage-specific floral organ abortion. Cell 1993, 74, 757–768.
161. Wu, X.; Knapp, S.; Stamp, A.; Stammers, D.K.; Jornvall, H.; Dellaporta, S.L.; Oppermann, U. Biochemical characterization of TASSELSEED 2, an essential plant short-chain dehydrogenase/reductase with broad spectrum activities. FEBS J. 2007, 274, 1172–1182.
162. Hartwig, T.; Chuck, G.S.; Fujioka, S.; Klempien, A.; Weizbauer, R.; Potluri, D.P.; Choe, S.; Johal, G.S.; Schulz, B. Brassinosteroid control of sex determination in maize. Proc. Natl. Acad. Sci. USA 2011, 108, 19814–19819.

CHAPTER 15

Analysis of Rice *RNA-Dependent RNA Polymerase 1* (*OsRDR1*) in Virus-Mediated RNA Silencing after Particle Bombardment

Hui Chen[1], Atsushi Tamai[2], Masashi Mori[2], Masashi Ugaki[3], Yoshikazu Tanaka[4 6], Partha P. Samadder[45], Akio Miyao[4], Hirohiko Hirochika[4], Naoto Yamaoka[1], Masamichi Nishiguchi[1]

[1.]*Faculty of AgricultureEhime UniversityMatsuyamaJapan*

[2.]*Research Institute for Bioresources and BiotechnologyIshikawa Prefectural UniversityIshikawaJapan*

[3.]*Department of Integrated BiosciencesThe University of TokyoKashiwaJapan*

[4.]*National Institute of Agrobiological SciencesTsukubaJapan*

[5.]*Department of Immunobiology, College of MedicineUniversity of ArizonaTucsonUSA*

[6.]*Wakasa Wan Energy Research CenterTsurugaJapan*

ABSTRACT

RNA-dependent RNA polymerases (RDRs) play a key role in various RNA silencing pathways in many organisms. Using the nucleotide sequence of *SGS2/SDE1/RDR6* in *Arabidopsis* as the search query for sequences that flank the insertions of rice retrotransposon *Tos17*, we selected rice mutant lines (*OsRDR1*). RT-PCR analysis showed that *OsRDR1* mRNA was undetectable in leaves and calli of the mutants, while it was detected in wild type. RNA silencing was induced by particle bombardment to investigate any effects of *OsRDR1* on RNA silencing with β-glucuronidase or green fluorescent protein DNA/RNA in the mutant lines. The results showed that RNA silencing was impaired in these mutant lines by inverted repeat (IR) DNA or in vitro transcribed double-stranded RNA. Further, the mutant lines were bombarded with *Brome mosaic bromovirus* (BMV, a ssRNA virus) or *Wheat dwarfgeminivirus* (WDV, a ssDNA virus), each carrying the IR sequence of a reporter gene. As a result, RNA silencing was impaired by BMV. Interestingly, however, it was not impaired by WDV. Thus we propose that *OsRDR1* is required for RNA silencing mediated by *Bromovirus*, but not by *Geminivirus* in this system.

Keywords: Rice *RNA-dependent RNA polymerase 1* RNA silencing *Brome mosaic bromovirus Wheat dwarf geminivirus*

INTRODUCTION

RNA-dependent RNA polymerases (RDRs) play a fundamental role in various RNA silencing pathways in numerous organisms (Curaba and Chen 2008; Wassenegger and Krczal 2006; Yang et al. 2008). The first *RDR* was discovered about 40 years ago in a search for enzymes that catalyze plant RNA viral replication (Astier-Manifacier and Cornuet 1971). The first plant-encoded RDR was isolated from tomato (*Lycopersicon esculentum*), and is now called *LeRDR1* (Schiebel et al. 1993). In the fungus *Neurospora crassa*, *NcRDR6* (*QDE-1*) is an essential component of quelling and can convert a single-stranded RNA (ssRNA) to a double-stranded (dsRNA) molecule in a primer-dependent/independent fashion (Makeyev and Bamford 2002). In *Arabidopsis thaliana*, six *RDRs* were identified as *AtRDR1, AtRDR2, AtRDR3a, AtRDR3b, AtRDR3c* and *AtRDR6*, three of which (*RDR3a, RDR3b, RDR3c*) share the unique DFDGD amino acid motif in the catalytic domain (Voinnet 2008; Wassenegger and Krczal 2006). Recent findings indicate that *RDR1*, *RDR2* and *RDR6* act in different biological processes of RNA silencing (Dalmay et al. 2000; Mourrain et al. 2000; Xie et al. 2004). *RDR1* expression is induced by salicylic acid (SA), a defense-signaling hormone that accumulates during infection by many viruses and viroids (Dorssers et al. 1984; Khan et al. 1986; Xie et al. 2001). More recent studies with *Cucumber mosaic virus* (CMV) demonstrate that the production of small interfering RNAs (siRNAs) is strongly dependent on *RDR1* without interference by the virus-encoded 2b silencing suppressor protein called viral suppressors of RNA silencing, and *RDR1*-dependent secondary virus-derived siRNAs (viRNAs) play a key role in non-cell-autonomous antiviral silencing (Diaz-

Pendon et al. 2007). It also implies that virus encoded proteins may inhibit one or more RDR pathways of siRNA biogenesis associated with silencing suppressor proteins (Donaire et al. 2008), while other reports indicate that *RDR1* mediates defense against herbivore attacks in *Nicotiana attenuata* (Pandey and Baldwin 2007). *RDR2* transcripts in *N. attenuata* are up-regulated by both UV and herbivore elicitation for protection (Pandey and Baldwin 2008), while *RDR6* is known to be involved in initiation and maintenance of post-transcriptional gene silencing (PTGS) and virus-induced gene silencing (VIGS) (Wassenegger and Krczal 2006). *RDR6* is required for VIGS induced by *Cabbage leaf curl virus* (CaLCuV) and CMV, but not by *Turnip crinkle virus* (TCV), *Tobacco rattle virus* (TRV), *Turnip mosaic virus* or *Tobacco mosaic virus* (Diaz-Pendon et al. 2007; Muangsan et al. 2004; Vaucheret 2006; Voinnet 2005).

In rice, five *RDR*s have been identified: *OsRDR1, OsRDR2, OsRDR3a, OsRDR3b* and *OsRDR6* (Wassenegger and Krczal 2006). A recent study indicates that the OsRDR6-dependent siRNA biogenesis that derives from an isocitrate lyase transcript was significantly up-regulated by the phytohormone abscisic acid (ABA), and *OsRDR6* may participate in the amplification of PTGS mediated by ABA in rice (Yang et al. 2008). The function of other *RDR* enzymes remains a key question. Here we isolated and identified several *OsRDR1* mutant lines using the nucleotide sequence of *SGS2/SDE1/RDR6* in *Arabidopsis* as the query to search for the sequences that flank the insertions of rice retrotransposon *Tos17*. We used particle bombardment to establish an RNA silencing system to investigate any effects of *OsRDR1* on RNA silencing with various silencing inducers, including inverted repeat (IR) DNA or dsRNA, in these mutant lines. We tested the role of *OsRDR1* in RNA silencing mediated by *Wheat dwarf virus* (a ssDNA geminivirus, WDV) or *Brome mosaic virus* (a ssRNA bromovirus, BMV), each carrying the IR sequence of a reporter gene. Here we used either β-glucuronidase (*GUS*) or green fluorescent protein (*GFP*) to make the data more versatile and to use the viral vectors available, pWI-GUS (Ugaki et al. 1991) and BMV-GFP. Our analyses show that *OsRDR1* is required for RNA silencing mediated by BMV, but not by WDV.

MATERIALS AND METHODS

Rice plants

Oryza sativa L. cv. Nipponbare and its knockout mutant lines (ND2001 and ND2059) were used. The homozygous individuals from M1 and M2 generation were screened by Southern blot analyses. Seedlings and calli of selected M2 plants were grown and used for respective experiments.

Southern blot analysis

Employing the cetyltrimethyl ammonium bromide (CTAB) method, we isolated rice genomic DNA (20 μg) and transferred it to a nylon membrane as described previously (Chen et al. 2008). Using the PCR DIG probe Synthesis Kit (Roche, Basel, Switzerland), the probe was prepared as described previously (Chen et al. 2008). A plasmid carrying the flanking sequences in each mutant was constructed as template DNA in the following protocol: a PCR product amplified from the coding region of *OsRDR1* by using forward primer (5′-AAACCCTTGCCTCCACCCAGGTG-3′) and reverse primer (5′-ATGGACTATACTCCAGCACCAA-3′), and the fragment amplified was then inserted into pUC19 (Stratagene, La Jolla, CA, USA), which had been digested with *Sma*I. The probes were labeled according to the manufacturer's instructions. DNA was then fixed to the membrane by UV irradiation, and the prehybridization and hybridization were done as described (Engler-Blum et al. 1993) for subsequent exposure to X-ray film (Boehringer Mannheim, Mannheim, Germany) for 12–18 h. Hybridizations were performed with DIG-dUTP-labeled (digoxigenin-labeled) DNA probes.

DNA sequencing and sequence analysis

DNA sequencing, sequence analysis and homologue searching were carried out as described previously (Chen et al. 2008).

Plasmids used for transient expression

The *GFP* expression plasmid p35S-GFP and the *DsRed* (*Discosoma* red fluorescent protein) expression plasmid pUbq-DsRed driven by an enhanced 35S promoter of *Cauliflower mosaic virus* (CaMV) and an ubiquitin promoter of maize polyubiquitin, respectively, were provided by Miki and Shimamoto (2004). Plasmid *GFP* RNAi was transcribed into an RNA predicted to form a double-stranded hairpin RNA structure (Miki and Shimamoto 2004). Plasmid pIG121 contains *GUS*, driven by the CaMV 35S promoter and 35S terminator, respectively (Ohta et al. 1990). Plasmid pUbq-DsRed was used as an internal control (Miki and Shimamoto 2004). The p35S-GUS RNAi was described previously (Chen et al. 2008).

The plasmid of pWI-GUS is a plant–bacterium shuttle vector derived from WDV having *GUS* under the control of the 35S promoter and terminator

(Ugaki et al. 1991). The pWI-GUS RNAi was constructed as follows (Figure 1a): a fragment containing the *GUS* sense fragment and λ phage sequence was derived from p35S-GUS RNAi through digestion with *Eco*RI and *Sac*I, cloned to pBluescriptII KS+ (Stratagene), creating pBS-GUS + λ. A PCR product containing a 831-bp linker of λ phage and 1300 bp of the inverted-repeat *GUS* fragment was amplified with pBS-GUS + λ as a template using forward primer 5′-GCCGCATGCGAATTCGGACAGACAGTG-3′ containing the *Sph*I site (underlined) and reverse primer 5′-GCCGCATGCGGGAATGGTGATTACCGAC-3′ containing the *Sph*I site (underlined). The amplified fragment was digested with *Sph*I and inserted into the end of the *GUS* fragment of pWI-GUS, which had been digested partially with *Sph*I (5 min at 37°C), so that the plasmid pWI-GUS RNAi was constructed after the orientation was selected and verified by sequencing.

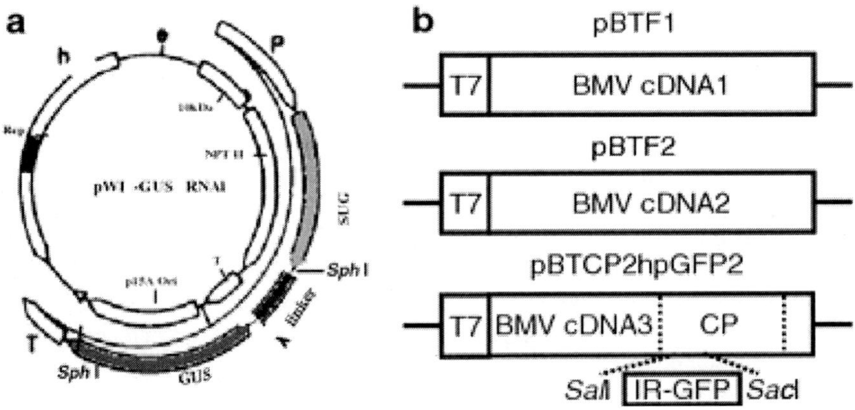

Figure 1. Diagram of pWI-GUS RNAi (**A**), which carries IR-GUS; *e* one of five Eco0109I sites, *NPT II* neomycin phosphotransferase II gene-coding region, *p15A ori* replication origin of plasmid p15A, *T* terminator of CaMV35S transcript, *Rep* replication-associated protein gene, and BMV RNAi (**B**), which carries IR-GFP in place of the 5′-half of the CP, *T7* T7 RNA polymerase promoter

The mixture of BMV-GFP or BMV-GFP RNAi (Figure 1b) was derived from in vitro transcription of pBTF1, pBTF2 and pBTFCP2smGFP or pBTFCP2hpGFP2, respectively (Mori et al. 1993). pBTFCP2smGFP and pBTFCP2hpGFP2 were constructed as follows (Figure 1b): the smGFP gene was amplified by PCR using primers 5′-GGGATGCATGA GTAAAGGAGAAGAACT-3′ and 5′-GGGGAGCTCTTAT TTGTATAG TTCATCCATGCC-3′ from clone smGFP from the Arabidopsis Biological Resource Center at Ohio State University. The PCR product was digested with *Nsi*I and *Sac*I and inserted between the *Nsi*I and *Sac*I site of pBTFCP2pIFN (Mori et al. 1993). The resulting plasmid was digested with *Nsi*I, blunt-ended with T4 DNA polymerase, then self-ligated, creating pBTFCP2smGFP. Then two DNA fragments, the one encoding 0.1 kbp from the C-terminus-, and the other encoding 0.4 kbp from the C-terminus of the *GFP* gene, was amplified using primer sets #1554 (5′-ATAACTAGTTT TCGAAAGATCCCAACGAA-3′)–#1553 (5′-TATGAGCTCGTC GACTAAAG CTCA TCATGTTTGT-3′), and #1555 (5′-ATAACTAGTCAAGACACGT-GCTGAAGTCA-3′)–#1553, respectively. As a PCR template, pBI-erG3 (Tamai et al. 2003) was used. The #1554–#1553 and the #1555–#1553 fragments were digested by *Sac*I/*Spe*I, and *Sal*I/*Spe*I, respectively, ligated with *Sal*I/*Sac*I-digested pBTFCP2smGFP. The resulting plasmid was named pBTFCP2hpGFP2, which was engineered to carry the *GFP* 301 bp as a spacer between an IR of *GFP* 98 bp.

As a complementation study, the p35S-OsRDR1 containing a whole ORF-coding region of cDNA from *OsRDR1* (3342 bp) was constructed as follows: two RT-PCR fragments (cDNA1, 1769 bp and cDNA2, 1583 bp) were amplified with forward primer 5′-GGATCCTGCGCCATGGGTGTCAA-3′ containing the *Bam*HI site (underlined) and reverse primer 5′-GCATGCTTTTCTTAAGGAAAG-3′ containing the *Sph*I site (underlined) and using forward primer 5′-GCATGCTAAAGTTTCAGTCAGA-3′ containing the *Sph*I site (underlined) and reverse primer 5′-GAGCTCAAAC ATGCTGGCCACA-3′ containing the *Sac*I site (underlined), respectively. The cDNA1 fragment was inserted into the pdTA2 vector (Toyobo, Osaka, Japan). This resulted in pdTA2-cDNA1. The cDNA2 fragment was digested with *Sph*I and *Sac*I and inserted into pdTA2-cDNA1, which had been digested with *Sph*I and *Sac*I, creating pdTA2-cDNA1 + cDNA2, then this plasmid was digested with *Bam*HI and *Sac*I to isolate a fragment of cDNA1 + cDNA2, which was gel-purified and inserted into pBI221, which had been digested with *Bam*HI and *Sac*I (to remove *GUS*). This resulted in p35S-OsRDR1 containing a whole ORF-coding region of cDNA from *OsRDR1* (3342 bp). The construct was verified by sequencing.

In vitro synthesis of dsRNA

In vitro synthesis of dsRNA for *GUS* was carried out as described previously (Chen et al. 2008).

Transient expression after particle bombardment

Gold particles (0.6 μm diameter, particle density 31 mg/ml; Sigma-Aldrich, St. Louis, MO, USA) were coated with the respective mixture of plasmids, then delivered biolistically into leaf segments through a particle inflow gun (IDERA GIE-III, Tanaka, Sapporo, Japan). The samples were bombarded (2.5 kg/cm^2 force and 500 mmHg pressure) with the gold particles coated with the following amounts of plasmids or dsRNA per shot: for RNA silencing with *GUS* and WDV, (i) pIG121 (5 µg) and pBIΔGUS (control vector) (5 µg), (ii) pIG121 (5 µg) and p35S-GUS RNAi (10 µg), (iii) pIG121 (5 µg) and ds*GUS* RNA (5 µg), (iv) p35S-GUS (5 µg), p35S-GUS RNAi (10 µg) and p35S-OsRDR1 (10 µg), (v) pIG121 (5 µg) and pWI-GUS RNAi (10 µg); for RNA silencing with *GFP* gene and BMV, (i) pUbq-DsRed (5 µg), p35S-GFP (5 µg) and pBI221 (control vector) (5 µg), (ii) pUbq-DsRed (5 µg), p35S-GFP (5 µg) and *GFP* RNAi (10 µg); (iii) pUbq-DsRed (5 µg), p35S-GFP (5 µg), *GFP* RNAi (10 µg) and p35S-OsRDR1 (10 µg), (iv) pUbq-DsRed (5 µg), p35S-GFP (5 µg) and BMV-GFP RNAi (10 µg); for BMV replication, pUbq-DsRed (5 µg) and BMV-GFP (10 µg); for WDV replication, pUbq-DsRed (5 µg) and pWI-GUS RNAi (10 µg). In the case of real-time RT-PCR analysis with *GUS* silencing, *DsRed* was used for normalization as a reference gene in all treatments.

After bombardment, the expression of GFP and GUS was detected as described previously (Chen et al. 2008).

RNA isolation and analysis

Following the manufacturer's instructions (Tri reagent kit, Molecular Research Center, Cincinnati, OH, USA), total RNA was extracted from rice leaves and calli, mRNA was purified using an Oligotex™-dT30 <Super> mRNA Purification Kit (Takara, Shiga, Japan), and mRNA treated with RNase-free DNase (Stratagene) was used as a template for RT-PCR with reverse transcriptase (MonsterScript™ 1st-Strand cDNA Synthesis Kit, Epicentre, Madison, WI, USA). RT-PCR was carried out using RT primer 5′-ATACGAGGCGGAATGAGAG-3′. The cDNA products of RT-PCR were then amplified by the following pairs of primers: 5′-GCTTGGATGAAACACGTAC-3′ (forward) and 5′-ATACGAGGCGGAATGAGAG-3′ (reverse). Rice ubiquitin was used as an internal standard (Miki and Shimamoto 2004).

Real-time RT-PCR experiments for detecting the expression of *GFP* and *GUS* were carried out as described previously (Chen et al. 2008), and real-time PCR experiments were done with the sets of primers specific to 10 kDa protein of WDV (5′-GGAGGCTTTTGGA CCACATCTTTT-3′ and 5′-TTCCTCGGTCCTCCTTTGCTTCTT-3′). The amount of *DsRed* transcripts was used for normalization as an internal control.

RESULTS

Isolation of homozygous *OsRDR1* knockout mutant lines

Homozygous *OsRDR1* knockout mutant lines from the *Tos17* insertion populations available at the Rice Tos17 Insertion Mutant Database (http://tos.nias.affrc.go.jp/~miyao/pub/tos17/index.html.en) were selected based on the nucleotide sequence of *SGS2/SDE1/RDR6* in *Arabidopsis*. Two lines (ND2001 and ND2059) had the highest homology to that sequence (referred to as *AtRDR6*). In these lines, *Tos17* is inserted at the same site as *OsRDR1*, suggesting that these two were each a callus clone derived from one transformation event because these mutants were selected by tissue culture (Hirochika 2001). The phylogenetic analysis of *OsRDR1* and other *RDR*s showed that this gene is most closely related to *AtRDR1*, and thus it is referred to as *OsRDR1* (Wassenegger and Krczal 2006). The flanking sequence of each mutant line was used to design the primers and to make the specific DNA probes as shown in the materials and methods. The co-segregation analysis of *Xba*I-digested genomic DNA, isolated from each of the individual mutant lines, was carried out by Southern blot analysis; one of three ND2001 and one out of three ND2059 plants were homozygous (Figure 2a). Furthermore, no significant differences between the wild type (WT) and the homozygous knockout mutant lines have been observed from germination through flowering (data not shown), indicating that the *OsRDR1* gene is not required for essential biological processes for morphology.

Identification of *OsRDR1* homologues in rice

To identify *OsRDR1* in rice, we searched for the *SGS2/SDE1 /RDR6* sequence in the rice genome sequence database from DNA Bank at the National Institute of Agrobiological Sciences (http://www.dna.affrc.go.jp) and from the chromatin database (http://www.chromdb.org) at the University of Arizona. The flanking sequence of lines ND2001 and ND2059 each had high identity to one genomic clone from chromosome 2, AP004880, one cDNA clone from AK101440 and one transcript sequence from RDR704. Using this information, one putative cDNA of rice *RDR* was cloned by RT-PCR as described in the materials and methods. Combining these results, AK101440 might be an incomplete clone including a missing transcript in exon 4. Comparison of the putative cDNA and its corresponding genomic sequences revealed

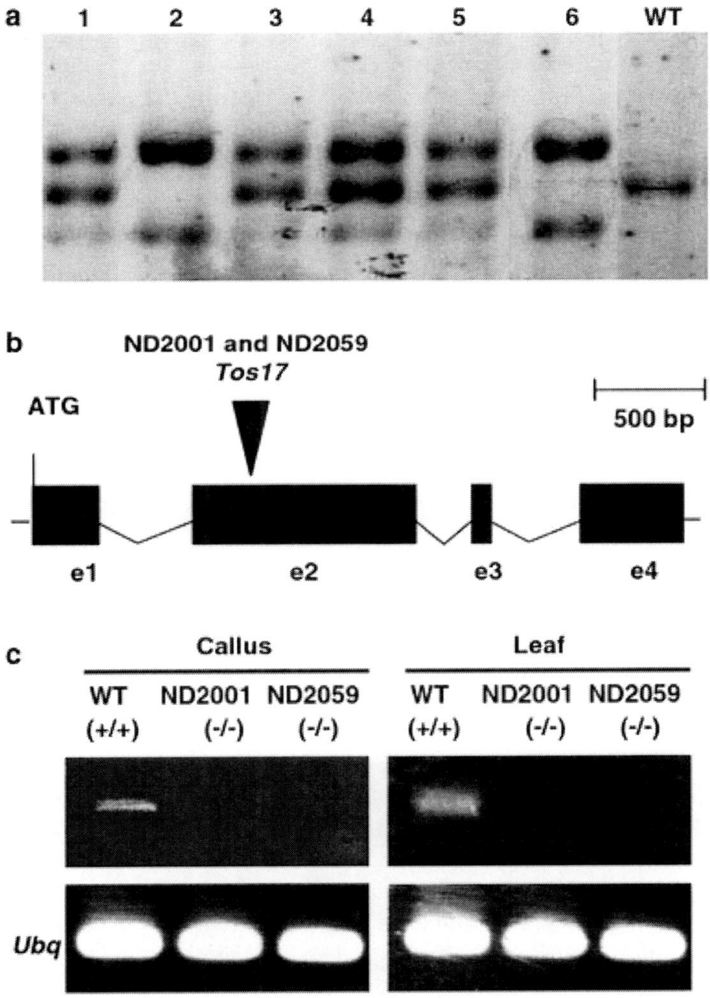

Figure 2. **A** Southern blot analysis of *OsRDR1* mutant lines. *Lanes 1–3* and *lanes 4–6* are M1 progeny of lines ND2001 and ND2059, respectively. *Lanes 2* and *6* are homozygous in terms of the *OsRDR1* knockout. **B** Genomic structure of functional *OsRDR1* gene and *Tos17* insertion site in the *OsRDR1* knockout mutant lines. Exons (e1–e4) are shown as *black boxes*. Introns are indicated by *broken lines*. *Black triangles* indicate the *Tos17* insertion position in knockout mutant lines. **C** Comparative RT-PCR of *OsRDR1* mRNA from different tissues between wild-type and mutant lines. *OsRDR1* mRNA was amplified with 30 PCR cycles. +/+ and −/− are the wild-type and mutant line, respectively. Rice ubiquitin (*Ubq*) serves as a normalization control for RT-PCR efficiency

that *OsRDR1* consists of four exons interrupted by three introns (Figure 2b). Two mutant lines have the same *Tos17* insertions in exon 2 (Figure 2b). A phylogenetic relationship between RDR homologues in rice and *Arabidopsis* was established. A Neighbor-Joining tree from bootstrap analysis (1000 replicates) was constructed using MEGA software version 3.1. These results indicate that *OsRDR1* is in the same clade as *AtRDR1* with a 99% consensus value of bootstrap in Figure S1 in Electronic Supplementary Material.

The expression of *OsRDR1* in WT and mutant lines (ND2001 and ND2059) was examined with RT-PCR analysis using specific primers for *OsRDR1*. *OsRDR1* mRNA was not detected in the young leaves or calli of the mutants, while a significant level was found in the WT after running 30 PCR cycles (Figure 2c). These results indicate that *OsRDR1* was not expressed in the mutant lines due to the insertion of *Tos17*.

Effects of *OsRDR1* disruption on RNA silencing induced by particle bombardment with IR DNA or dsRNA

As we previously reported (Chen et al. 2008), a transient RNA silencing system targeted to *GUS* was examined. The different mixtures of plasmids were used. As shown in Figure 3a, b, substantial levels of *GUS* were expressed at 48 h post-bombardment in the histochemical GUS assay of the WT and mutant line ND2001 when bombarded with plasmid (i) pIG121. The number of *GUS* expressing cells per shot was counted for quantitative analysis in three to six independent experiments ($n = 3–6$ shots). When using mixture (ii) pIG121 and p35S-GUS RNAi, the number of *GUS*-expressing cells was reduced in the WT but not in the mutant line. When using mixture (iii) pIG121 and ds*GUS*, a dsRNA for *GUS* silencing, the number of *GUS*-expressing cells in the WT clearly decreased but not in the mutant line (Figure 3a, b). The specific silencing of *GUS* was recovered by co-bombardment with mixture (iv) including plasmid p35S-OsRDR1 in the WT and mutant line (ND2001) (Figure 3a, b). Moreover, the accumulation of *GUS* transcripts versus *DsRed* transcripts (as an internal control) was examined in the leaf discs from the WT and the mutant line at 48 h post-bombardment with the mixture of those plasmids by real-time RT-PCR (Figure 3c). Similar results were obtained with three independent experiments essentially as levels of *GUS*-expressing cells (Figure 3b). These results were similar to those from the other mutant line (ND2059) (data no shown). Thus, in the mutant lines, RNA silencing was impaired by IR-DNA and dsRNA of *GUS*.

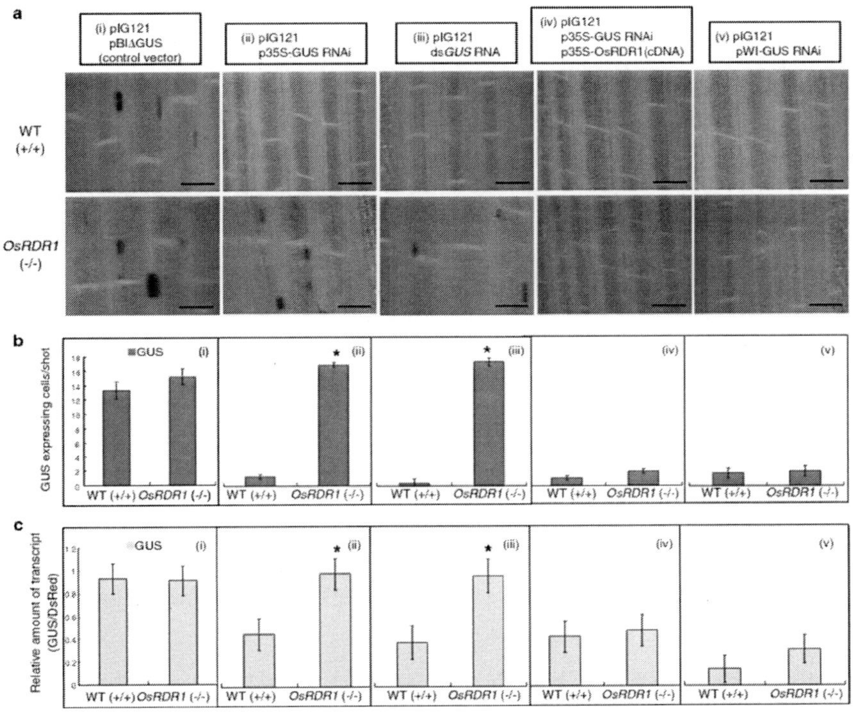

Figure 3. RNA silencing of *GUS* in the epidermal cells of rice leaf discs of wild-type (*WT*) and mutant line after cobombardment with different mixtures of plasmids. **A** Micrographs of *GUS* expression. *Bar* 30 μm. **B** Quantitative analysis of GUS-expressing rice cells. Data are mean ± SE ($n = 3$–6 shots). **C** Real-time RT-PCR analysis of accumulation of *GUS* transcripts 48 h after bombardment. Data were normalized using *DsRed* transcripts as an internal control. Data are mean ± SE ($n = 3$). *$P < 0.05$ (Fisher's test) versus value of WT

Effects of *OsRDR1* disruption on RNA silencing mediated by plant ssDNA/ssRNA viruses

It is worthwhile examining whether the *OsRDR1* gene is required for virus-mediated RNA silencing by particle co-bombardment. There are no appropriate viral vectors that infect rice plants available; thus, we selected WDV (ssDNA) and BMV (ssRNA), both of which are reported to infect rice protoplasts (Huntley and Hall 1996; Laufs et al. 1990). Here, experiments with this co-expression system were performed to detect RNA silencing with pIG121 and pWI-GUS RNAi constructed as described in the materials and methods. Specific silencing of *GUS* was observed after co-bombardment with mixture (v) including plasmid pWI-GUS RNAi in the WT and the mutant line (ND2001) (Figure 3a). This result indicates that *OsRDR1* is not required for WDV-mediated RNA silencing. To confirm bombardment efficiency, we also examined accumulation of *GUS* transcripts versus *DsRed* transcripts (as an internal control) in the leaf discs from the WT and the mutant line at 48 h post-bombardment with the mixture of those plasmids by real-time RT-PCR (Figure 3c). These results indicate that the levels of *GUS* transcript in the leaf discs bombarded were accordant with the levels *GUS*-expressing cells (Figure 3a). To check the possibility that WDV interferes with the transcription of *GUS*, an additional experiment has been done with pWI-GUS instead of pIG121. Figure S2a in ESM reveals that both the WT and mutant line (ND2001) expressed *GUS* (WDV driven) when bombarded with plasmid mixture (i) pWI-GUS and control DNA(pBIΔGUS). With mixture (iii) pWI-GUS and pWI-GUS RNAi, the number of cells that expressed *GUS* decreased markedly in the WT and in the mutant line, similar to the results shown in Figure 3a (v). Furthermore, Figure S2b (i) indicates that *DsRed* expression remained at a stable level as with *GUS* (WDV driven) expression. All these results showed that the WDV did not affect expression of *GUS* or *DsRed* gene.

We also examined whether BMV affected RNA silencing in the mutant line in this system. For RNA silencing induction, we used BMV-GFP RNAi, in vitro transcripts from BMV cDNA clones. These cDNA clones were the same as reported previously for RNA1 and 2 (Mori et al. 1993). For RNA3, pBTFCP2hpGFP2 was constructed to carry IR-GFP sequence in place of the 5′-half of coat protein gene (CP). When mixture (i) or (ii) in Figure 4 was used for bombardment, respectively, again the same results as in Figure 3 were obtained, showing that RNA silencing of *GFP* was impaired in the mutant line (Figure 4a). The number of GFP- or DsRed-fluorescing expressing cells was also counted for quantitative analysis from three independent experiments ($n = 3$ shots). With p35S-OsRDR1 in mixture (iii), which contains a whole ORF-coding region of *OsRDR1* cDNA, specific silencing of *GFP* was distinctly recovered in the mutant line, with the number of GFP-fluorescing cells decreasing as markedly in the mutant line as in the WT (Figure 4a). When using mixture (iv), which contained BMV-GFP RNAi instead of *GFP* RNAi, there was an obvious decrease in the number of GFP-fluorescing cells in the WT but not in the mutant line, while the number of DsRed-fluorescing cells remained constant in all treatments. To verify bombardment efficiency, we also used real-time RT-PCR to compare the accumulation of *GFP* transcripts to that of *DsRed* transcripts in leaf discs from the WT and the mutant line at 36 h post-bombardment with the various mixtures (Figure 4b). Similar results were obtained with mutant line ND2059 (data no shown). These results showed that *OsRDR1* was required for BMV-mediated RNA silencing.

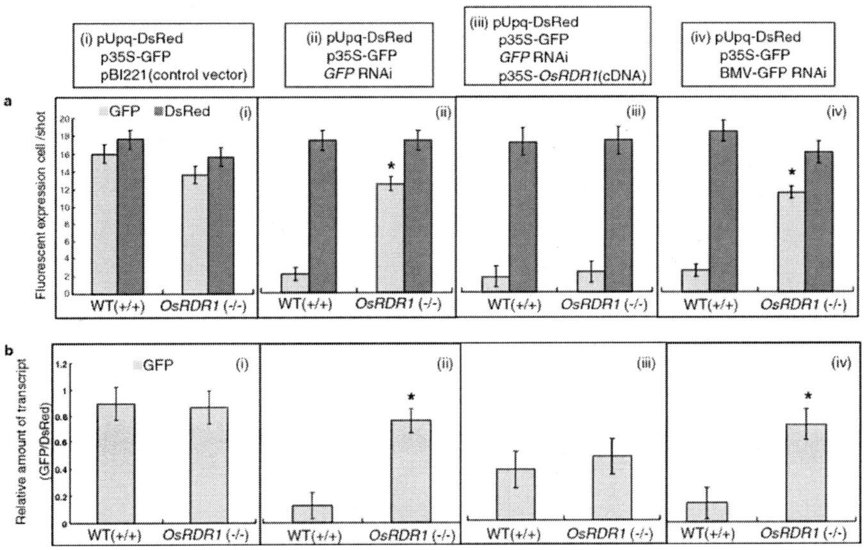

Figure 4. Effect on RNA silencing mediated by RNA virus in rice leaf discs of wild-type (*WT*) and mutant line after cobombardment with different mixtures of plasmids. **A** Quantitative analysis of GFP-fluorescing cells per shot. Data are mean ± SE ($n = 3$). **B** Real-time RT-PCR analysis of accumulation of *GFP* transcripts in the leaf discs at 36 h after bombardment. Data were normalized using *DsRed* transcripts as an internal control. Data are mean ± SE ($n = 3$). The relative amount of transcript (*GFP/DsRed*) is shown in the *vertical line*. *$P < 0.05$ (Fisher's test) versus value of WT

To make sure that both viruses are able to replicate in bombarded rice leaf discs, we checked for BMV and WDV replication at different times post-bombardment by real-time RT-PCR and PCR, respectively (Figure 5). The results show that BMV and WDV were able to replicate in the bombarded leaf discs of both WT and mutant rice.

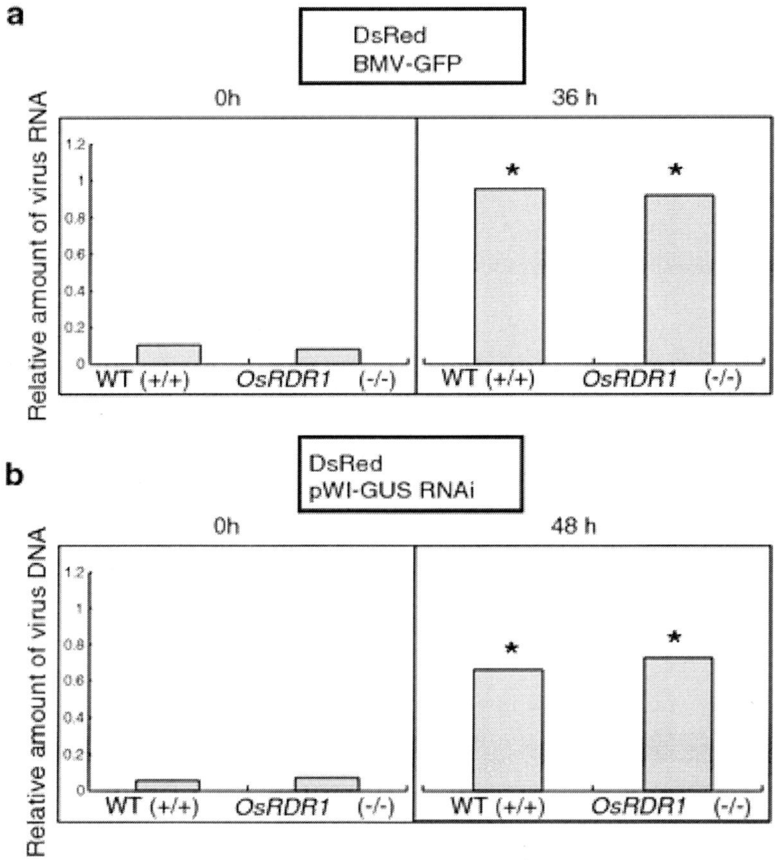

Figure 5. Analysis of BMV (BMV-GFP) (**A**) and WDV (pWI-GUS RNAi) (**B**) replication in bombarded rice leaf discs. Total RNA (**A**) or DNA (**B**) was extracted at 36 h (BMV) or 48 h (WDV) post-bombardment and used as a template for real time RT-PCR or PCR, respectively. Data were normalized using *DsRed* as an internal control. *$P < 0.05$ (Fisher's test) versus value at 0 h

DISCUSSION

In fungi, plants and several animals, RDRs are an independent class of enzymes with an important role in RNA silencing, heterochromatin formation and natural gene regulation. Previous findings show that *RDR6*, *SDE3*, *SGS3* and *AGO1* are required for sense transgene-mediated PTGS in *Arabidopsis*, whereas these genes are not involved in IR transgene-mediated PTGS (Dalmay et al. 2000; Diaz-Pendon et al. 2007;

Mourrain et al. 2000; Wassenegger and Krczal 2006). In our research, RNA silencing was induced not only by IR-DNAs but also by dsRNA in the WT. However, RNA silencing was impaired by introduction of either IR-DNA or dsRNA in the *OsRDR1*-deficient mutant lines. These results suggest that *OsRDR1* is required at the RNA level after transcription for RNA silencing.

Previous studies show that *RDR1* is induced by SA treatment, virus and viroid infection (Dorssers et al. 1984; Khan et al. 1986; Wassenegger and Krczal 2006). The loss-of-function in the *RDR1* mutant was expressed as enhanced disease susceptibility to various viruses (Xie et al. 2001; Yu et al. 2003). In *Arabidopsis*, the accumulation of TRV RNA in the *RDR1* mutant was significantly higher than in the WT, and disease symptoms in the mutant appeared to be more severe than in the WT (Yu et al. 2003). In *N. tabacum*, a *potato virus X* strain could spread and cause a systemic infection in the *RDR1* mutant line but not in the WT (Xie et al. 2001). Most recently, the finding revealed that the biogenesis of CMV siRNAs was found to be largely dependent on *RDR1* in plants infected with CMV-Δ2b, and genes such as *SGS3*, *SDE3* and *AGO1* might also be essential for *RDR1*-dependent silencing amplification in relation to the production of second siRNAs of CMV (Diaz-Pendon et al. 2007).

In our study, the results from the BMV-GFP RNAi experiments show that *OsRDR1* might be attributed to the amplification level of silencing related to the production of second BMV siRNA (Figure 4a, b). However, WDV-mediated RNA silencing was not impaired in the mutant lines (Figure 3a, b), possibly because geminivirus transcription depends on host-DNA-dependent RNA polymerase (Hanley-Bowdoin et al. 2000). VIGS by geminivirus might be amplified by other host *RDR*s such as *RDR2* and *RDR6* (Muangsan et al. 2004). Geminivirus DNA was suggested to be methylated by the RNA-directed DNA methylation pathway requiring *PolIVa*, *RDR2*, *DCL3* and *AGO4* in *Arabidopsis* (Donaire et al. 2008). Thus, WDV in this study is also presumed to be affected in rice by *OsRDR2* instead of *OsRDR1*.

In this RNA-silencing system, we have no data on siRNA, a hallmark of RNA silencing. After co-bombarded with the mixture containing BMV-GFP, *GFP* RNAi and pUbqDsRed, however, BMV-GFP was clearly reduced (data not shown) like the result shown in Figure 4, mixture (ii). This finding confirmed that RNA silencing functioned in this system as reported by English et al. (1996). Thus, the siRNA must be produced in each bombarded cell although it is difficult to check at the single cell level. The siRNA could be detected using protoplasts as shown by Qi et al. (2004).

Although rice is not a host for either BMV or WDV (http://www.agls.uidaho.edu/ebi/vdie/sppindex.htm), in some reports, each of these viruses infected and replicated in rice protoplasts (Huntley and Hall 1996; Laufs et al. 1990). Our results in Figure 5 showed that each virus could at least replicate in the leaf discs after co-bombardment. In the BMV vector used, the 5′-half of CP was replaced by the GFP sequence, which might possibly leave the virus unable to move from cell to cell (Schmitz and Rao 1996). In the WDV vector, the region covering the CP and part of the movement protein genes was replaced by a selectable marker (NPTII) and *GUS* (Ugaki et al. 1991). Thus, its cell-to-cell movement would be impaired. In Figure S2a (i) experiment in ESM, only a single cell expressed *GUS* (data not shown), suggesting that the virus did not spread from cell to cell. Similarly, with BMV vector, *GFP* was expressed in only one cell (data not shown). The spread *GUS/GFP* expression after co-bombardment with the wild-type virus would also be worth checking.

ACKNOWLEDGMENTS

The authors wish to thank K. Shimamoto and D. Miki for plasmids pUbq-DsRed, p35S-GFP and *GFP* RNAi and K. Nakamura for plasmid pIG121. They are also grateful to D. Murphy for his English correction. This work was supported by the Ministry of Agriculture, Forestry and Fisheries of Japan (Green Technology Project, MP2137). Scholarship for H. C. was supported by the Ministry of Education, Culture, Sports, Science and Technology of Japan.

REFERENCES

1. Astier-Manifacier S, Cornuet P (1971) RNA-dependent RNA polymerase in Chinese cabbage. Biochim Biophys Acta 232:484–493
2. Chen H, Samadder PP, Tanaka Y, Ohira T, Okuizumi H, Yamaoka N, Miyao A, Hirochika H, Ohira T, Tsuchimoto S, Ohtsubo H, Nishiguchi M (2008) OsRecQ1, a QDE-3 homologue in rice, is required for RNA silencing induced by particle bombardment for inverted repeat DNA, but not for double-stranded RNA. Plant J 56:274–286
3. Curaba J, Chen X (2008) Biochemical activities of Arabidopsis RNA-dependent RNA polymerase 6. J Biol Chem 283:3059–3066
4. Dalmay T, Hamilton A, Rudd S, Angell S, Baulcombe DC (2000) An RNA-dependent RNA polymerase gene in Arabidopsis is required for posttranscriptional gene silencing mediated by a transgene but not by a virus. Cell 101:543–553

5. Diaz-Pendon JA, Li F, Li WX, Ding SW (2007) Suppression of antiviral silencing by cucumber mosaic virus 2b protein in Arabidopsis is associated with drastically reduced accumulation of three classes of viral small interfering RNAs. Plant Cell 19:2053–2063
6. Donaire L, Barajas D, Martínez-García B, Martínez-Priego L, Pagán I, Llave C (2008) Structural and genetic requirements for the biogenesis of Tobacco rattle virus-derived small interfering RNAs. J Virol 82:5167–5177
7. Dorssers L, van der Krol S, van der Meer J, van Kammen A, Zabel P (1984) Purification of cowpea mosaic virus RNA replication complex: identification of a virus-encoded 110,000-dalton polypeptide responsible for RNA chain elongation. Proc Natl Acad Sci USA 81:1951–1955
8. Engler-Blum G, Meier M, Frank J, Müller GA (1993) Reduction of background problems in nonradioactive northern and Southern blot analyses enables higher sensitivity than 32P-based hybridizations. Anal Biochem 210:235–244
9. English JJ, Mueller E, Baulcombe DC (1996) Suppression of virus accumulation in transgenic plants exhibiting silencing of nuclear genes. Plant Cell 8:179–188
10. Hanley-Bowdoin L, Settlage SB, Orozco BM, Nagar S, Robertson D (2000) Geminiviruses: models for plant DNA replication, transcription, and cell cycle regulation. Crit Rev Biochem Mol Biol 35:105–140
11. Hirochika H (2001) Contribution of the Tos17 retrotransposon to rice functional genomics. Curr Opin Plant Biol 4:118–122
12. Huntley CC, Hall TC (1996) Interference with brome mosaic virus replication in transgenic rice. Mol Plant Microbe Interact 9:164–170
13. Khan ZA, Hiriyanna KT, Chavez F, Fraenkel-Conrat H (1986) RNA-directed RNA polymerases from healthy and from virus-infected cucumber. Proc Natl Acad Sci USA 83:2383–2386
14. Laufs J, Wirtz U, Kammann M, Matzeit V, Schaefer S, Schell J, Czernilofsky AP, Baker B, Gronenborn B (1990) Wheat dwarf virus Ac/Ds vectors: expression and excision of transposable elements introduced into various cereals by a viral replicon. Proc Natl Acad Sci USA 87:7752–7756
15. Makeyev EV, Bamford DH (2002) Cellular RNA-dependent RNA polymerase involved in posttranscriptional gene silencing has two distinct activity modes. Mol Cell 10:1417–1427
16. Miki D, Shimamoto K (2004) Simple RNAi vectors for stable and transient suppression of gene function in rice. Plant Cell Physiol 45:490–495
17. Mori M, Zhang GH, Kaido M, Okuno T, Furusawa I (1993) Efficient production of human gamma interferon in tobacco protoplasts by genetically engineered brome mosaic virus RNAs. J Gen Virol 74:1255–1260
18. Mourrain P, Béclin C, Elmayan T, Feuerbach F, Godon C, Morel JB, Jouette D, Lacombe AM, Nikic S, Picault N, Rémoué K, Sanial M, Vo TA, Vaucheret H (2000) Arabidopsis SGS2 and SGS3 genes are required for posttranscriptional gene silencing and natural virus resistance. Cell 101:533–542
19. Muangsan N, Beclin C, Vaucheret H, Robertson D (2004) Geminivirus VIGS of endogenous genes requires SGS2/SDE1 and SGS3 and defines a new branch in the genetic pathway for silencing in plants. Plant J 38:1004–1014
20. Ohta S, Mita S, Hattori T, Nakamura K (1990) Construction and expression in tobacco of a β-glucuronidase (GUS) reporter gene containing an intron within the coding sequence. Plant Cell Physiol 31:805–813
21. Pandey SP, Baldwin IT (2007) RNA-directed RNA polymerase 1 (RdR1) mediates the resistance of Nicotiana attenuata to herbivore attack in nature. Plant J 50:40–53
22. Pandey SP, Baldwin IT (2008) Silencing RNA-directed RNA polymerase 2 increases the susceptibility of Nicotiana attenuata to UV in the field and in the glasshouse. Plant J 54:845–862
23. Qi Y, Zhong X, Itaya A, Ding B (2004) Dissecting RNA silencing in protoplasts uncovers novel effects of viral suppressors on the silencing pathway at the cellular level. Nucleic Acids Res 32:e179

24. Schiebel W, Haas B, Marinković S, Klanner A, Sänger HL (1993) RNA-directed RNA polymerase from tomato leaves. II. Catalytic in vitro properties. J Biol Chem 268:11858–11867
25. Schmitz I, Rao ALN (1996) Molecular studies on bromovirus capsid protein. I. Characterization of cell-to-cell movement-defective RNA3 variants of brome mosaic virus. Virology 226:281–293
26. Tamai A, Kubota K, Nagano H, Yoshii M, Ishikawa M, Mise K, Meshi T (2003) Cucumovirus- and bromovirus-encoded movement functions potentiate cell-to-cell movement of tobamo- and potexviruses. Virology 315:56–67
27. Ugaki M, Ueda T, Timmermans MCP, Vieira J, Elliston KO, Messing J (1991) Replication of a geminivirus derived shuttle vector in maize endosperm cells. Nucleic Acids Res 19:371–377
28. Vaucheret H (2006) Post-transcriptional small RNA pathways in plants: mechanisms and regulations. Genes Dev 20:759–771
29. Voinnet O (2005) Induction and suppression of RNA silencing: insights from viral infections. Nat Rev Genet 6:206–220
30. Voinnet O (2008) Use, tolerance and avoidance of amplified RNA silencing by plants. Trends Plant Sci 13:317–328
31. Wassenegger M, Krczal G (2006) Nomenclature and functions of RNA-directed RNA polymerases. Trends Plant Sci 11:142–151
32. Xie Z, Fan B, Chen C, Chen Z (2001) An important role of an inducible RNA-dependent RNA polymerase in plant antiviral defense. Proc Natl Acad Sci USA 98:6516–6521
33. Xie Z, Johansen LK, Gustafson AM, Kasschau KD, Lellis AD, Zilberman D, Jacobsen SE, Carrington JC (2004) Genetic and functional diversification of small RNA pathways in plants. PLoS Biol 2:e104
34. Yang JH, Seo HH, Han SJ, Yoon EK, Yang MS, Lee WS (2008) Phytohormone abscisic acid control RNA-dependent RNA polymerase 6 gene expression and post-transcriptional gene silencing in rice cells. Nucleic Acids Res 36:1220–1226
35. Yu D, Fan B, MacFarlane SA, Chen Z (2003) Analysis of the involvement of an inducible Arabidopsis RNA-dependent RNA polymerase in antiviral defense. Mol Plant Microbe Interact 16:206–216

CHPATER 16

Control of Carbon Assimilation and Partitioning by Jasmonate: An Accounting of Growth–Defense Tradeoffs

Nathan E. Havko [1,†], Ian T. Major [2,†], Jeremy B. Jewell [1], Elham Attaran [1], John Browse [1,*] and Gregg A. Howe [2,*]

[1]*Institute of Biological Chemistry, Washington State University, Pullman, WA 99164, USA*

[2]*Department of Energy-Plant Research Laboratory and Department of Biochemistry and Molecular Biology, Michigan State University, East Lansing, MI 48824, USA*

ABSTRACT

Plant growth is often constrained by the limited availability of resources in the microenvironment. Despite the continuous threat of attack from insect herbivores and pathogens, investment in defense represents a lost opportunity to expand photosynthetic capacity in leaves and absorption of nutrients and water by roots. To mitigate the metabolic expenditure on defense, plants have evolved inducible defense strategies. The plant hormone jasmonate (JA) is a key regulator of many inducible defenses. Synthesis of JA in response to perceived danger leads to the deployment of a variety of defensive structures and compounds, along with a potent inhibition of growth. Genetic studies have established an important role for JA in mediating tradeoffs between growth and defense. However, several gaps remain in understanding of how JA signaling inhibits growth, either through direct transcriptional control of JA-response genes or crosstalk with other signaling pathways. Here, we highlight recent progress in uncovering the role of JA in controlling growth-defense balance and its relationship to resource acquisition and allocation. We also discuss tradeoffs in the context of the ability of JA to promote increased leaf mass per area (LMA), which is a key indicator of leaf construction costs and leaf life span.

Keywords: jasmonate; growth–defense tradeoff; defense response; growth inhibition; leaf mass per area; carbon partitioning

1. INTRODUCTION

As the primary photosynthetic organ in plants, leaves are the major source of reduced carbon skeletons that fuel the biosynthesis of energy-rich macromolecules. However, green leaves in general and protein-rich mesophyll cells in particular, are also a nutritional wellspring for arthropod herbivores and pathogens whose life cycles depend on plants as a food source. Plants effectively counter these biotic threats through constitutive or induced synthesis of myriad defense compounds, as well as development of specialized cells in which these compounds are produced, stored and secreted [1]. The expression of defense traits may exact allocation costs that negatively impact plant growth, particularly when resources such as light and nitrogen are limiting in the environment. Diversion of resources to defense at the expense of construction of new leaf tissue, which otherwise returns revenue in the form of photosynthate over the lifetime of the organ, also imposes so-called opportunity costs that are compounded as plants age [2,3]. These direct costs of defense may also impose ecological costs, including reduced ability to compete with neighboring plants or increased susceptibility to enemies that are not targeted by the induced defense [4,5]. The allocation of limited resources to plant growth and defense has therefore been described as a "dilemma" because growth rates must be sufficient to compete with neighboring plants for light capture while not neglecting investment in defense against plant-eating organisms [3].

Among the biochemical components that must be budgeted between growth and defense are carbon skeletons, ATP and reducing equivalents from photosynthesis, as well as assimilated nitrogen, phosphorous, sulfur, and trace nutrients. Investment of these resources into growth increases the area of resource-acquiring tissues (*i.e.*, photosynthetic structures and roots) and thus accelerates growth capacity. Conversely, investment in defense reduces the ability of herbivores and pathogens to destroy tissues that are needed for resource acquisition. To mitigate the cost of defense on reproductive fitness, plants have evolved inducible defenses, which are deployed upon perception of attack [6]. The plant hormone jasmonate (JA) has become recognized as a chief mediator of inducible defenses. JA is synthesized within minutes of the perception of a threat and exerts transcriptional control over thousands of genes to effect resistance to herbivory and disease [7,8,9].

The expression of defense traits in response to increased JA levels is accompanied by a potent inhibition of growth [10,11]. This has been demonstrated through the use of mutants that lack the ability to produce or perceive JA; such mutants are not only susceptible to biotic aggressors but also show reduced ability to respond to stress cues with growth arrest. In the last decade, much progress has been made toward understanding the core components of JA signaling and the way in which they are integrated into the wider hormone response network [12,13,14]. Despite these advances, however, our understanding of how JA signaling exerts control over growth, either through crosstalk with other growth-signaling pathways or as a consequence of resource allocation to defense (Figure 1), is still in its infancy. In the simplest scenario, carbon and energy diverted from growth would provide for a stoichiometric increase in defense compounds. However, as discussed below, there is a net loss in dry-weight gain within a day or so of the induction of defense signaling, suggesting a more complex scenario. Here, we discuss current opinions regarding the role of JA in governing metabolic tradeoffs between growth and defense, with particular emphasis on how JA modulates resource acquisition and allocation to achieve growth-defense balance.

2. JASMONATE IS CENTRAL TO THE GROWTH-DEFENSE TRADEOFF

Jasmonate was first identified more than half a century ago as a component of the essential oil of jasmine flowers [15]. Decades later, it was recognized that JA serves a physiological role in growth inhibition of a wide variety of tissues in both monocot and dicot plants [16,17,18,19,20]. The ability of exogenous JA to inhibit root elongation in a dose-dependent manner is now widely used as a simple quantitative bioassay of JA sensitivity [21,22,23]. A role for JA in defense was recognized by Farmer and Ryan, who demonstrated that airborne JA (JA methyl ester) induces the expression of proteinase inhibitors that contribute to resistance to herbivory [24]. Pioneering work in Zenk's laboratory further showed that JA is a potent elicitor of defense-related secondary metabolites in suspension-cultured plant cells [25]. Early genetic studies established that JA biosynthesis is essential for insect defense in both tomato and Arabidopsis [26,27]. Subsequently, numerous studies have confirmed the central role of JA in plant defense against a variety of herbivores and necrotrophic pathogens [9,12,28,29].

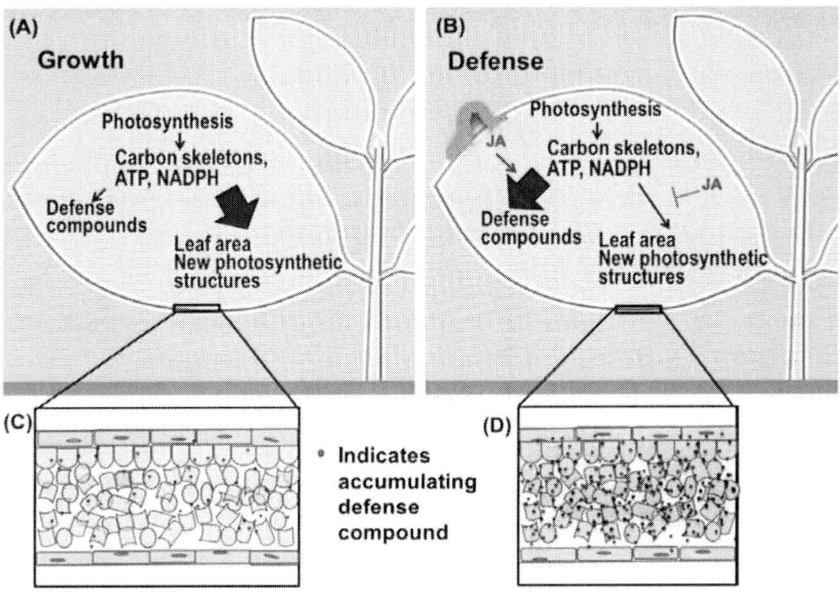

Figure 1. Jasmonate (JA) induces the re-budgeting of resources from tissue expansion to the production of defense compounds. (**A**) Plant growth is achieved using carbon skeletons, ATP, and NADPH from photosynthesis; (**B**) During the JA-mediated defense response, carbon skeletons, ATP, and NADPH that could otherwise contribute to growth are used for *de novo* synthesis of defense compounds; (**C**) In the absence of biotic attack, defense compounds are produced at a low basal level; (**D**) JA triggers the accumulation of defense compounds accompanied by an arrest of tissue expansion. In defended leaf tissue, cell size is similar to that in undefended leaves but the leaf mass per area (LMA) may increase as a consequence of increase carbon deposition into defense compounds.

The dual roles of JA in growth inhibition and defense have mostly been studied as independent phenomena but, in light of the tradeoffs mentioned above, there is a need to understand how these two processes are integrated. Recent work is beginning to uncover mechanisms that tip the growth-defense balance through crosstalk between JA signaling and other hormone response pathways [14,30,31]. A particularly important node of crosstalk is signal antagonism between JA and the growth-promoting gibberellin (GA) pathway (Figure 2). The core JA signaling module includes the JAZ transcriptional repressors that inhibit JA responses by blocking the activity of transcription factors, such as MYC2. The bioactive form of JA, jasmonoyl-L-isoleucine (JA-Ile), promotes the interaction between JAZ proteins and the CORONATINE INSENSITIVE 1 (COI1)-containing Skp1/Cullin/F-box (SCF^{COI1}) E3 ubiquitin ligase complex, leading to the ubiquitination and 26S proteasome-dependent degradation of JAZs [32,33,34]. Mutants lacking a functional COI1 are compromised in all known JA-regulated processes and do not respond to the effects of the exogenous hormone [35,36,37]. The JAZ proteins modulate growth responses, at least in part, through direct interaction with the DELLA repressors of the GA pathway (Figure 2). Current evidence indicates that JAZ-DELLA interactions competitively inhibit the ability of each repressor to exert repression on cognate transcription factors [38,39,40,41]. For example, GA-triggered degradation of DELLA proteins by the GID1-GA-SCF^{SLY1} complex frees JAZ and results in greater inhibition of MYC transcription factors and, consequently, repression of defense responses. JAZ-DELLA interactions also favor growth by interfering with the ability of DELLAs to repress PHYTOCHROME-INTERACTING FACTOR (PIF) transcription factors that promote growth responses (Figure 2) [42]. As discussed below, this conserved mechanism of growth-defense balance is modulated by the photoreceptor phytochrome B (phyB) in response to changes in light quality. These collective findings highlight the importance of JAZ-DELLA coupling in transcriptional networks that tune the delicate balance between growth and defense.

In addition to crosstalk between the JA and GA pathways, other hormones have received attention for modulating the effect of JA on growth. Evidence suggests that part of the growth repression arm of JA signaling involves antagonism between JA and the growth-promoting brassinosteroids. A leaky mutation in *DWARF4* (*DWF4*), a gene encoding a key enzyme in brassinosteroid biosynthesis, partially restores the sensitivity of the hypomorphic *coi1-2* allele to the growth inhibitory effects of exogenous JA [43]. Moreover, JA treatment represses *DWF4* transcript accumulation in a *COI1*-dependant manner. Together these results suggest that the growth repression limb of the JA pathway functions in part by blocking brassinosteroid production and downstream activation of cell expansion, a possibility supported by the identification of *DWF4* over-expressing Arabidopsis lines in which root sensitivity to JA is decreased [44]. Growth inhibition by JA may also involve repression of the cytokinin signaling pathway, which is a major positive regulator of cell division activity [17,45].

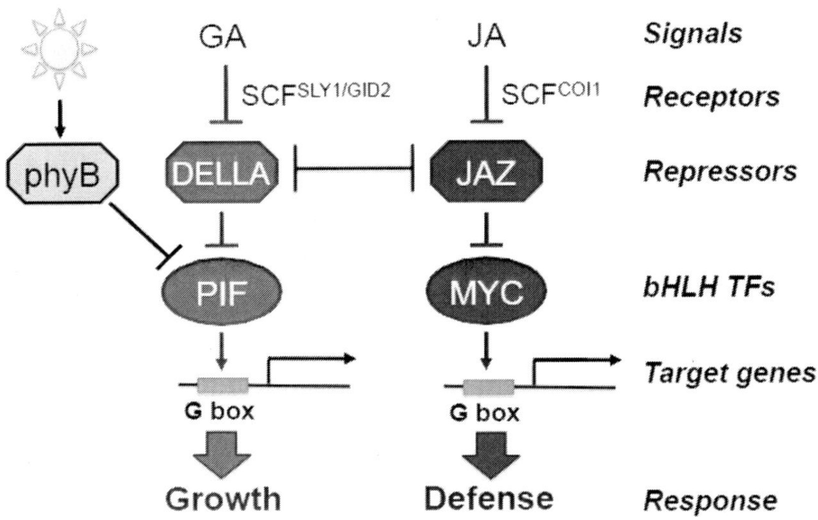

Figure 2. Simplified model depicting interactions between the JA, GA, and phyB signaling pathways. Points of positive and negative regulation are indicated by arrows and perpendicular lines, respectively. bHLH-TFs, basic helix-loop-helix transcription factors that bind G-box *cis*-regulatory elements typically located in the promoter region of response genes. In full sunlight, PIF transcription factors are inhibited by the active conformer of phyB. See text for details.

3. GROWTH INHIBITION: CELL ELONGATION OR CELL DIVISION?

Observations of JA-treated plants suggest that JA signaling interferes with processes that promote growth through both cell division and cell elongation. In Arabidopsis leaves, wound-induced JA biosynthesis resulted in a 50% decrease in the calculated number of palisade mesophyll cells per leaf without noticeable effects on the cell size, suggesting that JA inhibits leaf expansion primarily through cell cycle arrest [10]. This finding was supported by decreased activity of a cell-cycle reporter, *CycB1;2pro-GUS*, and is consistent with earlier work with tobacco suspension cultured cells in which exogenous JA arrested the cell cycle in the G2 phase [46,47]. An independent study of Arabidopsis leaves at various developmental stages also showed that JA-induced growth inhibition is associated with a strong reduction in cell number. However, delayed cell expansion also contributed to growth inhibition in this experimental system [48].

A number of studies have demonstrated JA-dependent inhibition of Arabidopsis hypocotyl elongation [49,50,51] and rice internode elongation [39]. In studies where cell length was determined, the JA-dependent reduction in cell length could account for differences in internode elongation [39]. Given that extension of Arabidopsis hypocotyls results primarily from cell elongation rather than division [52,53,54], it is likely that hypocotyl growth inhibition by JA is caused by a reduction in cell elongation as well.

Primary root growth inhibition appears to result from a combined reduction in both cell division and cell elongation. In Arabidopsis and rice, exogenous JA inhibits root growth via decreased cell elongation and cell division, although at low JA concentrations decreased cell elongation appears to be primarily responsible for growth inhibition [55,56]. Repeated wounding of Arabidopsis cotyledons triggered a JA-dependent shoot-to-root signal that also reduced root growth by decreasing cell elongation and cell division in the root differentiation zone [57]. Interestingly, loss-of-function mutations in the gene encoding the JA co-repressor NINJA result in reduced cell elongation in the differentiation zone, but not reduced cell division [58]. In summary, JA-dependent growth inhibition occurs through repression of both cell division and elongation, with the relative contribution of each depending on the tissue type.

4. MODULATION OF CARBON ASSIMILATION BY JASMONATE

Central to understanding the mechanistic basis of growth–defense tradeoffs is the question of how changes in endogenous JA levels modulate carbon assimilation and subsequent partitioning to various metabolic sinks. The effects of arthropod herbivore attack, simulated herbivory, and exogenous JA on photosynthesis have been examined in a variety of experimental systems, with mixed results. This variability likely reflects differences in damage type and intensity, location and type of tissue damaged, plant developmental stage, and environmental conditions [59]. A preponderance of studies report that treatments associated with activation of JA signaling reduce photosynthesis [59]. One interpretation of downregulation of growth and photosynthetic capacity in herbivore-damaged leaves is that these responses are part of the plant's defense strategy to limit the availability of food for herbivores [60].

However, robust plant defense responses also depend on a continuous supply of photosynthetic products to fuel local and systemic biosynthesis of defense compounds. Reduction in photosynthetic capacity by silencing of RuBisCO activase in coyote tobacco, for example, attenuated JA-mediated resistance to herbivory [61]. Evidence for positive or compensatory effects further suggests that foliage loss and induced expression of defenses can drive increased photosynthesis in remaining and new leaves, perhaps through altered source-sink relationships [62,63]. These observations are in general agreement with metabolic labeling studies showing that simulated herbivory and exogenous JA stimulate carbon partitioning to belowground tissues [64,65,66]. Zangerl and coworkers [67] reported a positive correlation between herbivore-induced furanocoumarin synthesis and the amount of leaf area having a reduced photosynthetic efficiency, suggesting that the production of this defense compound is relatively insensitive to herbivore-induced reduction in photosynthesis. This study also introduced the use of chlorophyll fluorescence imaging to spatially map the photosynthetic efficiency of whole leaves subjected to herbivory. This approach demonstrated decreases in photosynthesis in damaged tissue at the insect bite zone, which may result from water loss, as well as in undamaged regions of the leaf located a considerable distance from the wound site [59,67].

Downregulation in the expression of photosynthetic genes and, in some cases the corresponding proteins, in response to herbivory or JA treatment was initially observed in barley and since then reported in many other plant species [61,68,69,70,71,72]. Given that photosynthesis-related transcripts and proteins are highly abundant in leaves, this reprogramming of transcription may provide metabolic building blocks needed to redirect the cell's biosynthetic capacity from growth to defense without short-term negative effects on photosynthesis [70,71]. A recent study by Attaran *et al.* [71] showed that treatment of Arabidopsis with coronatine (COR), a potent agonist of the JA receptor, triggered a strong reduction of photosynthesis-associated gene expression without an accompanying decrease in photochemical efficiency. This experimental system allows for strong and rapid activation of the JA pathway without the complicating effects of tissue damage that accompany herbivory or other wound treatments. In summary, the reported effects of JA on multiple aspects of photosynthesis make it an important consideration for understanding the physiological basis of growth–defense tradeoffs.

5. JASMONATE SIGNALING PROMOTES INCREASED LEAF MASS PER AREA (LMA)

Whereas the effects of JA on photosynthesis have been well-examined, less is understood about how JA modulates the partitioning of fixed carbon to drive various aspects of growth, including leaf area expansion, leaf thickening, petiole elongation, and the construction of new leaves. Modeling of photosynthetic rate and carbon partitioning in relation to growth showed that carbon is partitioned to leaf area expansion and biomass at different rates, and that the relationship between leaf area and biomass is important to consider since leaf area determines light interception [73]. In particular, this study showed that partitioning of carbon to leaf thickness impacts growth more strongly than photosynthesis. Although it is known that activation of the JA defense response reduces leaf expansion [26,71] and biomass accumulation [10,11,49,50], we are not aware of studies that examined how JA affects growth rates of both leaf area and biomass.

Our early experiments showed that COR slows leaf area growth almost immediately after treatment (Figure 3A) [71]. COR treatment also arrested rosette diameter expansion (Figure 3B). This rapid effect partly reflects an arrest of petiole elongation and is consistent with the ability of JA to suppress cell elongation [39]. Leaf area and rosette expansion were arrested for several days before increasing again, though at an apparent reduced rate. The resumption in expansion appeared to be due to leaves that emerged after treatment, and which perhaps are less sensitive to the COR treatment. By comparison, dry mass accumulation remained similar to mock-treated plants for at least one day after treatment (Figure 3C). The absence of a corresponding, immediate reduction in dry mass accumulation supports our previous conclusion that activation of JA signaling does not have an immediate repressive effect on photosynthesis in Arabidopsis leaves [71]. These findings are consistent with the idea that reduced leaf area growth in response to JA results in opportunity costs as a consequence of lower total light interception and carbon assimilation on a per plant basis. However, that JA-elicited plants can "pay back" such costs in ways that enhance fitness in the presence of herbivores [74] suggests that a more in-depth analysis of the effects of JA on photosynthetic and non-photosynthetic leaf traits is warranted.

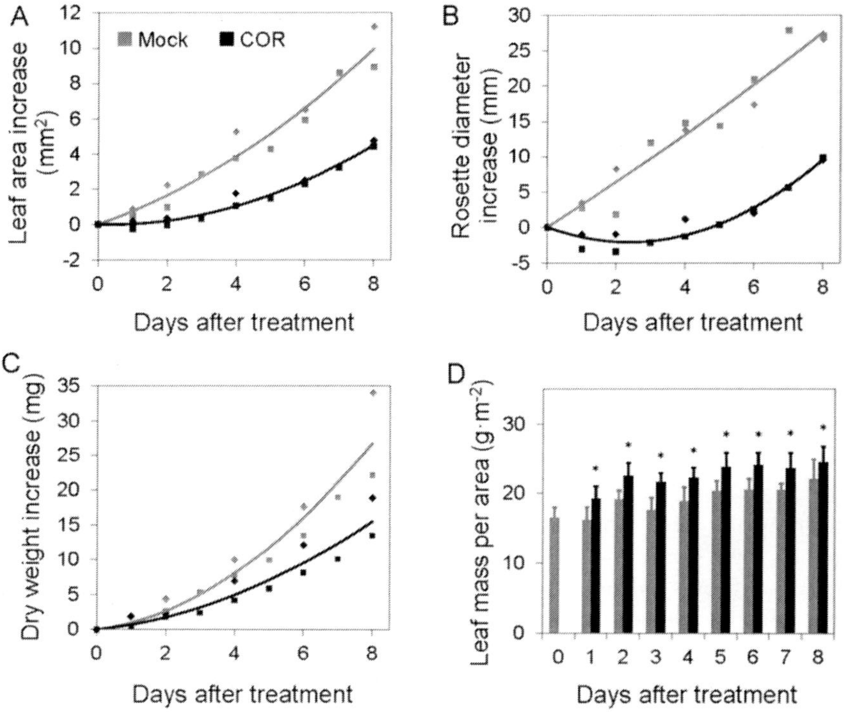

Figure 3. Jasmonate-mediated suppression of leaf area and biomass growth in *Arabidopsis thaliana* plants. Soil-grown Columbia-0 plants treated with mock (grey) or with 5 μM coronatine (COR, black) solutions [71] were measured for leaf area (**A**), rosette diameter (**B**), and dry weight (**C**) at indicated times after treatment. Data are the mean increase in growth (n = 12 plants) relative to the day of treatment (day 0) from two independent experiments (diamonds and squares distinguish experiments) and lines are second-order polynomial regressions of combined data from both experiments. Projected leaf area and rosette diameter were determined from overhead images, and dry weight was determined from rosettes (without roots) freeze-dried in a lyophilizer. In (**D**), leaf mass per area (LMA) was calculated from dry weight/leaf area from one experiment in A and C (square points). Error bars are standard deviation and asterisks indicate $P < 0.05$ between mock and COR treatment from a Student's t-Test.

The effects of JA on altered partitioning of resources to leaf area and biomass can be further examined from the ratio of dry mass to leaf area (Leaf Mass per Area, LMA). LMA is useful to describe the resources invested in a given area of leaf and may change as a consequence of variability in leaf thickness, leaf density, or both [75]. LMA has been widely used in studies of plant ecology and agronomy to describe different plant strategies for optimizing fitness to particular environments, as well as for analysis of chemical, physiological, and structural traits that together define the leaf economics spectrum [76]. In general, plants with low LMA tend to acquire resources rapidly and have higher growth rates, whereas plants with high LMA tend to conserve resources and have lower growth rates, but have higher fitness in stressful environments [75]. With their increased resource investment, high-LMA leaves tend to be more durable (e.g., as a consequence of increased thickness), and in terms of defense, are less preferred by herbivores to leaves with low LMA [75]. Thus, it would appear that leaf traits associated with increased LMA have striking similarity to known effects of JA on leaf development and function.

In the experiment described in Figure 3, LMA was significantly increased the day after treatment with COR, and remained significantly higher for the duration of the experiment (Figure 3D). Following day 1, the average difference in LMA between mock- and COR-treated leaves at each time point remained remarkably constant at approximately 3 g/m^2; this plateau effect may be a consequence of negative feedback loops that rapidly restrain JA responses [9]. One interpretation of the higher LMA of COR-treated leaves is that it reflects a concerted change in strategy from rapid leaf area-based growth to protection of energy acquiring-foliage through investment of resources in leaf defense and structural integrity. Thus, the opportunity costs associated with JA-mediated inhibition of leaf area growth may be counterbalanced by other features of high-LMA leaves, such as increased leaf life span [75]. These areas deserve future attention to better understand the physiological determinants of JA-mediated growth suppression and its relationship to solar energy capture, conversion, and deposition. Studies employing mutants that are impaired in growth-defense balance may provide additional insight into this question. In this context, it is interesting to note that mutants (e.g., *phyB*) exhibiting rapid growth and compromised defense have reduced LMA, whereas mutants with reduced responsiveness to GA grow slowly and have increased LMA [38,75,77]. These observations support the emerging paradigm that crosstalk between the JA and GA signaling pathways plays a significant role in controlling the nexus between leaf growth and defense. Further study of how JA-mediated changes in leaf chemistry are integrated with plasticity in leaf architecture may help to unravel the complexities of growth-defense balance.

6. AUDITING THE GROWTH-DEFENSE BUDGET

A comprehensive, quantitative understanding of the metabolic costs associated with partitioning of carbon into growth and defense pathways is still lacking. Efforts to meet this challenge should recognize that the growth-defense dichotomy oversimplifies the complexities of carbon budgeting because many other physiological processes, including reproduction, respiration, exudation, and volatile emission also consume photoassimilates [73]. Proximal analysis provides one approach to estimate the relative cost of producing defense metabolites from precursors in primary metabolism [78,79,80]. Also called detailed biochemical pathway analysis, proximal analysis estimates the energy required to produce biochemical constituents of a tissue, with the sum-total energy requirement for the synthesis, transport and maintenance of all compounds in a tissue equaling the tissue construction cost. This approach requires detailed information about tissue components and the biochemistry of their synthesis. Generally, more reduced secondary metabolites such as terpenoids have a higher metabolic cost than carbohydrate and phenolic metabolites. For example, production of one gram of the monoterpene camphor is calculated to cost 3.1 g glucose equivalents based on the requirements of acetyl-CoA, ATP and NADPH, whereas 1 g of phenol glycoside costs approximately 1.5 to 2 g glucose equivalents [78].

Flux balance analysis may also be used to estimate energy consumption required to deploy induced defenses. Flux balance analysis is performed by constructing, *in silico*, a metabolic network of all of the known or predicted biochemical conversions between compounds within a tissue. The network may be informed by measurements of biochemical activities and/or gene expression/predicted function. Glucosinolate production in Arabidopsis provides an example for which there is sufficient biochemical knowledge to undertake such an analysis [81]. Flux-balance modeling has estimated, for example, that production of glucosinolates in Arabidopsis increases the photosynthetic requirement by 15% [82].

Nitrogen accounts for a relatively small proportion of plant dry weight but is vitally important for plant growth and photosynthesis, and is often a limiting factor for growth in natural environments. Thus, JA-dependent changes in nitrogen assimilation or allocation are expected to have significant impacts on plant growth. Available data indicate that growth–defense tradeoffs involve not only the reallocation of reduced carbon but also substantial redistribution of nitrogen [3,5]. ^{15}N flux studies have leveraged nitrogen allocation to RuBisCO and small N-containing defense metabolites as a proxy for growth–defense tradeoffs [83,84]. In young rosette leaves of *Nicotiana attenuata*, simulated herbivory reduced nitrogen investment in RuBisCO and total soluble proteins by 89% [84]. Importantly, nitrogen investment in RuBisCO and total soluble protein was only reduced by 47% in transgenic plants with partially impaired JA biosynthesis. Thus, diversion of nitrogen in response to JA signaling may be a significant driver of growth inhibition.

Calorimetry offers a potentially informative yet simple approach for estimating the biochemical cost of defense. Combustion of plant tissue in a calorimeter gives a simple measure of the total energy stored within the system. Although not as informative as proximal analysis in assessing the contribution of individual compound classes to tissue construction costs, the two methods are typically in broad agreement [85,86]. Moreover, calorimetric estimates can be applied to whole plants. Although this approach has been used to assess altered carbon partitioning to lipid biosynthesis in Arabidopsis leaves [87], to our knowledge the method has not been applied to understanding growth–defense metabolic tradeoffs elicited by JA.

If increased LMA represents a significant component of JA-mediated changes in growth-defense balance as suggested by our preliminary studies (Figure 3D), it may be possible to use this trait as a starting point for quantitative analysis of the growth-defense budget. Variation in LMA as a function of genotype and environment often reflects differences in leaf anatomical traits [75]. Such features include mesophyll tissue volume, cellular volume occupied by the central vacuole and other organelles, volume of air spaces between cells, and leaf thickness. A second and complementary approach to explain variation in LMA is to deconstruct the trait into the chemical constituents of leaf biomass, including structural and non-structural carbohydrates, lignin, proteins, lipids, organic acids, minerals, and secondary metabolites. In-depth knowledge of central carbon metabolism and partitioning in leaves is expected to facilitate this approach [73,88], as is knowledge of how JA, and biotic stress in general, reconfigures primary and secondary metabolism [89].

7. GENETIC DISSECTION OF GROWTH-DEFENSE ANTAGONISM

Over the last twenty years, genetic investigations have revealed key steps in the synthesis and perception of JA in a variety of plant species [12,13]. Studies employing these mutants support the general conclusion that stress-induced JA biosynthesis leads to activation of transcriptional modules that simultaneously attenuate vegetative growth and heighten defense against a remarkable spectrum of plant-associated organisms, ranging from microbes to vertebrate herbivores [9]. More recently, molecular genetic investigations are beginning to untangle regulatory circuits through which JA governs growth and defense. The recent identification of mutants that disrupt either the growth or the defense arm of the JA response pathway suggests that the two processes involve interconnected but distinct regulatory networks.

Hu and coworkers [90] used an RNA interference (RNAi)-based screen to identify JAV1 as negative regulator of the defense limb of JA signaling. Silencing of *JAV1* dramatically reduced plant susceptibility to the necrotrophic fungus *Botrytis cinerea* and the lepidopteran herbivore *Spodoptera exigua*. Similar to JAZ repressors, JAV1 was degraded by a pathway that depends both on COI1 and the 26S proteasome. Unlike JAZs, however, JAV1 does not interact directly with COI1 in the presence of COR. The E3 ubiquitin ligase responsible for targeting JAV1 for destruction remains to be identified. Interestingly, *JAV1*-silenced plants had no obvious growth phenotypes despite their enhanced defense response [90]. These findings provide genetic evidence for distinct growth and defense branches of JA response pathway.

Additional evidence that JA-mediated growth and defense responses can be genetically separated comes from recent studies on the interaction between the JA and light signaling pathways that control shade avoidance responses. The shade avoidance syndrome is characterized by rapid growth of stems and petioles, which allows neighboring plants to compete for light [91]. Full sunlight has a high ratio of red (R) to far-red (FR) light, which maintains the phytochrome B (phyB) photoreceptor in its active Pfr form. In this state, phyB represses vegetative growth by inhibiting the activity of PIF transcription factors (Figure 2). Because canopy leaves preferentially absorb red light, light that is transmitted or reflected from neighboring plants has a decreased R:FR ratio. Perception of this shade signal by phyB converts the photoreceptor to its inactive (Pr) form, thereby relieving restraint on PIFs. The ensuing transcriptional responses promote the biosynthesis of auxin, which together with other growth hormones, stimulates cell elongation-type growth [91].

Recent studies have revealed that inactivation of phyB by shade light also suppresses JA-triggered defense responses by a mechanism that involves depletion of DELLA proteins, enhanced stability of JAZ repressors, and accelerated degradation of MYCs [50,77,92,93]. This mechanistic insight into growth-defense balance provides a compelling example of how crosstalk at the DELLA-JAZ interface controls a key aspect of plant phenotypic plasticity, and opens the way to identifying specific regulatory genes that mediate these interactions. *JAZ10*, for example, which encodes several alternative splice variants that strongly repress MYC2 [11,94,95,96], plays a central role in repressing JA-mediated defense responses during the shade avoidance response [50,78,93,94]. Advanced phenotypic profiling techniques have recently been used to establish new genetic connections between the JA and shade avoidance pathways [97].

Although phyB-JA crosstalk appears to be hardwired to ensure mutually exclusive expression of either growth- and defense-related transcriptional programs, there are emerging examples of how genetics can be used to tease apart the physiological basis of these tradeoffs. One example comes from studies with the *sav3* mutant that lacks the ability to synthesize auxin during the shade avoidance response [98]. *sav3* mutant plants lack shade-triggered growth responses but remain sensitive to the attenuating effects of low R:FR on defense responses, indicating that repression of defense does not result simply from diversion of resources to growth [77]. Conversely, a *phyB jaz10* double mutant was shown to retain a constitutive shade avoidance growth response (due to genetic inactivation of phyB) and simultaneously express higher levels of JA-induced secondary metabolites [50]. These collective studies provide clues as to how growth–defense tradeoffs may be genetically manipulated.

8. CONCLUSIONS AND FUTURE PERSPECTIVES

Tremendous progress in understanding JA synthesis and signaling in the past 25 years has placed the hormone at the nexus of a complex circuit that controls myriad plant stress responses [9,72]. Now, JA is receiving renewed attention as a growth regulator. The discovery of JAZ-DELLA antagonism has provided deep mechanistic insight into how JA-mediated defense operates in a dynamic balance with growth. Although this and other recent discoveries begin to frame the counterpoise of JA signaling with plant growth, additional effort is required to understand the relationships between JA signaling, deployed defenses, and modulation of growth in the context of leaf expansion, dry-weight gain and partitioning of photoassimilates. Available results indicate that the shift from growth to defense in response to JA signaling is not a simple reallocation of resources but instead involves a rapid and substantial decrease in the rate of dry-matter accumulation. The positive effect of JA signaling on LMA, however, suggests that the hormone may influence leaf architecture via increased leaf thickness and/or leaf density, potentially extending the lifetime photosynthetic output of challenged plants in natural environments. These insights are critically important for an understanding of the survival and fitness of plants in their environment and may reveal opportunities to optimize plant growth, especially in protected agricultural contexts.

ACKNOWLEDGMENTS

This work was supported by the Chemical Sciences, Geosciences and Biosciences Division, Office of Basic Energy Sciences, Office of Science, US Department of Energy through grants DE-FG02-91ER20021 (Gregg A. Howe) and DE-FG02-99ER20323 (John Browse). Ian T. Major was partially supported by a fellowship from the Natural Sciences and Engineering Research Council of Canada. Nathan E. Havko was partially supported by a NIH/NIGMS-funded predoctoral fellowship (T32 GM008336).

AUTHOR CONTRIBUTIONS

Ian T. Major and Gregg A. Howe designed the leaf growth experiments described in Figure 3. Ian T. Major performed the experiments and analyzed the data. All co-authors contributed to the writing of the paper.

CONFLICTS OF INTEREST

The authors declare no conflicts of interest.

REFERENCES

1. Howe, G.; Jander, G. Plant immunity to insect herbivores. Annu. Rev. Plant Biol. 2008, 59, 41–66.
2. Coley, P.D.; Bryant, J.P.; Chapin, F.S. Resource availability and plant antiherbivore defense. Science 1985, 230, 895–899.
3. Herms, D.A.; Mattson, W.J. The dilemma of plants—To grow or defend. Q. Rev. Biol. 1992, 67, 283–335.
4. Futuyma, D.J.; Agrawal, A.A. Macroevolution and the biological diversity of plants and herbivores. Proc. Natl. Acad. Sci. USA 2009, 106, 18054–18061.
5. Stamp, N. Out of the quagmire of plant defense hypotheses. Q. Rev. Biol. 2003, 78, 23–55.
6. Karban, R.; Baldwin, I.T. Induced Responses to Herbivory; University of Chicago Press: Chicago, IL, USA, 1997.

7. Glauser, G.; Dubugnon, L.; Mousavi, S.A.R.; Rudaz, S.; Wolfender, J.L.; Farmer, E.E. Velocity estimates for signal propagation leading to systemic jasmonic acid accumulation in wounded arabidopsis. J. Biol. Chem. 2009, 284, 34506–34513.
8. Koo, A.J.; Gao, X.; Jones, A.D.; Howe, G.A. A rapid wound signal activates the systemic synthesis of bioactive jasmonates in Arabidopsis. Plant J. 2009, 59, 974–986.
9. Campos, M.L.; Kang, J.H.; Howe, G.A. Jasmonate-triggered plant immunity. J. Chem. Ecol. 2014, 40, 657–675.
10. Zhang, Y.; Turner, J. Wound-induced endogenous jasmonates stunt plant growth by inhibiting mitosis. PLoS ONE2008, 3, e3699.
11. Yan, Y.; Stolz, S.; Chetelat, A.; Reymond, P.; Pagni, M.; Dubugnon, L.; Farmer, E.E. A downstream mediator in the growth repression limb of the jasmonate pathway. Plant Cell 2007, 19, 2470–2483.
12. Browse, J. Jasmonate passes muster: A receptor and targets for the defense hormone. Annu. Rev. Plant Biol. 2009, 60, 183–205.
13. Wasternack, C.; Hause, B. Jasmonates: Biosynthesis, perception, signal transduction and action in plant stress response, growth and development. Ann. Bot. 2013, 111, 1021–1058.
14. Huot, B.; Yao, J.; Montgomery, B.L.; He, S.Y. Growth-defense tradeoffs in plants: A balancing act to optimize fitness. Mol. Plant 2014, 7, 1267–1287.
15. Demole, E.; Lederer, E.; Mercier, D. Isolement et détermination de la structure du jasmonate de méthyle, constituant odorant caracteristique de l'essence de jasmin. Helv. Chim. Acta 1962, 45, 675–685.
16. Aldridge, D.C.; Galt, S.; Giles, D.; Turner, W.B. Metabolites of Lasiodiplodia theobromae. J. Chem. Soc. C 1971.
17. Ueda, J.; Kato, J. Inhibition of cytokinin-induced plant-growth by jasmonic acid and its methyl-ester. Physiol. Plant.1982, 54, 249–252.
18. Dathe, W.; Ronsch, H.; Preiss, A.; Schade, W.; Sembdner, G.; Schreiber, K. Endogenous plant hormones of the broad bean, Vicia faba L. (−)-jasmonic acid, a plant-growth inhibitor in pericarp. Planta 1981, 153, 530–535.
19. Corbineau, F.; Rudnicki, R.M.; Come, D. The effects of methyl jasmonate on sunflower (Helianthus annuus L.) seed germination and seedling development. Plant Growth Regul. 1988, 7, 157–169.
20. Nojiri, H.; Sugimori, M.; Yamane, H.; Nishimura, Y.; Yamada, A.; Shibuya, N.; Kodama, O.; Murofushi, N.; Omori, T. Involvement of jasmonic acid in elicitor-induced phytoalexin production in suspension-cultured rice cells. Plant Physiol. 1996, 110, 387–392.
21. Staswick, P.E.; Su, W.; Howell, S.H. Methyl jasmonate inhibition of root growth and induction of a leaf protein are decreased in an Arabidopsis thaliana mutant. Proc. Natl. Acad. Sci. USA 1992, 89, 6837–6840.
22. Feys, B.; Benedetti, C.E.; Penfold, C.N.; Turner, J.G. Arabidopsis mutants selected for resistance to the phytotoxin coronatine are male sterile, insensitive to methyl jasmonate, and resistant to a bacterial pathogen. Plant Cell 1994, 6, 751–759.
23. Lorenzo, O.; Chico, J.M.; Sanchez-Serrano, J.J.; Solano, R. JASMONATE-INSENSITIVE1 encodes a MYC transcription factor essential to discriminate between different jasmonate-regulated defense responses in arabidopsis. Plant Cell2004, 16, 1938–1950.
24. Farmer, E.E.; Ryan, C.A. Interplant communication: Airborne methyl jasmonate induces synthesis of proteinase inhibitors in plant leaves. Proc. Natl. Acad. Sci. USA 1990, 87, 7713–7716.
25. Gundlach, H.; Muller, M.J.; Kutchan, T.M.; Zenk, M.H. Jasmonic acid is a signal transducer in elicitor-induced plant cell cultures. Proc. Natl. Acad. Sci. USA 1992, 89, 2389–2393.
26. McConn, M.; Creelman, R.A.; Bell, E.; Mullet, J.E.; Browse, J. Jasmonate is essential for insect defense in Arabidopsis. Proc. Natl. Acad. Sci. USA 1997, 94, 5473–5477.

27. Howe, G.A.; Lightner, J.; Browse, J.; Ryan, C.A. An octadecanoid pathway mutant (JL5) of tomato is compromised in signaling for defense against insect attack. Plant Cell 1996, 8, 2067–2077.
28. Glazebrook, J. Contrasting mechanisms of defense against biotrophic and necrotrophic pathogens. Annu. Rev. Phytopathol. 2005, 43, 205–227.
29. Kessler, A.; Baldwin, I.T. Defensive function of herbivore-induced plant volatile emissions in nature. Science 2001, 291, 2141–2144.
30. Ballare, C.L. Light regulation of plant defense. Annu. Rev. Plant Biol. 2014, 65, 335–363.
31. Kazan, K.; Manners, J.M. JAZ repressors and the orchestration of phytohormone crosstalk. Trends Plant Sci. 2012, 17, 22–31.
32. Thines, B.; Katsir, L.; Melotto, M.; Niu, Y.; Mandaokar, A.; Liu, G.; Nomura, K.; He, S.Y.; Howe, G.A.; Browse, J. JAZ repressor proteins are targets of the SCFCOI1 complex during jasmonate signalling. Nature 2007, 448, 661–665.
33. Chini, A.; Fonseca, S.; Fernandez, G.; Adie, B.; Chico, J.M.; Lorenzo, O.; Garcia-Casado, G.; Lopez-Vidriero, I.; Lozano, F.M.; Ponce, M.R.; et al. The JAZ family of repressors is the missing link in jasmonate signalling. Nature 2007, 448, 666–671.
34. Sheard, L.B.; Tan, X.; Mao, H.B.; Withers, J.; Ben-Nissan, G.; Hinds, T.R.; Kobayashi, Y.; Hsu, F.F.; Sharon, M.; Browse, J.; et al. Jasmonate perception by inositol-phosphate-potentiated COI1-JAZ co-receptor. Nature 2010, 468, 400–405.
35. Xie, D.X.; Feys, B.F.; James, S.; Nieto-Rostro, M.; Turner, J.G. COI1: An Arabidopsis gene required for jasmonate-regulated defense and fertility. Science 1998, 280, 1091–1094.
36. Li, L.; Zhao, Y.; McCaig, B.C.; Wingerd, B.A.; Wang, J.; Whalon, M.E.; Pichersky, E.; Howe, G.A. The tomato homolog of CORONATINE-INSENSITIVE1 is required for the maternal control of seed maturation, jasmonate-signaled defense responses, and glandular trichome development. Plant Cell 2004, 16, 126–143.
37. Paschold, A.; Halitschke, R.; Baldwin, I.T. Co(i)-ordinating defenses: NaCOI1 mediates herbivore-induced resistance in nicotiana attenuata and reveals the role of herbivore movement in avoiding defenses. Plant J. 2007, 51, 79–91.
38. Hou, X.; Lee, L.Y.; Xia, K.; Yan, Y.; Yu, H. DELLAs modulate jasmonate signaling via competitive binding to JAZs. Dev. Cell 2010, 19, 884–894.
39. Yang, D.L.; Yao, J.; Mei, C.S.; Tong, X.H.; Zeng, L.J.; Li, Q.; Xiao, L.T.; Sun, T.P.; Li, J.G.; Deng, X.W.; et al. Plant hormone jasmonate prioritizes defense over growth by interfering with gibberellin signaling cascade. Proc. Natl. Acad. Sci. USA 2012, 109, 1192–1200.
40. Wild, M.; Daviere, J.M.; Cheminant, S.; Regnault, T.; Baumberger, N.; Heintz, D.; Baltz, R.; Genschik, P.; Achard, P. The Arabidopsis DELLA RGA-like3 is a direct target of MYC2 and modulates jasmonate signaling responses. Plant Cell 2012, 24, 3307–3319.
41. Qi, T.; Huang, H.; Wu, D.; Yan, J.; Qi, Y.; Song, S.; Xie, D. Arabidopsis DELLA and JAZ proteins bind the WD-repeat/bHLH/MYB complex to modulate gibberellin and jasmonate signaling synergy. Plant Cell 2014, 26, 1128–1133.
42. Leivar, P.; Quail, P.H. PIFs: Pivotal components in a cellular signaling hub. Trends Plant Sci. 2011, 16, 19–28.
43. Ren, C.M.; Han, C.Y.; Peng, W.; Huang, Y.; Peng, Z.H.; Xiong, X.Y.; Zhu, Q.; Gao, B.D.; Xie, D.X. A leaky mutation in DWARF4 reveals an antagonistic role of brassinosteroid in the inhibition of root growth by jasmonate in Arabidopsis. Plant Physiol. 2009, 151, 1412–1420.
44. Kim, B.; Fujioka, S.; Kwon, M.; Jeon, J.; Choe, S. Arabidopsis brassinosteroid-overproducing gulliver3-D/DWARF4-D mutants exhibit altered responses to jasmonic acid and pathogen. Plant Cell Rep. 2013, 32, 1139–1149.
45. Schaller, G.E.; Street, I.H.; Kieber, J.J. Cytokinin and the cell cycle. Curr. Opin. Plant Biol. 2014, 21, 7–15.

46. Świątek, A.; Lenjou, M.; van Bockstaele, D.; Inzé, D.; van Onckelen, H. Differential effect of jasmonic acid and abscisic acid on cell cycle progression in tobacco BY-2 cells. Plant Physiol. 2002, 128, 201–211.
47. Świątek, A.; Azmi, A.; Stals, H.; Inzé, D.; van Onckelen, H. Jasmonic acid prevents the accumulation of cyclin B1; 1 and CDK-B in synchronized tobacco BY-2 cells. FEBS Lett. 2004, 572, 118–122.
48. Noir, S.; Bomer, M.; Takahashi, N.; Ishida, T.; Tsui, T.L.; Balbi, V.; Shanahan, H.; Sugimoto, K.; Devoto, A. Jasmonate controls leaf growth by repressing cell proliferation and the onset of endoreduplication while maintaining a potential stand-by mode. Plant Physiol. 2013, 161, 1930–1951.
49. Robson, F.; Okamoto, H.; Patrick, E.; Harris, S.R.; Wasternack, C.; Brearley, C.; Turner, J.G. Jasmonate and phytochrome a signaling in Arabidopsis wound and shade responses are integrated through JAZ1 stability. Plant Cell2010, 22, 1143–1160.
50. Leone, M.; Keller, M.M.; Cerrudo, I.; Ballare, C.L. To grow or defend? Low red : Far-red ratios reduce jasmonate sensitivity in Arabidopsis seedlings by promoting della degradation and increasing JAZ10 stability. New Phytol. 2014, 204, 355–367.
51. Chen, H.J.; Chen, C.L.; Hsieh, H.L. Far-red light-mediated seedling development in Arabidopsis involves FAR-RED INSENSITIVE 219/JASMONATE RESISTANT 1-dependent and -independent pathways. PLoS ONE 2015, 10, e0132723.
52. Smalle, J.; Haegman, M.; Kurepa, J.; van Montagu, M.; van der Straeten, D. Ethylene can stimulate Arabidopsis hypocotyl elongation in the light. Proc. Natl. Acad. Sci. USA 1997, 94, 2756–2761.
53. Cowling, R.J.; Harberd, N.P. Gibberellins control Arabidopsis hypocotyl growth via regulation of cellular elongation. J. Exp. Bot. 1999, 50, 1351–1357.
54. Szekeres, M.; Nemeth, K.; KonczKalman, Z.; Mathur, J.; Kauschmann, A.; Altmann, T.; Redei, G.P.; Nagy, F.; Schell, J.; Koncz, C. Brassinosteroids rescue the deficiency of CYP90, a cytochrome P450, controlling cell elongation and de-etiolation in Arabidopsis. Cell 1996, 85, 171–182.
55. Toda, Y.; Tanaka, M.; Ogawa, D.; Kurata, K.; Kurotani, K.-I.; Habu, Y.; Ando, T.; Sugimoto, K.; Mitsuda, N.; Katoh, E.; et al. RICE SALT SENSITIVE3 forms a ternary complex with JAZ and class-C bHLH factors and regulates jasmonate-induced gene expression and root cell elongation. Plant Cell 2013, 25, 1709–1725.
56. Chen, Q.; Sun, J.Q.; Zhai, Q.Z.; Zhou, W.K.; Qi, L.L.; Xu, L.; Wang, B.; Chen, R.; Jiang, H.L.; Qi, J.; et al. The basic helix-loop-helix transcription factor MYC2 directly represses PLETHORA expression during jasmonate-mediated modulation of the root stem cell niche in Arabidopsis. Plant Cell 2011, 23, 3335–3352.
57. Gasperini, D.; Chetelat, A.; Acosta, I.F.; Goossens, J.; Pauwels, L.; Goossens, A.; Dreos, R.; Alfonso, E.; Farmer, E.E. Multilayered organization of jasmonate signalling in the regulation of root growth. PLoS Genet. 2015, 11, e1005300.
58. Acosta, I.F.; Gasperini, D.; Chetelat, A.; Stolz, S.; Santuari, L.; Farmer, E.E. Role of NINJA in root jasmonate signaling. Proc. Natl. Acad. Sci. USA 2013, 110, 15473–15478.
59. Délano Frier, J.P.; Sánchez Hernández, C.V.; Tiessen, A. Friend or foe? Exploring the factors that determine the difference between positive and negative effects on photosynthesis in response to insect herbivory. In Artificial Photosynthesis; Najafpour, M.M., Ed.; InTech: Rijeka, Croatia, 2012; pp. 155–206.
60. Hermsmeier, D.; Schittko, U.; Baldwin, I.T. Molecular interactions between the specialist herbivore Manduca sexta (Lepidoptera, Sphingidae) and its natural host Nicotiana attenuata. I. Large-scale changes in the accumulation of growth- and defense-related plant mRNAs. Plant Physiol. 2001, 125, 683–700.
61. Mitra, S.; Baldwin, I.T. Independently silencing two photosynthetic proteins in Nicotiana attenuata has different effects on herbivore resistance. Plant Physiol. 2008, 148, 1128–1138.
62. Thomson, V.P.; Cunningham, S.A.; Ball, M.C.; Nicotra, A.B. Compensation for herbivory by Cucumis sativus through increased photosynthetic capacity and efficiency. Oecologia 2003, 134, 167–175.

63. Halitschke, R.; Hamilton, J.G.; Kessler, A. Herbivore-specific elicitation of photosynthesis by mirid bug salivary secretions in the wild tobacco Nicotiana attenuata. New Phytol. 2011, 191, 528–535.
64. Ferrieri, A.P.; Agtuca, B.; Appel, H.M.; Ferrieri, R.A.; Schultz, J.C. Temporal changes in allocation and partitioning of new carbon as 11C elicited by simulated herbivory suggest that roots shape aboveground responses in Arabidopsis. Plant Physiol. 2013, 161, 692–704.
65. Babst, B.A.; Ferrieri, R.A.; Gray, D.W.; Lerdau, M.; Schlyer, D.J.; Schueller, M.; Thorpe, M.R.; Orians, C.M. Jasmonic acid induces rapid changes in carbon transport and partitioning in Populus. New Phytol. 2005, 167, 63–72.
66. Schwachtje, J.; Minchin, P.E.; Jahnke, S.; van Dongen, J.T.; Schittko, U.; Baldwin, I.T. SNF1-related kinases allow plants to tolerate herbivory by allocating carbon to roots. Proc. Natl. Acad. Sci. USA 2006, 103, 12935–12940.
67. Zangerl, A.R.; Rutledge, C.E. The probability of attack and patterns of constitutive and induced defense: A test of optimal defense theory. Am. Nat. 1996, 147, 599–608.
68. Reymond, P.; Bodenhausen, N.; van Poecke, R.M.; Krishnamurthy, V.; Dicke, M.; Farmer, E.E. A conserved transcript pattern in response to a specialist and a generalist herbivore. Plant Cell 2004, 16, 3132–3147.
69. Giri, A.P.; Wunsche, H.; Mitra, S.; Zavala, J.A.; Muck, A.; Svatos, A.; Baldwin, I.T. Molecular interactions between the specialist herbivore Manduca sexta (Lepidoptera, Sphingidae) and its natural host Nicotiana attenuata. VII. Changes in the plant's proteome. Plant Physiol. 2006, 142, 1621–1641.
70. Bilgin, D.D.; Zavala, J.A.; Zhu, J.; Clough, S.J.; Ort, D.R.; DeLucia, E.H. Biotic stress globally downregulates photosynthesis genes. Plant Cell Environ. 2010, 33, 1597–1613.
71. Attaran, E.; Major, I.T.; Cruz, J.A.; Rosa, B.A.; Koo, A.J.; Chen, J.; Kramer, D.M.; He, S.Y.; Howe, G.A. Temporal dynamics of growth and photosynthesis suppression in response to jasmonate signaling. Plant Physiol. 2014, 165, 1302–1314.
72. Wasternack, C. How jasmonates earned their laurels: Past and present. J. Plant Growth Regul. 2015, 34, 761–794.
73. Weraduwage, S.M.; Chen, J.; Anozie, F.C.; Morales, A.; Weise, S.E.; Sharkey, T.D. The relationship between leaf area growth and biomass accumulation in Arabidopsis thaliana. Front. Plant Sci. 2015.
74. Baldwin, I.T. Jasmonate-induced responses are costly but benefit plants under attack in native populations. Proc. Natl. Acad. Sci. USA 1998, 95, 8113–8118.
75. Poorter, H.; Niinemets, U.; Poorter, L.; Wright, I.J.; Villar, R. Causes and consequences of variation in leaf mass per area (LMA): A meta-analysis. New Phytol. 2009, 182, 565–588.
76. Wright, I.J.; Reich, P.B.; Westoby, M.; Ackerly, D.D.; Baruch, Z.; Bongers, F.; Cavender-Bares, J.; Chapin, T.; Cornelissen, J.H.C.; Diemer, M.; et al. The worldwide leaf economics spectrum. Nature 2004, 428, 821–827.
77. Moreno, J.E.; Tao, Y.; Chory, J.; Ballare, C.L. Ecological modulation of plant defense via phytochrome control of jasmonate sensitivity. Proc. Natl. Acad. Sci. USA 2009, 106, 4935–4940.
78. Gershenzon, J. The cost of plant chemical defense against herbivory: A biochemeical perspective. In Insect-Plant Interactions; CRC Press: Boca Raton, FL, USA, 1994; pp. 105–173.
79. Gershenzon, J. Metabolic costs of terpenoid accumulation in higher plants. J. Chem. Ecol. 1994, 20, 1281–1328.
80. De Vries, F.P.; Brunsting, A.; van Laar, H. Products, requirements and efficiency of biosynthesis a quantitative approach. J. Theor. Biol. 1974, 45, 339–377.
81. Halkier, B.A.; Gershenzon, J. Biology and biochemistry of glucosinolates. Annu. Rev. Plant Biol. 2006, 57, 303–333.
82. Bekaert, M.; Edger, P.P.; Hudson, C.M.; Pires, J.C.; Conant, G.C. Metabolic and evolutionary costs of herbivory defense: Systems biology of glucosinolate synthesis. New Phytol. 2012, 196, 596–605.

83. Ullmann-Zeunert, L.; Muck, A.; Wielsch, N.; Hufsky, F.; Stanton, M.A.; Bartram, S.; Bocker, S.; Baldwin, I.T.; Groten, K.; Svatos, A. Determination of 15N-incorporation into plant proteins and their absolute quantitation: A new tool to study nitrogen flux dynamics and protein pool sizes elicited by plant-herbivore interactions. J. Proteome Res. 2012, 11, 4947–4960.
84. Ullmann-Zeunert, L.; Stanton, M.; Wielsch, N.; Bartram, S.; Hummert, C.; Svatoš, A.; Baldwin, I.; Groten, K. Quantification of growth-defense trade-offs in a common currency: Nitrogen required for phenolamide biosynthesis is not derived from ribulose-1,5-bisphosphate carboxylase/oxygenase turnover. Plant J. 2013, 75, 417–429.
85. Williams, K.; Percival, F.; Merino, J.; Mooney, H.A. Estimation of tissue construction cost from heat of combustion and organic nitrogen content. Plant Cell Environ. 1987, 10, 725–734.
86. Griffin, K. Calorimetric estimates of construction cost and their use in ecological studies. Funct. Ecol. 1994, 8, 551–562.
87. Sanjaya; Miller, R.; Durrett, T.P.; Kosma, D.K.; Lydic, T.A.; Muthan, B.; Koo, A.J.K.; Bukhman, Y.V.; Reid, G.E.; Howe, G.A.; et al. Altered lipid composition and enhanced nutritional value of Arabidopsis leaves following introduction of an algal diacylglycerol acyltransferase 2. Plant Cell 2013, 25, 677–693.
88. Stitt, M.; Hurry, V. A plant for all seasons: Alterations in photosynthetic carbon metabolism during cold acclimation in Arabidopsis. Curr. Opin. Plant Biol. 2002, 5, 199–206.
89. Zhou, S.; Lou, Y.R.; Tzin, V.; Jander, G. Alteration of plant primary metabolism in response to insect herbivory. Plant Physiol. 2015, 169, 1488–1498.
90. Hu, P.; Zhou, W.; Cheng, Z.; Fan, M.; Wang, L.; Xie, D. JAV1 controls jasmonate-regulated plant defense. Mol. Cell2013, 50, 504–515.
91. Casal, J.J. Photoreceptor signaling networks in plant responses to shade. Annu. Rev. Plant Biol. 2013, 64, 403–427.
92. Chico, J.M.; Fernandez-Barbero, G.; Chini, A.; Fernandez-Calvo, P.; Diez-Diaz, M.; Solano, R. Repression of jasmonate-dependent defenses by shade involves differential regulation of protein stability of MYC transcription factors and their JAZ repressors in Arabidopsis. Plant Cell 2014, 26, 1967–1980.
93. Cargnel, M.D.; Demkura, P.V.; Ballare, C.L. Linking phytochrome to plant immunity: Low red: Far-red ratios increase Arabidopsis susceptibility to Botrytis cinerea by reducing the biosynthesis of indolic glucosinolates and camalexin. New Phytol. 2014, 204, 342–354.
94. Chung, H.S.; Howe, G.A. A critical role for the tify motif in repression of jasmonate signaling by a stabilized splice variant of the JASMONATE ZIM-DOMAIN protein JAZ10 in Arabidopsis. Plant Cell 2009, 21, 131–145.
95. Chung, H.S.; Cooke, T.F.; DePew, C.L.; Patel, L.C.; Ogawa, N.; Kobayashi, Y.; Howe, G.A. Alternative splicing expands the repertoire of dominant JAZ repressors of jasmonate signaling. Plant J. 2010, 63, 613–622.
96. Moreno, J.E.; Shyu, C.; Campos, M.L.; Patel, L.C.; Chung, H.S.; Yao, J.; He, S.Y.; Howe, G.A. Negative feedback control of jasmonate signaling by an alternative splice variant of JAZ10. Plant Physiol. 2013, 162, 1006–1017.
97. Nozue, K.; Tat, A.V.; Kumar Devisetty, U.; Robinson, M.; Mumbach, M.R.; Ichihashi, Y.; Lekkala, S.; Maloof, J.N. Shade avoidance components and pathways in adult plants revealed by phenotypic profiling. PLoS Genet. 2015, 11, e1004953.
98. Tao, Y.; Ferrer, J.L.; Ljung, K.; Pojer, F.; Hong, F.; Long, J.A.; Li, L.; Moreno, J.E.; Bowman, M.E.; Ivans, L.J.; et al. Rapid synthesis of auxin via a new tryptophan-dependent pathway is required for shade avoidance in plants. Cell 2008, 133, 164–176.

CHAPTER 17

Characterization of *Phytophthora Infestans* Population Diversity in Canada Reveals Increased Migration and Genotype Recombination

R. D. Peters[1], K. I. Al-Mughrabi[2], M. L. Kalischuk[3], K. F. Dobinson[4], K. L. Conn[4], H. Alkher[5], M. R. Islam[5], F. Daayf[5], J. Lynn[3], B. Bizimungu[6], D. De Koeyer[6], C. A. Lévesque[7] And L. M. Kawchuk[3]

[1]Agriculture and Agri-Food Canada, 440 University Avenue, Charlottetown, PE C1A 4N6, Canada

[2]New Brunswick Department of Agriculture, Aquaculture and Fisheries, 39 Barker Lane, Wicklow, NB E7L 3S4, Canada [3]Agriculture and Agri-Food Canada, 5403-1st Avenue South, Lethbridge, AB T1J 4B1, Canada

[4]Agriculture and Agri-Food Canada, 1391 Sandford Street, London, ON N5V 4T3, Canada [5]University of Manitoba, 66 Chancellors Circle, Winnipeg, MB R3T 2N2, Canada [6]Agriculture and Agri-Food Canada, 850 Lincoln Road, Fredericton, NB E3B 4Z7, Canada [7]Agriculture and Agri-Food Canada, 960 Carling Avenue, Ottawa, ON K1A 0C6, Canada

ABSTRACT

Late blight, caused by Phytophthora infestans, is a devastating disease of potatoes and tomatoes that causes significant crop losses worldwide. Late blight diversity has been increasing since 2009 in many areas of Canada on both potatoes and tomatoes. Although the US-8 genotype of P. infestans has dominated pathogen populations in much of Canada for the past 15 years, novel genotypes have emerged in recent years. A comprehensive survey of late blight samples collected across Canada in 2011 indicated that although the US-8 genotype of P. infestans still dominated pathogen populations in Prince Edward Island, new US-23 and US-24 genotypes had become established in other parts of Canada. The US-11 and US-22 genotypes dominated populations of the pathogen from specific areas of Alberta and Ontario, respectively. Clonal propagation and migration contributed to long-distance transport of P. infestans genotypes and in some situations, both A1 and A2 mating types of the pathogen were found in the same production region. This finding is of concern because it increases the potential for sexual recombination within pathogen populations, leading to the production of oospores that can survive in the absence of host tissue and create new strains of P. infestans. Evidence for recombination in the P. infestans population was detected in Ontario on tomato. Segregation of Gpi allozymes and mating type appeared to occur independently of the RG57 loci, producing several new P. infestans genotypes. Prediction and management of late blight will need to be evaluated and modified to accommodate these new potentially aggressive P. infestans genotypes.

Keywords: allozyme, genetic diversity, late blight, metalaxyl, microsatellite, mtDNA, RFLP

INTRODUCTION

Late blight is the most damaging disease of potato. Worldwide losses each year to late blight are conservatively estimated at more than $6 billion (Haas et al. 2009). The disease is caused by the oomycete Phytophthora infestans (Mont.) de Bary and occurs in all major potato and tomato producing areas (Fry et al. 1991; Spielman et al. 1991; Goodwin et al. 1994a; Sujkowski et al. 1994; Goodwin et al. 1996; Day & Shattock 1997; Forbes et al. 1997). Genetic diversity of P. infestans has increased in Canada and the United States, as reflected by the development of strains with different virulence and reaction to fungicides such as metalaxyl (Deahl & Inglis 1995; Peters et al. 1998; Daayf & Platt 2000; Daayf et al. 2000). A recent resurgence of the disease in Canada and the United States has been largely attributed to clonal propagation and long-distance migration of more aggressive strains of the pathogen (Hu et al. 2012; Kalischuk et al.

2012). Whole-genome sequen-cing of P. infestans revealed that it has a relatively large and complex genome that contributes to the infamous complex-ity of the pathogen and its pathogenicity (Haas et al. 2009).

Phytophthora infestans is a heterothallic pathogen with two mating types, A1 and A2. Prior to 1980, disease losses in the United States and Europe were caused by A1 lineages. Both A1 and A2 mating types were introduced into the United States and Europe from Mexico during the 1980s (Deahl et al. 1991; Goodwin et al. 1994b). As a result of this migration in many areas, the A1 mating type was often replaced by more aggressive strains that were often A2 mat-ing type and metalaxyl-resistant (Goodwin 1997). Distinctive North American genotypes that share properties such as host preference, fungicide sensitivity, mating type and aggressive-ness, have since been identified and classified as CA-1 to CA-8 or US-1 to US-24 based on various genetic markers (Deahl et al. 1993; Legard et al. 1995; Kato et al. 1997; Forbes et al. 1998; Goodwin et al. 1998; Danies et al. 2013).

Previous increases in incidence and severity of late blight have coincided with major genetic changes in the pathogen populations (Spielman et al. 1991; Deahl & Inglis 1995; Goodwin et al. 1995a, 1995b; Marshall-Farrar et al. 1998; Kawchuk et al. 2011; Danies et al. 2013). In Canada and the United States, the emergence of new P. infestans genotypes suggests that sexual recombination is now contributing to the increased diver-sity in genetic composition of the pathogen (Goodwin et al. 1995a; Kalischuk et al. 2012). Sexual recombination produces thick-walled oospores capable of surviving extreme temperatures and extended periods in the absence of host tissue (Anderson et al. 2009). However, evidence of sexual reproduction in Canada and the United States remained limited, probably because both mating types were found together only in a small number of sites in this area (Goodwin et al. 1995a; Chycoski & Punja 1996). As the genetic composition of the P. infestans populations continues to diversify, characterization of the pathogen and management of late blight will become more challenging. Increased genetic diversity in the P. infestans populations already requires the implementation of several routinely applied diagnostic procedures includ-ing allozymes, restriction fragment length polymorphisms (RFLPs) and simple sequence repeats (SSRs) to produce results that correspond with critical phenotypic character-istics (Goodwin et al. 1992, 1995c; Lees et al. 2006; Hu et al. 2012). Understanding the variation in P. infestans populations and potential for sexual recombination will be critical in tracking pathogen movement and develop-ing effective control strategies.

The objectives of this study were to analyze the genetic diversity of P. infestans populations in Canada, to identify changes in overall population composition and to determine the frequency of migration and recombination events. These results would help predict the prevalent P. infestans genotype in subsequent years and establish a linkage between genotype specific loci frequently applied in isolate identification and characterization. Genetic analyses of pathogen diversity will advance our understanding of P. infestans epidemiology, providing information regarding migration and recombina-tion that influences the pathogen population and effectiveness of recommended disease prevention strategies.

MATERIALS AND METHODS

Isolate collection

During 2011, isolates of P. infestans from across Canada were obtained from potato or tomato plants, tubers and infected fruits. Sources included field and greenhouse grown crops from commercial growers or home gardens. Individual isolates were obtained by transferring sporan-gia from sporulating tissues using an approximately 1-mm rye A agar cube, as described previously (Caten & Jinks 1968). Tissues that were not sporulating were placed on moistened filter paper in a Petri dish and incubated at 18°C for a few days to induce sporulation. Infected tomato fruit were surface-sterilized with 1.5% sodium hypochlorite for 2 min, rinsed in sterile water and a 3-mm slice removed and placed cut side up in a Petri dish to promote sporulation. Alternatively, P. infestans was rescued from tissues that were beginning to decay by placing them in a Petri dish beneath a fresh slice of potato 'Shepody' and isolating from developing lesions. Isolated sporangia and growing cultures were maintained at 15–22°C on rye A (Caten & Jinks 1968) or V-8 juice agar (Miller 1955). Haplotype characterization was per-formed using P. infestans isolates produced from single zoospores.

Isolate characterization

Phytophthora infestans mycelium used in allozyme and DNA RFLP analyses was grown as described previously (Goodwin et al. 1992, 1995c). Genetic analyses were performed in replicate on 138 isolates derived from single zoospore or sporangia initially plated onto rye A agar. Each isolate was grown in standing culture (20 mL V-8 broth) in 9-cm disposable Petri dishes. Petri dishes were inoculated with a mycelial plug cut from the margin of an actively growing colony and incubated at 18°C in the dark for 1–3 weeks. Isolates of P. erythroseptica,

P. brassicae, P. nicotianae and Pythium ultimum were used as negative controls.

Isolates were examined at the Glucose-6-phosphate isomerase (Gpi) locus by cellulose-acetate electrophoresis (CAE) to determine allozyme genotypes (Oudemans & Coffey 1991; Goodwin et al. 1996). Reference isolates (US-6 (100/100), US-8 (100/111/122), US-11 (100/100/ 111) and US-22 (100/122)) were inoculated on each CAE plate. Migration distances of Gpi proteins from unchar-acterized isolates were compared to the proteins from the reference genotypes.

Procedures for DNA RFLP analyses were as described previously (Goodwin et al. 1992). Pathogen DNA was extracted from mycelium and digested with EcoRI. To prepare the multilocus P. infestans specific RFLP probe RG57, a genomic DNA fragment was amplified by polymerase chain reaction (PCR) from P. infestans US-11 using the primers RG57P1F (5'ACCATG CAGCTGAATTGCCAT3') and RG57P1R (5'CTCTCATAACCGACGTTTTC3') that are specific to the RG57 sequence (GenBank Accession AF495861) (Goodwin et al. 1992). The PCR was performed in 100 μL of 10 mM Tris-HCl, pH 8.3, 50 mM KCl, 1.5 mM MgCl2, 0.001% gelatin, 0.25 mM of each dNTP, 2 μM of each primer, 0.5 U Vent Taq DNA polymerase (New England Biolabs, Inc., Mississauga, ON) and 200 ng template DNA as previously described (Khadhair et al. 1998; Entz et al. 2005). The PCR product was used as a template for random labelling with 32P. Blotting of the digest on the Nytran SPC membranes, hybridization and post-hybridization washing were performed as recommended by the membrane manufacture (GE Healthcare, Baie d'Urfe, QC).

To determine mitochondrial (mt) DNA haplotypes, DNA was extracted from the P. infestans isolates, PCR amplified and restricted as previously described using the identical primers and restriction enzymes (Griffith & Shaw 1998). Each analysis was replicated at least twice. Haplotypes were determined by comparing the DNA profiles to those of reference isolates having Ia, Ib, IIa and IIb profiles of the mitochondrial genome and con-firmed by amplification and sequencing of P2 and P4 fragments as previously described (Griffith & Shaw 1998). The DNA sequence was determined with a BigDye Terminator Cycle Sequencing Kit and an ABI 377 automated sequencer (Life Technologies, Burlington, ON) using the same PCR primers.

Metalaxyl sensitivity of the P. infestans isolates was determined by measuring growth on metalaxyl (Ridomil Gold; Syngenta Crop Protection, Guelph, ON) – amended rye A agar as previously described (Goodwin et al. 1996; Holley & Kawchuk 1996; Peters et al. 1998). This assay was performed at least twice with three replicate plates per isolate. Isolates were grown on media containing 0 or 5 μg of metalaxyl per mL and colony diameters measured after 7 days of growth. Isolates that produced more than 40% of the growth on the amended medium relative to that on the metalaxyl-free medium were designated as resistant to metalaxyl. Those that produced less than 40% growth were considered metalaxyl-sensitive.

Mating type of each isolate was determined in dupli-cate on RSA (Peters et al. 1998). Isolates were paired with characterized A1 and A2 isolates (P183A and P185A, respectively; provided by K. L. Deahl, USDA, Beltsville, MD). Plates were incubated at 18°C in the dark for 2 weeks and oospore formation for each isolate pairing was assessed microscopically. Isolates that produced oospores when paired with the A1 isolate but not with the A2 isolate, were labelled as A2 mating type, and those that produced oospores only when paired with the A2 isolate, were considered to be of the A1 mating type.

Multilocus P. infestans genotypes were determined from mating type, Gpi allozyme loci, and RG57 RFLP profiles. Recombinant P. infestans isolates from ON were also examined with microsatellite markers Pi4B, PiG11, D13, Pi04, Pi06, Pi56, Pi63, and Pi89 following the procedure described by Lees et al. (2006). Each locus was treated as a dichotomous variable, with values con-verted to a binary format by assigning the values 0 or 1 (absence or presence) for each known allele (two for mating type, seven for Gpi and 26 for RG57). Similarity coefficients for the P. infestans multilocus genotypes were determined with the Jaccard index and subtracted from unity to provide estimates of genetic distances. Phylogenetic trees were produced from the distance matrix with the neighbor-joining clustering algorithm. All analyses were performed using SAS (release 6.12, SAS Institute, Inc., Cary, NC).

RESULTS

Pathogen collection

One hundred and thirty-eight isolates of P. infestans were collected from potato and tomato in 2011 from eight provinces in Canada including British Columbia (BC), Alberta (AB), Saskatchewan (SK), Manitoba (MB), Ontario (ON), Quebec (QC), New Brunswick (NB) and Prince Edward Island (PE) (Table 1). Only tomato tissues infected with P. infestans were identified in SK and ON, and with the exception of single isolates from infected tomato in MB and PE, all other diseased samples were from potato. All samples originated from commercial fields, greenhouses or backyard gardens. P. infestans was not observed in any samples collected from commer-cial garden retail centres.

Allozyme, mtDNA haplotype and DNA restriction frag-ment length polymorphism analyses Allozyme analysis of the 138 P. infestans isolates identified four Gpi allozyme profiles including Gpi 100/100 (US-23), Gpi 100/111/122

(US-8), Gpi 100/122 (US-22, CA-9, CA-10), and Gpi 100/100/111 (US-11, US-24 and CA-11) (Figure 1). Sequences of P2 and P4 regions of the mitochondrial genome confirmed the identity of the mtDNA haplotypes, and indicated that only two (Ia and IIb) of the four haplotypes were represented among the 138 P. infestans isolates collected in Canada during 2011 (Figure 2). Both haplotypes were isolated from potato and tomato. The Ia haplotype was the most abundant (127 isolates) and widely distributed and the IIb haplotype was only found in the US-11 isolates (Table 1).

The RFLP analyses confirmed the genotype identifica-tions made by the Gpi allozyme analyses and distinguished US-11 from US-24 and US-6 from US-23 (Figure 3). The US-23 genotype was identified only in BC, MB, NB, PE and SK. Only the P. infestans US-11 and US-23 genotypes were recovered from both potato and tomato samples. With the exception of a single isolate from ON, the US-11 genotype was isolated only from central AB and the US-8 and US-24 genotypes were only isolated from diseased potato samples. Many P. infestans isolates from AB, MB, NB and QC were genotype US-24, revealing a migration and proliferation of this geno-type. Isolates of the P. infestans US-22 genotype were identified only in ON, where they were the dominant genotype and may have produced the new P. infestans genotypes CA-9, CA-10 and CA-11. The RFLP profile of P. infestans genotype CA-9 was identical to that of US-22, while P. infestans genotypes CA-10 and CA-11 both had RFLP profiles resembling that of US-22 (Figure 3) but with an additional DNA product (DNA 18) and missing a second allele (DNA 16). Although CA-9 and CA-10 had a Gpi profile 100/122 identical to that of US-22, CA-11 had a Gpi 100/100/111 banding pattern identical to that of US-11 and US-24. Under the high stringency annealing and washing conditions used in this study, the RG57 sequence showed no hybridization to DNA from other oomycetes including P. ultimum, P. erythroseptica, P. brassicae and P. nicotianae.

Metalaxyl sensitivity analyses

All isolates of the US-8 and US-11 genotypes assayed in this study were metalaxyl insensitive (average 83 ± 6 mm and 82 ± 9 mm, respectively) (Table 1). Isolates of all other P. infestans genotypes were sensi-tive to metalaxyl. Although the P. infestans US-24 isolates are metalaxyl sensitive, they do exhibit some level of metalaxyl tolerance compared to the sensitive genotypes, producing some growth (average 8 ± 2 mm). The P. infestans

US-8 and US-23 genotypes have been the most common P. infestans strains causing potato late blight in Canada during 2010 (Kalischuk et al. 2012) but in 2011 were largely replaced by the metalaxyl sensitive US-24. Surprisingly, in the province of NB, the P. infestans genotype US-8 was completely displaced within a sin-gle year by the US-23 and US-24 genotypes.

Table 1. Source and genotypes of Phytophthora infestans isolates collected in Canada during 2011.

Province	Host	Mating type	Allozyme[a]	RG57 RFLP[b]	Genotype	mtDNA[c]	Metalaxyl[d]	Isolates
British Columbia	Potato	A1	*100/100*	11001100010011001001100111	US-23	Ia	S	9
Alberta	Potato	A1	*100/100/111*	10101110010011010101100101	US-11	IIb	R	10
	Potato	A1	*100/100/111*	10101010010011010001101101	US-24	Ia	S	12
Saskatchewan	Tomato	A1	*100/100*	11001100010011001001100111	US-23	Ia	S	1
Manitoba	Potato	A1	*100/100/111*	10101010010011010001101101	US-24	Ia	S	20
	Tomato	A1	*100/100*	11001100010011001001100111	US-23	Ia	S	1
Ontario	Tomato	A1	*100/100/111*	10101110010011010101100101	US-11	IIb	R	1
	Tomato	A1	*100/122*	10001000000011010001100101	CA-9	Ia	S	1[e]
	Tomato	A2	*100/122*	10001000000011010001100101	US-22	Ia	S	10
	Tomato	A1	*100/122*	10001000000011000101100101	CA-10	Ia	S	4[e]
	Tomato	A2	*100/100/111*	10001000000011000101100101	CA-11	Ia	S	1[e]
Quebec	Potato	A2	*100/111/122*	10001000010011010001101101	US-8	Ia	R	1
	Potato	A1	*100/100/111*	10101010010011010001101101	US-24	Ia	S	4
New Brunswick	Potato	A1	*100/100*	11001100010011001001100111	US-23	Ia	S	7
	Potato	A1	*100/100/111*	10101010010011010001101101	US-24	Ia	S	5
Prince Edward Island	Potato	A2	*100/111/122*	10001000010011010001101101	US-8	Ia	R	50
	Tomato	A1	*100/100*	11001100010011001001100111	US-23	Ia	S	1

[a]*Glucose-6-phosphate isomerise* (*Gpi*) locus detected by cellulose-acetate electrophoresis (CAE).
[b]Restriction fragment lengthpolymorphic (RFLP) pattern produced with *Eco* RI and RG57.
[c]Mitochondrial DNA haplotype as determined by polymerase chain reaction and restriction digests.
[d]S = sensitive and *R* = resistant to metalaxyl.
[e]Recombinants based on mating type, *Gpi*, and RG57 RFLP profile.

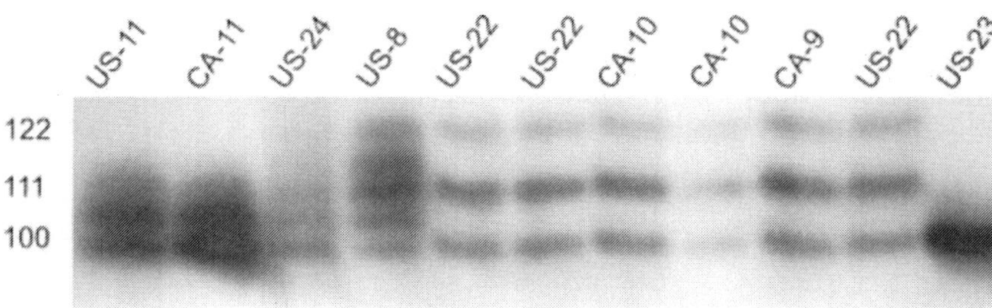

Figure 1 (Colour online) Glucose-6-phosphate isomerase (Gpi) allo-zyme profile of Phytophthora infestans isolates from potato and tomato samples collected in Canada in 2011. Allozymes were separated on cellulose-acetate gels and stained for Gpi activity. The Gpi profiles of the P. infestans genotypes shown include US-8, Gpi 100/111/122; US-11, Gpi 100/100/111; US-22, Gpi 100/122; US-23, Gpi 100/100; US-24, Gpi 100/100/111; CA-9, Gpi 100/122; CA-10, Gpi 100/122, and CA-11, Gpi 100/100/111. The Gpi geno-type of US-11 was identical to that of US-24 and CA-11 (Gpi 100/ 100/111) and US-22 was indistinguishable from CA-9 and CA-10 (Gpi 100/122).

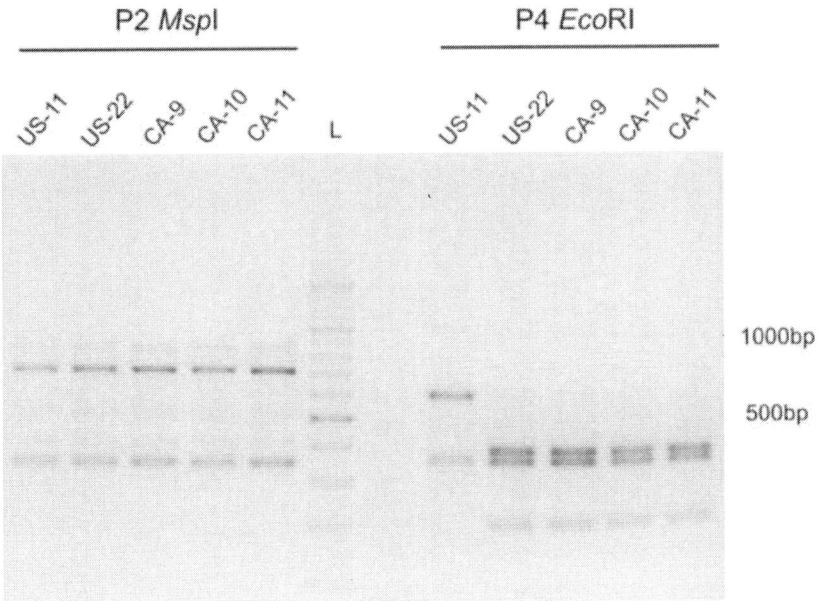

Figure 2 (Colour online) Amplified DNA products P2 and P4 contain variable sequences of the mitochondrial genome of Phytophthora infestans isolates from potato and tomato samples collected in Canada in 2011. Amplicons are shown after restriction digestion with MspI or EcoRI and gel electrophoresis. Four mitochondrial DNA haplotypes of P. infestans designated as Ia, Ib, IIa or IIb have been previously identified through DNA amplification and restric-tion. Isolates of P. infestans US-22, CA-9, CA-10 and CA-11 genotypes produced the mtDNA haplotype Ia and US-11 isolates were mtDNA haplotype IIb. Lane L shows a 100 bp DNA ladder (Fisher Scientific, Ottawa, ON).

Mating type analyses

Only P. infestans mating type A1 was isolated from western Canadian (BC, AB, SK and MB) late blight samples obtained in 2011, while isolates from eastern Canada (ON, QC, NB and PE) were either A1 or A2 mating types (Figure 4A). Although the close proximity in QC, PE and NB of both mating types may have increased opportunities for recombination, there were no atypical P. infestans genotypes recovered from these provinces. However, P. infestans genetic diversity was highest in ON where both mating types occur in extremely close proximity (Figure 4B). Evidence for recombination was observed for the US-11 A1 and US-22 A2 genotypes on tomato in ON, but not the US-8 A2 and US-24 A1 genotypes on potato in QC.

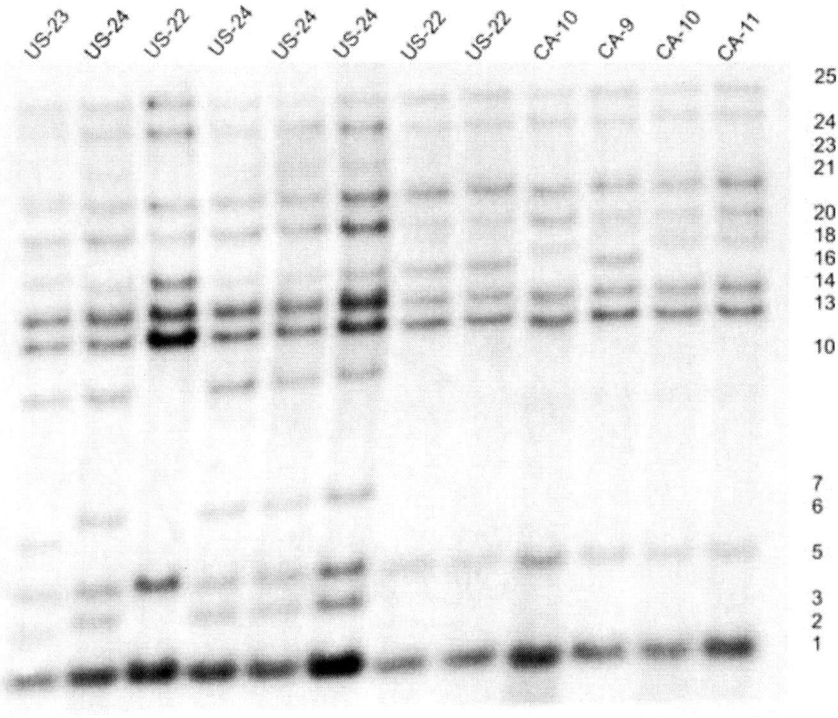

Figure 3 Restriction fragment length polymorphism (RFLP) profile of dominant and recombinant Phytophthora infestans genotypes from representative isolates recovered in 2011 on potato and tomato samples (n = 138) collected across Canada. Genotype designations are indicated above each lane and the RG57 products are shown to the right of the blot. Genotype CA-10 and CA-11 were distinguished by the unique DNA 1,5,13,14,18,20,21,24,25 RG57 RFLP profile. Origins of each P. infestans genotype are US-22, ON; US-23, NB; US-24, AB and MB; and CA-9, CA-10, and CA-11, ON.

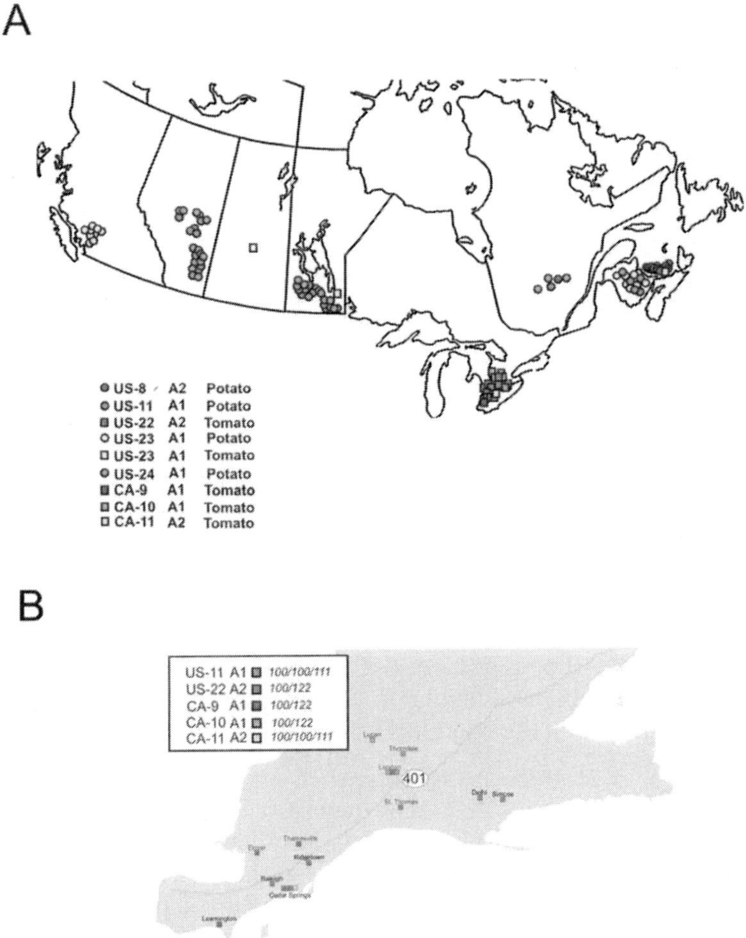

Figure 4 Locations of Phytophthora infestans populations and mating type distribution in major potato production areas of Canada sampled during 2011 from which samples with symptoms of late blight were collected. Potato samples (US-8, n = 51; US-11, n = 10; US-23, n = 16; US-24, n = 41) are indicated by circles and tomato samples (US-11, n = 1; US-22, n = 11; US-23, n = 3; CA-9, n = 1; CA-10, n = 4; CA-11, n = 1) are shown as squares. (A) Areas show general distribution and diversity of genotypes and geographic proximity of the P. infestans mating types A1 (in BC, AB, SK, MB and NB), A1 and A2 (ON and QC), and A2 (PE). (B) Distribution of the P. infestans isolates in ON showing the proximity of the genotypes US-11 A1 and US-22 A2 relative to the recombinant genotypes CA-9 A1, CA-10 A1 and CA-11 A2. Although the P. infestans genotypes occur in extremely close proximity, the A1 mating types occur to the north and extreme south of the A2 mating types.

Isolates of P. infestans recovered from tomato in ON in 2011 were determined to be mostly genotype US-22 mating type A2 (10/17 isolates). This genotype has not been reported in any other areas of Canada. Only a single isolate of US-11 mating type A1 was collected from ON tomato samples (Figure 4 and Table 1). Interestingly, several P. infestans isolates producing the unique RG57 RFLP pattern 1,5,14,18, 20,21,24,25 or unique combinations of associated Gpi allozyme and mating type, were also isolated from tomato in ON and designated CA-9, CA-10 and CA-11 following the recommendations by Forbes et al. (1998) (Table 1). The new RFLP profiles resemble that of P. infestans genotype US-22 but other characteristics correspond to the US-11 genotype. Additional evi-dence indicating similarity of the P. infestans CA genotypes to the US-11 genotype include the occurrence of the Gpi allozyme 100/100/111 profile in CA-11 and the same RG57 RFLP product 18 in US-11, CA-10 and CA-11 (Table 1).

Of the 138 P. infestans isolates recovered from across Canada in 2011, the six ON isolates were the only ones that showed independent segregation between the loci for mating type, Gpi allozyme and the multilocus RG57 sequence RFLP profile (Table 1). These P. infestans iso-lates from ON resembled US-22 (A2 mating type, Gpi 100/122 and RG57 RFLP 1,5,14,16,20,21,24,25 profile) but had the A1 mating type (CA-9) or produced the new 1,5,14,18,20,21,24,25 RG57 profile with either the A1 mating type (CA-10) or the US-11 Gpi 100/100/111 pro-file (CA-11). The CA-9 P. infestans isolate is unique in that the mating type has segregated independently from all the other loci used to determine genotype.

All recombinant P. infestans genotypes were recovered from infected tomato and resembled US-22 in that they were sensitive to metalaxyl and have the Ia mtDNA haplotype (Table 1). Isolation of the new P. infestans genotypes from the same region suggests that they are related and capable of migration to cause subsequent infections in surrounding areas. Results with the micro-satellite analysis showed additional polymorphisms within the recombinant P. isolates CA-10.1 (Pi89 179/ 179) and CA-11 (Pi4B 213/226), as compared to the US-22 genotype products Pi4B 213/213, PiG11 131/155, D13 106/106, Pi04 166/170, Pi06 177/177, Pi56 230/250/255, Pi63 279/279, and Pi89 177/179. Cluster analysis of the data for the genotypic specific markers from all the P. infestans isolates further indicated that the recombinants CA-9, CA-10 and CA-11 were most closely related to US-11 and US-22 (Figure 5).

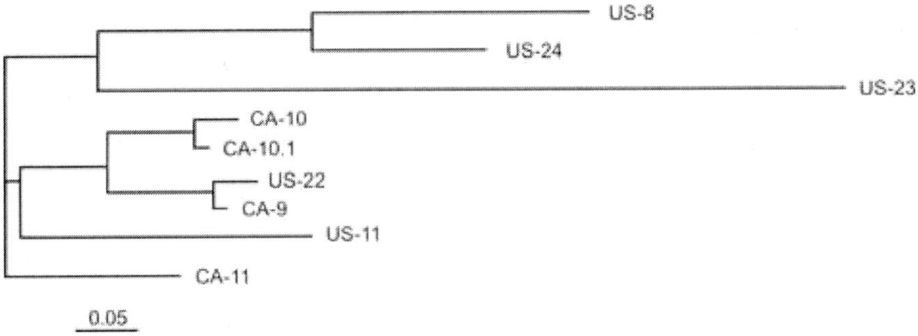

Figure 5 Cluster analysis of the genotype specific markers from the Phytophthora infestans isolates in this study based on a Jaccard distance coefficient of restriction fragment polymorphism, mating type, and allozyme allelic diversity. Genotypes are indicated by the International Organization for Standardization (ISO) two-letter country code followed by a unique number (see Table 1).

DISCUSSION

Genetic changes in the late blight pathogen population have occurred rapidly on potato and tomato in Canada during 2011 compared to 2010 (Kalischuk et al. 2012). This change is primarily due to clonal propagation of the P. infestans US-8, US-11, US-22, US-23 and US-24 genotypes (Table 1). Previous studies had shown the occurrence of these genotypes in specific geographical areas. Therefore, the rapid increase in genetic complexity and composition seen in this study was unexpected based on the only relatively recent emergence in the United States and Canada of P. infestans genotypes such as US-22, US-23 and US-24 (Kawchuk et al. 2011; Hu et al. 2012; Kalischuk et al. 2012). Clonal propagation and movement across the continent appears to still dom-inate P. infestans population dynamics but we now have evidence of recombination in an area of Canada (ON) where the A1 and A2 mating types occurred in close proximity. In particular, several new multiloci P. infestans genotypes were isolated from tomato, which show inde-pendent segregation of mating type and genotype specific loci. This development may further contribute to the increasing P. infestans population diversity in Canada.

Prior to 2011, the dominant P. infestans genotype in western Canadian provinces (BC, AB and SK) was reported to be US-23 (Kawchuk et al. 2011; Kalischuk et al. 2012). Initially isolated from tomato in retail garden centers and later found on field tomatoes and potatoes, an extensive awareness program for growers during 2011 probably helped to dramatically reduce the occurrence of US-23 in the Prairie provinces during 2011 (Kalischuk et al. 2012). Nevertheless, all of the P. infestans isolates obtained from BC in 2011 were US-23 and the first occurrence of the P. infestans genotype US-23 in eastern Canada was detected in NB during 2011 in several dis-eased potato samples. Many samples collected in MB in 2010 were infected with the US-24 genotype, which had not been previously reported in any other Canadian pro-vinces (Kalischuk et al. 2012). However, migration and proliferation of US-24 into AB and NB has now made this one of the dominant potato-infecting genotypes in Canada. In AB and ON, genotype US-11 was isolated from both potato and tomato. Although all of the P. infestans genotypes from western Canada were still of A1 mating type in 2011, migrations have apparently introduced this mating type into the eastern Canadian provinces of ON, QC and PE where the A2 genotypes were already established. This development would increase the opportunity for sexual recombination and development of new genotypes due to having both mat-ing types co-exist in an area (Table 1).

The US-8 A2 genotype that dominated P. infestans populations in Canada during the 1990s (Peters et al. 1998, 2001; Platt et al. 1999; Chen et al. 2003) was still dominant in PE in 2011. Remarkably, this genotype dominated the province of NB in 2010 but was comple-tely replaced by the US-23 and US-24 genotypes in 2011. Introduction of P. infestans genotypes via infected potato seed, combined with environmental conditions conducive to the development of late blight, probably contributed to this rapid and province-wide displacement of pathotypes in NB as has been observed elsewhere (Fry et al. 1991; Fry et al. 2013; Goodwin et al. 1994a; Sujkowski et al. 1994; Day & Shattock 1997; Forbes et al. 1997). Isolates of the US-24 genotype were also recovered from diseased plant samples from QC, raising concerns that the occur-rence of both mating types in this area may again facil-itate sexual recombination. The resulting oospores would allow the pathogen to overwinter in the absence of the host and potentially produce new recombinant genotypes. However, previous studies have reported the presence of the opposite mating types of P. infestans in close associa-tion without any evidence for sexual recombination (Deahl et al. 1991; Goodwin et al. 1995a; Chycoski & Punja 1996).

In ON, while most of the P. infestans isolates recovered from diseased tomato tissues were identified by allozyme and RFLP analyses to be the US-22 A2 P. infestans genotype, one US-11 A1 isolate was recovered in close proximity to the isolates of the opposite mating type. This could potentially provide an opportunity for sexual propagation. Examination showed that isolates of the ON and AB US-11 P. infestans genotype shared the unique IIb mtDNA haplotype (Griffith & Shaw 1998). A recent analysis of P. infestans populations in the United States found that all isolates were of the mtDNA Ia haplotype (Hu et al. 2012).

Evidence of recombination in ON was first detected through the recovery of a P. infestans isolate designated CA-9. Although this isolate had a Gpi profile 100/122 and an RG57 RFLP 1,5,14,16,20,21,24,25 pattern iden-tical to that seen in P. infestans US-22 isolates, the CA-9 genotype was not an A2 but rather an A1 mating type and therefore capable of sexually recombining with US-22 A2 isolates. To our knowledge, this is the first reported independent segregation between the mating type and the RG57 loci. These results indicate that although link-age distances are probably small, they are sufficient for recombination and thus independent segregation to occur.

Additional evidence for recombination in the ON P. infestans population was observed in several other unique isolates obtained from diseased tomato plants. Isolates having the Gpi 100/122 profile of the P. infestans US-22 genotype were recovered that were of the A1 mating type and had the new RG57 RFLP profile 1,5,14,18,20,21,24,25. These P. infestans isolates were designated CA-10 following the nomenclature recom-mendation that there be a minimum of two polymorphic allozyme or DNA loci (Goodwin et al. 1994b; Forbes et al. 1998). These guidelines will need to be expanded and include a change in mating type to accommodate P. infestans genotypes such as CA-9. Finally, another isolate of P. infestans, designated CA-11, was also recov-ered from diseased tomato samples in ON. This isolate displayed the RG57 RFLP profile of the CA-10 genotype but was an A2 mating type and had the Gpi profile 100/ 100/111 associated with the US-11 and US-24 genotypes. This is similar to a P. infestans isolate not assigned a genotype from the Columbia Basin (Miller et al. 1997). As with CA-9, where mating type segregated indepen-dently of the Gpi profile and RG57 RFLP loci, CA-11 data shows that the Gpi profile may segregate indepen-dently of the mating type and RG57 RFLP loci.

Because the isolated CA-genotypes cannot always be determined from the RG57 RFLP profile alone, our data show that identification of future P. infestans genotypes needs to also include the Gpi allozyme pattern and mating type. Microsatellites can produce additional polymorph-isms to further differentiate closely related P. infestans isolates such as CA-10 and CA-10.1. Several factors suggest that the recombinant isolates were produced from sexual recombination between P. infestans geno-types US-11 mating type A1 and US-22 mating type A2. These particular genotypes were found in relatively close proximity to each other in ON, and the recombinant isolates had RG57 RFLP and Gpi allozyme profiles like those of P. infestans US-11 and US-22 isolates. Such recombination is not without precedent, as recombination between US-6 and US-8 may have been the source for several P. infestans genotypes, including US-11 (Goodwin et al. 1998). In Canada and the United States, the potential recombinant genotypes have always made up only a small proportion of the total samples. In con-trast, nearly 100% of the isolates from Europe that were tested were recombinant in origin (Drenth et al. 1993; Sujkowski et al. 1994; Anderson et al. 2009). Continued monitoring will be required to determine whether sexu-ally reproducing populations of P. infestans have now become established in Canada and how this development might affect late blight epidemiology in subsequent years.

In conclusion, our analyses revealed that the P. infes-tans population complexity in Canada has increased since a survey that was taken as recently as 2010 (Kalischuk et al. 2012). This change was predominantly through clonal propagation and migration. The increased com-plexity has contributed to the occurrence of both mating types in similar geographical areas and appears to have generated recombinants that independently segregate for various loci, including mating type. The presence of new genotypes in different regions of Canada increases con-cern regarding disease control measures, as new geno-types may have unique epidemiological characteristics such as host and environmental preferences. Further glo-bal movement of potato seed tubers and transplants of susceptible hosts, such as tomato, is providing an unpre-cedented opportunity for rapid migration of P. infestans genotypes into new areas within a single season, and is contributing to increased opportunities for sexual repro-duction. Models predicting the onset and development of late blight will need to be evaluated and adjusted to cope with the characteristics of new or dominant genotypes. Monitoring the distribution and evolution of P. infestans should provide early warning of the anticipated genotypes influencing late blight epidemiology and facilitate the development of more effective integrated methods for disease management.

ACKNOWLEDGMENTS

The authors wish to thank the many collaborators that supplied infected tissues or P. infestans isolates. This pro-ject was supported in part by the Canadian Horticultural Council, Alberta Crop Industry Development Fund Ltd. Project 2010C001R and AAFC Project 1751.

REFERENCES

1. Anderson PK, Widmark A-K, Yuen JE, Nielsen B, Rovnskov S, Kessel GJT, Evenhuis A, Turkensteen IJ, Hansen JG, Lehtinen A et al. 2009. The role of oospores in the epidemiology of potato late blight. Acta Hort. 834:61–68.
2. Caten CE, Jinks JL. 1968. Spontaneous variability of single isolates of Phytophthora infestans. I. Cultural variation. Can J Bot. 46:329–348.
3. Chen Q, Kawchuk LM, Lynch DR, Goettel MS, Fujimoto DK. 2003. Identification of late blight, Colorado potato beetle, and blackleg resistance in three Mexican and two South American wild 2x (1EBN) Solanum species. Am J Potato Res. 80:9–19.
4. Chycoski CI, Punja ZK. 1996. Characteristics of populations of Phytophthora infestans from potato in British Columbia and other regions of Canada during 1993 to 1995. Plant Dis. 80:579–589.
5. Daayf F, Platt HW. 2000. Changes in metalaxyl resistance among glucose phosphate isomerase genotypes of Phytophthora infestans in Canada during 1997 and 1998. Am J Potato Res. 77:311–318.
6. Daayf F, Platt HWB, Peters RD. 2000. Changes in mating types, resistance to metalaxyl, and Gpi-allozyme genotypes of Phytophthora infestans in Canadian provinces from 1996 to 1998. Can J Plant Pathol. 22:110–116.
7. Danies G, Small IM, Myers K, Childers R, Fry WE. 2013. Phenotypic characterization of recent clonal lineages of Phytophthora infestans in the United States. Plant Dis. 97:873-881.
8. Day JP, Shattock RC. 1997. Aggressiveness and other factors relating to displacement of populations of Phytophthora infestans in England and Wales. Eur J Plant Pathol. 103:379–391.
9. Deahl KL, Goth RW, Young R, Sinden SL, Gallegly ME. 1991. Occurrence of the A2 mating type of Phytophthora infestans in potato fields in the United States and Canada. Am Potato J. 68:717–725.
10. Deahl KL, Inglis DA. 1995. Occurrence of Metalaxyl-insensitive Phytophthora infestans on Solanum sarachoides in Northwestern Washington. Plant Dis. 79:540.
11. Deahl KL, Inglis DA, DeMuth SP. 1993. Testing for resistance to metalaxyl in Phytophthora infestans isolates from northwestern Washington. Am Potato J. 70:779–795.
12. Drenth A, Goodwin SB, Fry WE, Davidse LC. 1993. Genotypic diversity of Phytophthora infestans in the Netherlands revealed by DNA polymorphisms. Phytopathology. 83:1087–1092.
13. Entz SC, Johnson DL, Kawchuk LM. 2005. Development of a PCR-based diagnostic assay for the specific detection of the entomopathogenic fungus Metarhizium anisopliae var. acridum. Mycol Res. 109:1302–1312.
14. Forbes GA, Escobar XC, Ayala CC, Revelo J, Ordoñez ME, Fry BA, Doucett K, Fry WE. 1997. Population genetic structure of Phytophthora infestans in Ecuador. Phytopathology. 87:375–380.
15. Forbes GA, Goodwin SB, Drenth A, Oyarzun P, Ordoñez ME, Fry WE. 1998. A global marker database for Phytophthora infestans. Plant Dis. 82:811–818.
16. Fry WE, Drenth A, Spielman LJ, Mantel BC, Davidse LC, Goodwin SB. 1991. Population genetic structure of Phytophthora infestans in the Netherlands. Phytopathology. 81:1330–1336.
17. Fry WE, McGrath MT, Seaman A, Zitter TA, McLeod A, Danies G, Small IM, Myers K, Everts K, Gevens AJ, Gugino B.K., Johnson S.B., Judelson H., Ristaino J., Roberts P., Secor G., Seebold K., Snover-Clift K., Wyenandt A., Grünwald N.J., Smart C.D.. 2013. The 2009 late blight pandemic in the eastern United States – causes and results. Plant Disease. 97:296–306.
18. Goodwin SB. 1997. The population genetics of Phytophthora. Phytopathology. 87:462–473.

19. Goodwin SB, Cohen BA, Deahl KL, Fry WE. 1994b. Migration from northern Mexico as the probable cause of recent genetic changes in populations of Phytophthora infestans in the United States and Canada. Phytopathology. 84:553–558.
20. Goodwin SB, Cohen BA, Fry WE. 1994a. Panglobal distribution of a single clonal lineage of the Irish potato famine fungus. Proc Natl Acad Sci. 91:11591–11595.
21. Goodwin SB, Drenth A, Fry WE. 1992. Cloning and genetic analyses of two highly polymorphic, moderately repetitive nuclear DNAs from Phytophthora infestans. Curr Genet. 22:107–115.
22. Goodwin SB, Schneider RE, Fry WE. 1995c. Use of cellulose acetate electrophoresis for rapid identification of allozyme genotypes of Phytophthora infestans. Plant Dis. 79:1181–1185.
23. Goodwin SB, Smart CD, Sandrock RW, Deahl KL, Punja ZK, Fry WE. 1998. Genetic change within populations of Phytophthora infestans in the United States and Canada during 1994 to 1996: role of migration and recombination. Phytopathology. 88:939–949.
24. Goodwin SB, Sujkowski LS, Dyer AT, Fry BA, Fry WE. 1995a. Direct detection of gene flow and probable sexual reproduction of Phytophthora infestans in northern North America. Phytopathology. 85:473–479.
25. Goodwin SB, Sujkowski LS, Fry WE. 1995b. Rapid evolution of pathogenicity within clonal lineages of the potato late blight disease fungus. Phytopathology. 85:669–676.
26. Goodwin SB, Sujkowski LS, Fry WE. 1996. Widespread distribution and probable origin of resistance to metalaxyl in clonal genotypes of Phytophthora infestans in the United States and western Canada. Phytopathology. 86:793–800. 10.1094/Phyto-86-793
27. Griffith GW, Shaw DS. 1998. Polymorphisms in Phytophthora infestans: four mitochondrial haplotypes are detected after PCR amplification of DNA from pure cultures or from host lesions. Appl Environ Microbiol. 64:4007–4014.
28. Haas BJ, Kamoun S, Zody MC, Jiang RHY, Handsaker RE, Cano LM, Grabherr M, Kodira CD, Raffaele S, Torto-Alalibo T et al. 2009. Genome sequence and analysis of the Irish potato famine pathogen Phytophthora infestans. Nature. 461:393–398.
29. Holley JD, Kawchuk LM. 1996. Distribution of thiabendazole and thiophanate-methyl resistant strains of Helminthosporium solani and Fusarium sambucinum in Alberta potato storages. Can Plant Dis Surv. 76:21–27.
30. Hu C-H, Perez F, Donahoo R, McLeod A, Myers K, Ivors K, Secor G, Roberts P, Deahl K, Fry WE, Ristaino JB. 2012. Recent Genotypes of Phytophthora infestans in the Eastern United States Reveal Clonal Populations and Reappearance of Mefenoxam Sensitivity. Plant Dis. 96:1323–1330.
31. Kalischuk M, Al-Mughrabi KI, Peters RD, Howard RJ, Platt HW, Kawchuk LM. 2012. Genetic composition of Phytophthora infestans in Canada reveals migration and increased diversity. Plant Dis. 96:1729–1735.
32. Kato M, Mizubuti ES, Goodwin SB, Fry WE. 1997. Sensitivity to protectant fungicides and pathogenic fitness of clonal lineages of Phytophthora infestans in the United States. Phytopathology. 87:973–978.
33. Kawchuk LM, Howard RJ, Peters RD, Al-Mughrabi KI. 2011. First report of Phytophthora infestans genotype US23 causing late blight in Canada. Plant Dis. 95:873.
34. Khadhair A-H, Kawchuk LM, Taillon RC, Botar G. 1998. Detection and molecular characterization of an aster yellows phytoplasma in parsley. Can J Plant Pathol. 20:55–61.
35. Lees AK, Wattier R, Shaw DS, Sullivan L, Williams NA, Cooke DEL. 2006. Novel microsatellite markers for the analysis of Phytophthora infestans populations. Plant Pathol. 55:311–319.
36. Legard DE, Lee TY, Fry WE. 1995. Pathogenic specialization in Phytophthora infestans: Aggressiveness on tomato. Phytopathology. 85:1356–1361.
37. Marshall-Farrar KD, McGrath M, James RV, Stevenson WR. 1998. Characterization of Phytophthora infestans in Wisconsin from 1993 to 1995. Plant Dis. 82:434–436.

38. Miller JS, Hamm PB, Johnson DA. 1997. Characterization of the Phytophthora infestans population in the Columbia Basin of Oregon and Washington from 1992 to 1995. Phytopathology. 87:656–660.
39. Miller PM. 1955. V-8 juice agar as a general-purpose medium for fungi and bacteria. Phytopathology. 45:461–462.
40. Oudemans P, Coffey MD. 1991. A revised systematics of twelve papillate Phytophthora species based on isozyme analysis. Mycol Res. 95:1025–1046.
41. Peters RD, Förster H, Platt HW, Hall R, Coffey MD. 2001. Novel genotypes of Phytophthora infestans in Canada during 1994 and 1995. Am J Potato Res. 78:39–45.
42. Peters RD, Platt HW, Hall R. 1998. Characterization of changes in populations of Phytophthora infestans in Canada using mating type and metalaxyl sensitivity markers. Can J Plant Pathol. 20:259–273.
43. Platt HW, Peters RD, Medina M, Arsenault W. 1999. Impact of seed potatoes infected with Phytophthora infestans (US-1 or US-8 genotypes) on crop growth and disease risk. Am J Potato Res. 76:67–73.
44. Spielman LJ, Drenth A, Davidse LC, Sujkowski LJ, Gu WK, Tooley PW, Fry WE. 1991. A second world-wide migration and population displacement of Phytophthora infestans?. Plant Pathol. 40:422–430.
45. Sujkowski LS, Goodwin SB, Dyer AT, Fry WE. 1994. Increased genotypic diversity via migration and possible occurrence of sexual reproduction of Phytophthora infestans in Poland. Phytopathology. 84:201–207.

CHAPTER 18

Recessive and Dominant Genes Confer Resistance to *Colletotrichum Truncatum* in Cultivated Lentil

Lone Buchwaldt[1], Rubeena Shaikh[1], Jennifer Adam[1], Abebe Tullu[2] And Alfred E. Slinkard[2]

[1]*Agriculture and Agri-Food Canada, 107 Science Place, Saskatoon, SK, S7N 0X2, Canada*

[2]*Crop Development Centre, University of Saskatchewan, 51 Campus Drive, Saskatoon, SK, S7N 5A8, Canada*

ABSTRACT

Anthracnose caused by the fungal pathogen *Colletotrichum truncatum* is one of the most damaging diseases of lentil in western Canada. Two physiological races, Ct0 and Ct1, exist of this pathogen and lentil accessions have been identified with resistance to one or both races. In this study, the genetics of resistance to race Ct1 was examined in cultivar 'Indianhead' (Czechoslovakia) and two plant introductions PI320937 (Germany) and PI345629 (former Soviet Union). Segregating populations were generated from crosses between each resistant line and a susceptible cultivar 'Eston' and from crosses among the three resistant lines to determine gene allelism. Populations of F_1, F_2, BC_1R and BC_1S were screened with two isolates, 91IH and 95B36, separately, both characterized as race Ct1 while F_3 families were screened with one isolate, 91IH. Plants were grown under controlled conditions in a growth chamber and inoculated at early flowering with a conidial suspension of *C. truncatum*. Disease severity was rated 14 days after inoculation as number of lesions on the main stem, depth of penetration and degree of shoot die-back or plant wilt. Chi-square analyses of observed segregation ratios of resistant and susceptible plants in each population revealed that resistance is governed by two recessive genes, *ctr1* and *ctr2*, and three closely linked dominant genes, *CtR3*, *CtR4* and *CtR5*. The resistance genes *ctr2CtR5* in PI345629 and *CtR4* in PI320937 were detected by both isolates, while *ctr1* and *CtR3* in 'Indianhead' were detected by isolate 91IH and 95B36, respectively. These data demonstrate that race Ct1 consists of different genotypes, which however cannot be differentiated using the seven resistant lentil lines initially used to characterize race Ct0 and Ct1. A set of lentil lines with single resistance genes is needed to capture more of the pathogenic variability in *C. truncatum*.

Keywords: allelism, anthracnose, inheritance, *Lens culinaris* subsp. *culinaris*, races

INTRODUCTION

Canada is the second largest producer of lentil (*Lens culinaris* Medikus subsp. *culinaris*) after India and has become the world's major exporter. A total of 970,000 ha were planted to lentil in 2009 with 95% in the province of Saskatchewan and the remainder in Alberta and Manitoba. Anthracnose, caused by the fungal pathogen

Colletotrichum truncatum (Schwein.) Andrus & W.D. Moore, is widespread in western Canada and has resulted in yield losses up to 60% (Morrall, 1997). Areas in Manitoba with intensive lentil production were devastated by widespread anthracnose in the 1990s when the dis-ease was found in 40–50% of commercial lentil fields (Buchwaldt *et al.*, 1992; Platford & McCallum, 1996). The disease spread to North Dakota (Venette *et al.*, 1994) and Saskatchewan, where anthracnose was identified in 30–60% of fields surveyed in 1998 and 1999 (Anderson *et al.*, 1999, 2000).

The first anthracnose symptoms in lentil appear at early flowering when tan necrotic lesions develop on leaflets, leading to premature defoliation. Necrotic lesions with a typical black margin begin forming at the base of the plants and spread to the stems over the growing season. The pathogen penetrates into the vascular tissue, caus-ing the plant to wilt, and as a result large patches of drying plants develop in the field. The pathogen forms numerous microsclerotia on infected tissues which are the primary source of inoculum for subsequent lentil crops (Buchwaldt *et al.*, 1996). During the first anthracnose epi-demics in western Canada, all lentil cultivars were highly susceptible except the cultivar 'Indianhead' (Gibson, 1994). However, this cultivar was not widely grown due to its black seed coat colour which is not one of the mar-ketable seed classes. Resistance from 'Indianhead' was therefore crossed into breeding lines with commercially desirable seed characteristics (Buchwaldt *et al.*, 1995), which eventually resulted in development of cultivars such as 'CDC Robin' and 'CDC Redberry' with improved anthracnose resistance (Vandenberg *et al.*, 2002, 2006). Lack of other sources of resistance prompted screen-ing of more than 2300 accessions of cultivated lentil (*L. culinaris*) obtained from gene bank collections in USA, Germany, Russia, Bulgaria, Poland and Hungary (Buchwaldt *et al.*, 2004; Shaikh *et al.*, 2012). The search for resistance simultaneously revealed the presence of physiological races of *C. truncatum* (Buchwaldt *et al.*, 2004). The race designation was used to denote a spec-trum of pathogenicity against differential varieties as described by Flor (1970), which can be used whether or not the genetic basis of resistance has been elucidated. The two races were identified by inoculation of lentil lines, some of which were resistant to isolates of race Ct0, while others were resistant to isolates of race Ct1. Lines that are resistant to race Ct1 are susceptible to race Ct0 and vice versa, except for line VIR2633 which is resistant to both races (Buchwaldt *et al.*, 2004; Shaikh *et al.*, 2012). The majority of the resistant lentil accessions are lan-draces originating primarily in Czechoslovakia and other East European countries. The objective of the present study was to examine the inheritance of anthracnose resistance in three *L. culinaris* accessions from different geographic regions using two *C. truncatum* isolates both characterized as race Ct1.

MATERIALS AND METHODS

Lentil populations

Three *L. culinaris* lines were selected from a pre-vious study based on their resistance to *C. trunca-tum* race Ct1 and country of origin (Buchwaldt *et al.*, 2004). These were cultivar 'Indianhead' originally col-lected in Czechoslovakia and two plant introductions, PI320937 and PI345629, originating in Germany and the former Soviet Union, respectively. The two PI accessions were obtained from the National Germplasm System, USDA-ARS (Pullman, WA, USA). The Canadian cultivar 'Eston' was selected as the susceptible parent.

The number of resistance genes in each lentil line was determined by screening of F_1, BC_1R, BC_1S, F_2 and F_3 plants derived from crosses between the sus-ceptible parent 'Eston' and each of the three resistant lines, 'Eston' × 'Indianhead', 'Eston' × PI320937 and 'Eston' × PI345629 (Tables 1–3). Two isolates, 91IH and 95B36, were used to screen these populations, except in 'Eston' × PI345629, which was screened with one isolate, 91IH, due to the lack of seed. Dominance of the resistant trait was determined in F_1 in all cases, and confirmed in back crosses in four of the five populations. The number of genes conferring resistance was verified by screening three large F_3 populations using isolate 91IH. The allelic relationship among resistance genes was determined by screening of F_1 and F_2 progenies derived from crosses among the three resistant lines, 'Indianhead' × PI320937, 'Indianhead' × PI345629, and PI320937 × PI345629 (Tables 4–6). A large number of F_2 plants were screened with isolate 91IH and 95B36, separately, to determine whether two or three resistance genes were segregating. Dominance was determined by inoculation of F_1 plants with either isolate, except in one case when not enough seed was available.

Each successful cross-pollinated bud produced one or two F_1 seeds. Morphological traits were used to deter-mine whether F_1 were true hybrids such as flower and cotyledon colour which segregated in subsequent generations. The parental entries 'Eston' and PI345629 have white flowers while 'Indianhead' and PI320937 have pur-ple flowers. 'Indianhead' has red cotyledons compared with yellow for the other entries; and finally, black seed coat colour is characteristic of both 'Indianhead' and PI320937.

The F_1 seeds were divided into three groups, one for disease screening, another for back crossing to the resistant (BC_1R) and susceptible (BC_1S) parents, and a third group was selfed to produce F_2 seeds. Likewise, the F_2 seeds were divided into two groups, one group for disease screening and another for selfing to produce F_3 families, since susceptible F_2 plants used for disease screening would not produce seed and therefore result in a skewed F_3 segregation ratio.

Before sowing, all seeds were scarified by removing a piece of the seed coat with a scalpel. Seeds of F_1, BC_1 and F_2 were sown in 10 cm^2 pots. Each pot con-tained 4–6 test plants and two plants of the susceptible 'Eston' to assess whether disease pressure was adequate. In addition, 4–5 pots each of 'Eston' and the appropriate resistant parent were included at every test date. Batches of 150–200 F_1, BC_1 or F_2 plants derived from the same cross were planted and inoculated together on the same test date. Seeds of 50 F_3 families per cross were sown in trays (55 cm wide × 75 cm long × 15 cm deep) each with 10 rows and 10 seeds per row. Twenty seeds of each F_3 family were sown in two different trays along with one row each of 'Eston' and the resistant parent of the particular cross. Both pots and trays were filled with a soil-less mixture (Sunshine Germination Mix 3, Sun Gro

Horticulture, Vancouver, Canada) and fertilized weekly with 20 N : 20 P : 20 K. Pots and trays were ran-domized and placed in a growth cabinet maintained at 21/16 ± 2 °C day/night temperature and 16 h photope-riod provided by a mixture of cool-white fluorescent and incandescent lamps.

C. truncatum isolates and disease screening

Two single-spore *C. truncatum* isolates, 91IH and 95B36, were selected from a previous study (Buchwaldt *et al.*, 2004). Isolate 91IH was collected from the resistant culti-var 'Indianhead' and 95B36 from the susceptible cultivar 'Eston'. Both isolates were characterized as race Ct1 since they were non-pathogenic on a set of seven lentil acces-sions, but pathogenic on cultivar 'Eston'.

To maintain infectivity, the two isolates were sepa-rately inoculated and re-isolated from 'Eston' and infected tissue plated on half-strength oat meal agar ($^1/_2$ OMA) (Difco) in 9-cm Petri dishes. Stock cultures were made by transferring conidia and microsclerotia from the grow-ing margin to sterilized cryo-freezer solution (10% skim milk, 40% glycerol) in 2 mL vials which were stored at −80 °C as previously described (Shaikh *et al.*, 2012). Inoculum was produced by transferring 30–50 μL of a stock solution to $^1/_2$ OMA in Petri dishes. The plates were incubated at 20/16 ± 1 °C day/night cycle in a 16 h photoperiod provided by 3 : 1 cool-white fluorescent light and black light. Conidial suspension for inoculation was prepared by flooding 7 to 10-day-old cultures with ster-ile distilled water and filtering the suspension through one layer of Miracloth (Calbiochem-Behring Corp., La Jolla, CA). The final concentration was adjusted to 10^5 conidia mL^{-1} as described previously for screening of lentil accessions (Buchwaldt *et al.*, 2004; Shaikh *et al.*, 2012). Inoculation was carried out 3–4 weeks after sowing when the majority of the lentil plants had 10–12 intern-odes and flower buds on the oldest branches. Plants were inoculated with a conidial suspension of either isolate 91IH or 95B36 at 1.5 mL per plant using an airbrush sprayer (Paasche Airbrush Company, IL, USA) set at 125–150 kPa. The plants were incubated for 24 h in a humidity chamber at 100 % relative humidity provided by a cool-mist humidifier. Light was reduced to a min-imum to maintain the temperature at 21 ± 2 °C. After incubation, the plants were transferred to a greenhouse maintained at 20–24/15–19 °C day/night temperature range with a supplement of high-pressure sodium light for a 16 h daylength.

Disease rating and data analysis

Symptoms on the main stem of individual plants were rated 14 days after inoculation. The symptoms were assessed both quantitatively and qualitatively as described previously (Buchwaldt *et al.*, 2004). The number of lesions on the main stem was recorded followed by a notation of the position on the stem, depth of penetration and the resulting amount of plant wilting. The position of lesions was noted as stem base, lower half of the stem or top half of the stem. Depth of penetration into the stem was noted as superficial, deep or a mixture of both superfi-cial and deep lesions. Finally, the amount of plant wilting was noted as shoot die-back, half of the plant wilted or the whole plant wilted. The resistant reaction ranged from zero or only superficial stem lesions to a mixture of up to 10 lesions confined to the lower part of the stem but not causing shoot die back or wilt. The susceptible reac-tion ranged from more than 10 deep lesions including the upper part of the plant to more than 30 lesions resulting in shoot die back or plant wilting. The observed resistant and susceptible segregation ratios were analysed using the chi-square (χ^2) test for goodness of fit to the expected ratios, and probability values (P) were calculated for goodness of fit to selected genetic models.

RESULTS

Disease reaction of parental lines

The disease severities of 'Eston' plants sown in each pot along with test plants were used to confirm that adequate infectivity was obtained and were not included in Tables 1 to 6. There were 21 test dates with appropriate levels of infectivity. Leaf lesions on 'Eston' were severe and caused high levels of defoliation. In contrast, leaves of PI320937 developed few lesions, while 'Indianhead' and PI345629 developed variable degrees of leaf symptoms and some defoliation. It was therefore not possible to reliably assess the degree of leaf symptoms on individual progeny plants.

Resistance in 'Indianhead' is controlled by ctr1CtR3

Inoculation of 17 F1 plants derived from 'Eston' × 'Indianhead' with isolate 91IH showed all were susceptible; screening of 264 F2 plants showed a segregation of 1 resistant : 3 susceptible, both showing the presence of a single recessive gene conferring resistance to C. truncatum (Table 1). This was confirmed in both back-crosses to the susceptible and resistant parent, respectively, since all BC1S plants were susceptible and BC1R plants segregated 1 resistant : 1 susceptible. The presence of a single gene was also evident in the segregation of 45 F3 families which complied satisfactorily to the ratio 1 homozygous resistant : 2 segregating : 1 homozygous susceptible when screened with isolate 91IH (Figure 1a). In contrast, inoculation with the other isolate, 95B36, showed that all F1 plants were resistant, indicating a dominant trait, which was confirmed by segregation of 1 resistant : 1 susceptible in BC1S plants, while all BC1R progenies were resistant (Table 1). The combined data obtained with two different C. truncatum isolates provided evidence that resistance in lentil cultivar 'Indianhead' is most likely controlled by a recessive and a dominant gene, ctr1CtR3. It is less likely that the gene is the same but showing different dominance relationship to each isolate

Resistance in PI320937 is controlled by CtR4

In the cross 'Eston' × PI320937, all F_1 plants were resistant when inoculated with either isolate 91IH or 95B36, pointing to the presence of dominant resistance

Table 1. Segregation of resistance to *C. truncatum* in *L. culinaris* 'Eston' × 'Indianhead' progenies providing evidence for one recessive resistance gene, *ctr1*, effective against isolate 91IH and one dominant gene, *CtR3*, effective against isolate 95B36 in 'Indianhead'.

Parents and progenies	R-gene	Total plants	Observed ratio R : S	Expected ratio R : S	χ^2	P
Isolate 91IH						
'Eston'		79				
'Indianhead'	*ctr1~~CtR3~~*	53				
F_1		17	0 : 17	All S		
BC_1S		9	0 : 9	All S		
BC_1R		11	5 : 6	1 : 1		
F_2		264	78 : 186	1 : 3	2.91	0.09
F_3 families (45)		816	10 : 29 : 6	1 : 2 : 1	3.92	0.14
Isolate 95B36						
'Eston'		20				
'Indianhead'	*~~ctr1~~CtR3*	19				
F_1		10	10 : 0	All R		
BC_1S		45	21 : 24	1 : 1	0.20	0.65
BC_1R		48	48 : 0	All R		

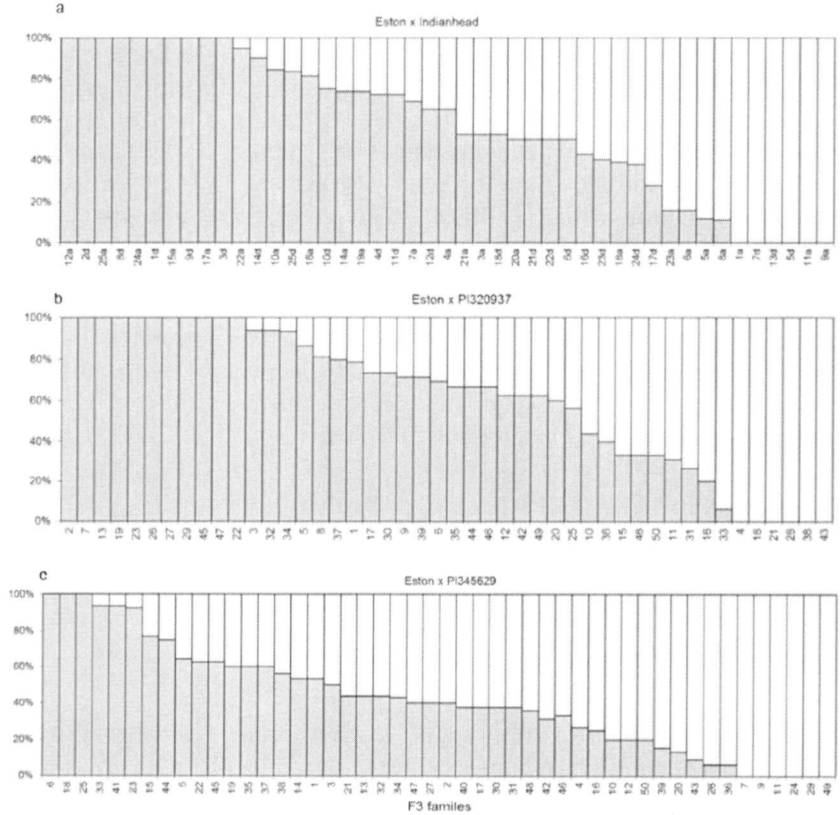

Figure 1. Per cent resistant (white bars) and susceptible (grey bars) lentil plants in three F_3 families resistant to *C. truncatum* isolate 91IH derived from crosses between the susceptible 'Eston' and each of (**a**) 'Indianhead', (**b**) PI320937 and (**c**) PI345629

(Table 2). Both groups of 192 and 126 F_2 plants seg-regated as 3 resistant : 1 susceptible when screened with either isolate, showing the presence of a single domi-nant gene (Table 2). Furthermore, all BC_1R plants were resistant when inoculated with isolate 95B36, while BC_1S progenies segregated as 1 resistant : 1 susceptible, thus demonstrating a single dominant gene for resistance. The presence of a single gene was also evident in the segregation of 46 F_3 families which conformed satisfac-torily to the ratio of 1 homozygous resistant : 2 segre-gating : 1 homozygous susceptible when inoculated with isolate 91IH (Figure 1*b*). The combined data showed that resistance to *C. truncatum* in lentil accession PI320937 is most likely governed by a single dominant gene, *CtR4*, effective against both isolates 91IH and 95B36. However, two genes effective against each isolate cannot be ruled out.

Resistance in PI345629 is controlled by ctr2CtR5

In the cross 'Eston' × PI345629, the 287 F_2 plants segre-gated as 13 resistant : 3 susceptible when inoculated with isolate 91IH, indicating the presence of one dominant and one recessive resistance gene, *ctr2CtR5* (Table 3). This was also evident in back-cross plants of F_1 to the susceptible parent, BC_1S, which segregated as 1 resis-tant : 1 susceptible, while at the same time, the back-cross to the resistant parent, BC_1R, were all resistant. However, the segregation of 48 F3 families did not fit the expected 7 homozygous resistant : 6 segregating : 3 homozygous susceptible ratio for one recessive and one dominant gene when inoculated with isolate 91IH (Figure 1c). Only three F3 families were homozygous resistant, indicating that the dominant gene, CtR5, may need the presence of the recessive gene, ctr2, for expression of the same level of resistance as in the parent PI345629. F3 families are less informative when resistance is conferred by two or more genes, particularly if epistasis is involved

Table 2. Segregation of resistance to *C. truncatum* in *L. culinaris* 'Eston' × PI320937 progenies providing evidence for one dominant resistance gene, *CtR4*, in PI320937 effective against two isolates, 91IH and 95B36.

Parents and progenies	R-gene	Total plants	Observed ratio R : S	Expected ratio R : S	χ^2	P
Isolate 91IH						
'Eston'		34				
PI320937	*CtR4*	37				
F_1		17	17 : 0	All R		
F_2		192	140 : 52	3 : 1	0.44	0.50
F_3 families (46)		698	11 : 29 : 6	1 : 2 : 1	4.22	0.12
Isolate 95B36						
'Eston'		18				
PI320937	*CtR4*	20				
F_1		6	6 : 0	All R		
BC_1S		43	22 : 21	1 : 1	0.00	1.00
BC_1R		32	32 : 0	All R		
F_2		126	90 : 36	3 : 1	0.67	0.41

Table 3. Segregation of resistance to C. truncatum in L. culinaris 'Eston' × PI345629 progenies providing evidence for one recessive and one dominant resistance gene, ctr2CtR5, in PI345629 effective against isolate 91IH.

Parents and progenies	R-gene	Total plants	Observed ratio R : S	Expected ratio R : S	χ^2	P
Isolate 91IH						
'Eston'		50				
PI345629	*ctr2CtR5*	40				
F_1		8	8 : 0	All R		
BC_1S		46	23 : 23	1 : 1		
BC_1R		42	42 : 0	All R		
F_2		287	207 : 80	13 : 3	15.42	0.00
F_3 families (48)		728	3 : 39 : 6	7 : 8 : 1	27.80	0.00

Allelism tests

To examine whether the resistance genes in 'Indianhead', PI 320937 and PI 345629 were allelic, the three resistant lines were inter-crossed to produce F1 and F2 populations. These progenies were inoculated separately with the same two isolates, 91IH and 95B36, as described for screening of progenies derived from the susceptible × resistant crosses. When screening of resistant × resistant crosses involving 'Indianhead', it was important to note that isolate 91IH interacted with the recessive gene ctr1 but not CtR3, while isolate 95B36 interacted with the dominant gene CtR3 but not ctr1. Both isolates detected the other resistance genes ctr2, CtR4 and CtR5 in crosses involving PI320937 and PI345629

All F1 plants derived from the resistant × resistant crosses were resistant, showing that dominant genes were involved in each case (Tables 4–6). Inoculation of 168 F2 plants derived from 'Indianhead' × PI320937 with isolate 91IH showed a 13 resistant : 3 susceptible ratio (Table 4), which is in agreement with segregation of one recessive gene, ctr1, in 'Indianhead' and one dominant gene, CtR4, in PI320937, as also determined in the susceptible × resistant crosses described above (Tables 1 and 2). Inoculation with the other isolate, 95B36, of 341 F2 plants from the same cross showed all were resistant, suggesting that the CtR3 gene in 'Indianhead' and CtR4 gene in PI320937 were very closely linked. Interestingly, the CtR3 and CtR4 genes are different, since CtR4 in PI320937 was detected by both isolates (Table 2), while CtR3 in 'Indianhead' was detected by one isolate, 95B36 (Table 1)

Table 4. Segregation of resistance to C. truncatum in L. culinaris 'Indianhead' × PI320937 progenies supporting the presence of ctr1CtR3 in 'Indianhead' and CtR4 in PI320937.

Parents and progenies	R-gene	Total plants	Observed ratio R : S	Expected ratio R : S	χ^2	P
Isolate 91IH						
'Indianhead'	*ctr1~~CtR3~~*	25				
PI320937	*CtR4*	39				
F_1		9	9 : 0	All R		
F_2		168	144 : 24	13 : 3	1.94	0.16
Isolate 95B36						
'Indianhead'	*~~ctr1~~CtR3*	20				
PI320937	*CtR4*	20				
F_1		27	27 : 0	All R		
F_2		341	341 : 0	15 : 1	22.37	0.00

Table 5. Segregation of resistance to C. truncatum in L. culinaris PI320937 × PI345629 progenies supporting the presence of ctr1CtR3 in 'Indianhead' and ctr2CtR5 in PI345629.

Parents and progenies	R-gene	Total plants	Observed ratio R : S	Expected ratio R : S	χ^2	P
Isolate 91IH						
'Indianhead'	*ctr1~~CtR3~~*	16				
PI345629	*ctr2CtR5*	31				
F_1		12	12 : 0	All R		
F_2		153	131 : 22	55 : 9	0.12	0.72
Isolate 95B36						
'Indianhead'	*~~ctr1~~CtR3*	20				
PI 345629	*ctr2CtR5*	19				
F_1		3	3 : 0	All R		
F_2		435	428 : 7	61 : 3	8.86	0.00

Inoculation with isolate 91IH of 153 F2 progenies derived from 'Indianhead' × PI345629 showed a 55 resistant : 9 susceptible ratio as expected for segregation of one dominant and two recessive resistance genes e.g. ctr1 in 'Indianhead' and ctr2CtR5 in PI345629 (Table 5). Inoculation with the other isolate, 95B36, of 435 F2 plants from the same cross would have shown segregation of 61 resistant : 3 susceptible if all genes were unlinked; however, there were 1.6% susceptible F2 plants compared with the expected 4.9%, which indicates that CtR3 in 'Indianhead' and CtR5 in PI345629 were linked, similar to the cross involving CtR3 in 'Indianhead' and CtR4 in PI320937 described above (Table 4).

Inoculation of 105 F2 plants derived from PI320937 × PI345629 with isolate 91IH showed a segregation of 61 resistant : 3 susceptible ratio as expected for one recessive and two dominant resistance genes (Table 6). The same segregation ratio was expected from inoculation of 491 F2 plants of the same cross with isolate 95B36. However, in this case, 3.2% susceptible plants were obtained instead of the expected 4.9%, which suggested a weak linkage between CtR4 in PI320937 and CtR5 in PI345629.

Based on the differential reaction of three lentil accessions with two pathogen isolates and extensive allelism tests, it was concluded that both recessive and dominant genes control resistance to C. truncatum in L. culinaris. The haploid gene notation for resistance to C. truncatum in the three lentil lines is ctr1CtR3 in 'Indianhead', CtR4 in PI320937 and ctr2CtR5 in PI345629. The data also revealed that the dominant genes CtR3, CtR4 and CtR5 are closely linked and likely different alleles at the same locus conferring resistance to C. truncatum.

DISCUSSION

This is the first comprehensive study of inheritance for resistance to anthracnose in cultivated lentil. Three resistant lentil accessions were selected from as wide a geographical area as possible to maximize the likelihood of identifying different resistance genes. Since the majority of anthracnose-resistant germplasm originate from Europe and Russia (Buchwaldt et al., 2004; Shaikh et al., 2012), resistant lines were chosen among germplasm from Germany (PI320937), Czechoslovakia ('Indianhead') and the former Soviet Union (PI345629).

Table 6. Segregation for resistance to C. truncatum in L. culinaris PI320937 × PI345629 progenies supporting the presence of CtR4 in PI320937 and ctr2CtR5 in PI345629.

Parents and progenies	R-gene	Total plants	Observed ratio R : S	Expected ratio R : S	χ^2	P
Isolate 91IH						
PI320937	*CtR4*	28				
PI345629	*ctr2CtR5*	20				
F_2		105	101 : 4	61 : 3	0.21	0.65
Isolate 95B36						
PI320937	*CtR4*	18				
PI345629	*ctr2CtR5*	20				
F_1		19	19 : 0	All R		
F_2		491	476 : 15	61 : 3	2.92	0.09

Colletotrichum truncatum isolates from lentil crops in western Canada comprise two physiological races, Ct0 and Ct1, which were identified by their differential reaction on the first resistant lentil lines that were identified (Buchwaldt et al., 2004). The current genetic study focused on inheritance of resistance to race Ct1. To begin elucidation of the pathogenic variability within C. truncatum, two isolates, both characterized as race Ct1, were used for screening of segregating generations. This strategy permitted identification of two recessive resistance genes, ctr1 and ctr2, and three closely linked dominant genes, CtR3, CtR4 and CtR5. The resistance genes ctr2CtR5 in PI345629 and CtR4 in PI320937 were detected by both isolates. In contrast, the recessive and dominant genes, ctr1 and CtR3, in 'Indianhead' were detected by separate isolates. The data revealed that C. truncatum isolates originally characterized as race Ct1 can be further divided based on their differential interaction with specific resistance genes, thereby demonstrating a gene-for-gene relationship between C. truncatum and L. culinaris. A set of differential lines with individual resistance genes is now needed to characterize the pathogen population further. Some lines could be developed by multiple selfing of F2 and F3 lines selected from the current study. Recently identified lentil accessions resistant to either race Ct1 and Ct0 could provide additional differential capability (Shaikh et al., 2012).

Pathogenic variability in C. truncatum may have resulted from asexual as well as sexual recombination. Although the sexual phase (Glomerella truncatum Armstrong-Cho & Banniza sp. nov.) of C. truncatum has only been induced in vitro (Armstrong-Cho & Banniza, 2006), it cannot be ruled out as a source of genetic variation under natural conditions. The sexual phase C. lindemuthianum (Sacc. & Magnus) Lams.-Scrib. of the related Glomerella lindemuthiana (Shear) on bean (Phaselous vulgaris L.) best explained the high genetic variability found in this fungal pathogen (Souza et al., 2010). Genetic variability could also have arisen as a result of conidial fusion which has been extensively studied in several Colletotrichum species (Roca et al., 2003; Ishikawa et al., 2012).

Our data showed the presence of both recessive and dominant resistance genes in L. culinaris. The presence of a dominant resistance gene, CtR4, in lentil line PI320937 is in agreement with previous results obtained during development of molecular markers linked to anthracnose resistance in this line (Tullu et al., 2003). In the study by Tullu et al. (2003), a F6 population derived from a cross between PI320937 and 'Eston' segregated as 1 resistant : 1 susceptible when inoculated with isolate 95B36, demonstrating a single resistance gene. The study also identified two RAPD and three AFLP markers flanking the resistance locus. Interestingly, one of the markers linked to resistance in PI320937 was later used to transfer resistance from the cultivar 'CDC Robin', which is derived from 'Indianhead', into various progenies (Tar'an et al., 2003), which points to a close linkage between resistance genes in PI320937 and 'Indianhead' similar to our findings.

Resistance to C. truncatum has been identified in several wild lentil accessions especially within L. ervoides (Brign.) Grande which belong to the tertiary gene pool (Tullu et al., 2006). An interspecific cross between L. culinaris 'Eston' and a resistant L. ervoides accession was made possible by embryo rescue techniques. Screening of F8 progenies under controlled conditions with two isolates representing race Ct1 and Ct0, respectively, showed that anthracnose resistance was likely conferred by two recessive genes effective against both races (Fiala et al., 2009). Subsequently, a resistant hybrid of L. ervoides × L. culinaris was crossed with both 'Eston' and the resistant 'CDC Redberry', another cultivar derived from 'Indianhead', confirmed that resistance to both C. truncatum races was conferred by the same or closely linked recessive genes in L. ervoides and 'CDC Redberry'. Further elucidation was difficult due to segregation distortion and effect of plant growth stage at the time of inoculation (Vail & Vandenberg, 2011). Nevertheless, these and our results show that recessive resistance genes are present in both L. ervoides and L. culinaris.

Colletotrichum species cause significant losses in yield and seed quality in both cool and warm season legume crops such as common bean, soybean (Glycine max (L.) Merr.), alfalfa (Medicago sativa L.), lupin (Lupinus angustifolius L.), peanut (Arachis hypogaea L.) and several Stylosanthes species (Lenné, 1992). The interaction between common bean and C. lindemuthianum is particularly well characterized. Nine Co resistance genes have been reported in bean and a set of 12 cultivars have been used to differentiate pathogen races, which are labelled using a binary numbering system (Kelly & Vallejo, 2004). The Co resistance genes are dominant except for co-8 which is recessive. Two genes, Co-3 and Co-9, are allelic, while the other genes are independently inherited with multiple alleles at the Co-1, Co-3 and Co-4 loci. A similar complexity in the L. culinaris–C. truncatum interaction has been demonstrated in the present study. Genetic studies of C. truncatum resistance in soybean showed that resistance in three cultivars was governed by single dominant and non-allelic genes (Kaushal & Sood, 2002). Partial resistance effective against the three known races of C. trifolii Bain & Essary has been found in some alfalfa cultivars, but the genetic basis of resistance has been difficult to determine due to heterozygosity of the autotetraploid cultivars used (Elgin & Ostazeski, 1985; Mackie et al., 2007).

Resistance genes in L. culinaris to C. truncatum do not confer complete resistance in the host since a mixture of superficial and deep lesions can sometimes be found on the lower half of the stems. Detailed measurements by Chongo & Bernier (1999) of disease symptoms in 'Indianhead', PI320937 and PI345629 showed that fewer and smaller lesions developed on the stems, which together with prolonged incubation and latent periods, resulted in a reduced rate of disease development. Under field conditions, the disease severities in plots planted to the resistant lines were significantly lower than in the susceptible 'Eston', and seed yields of these lines were comparable to 'Eston' plots that had received a fungicide application (Chongo et al., 1999). The trials were carried out in commercial lentil fields under natural infection by both C. truncatum races. Needless to say, incorporation of resistance to both races in future lentil cultivars is necessary to improve anthracnose resistance of lentil grown in western Canada. Fortunately, L. culinaris accessions have recently been identified with resistance to race Ct0 including one accession, VIR2633, with resistance to both races (Shaikh et al., 2012). These gene bank accessions went through several cycles of inoculation and single plant selection to obtain lines homozygous for anthracnose resistance. Genetic studies of resistance to race Ct0 similar to the present study are now needed. Concurrent evaluation of the pathogen population for pathogenicity on these lines will be needed to better deploy effective resistance genes over time.

ACKNOWLEDGEMENTS

Funding from Agriculture and Agri-Food Canada, Saskatchewan Pulse Growers, Manitoba Pulse Grower Association and the Western Grains Research Foundation is greatly appreciated.

REFERENCES

1. Anderson, K.L., Buchwaldt, L. and Gossen, B.D. 1999. Disease survey of lentil in Saskatchewan 1998. Can. Plant Dis. Surv., 79: 126–127.
2. Anderson, K.L., Buchwaldt, L., Chongo, G., Gossen, B.D., Morrall, R.A.A. and Pearse, P.G. 2000. Diseases of lentil in Saskatchewan 1999. Can. Plant Dis. Surv., 80: 96–98.
3. Armstrong-Cho, C.L. and Banniza, S. 2006. Glomerella truncata sp. nov., the teleomorph of Colletotrichum truncatum. Mycol Res., 110: 951–956.
4. Buchwaldt, L., Bernier, C.C. and Platford, R.G. 1992. Diseases of lentil in southern Manitoba in 1991. Can. Plant Dis. Surv., 72: 78–79.
5. Buchwaldt, L., Vandenberg, A. and Bernier, C.C. 1995. "Selection in lentil for anthracnose resistance and seed characteristics". In 2nd European Conference on Grain Legumes. Improving Production and Utilization of Grain Legumes 90–91. Copenhagen, , Denmark July 9–13, 1995
6. Buchwaldt, L., Morrall, R.A.A., Chongo, G. and Bernier, C.C. 1996. Windborne dispersal of Colletotrichum truncatum and survival in infested lentil debris. Phytopathology, 86: 1193–1198.
7. Buchwaldt, L., Anderson, K.L., Morrall, R.A.A., Bernier, C.C. and Gossen, B.D. 2004. Identification of lentil germplasm resistant to Colletotrichum truncatum and characterization of two pathogen races. Phytopathology, 94: 236–243.

8. Chongo, G. and Bernier, C.C. 1999. Field and growth chamber evaluation of components of partial resistance to Colletotrichum truncatum in lentil. Can. J. Plant Pathol., 21: 58–63.
9. Chongo, G., Bernier, C.C. and Buchwaldt, L. 1999. Control of anthracnose in lentil using partial resistance and fungicide applications. Can. J. Plant Pathol., 21: 16–22.
10. Elgin, J.H. JR. and Ostazeski, S.A. 1985. Inheritance of resistance to race 1 and race 2 anthracnose in Arc and Sarenac AR alfalfa. Crop Sci., 25: 861–885.
11. Fiala, J.V., Tullu, A., Banniza, S., Seguin-Swartz, G. and Vandenberg, A. 2009. Interspecies transfer of resistance to anthracnose in lentil (Lens culinaris Medic.). Crop Sci., 49: 825–830.
12. Flor, H.H. 1970. Designation of physiological races of plant pathogens. Nature, 227(5264): 1268–1269.
13. Gibson, R.J. 1994. "An investigation into the epidemiology and control of anthracnose (Colletotrichum truncatum) of lentil in Manitoba. M.Sc. Thesis". In University of Manitoba, Canada: Winnipeg.
14. Ishikawa, F.H., Souza, E.A., Shoji, J-Y., Connolly, L., Freitag, M., Read, N.D. and Roca, M.G. 2012. Heterokaryon incompatibility is suppressed following conidial anastomosis tube fusion in a fungal plant pathogen fusion in a fungal plant pathogen. PLoS ONE, 7(2): e31175
15. Kaushal, R.P. and Sood, O.P. 2002. Genetics of resistance in soybean to pod blight caused by Colletotrichum truncatum and Diaporthe phaseolorum. J. Mycol. Plant Path., 32: 219–221.
16. Kelly, J.D. and Vallejo, V.A. 2004. A comprehensive review of the major genes conditioning resistance to anthracnose in common bean. HortSci., 39: 1196–1207.
17. Lenné, J.M. 1992. "Colletotrichum diseases of legumes". In Colletotrichum: Biology, Pathology and Control, Edited by: Bailey, J.A. and Jeger, M.J. 134–166. Wallingford: CAB International.
18. Mackie, J.M., Musial, J.M., Armour, D.J., Phan, H.T.T., Ellwood, S.E., Aitken, K.S. and Irwin, J. A.G. 2007. Identification of QTL for reaction to three races of Colletotrichum truncatum trifolii and further analysis of inheritance of resistance in autotetraploid lucerne. Theor. Appl. Genet., 114: 1417–1426.
19. Morrall, R.A.A. 1997. Evolution of lentil diseases over 25 years in western Canada. Can. J. Plant Pathol., 19: 197–207.
20. Platford, R.G. and McCallum, B. 1996. Diseases of lentil in Manitoba in 1995. Can. Plant Dis. Surv., 76: 112–113.
21. Roca, M.G., Davide, L.C., Mendes-Costa, M.C., & Wheals, A. (2003). Conidial anastomosis tubes in Colletotrichum. Fungal Genet. Biol., 40, 138–145.
22. Shaikh, R., Diederichsen, A., Harrington, M., Adam, J., Conner, R.L. and Buchwaldt, L. 2012. "New sources of resistance to Colletotrichum truncatum race Ct0 and Ct1 in Lens culinaris Medikus. subsp. culinaris obtained by single plant selection in germplasm accessions". In Genet. Resour. Crop Evol
23. Souza, E.A., Camargo Jr, O.A. and Pinto, J.M.A. 2010. Sexual recombination in Colletotrichum lindemuthianum occurs on a fine scale. Gen. Mol. Res., 9: 1759–1769.
24. Tar'an, B., Buchwaldt, L., Tullu, A., Banniza, S., Warkentin, T.D. and Vandenberg, A. 2003. Using molecular markers to pyramid genes for resistance to ascochyta blight and anthracnose in lentil (Lens culinaris Medik.). Euphytica, 134: 223–230.
25. Tullu, A., Buchwaldt, L., Warkentin, T., Taran, B. and Vandenberg, A. 2003. Genetics of resistance to anthracnose and identification of AFLP and RAPD markers linked to the resistance gene in PI 320937 germplasm of lentil (Lens culinaris Medikus). Theor. Appl. Genet., 106: 428–434.
26. Tullu, A., Buchwaldt, L., Lulsdorf, M., Banniza, S., Barlow, B., Slinkard, A.E., Sarkar, A., Tar'an, B., Warkentin, T.D. and Vandenberg, A. 2006. Sources of resistance to anthracnose (Colletotrichum truncatum) in wild Lens species. Genet. Resour. Crop Evol., 53: 111–119.
27. Vandenberg, A., Kiehn, F.A., Vera, C., Gaudiel, R., Buchwaldt, L., Dueck, S., Wahab, J. and Slinkard, A.E. 2002. CDC Robin lentil. Can. J. Plant Sci., 82: 111–112.

28. Vandenberg, A., Banniza, S., Warkentin, T.D., Ife, S., Barlow, B., McHale, S., Brolley, B., Gan, Y., McDonald, C., Bandara, M. and Dueck, S. 2006. CDC Redberry lentil. Can J. Plant Sci., 86: 487–489.
29. Vail, S. and Vandenberg, A. 2011. Genetic control of interspecific-derived and juvenile resistance in lentil to Colletotrichum truncatum. Crop Sci., 51: 1481–1490.
30. Venette, J.R., Lamey, H.A. and Lamppa, L. 1994. First report of lentil anthracnose (Colletotrichum truncatum) in the United States. Plant Dis., 78: 1216

CHAPTER 19

Gray Mold of Castor: A Review

Dartanhã José Soares

Empresa Brasileira de Pesquisa Agropecuária,

Embrapa Algodão, Campina Grande Brazil

1. INTRODUCTION

Castor plant (*Ricinus communis* L.) is a non-edible oilseed crop with unique oil features for the chemistry industry. The crop was very important in the mid and late nineteenth century and also during WWI. After that the crop lost its importance in developed countries (Godfrey, 1923), but in India and Brazil it has remained as the most important non-edible oilseed crop of the arid and semi-arid regions (Dange et al., 2005; Santos et al., 2007). Nowadays, due the constant pressure for renewable fuels, castor has been investigated as a potential source of biofuel, mainly in Brazil due to governmental stimulus, and this has raised the crop importance once again. Regardless of the lack of a well established crop system, castor hosts several pests and diseases which cause heavy losses in the crop yield. One of the most destructive diseases of castor is gray mold, caused by the fungus *Botryotinia ricini* (Godfrey) Whetzel. Actually, it is the anamorphic phase of *B. ricini*, known as *Amphobotrys ricini* (N.F. Buchw.) Hennebert, that is responsible for disease epidemics and heavy yield losses frequently observed in castor crops. The first epidemic outbreak caused by this fungus was reported by H.E. Stevens of the Florida Experiment Station, Gainesville, Florida (Godfrey, 1919, 1923). At that time, a meticulous study was conducted and much of our knowledge regarding the disease and its causal agent was published in the classic work of Godfrey (1923). Subsequently, only sporadic works were conducted by other scientists around the world, consequently few advances have been made on management of gray mold. Breeding programs have failed in developing varieties with satisfactory resistance levels (Kolte, 1995), and chemical control is still ineffective and economically prohibitive, mainly due to the lack of basal information about the causal organism and its biology. In this chapter, the major aspects of castor gray mold will be reviewed.

2. GRAY MOLD OF CASTOR

2.1 Historic and economic importance

Castor gray mold was first reported in the USA in 1918, following pioneering investigations by H.E. Stevens and F. W. Patterson, who promptly suggested that the causal organism of castor gray mold was an unknown *Botrytis* species (Godfrey, 1919, 1923). This fungus had caused serious losses of castor crop in the summer of 1918 mainly in Florida and others southern States, where it was responsible for losses up to 100% of castor yield (Godfrey, 1923). Later, the disease was reported in almost all countries where castor has been cultivated (Kolte, 1995), having nowadays a worldwide distribution.

The first occurrence of this disease in USA was directly linked to seeds imported from Bombay (now Mumbai), India, even though until that time, such disease had not been described in that country (Godfrey, 1923). In his work, Godfrey (1923) did a detailed account of the destructive potential of the gray mold of castor under favourable condition. By attacking mainly reproductive organs of the castor plant, gray mold disease is implicated in direct losses of yield whatever the level of infection.

In India, today the major castor producer, gray mold is found in few states and is regarded as troublesome only in Andhra Pradesh and Tamil Nadu, in the South, where the weather conditions are more favourable for disease development where in 1987, an epidemic outbreak of gray mold occurred (Dange et al., 2005).

In Brazil, the disease was first reported in the São Paulo state in 1932. However, it was only in 1936 that any attention was given to the disease due to the serious losses which occurred that year (Gonçalves, 1936). Currently, gray mold is present in almost all Brazilian states and its importance has grown at the same time that the crop cultivation has been intensified, mainly in those regions where the weather conditions are favourable for disease development, including the Southern and South-eastern Brazilian states (Araújo et al., 2007; Freire et al., 2007). In the region of the "Brejo Paraibano", where the recommended sowing period is between mid-April to early-May (Amorim Neto et al., 2000), the flowering period (mid-June to early-August) usually coincides with highly favourable conditions for disease development (Moraes et al., 2009). Yield losses of up to 100% are quite frequent when highly susceptible cultivars are planted. Conversely, in Bahia the major castor producer in Brazil gray mold is not a problem because the weather conditions are usually not favourable for disease development.

2.2 Etiology, taxonomy and population structure

The causal agent of gray mold of castor was originally described by Godfrey (1919) as *Sclerotinia ricini* Godfrey, based on the holomorph. Later, Whetzel (1945) transferred the species *S. ricini* to the genus *Botryotinia*, which since then has been known as *Botryotinia ricini* (Godfrey) Whetzel. Subsequently, the anamorphic state of *Botryotinia ricini* was named as *Botrytis ricini* N.F. Buchw. (Buchwald, 1949). This led to general confusion between the non-mycologist communities, which adopted the name *Botrytis ricini* N.F. Buchw., instead of *Botryotinia ricini* (Godfrey) Whetzel. In 1973, Hennebert erected the genus *Amphobotrys* to accommodate the anamorphic state of *B. ricini*, based mainly on the distinctive pattern of conidiophore ramification, and since then the anamorphic state became known as *Amphobotrys ricini* (N.F. Buchw.) Hennebert (Hennebert, 1973). Even so, several authors used, and still use, the erroneous name "*Botrytis ricini*" attributing its authority to Godfrey (Barreto & Evans, 1998; Batista et al., 1998; Dange et al., 2005; Lima & Soares, 1990).

Although the correct name to be applied to the causal agent of gray mold of castor is *Botryotinia ricini*, only the anamorphic state is observed in the field by most authors, and thus the name of the anamorph, at the expense of the name of the holomorph, is preferred (Holcomb et al., 1989; Lima et al., 2008).

Botryotinia ricini belongs to Sclerotiniaceae (Helotiales, Ascomycota) and is characterized by its dark, plane-convex, elongated sclerotia (Figure 1), which give rise to cinnamon brown to chestnut brown, long stipitate apothecia, with cylindrical to cylindro-clavate asci, apex slightly thickened, 8-spored; ascopores ellipsoidal, often sub-fusoid, one-celled, bi-guttulate and hyaline; paraphyses hyaline, filiform, septate (Godfrey, 1919). Its anamorphic phase is characterized by cylindrical, straight, dichotomously branched, pale brown conidiophores, with conidiogeneous cell not inflated, thin-walled; conidia globose, maturing synchronically, on short denticles, smooth, one-celled, sub-hyaline to pale brown (Figure 1) (Godfrey, 1919; Hennebert, 1973; Lima et al., 2008). A synanamorph (*Myrioconium* sp. – spermatial state) may sometimes be present on culture media (Godfrey, 1923; Hennebert, 1973; Seifert et al., 2011). According to Kirk et al. (2008), the genus *Amphobotrys* remains monotypic.

Botryotinia ricini is regarded as a homothallic species (Beveer & Weeds, 2007), so sexual reproduction will readily take place. If sexual reproduction had taken place, a high degree of diversity would be expected within the *B. ricini* population; however, Bezerra (2007) shows evidence that, in the state of Paraíba (Northeast Brazil) populations of *B. ricini* are clonal, which means that sexual reproduction had not taken place within those populations. Nonetheless, this conclusion must be viewed with caution because the population sampled was relatively small. Unfortunately, there is no other work on this subject and, therefore, the population structure of *B. ricini* remains unknown.

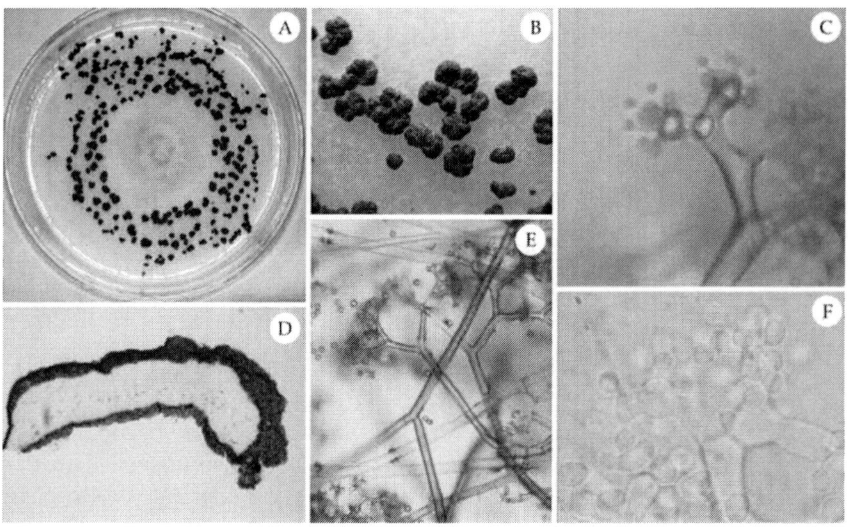

Figure 1. Dark sclerotia on culture medium (A); close-up view of the sclerotia (B); transversal section through a sclerotium to show its plane-convex form (D); dichotomous branch of the conidiophores (E); conidiogenous cells showing the synchronic conidiogenesis (C); and close-up the conidiogenous cell to show the denticles and globose conidia (F). Photos: D.J. Soares.

2.3 Host penetration and colonization

In his classical work, G.H. Godfrey also investigated the infection process of *B. ricini* on leaves of the castor plant and concluded that penetration occurs directly through the host cuticle, in a process similar to *Botrytis cinerea* (Godfrey, 1923). After penetrating the cuticle, the fungus quickly spreads over the host tissues leading to a complete disorganization and breakdown. Although Godfrey had made mention as to the possible role of an enzymatic action in the penetration process, his conclusion pointed out to a mechanical penetration of the germ-tube, without tissues dissolving, prior to infection. On the other hand, Thomas & Orellana (1963b) found that it was not possible to verify the direct germ-tube penetration of *B. ricini*, through the cuticle or stomata, on castor capsules, before tissue maceration by pectic enzymes action, suggesting that the fungus first degraded the cuticle and later penetrated the host tissues (Orellana & Thomas, 1962). Probably *B. ricini* uses both mechanical and chemical processes to penetrate undamaged host tissue, however, no further studies have been done to clarify these questions.

Although the infection process of *B. ricini* needs to be better understood, it is likely that enzymes, such as lipases and cutinases, play an important role in the infection process similar to several other *Botrytis*-host interactions (Kars & van Kan, 2007). Additionally, *Botrytis* species can also affect the 'redox' process in the host plants, during the colonization of host tissues, through the production of enzymes like superoxide dismutase (Lyon et al., 2007). Due the great biological similarity of *Botrytis* spp. and *B. ricini*, probably, such enzymes also have an important role in the infection process of the causal agent of castor gray mold, as already evidenced by Orellana & Thomas (1962). Hoffmann et al. (2004) had, for example, extracted alpha and beta esterase and superoxide dismutase from *B. ricini*, however they did not perform a study to determine the role of such enzymes in the infection process of this fungus.

It is important to highlight the great distinction between the penetration process of a fungus under controlled and highly favourable condition in contrast with the natural process in the field. In the latter case, all aerial parts of the host are potential targets for deposition and penetration of *B. ricini*, because not only the conidia, regarded as the major propagative unit, usually responsible for the epidemic outbreak, but also the ascospores, sclerotia and mycelia fragments can give rise to infection, as observed in *Botrytis* spp. (Jarvis, 1978). So, besides direct penetration, probably natural openings and wounds also serve as a point of entrance for the fungus. Growth of the fungus on the host surface and consequently its penetration in the host tissues will depend on factors such as inoculum type, free water and nutrient availability, cuticle features, presence of exudates on floral organs and other glands, besides the abundance of natural openings and the size and age of wounds, as pointed out by Holz et al., (2007) for *Botrytis* species.

2.4 Symptoms

The primary targets of the fungus are the inflorescence and the capsules, in any development stage (Figure 2) (Araújo et al., 2007; Dange et al., 2005; Gonçalves, 1936; Lima et al., 2001). Some authors (Drumond & Coelho, 1981; Batista et al., 1996) claim that the male flowers are the first to be infected, but it is not always the case because any part of the inflorescence can be infected, the female flowers being the preferential target. That claim came from the fact that the male flowers are the first to be exposed, at the earlier stage of inflorescence formation; consequently such flowers are exposed longer to the infection units of the fungus. However, as soon as the male flowers suffer anthesis they are no longer a target and are hardly infected, contradicting the statement of Drumond & Coelho (1981) that "the fungus attacks first the male flowers because the anthers, being soaked with the rain water or dew, easily retain the fungus spores carried by the wind".

Figure 2. Symptoms of gray mold attack on castor inflorescence and raceme. A-B. Symptoms on young inflorescences, before fertilization of female flowers. C-H. Symptoms on capsules at distinct development stages. Photos: D.J. Soares.

The first symptoms are visible as bluish spots on the inflorescences, on both female and male (before anthesis) flowers, and on developing fruits. On fruits, the symptoms can evolve to circular or elliptic, sunken, dark coloured spots that can result in rupture of the capsule (Figure 3 A-B) (Araújo et al., 2007). These symptoms are usually more frequent when a period of low relative humidity unfavourable to fungal sporulation occurs soon after the fungus penetrates the host tissues.

Depending on weather conditions (e.g. long periods with high relative humidity soon after the fungus penetrates the host), the occurrence of yellow ooze at the point of infection is frequent (Figure 3 C-D) (Batista et al., 1996; Dange et al., 2005) as a result of the rapid enzymatic tissue degradation. The symptoms on the male flowers, before anthesis, are small, pale brown, necrotic spots, which can evolve to larger brown spots with a darker edge (Figure 3 E). The infected flowers and young capsules became softened due the fungal colonization and mycelial growth is, at first, pale gray and later dark olivaceous. A profuse sporulation is usually observed in such stage (Figure 3 F). When the infection starts on immature capsules, they become rotten; if the infection starts later, with fully developed capsules, the seeds usually became hollow, with coat discoloration and weight loss (Dange et al., 2005). On the inflorescence, the male flowers can be infected first, but the fungus has a clear preference for the female flowers (Figure 3 G). Infection can lead to complete destruction of the raceme (Figure 3 H), particularly if it reaches the main stem and the weather conditions are favourable for the disease. Several other plant parts, e.g. leaves, petioles and stem can also be infected, mainly due to the deposition or fall of infected material from the inflorescence or racemes. On leaves, the lesions are usually irregular, but can assume an elliptic or circular pattern, the size is very variable, sometimes coalescing and resulting in a foliar blight (Figure 3 I-J). On petioles and stems, necrotic, sunken lesions usually are formed which can cause the strangulation and consequently death of the parts above the infection point (Figure K-N) (Batista et al., 1996; Dange et al., 2005).

2.5 Epidemiology

There are few studies of the epidemiological aspects of gray mold on castor. Godfrey (1923) mentioned that temperatures around 25ºC and high relative humidity are highly favourable to disease development. Such statements have been exhaustively repeated in almost all publications about the subject over the last decades (Araújo et al., 2007; Batista et al., 1998; Gonçalves, 1936; Kimati, 1980; Lima & Soares, 1990; Massola JR & Bedendo, 1997; Melhorança & Staut, 2005;). The minimum and maximum temperature for mycelial growth was established by Godfrey as 12 and 35ºC, respectively (Godfrey, 1923).

Figure 3. Symptoms of gray mold on castor plants. Dark-bluish spot (A) and capsule rupture (B) under unfavourable conditions; Gray-bluish spot with yellowish-brown ooze (C-D) under favourable conditions; Dark brown spot on male flowers before anthesis (E); Profuse sporulation on infected capsule (F); Infected inflorescence showing the fungus preference for female flowers (G); Completely destroyed inflorescence (H); Leaf spot (I) and blight (J); Apical infection (K); Petiole infection (L); Stem (M) and rachis infection (N). Photos: D.J. Soares.

Some complementary studies have confirmed that temperatures around 25ºC are favourable to fungal growth and disease development (Araújo et al., 2003; Suassuna et al., 2003; Sussel, 2008). At temperatures below 20ºC, the disease is little expressed and highly dependent on long periods of high relative humidity (Sussel, 2008). According to Sussel et al. (2011), there is a high correlation between the temperature and duration of leaf wetness with the disease incidence and severity. The same authors also concluded that the disease was more intense with a temperature of 28ºC and 72 hours of leaf wetness, and that for temperatures below 15º the fungus needs more than 6 hours of leaf wetness, otherwise the disease does not occur (Sussel et al., 2011). Even when under optimal temperatures (near 25ºC), the fungus appears to be highly dependent on periods of high humidity, since in a study developed by Esuruoso (1966) with 39 castor varieties in two distinct places, Ibadan and Ilora (West Nigeria), that share similar temperatures (24.4 to 27.9 and 24.2 to 27.2°C, respectively), but that rainfall was twofold higher in Ibadan than Ilora, the disease level in Ibadan was 1.5 fold higher than in Ilora. This dependence on high rainfall, and not only on high relative humidity, was also observed in a study conducted in the Paraiba state (Northeast Brazil) where a high correlation between the disease progress and the accumulated rainfall between the seventh and fifth day before the evaluation was observed (D.J. Soares, unpublished data). The correlation, however, was inversely proportional to the rainfall intensity which means that long periods of low-intensity rains are more favourable than short periods of high-intensity rains.

Sussel (2008) concluded that the disease shows a random distribution pattern, matching its airborne nature. However, under high rainfall, the disease assumes an aggregate pattern, typical of those dispersed by water splash. With reference to aerobiology, Sussel (2008) obtained a positive correlation between the average number of conidia collected daily in air and the weather variables: minimum temperature, mean relative humidity, mean precipitation and leaf wetness. There was an increase in the number of conidia collected with the elevation of the minimum temperature from 14.6 to 18.1ºC; the same occurred when the mean relative humidity was raised from 42 to 95% and when the daily precipitation increased from 1 to 20 hours (Sussel, 2008, 2009a).

Under controlled conditions, at 25ºC and relative humidity near saturation, Soares et al. (2010) had determined that the incubation period of *B. ricini* can vary from 44 to 88 hours (average 72 h) and the latent period from 72 to 144 hours (average 96 h), depending on the genotype.

Although in the recent years, there have been some advances in knowledge about the epidemiology of gray mold of castor, several issues remains unresolved. Further studies are needed to understand the role of each potential dispersal unit; how the fungus survives between the growth seasons, for example, as sclerotia, on secondary hosts, on spontaneous (volunteer) castor plants. Additionally, it is also crucial to determine whether it is possible to predict disease development based on environmental variables, like precipitation or surface wetness.

2.5.1 Live cycle and host range

The disease starts with spore deposition on the host surface, followed by penetration and colonization of the host tissues. Soon after colonization, the fungus, under favourable conditions, sporulates profusely on the dead tissues, and then the conidia became the main inoculum source for new infection sites. Although most authors recognized the major role of conidia after the disease had been established in the field, there is much speculation about the primary inoculum source of *B. ricini*.

Godfrey (1923) claims that the fungus survives on soil or crop residues as sclerotia and, under the right conditions, these can produce sexual structures, which will be responsible for the initial infection. However, there is no report of sexual reproduction under natural conditions, other than the original reports of Godfrey (1919, 1923), so the purported role of ascospores as the initial inoculum source remains unclear, despite the fact that apothecia are easily overlooked in the field.

Under tropical climate, the initial inoculum source are probably the conidia from wild castor plants which grows spontaneously near the crop areas all year (Gonçalves, 1936). Wild castor plants can produce flowers throughout the year and consequently new susceptible tissue will be available for the fungus to self perpetuate in its anamorphic state through the year. By infecting the first inflorescence under favourable conditions, the fungus produces abundant sporulation, thus allowing multiple rounds of re-infection, since this pathogen is easily spread by wind, rain splash and, probably, by insects (Figure 4 A-C) (Dange et al., 2005).

It was also mentioned by Godfrey (1923), that the fungus is seed-borne, the seeds being regarded as the primary inoculum source (Figure 4 D-E). However, the role of seeds as a primary inoculum source requires further study. Probably the seeds have no essential role at the beginning of the epidemic because there is usually an interval of almost two months between sowing and flowering, so the inoculum originating from the seeds will not be available to infect the flowers. This means that, although *B. ricini* is a seed-borne fungus, it is probably not seed transmitted, because it could hardly infect the crops which grow from them, as conceived by Maude (1996). Thus, the most important role of the seeds is to carry the pathogen to new areas, rather than to act as a primary inoculum source for epidemic on the crop season.

Initially, it was suspected that *B. ricini* was host-specific and that it has a very narrow host range (Godfrey, 1923). Through artificial inoculation, Godfrey (1923) showed that the fungus is extremely dependent on high humidity and it was not able to cause disease, at same levels observed on castor plant, when inoculated on several other hosts, including members of Euphorbiaceae. In field inspections, the same author did not detect any other plant species showing symptoms of natural infection in the surroundings of castor-growing areas which were severely affected by the disease. Nevertheless, since the 1980s several reports of natural infection of *B. ricini* on members of Euphorbiaceae have been made, including both weeds and ornamentals: such as *Caperonia palustris* (Whitney & Taber,1986), *Euphorbia supina* (Holcomb et al., 1989; Russo & Rossman, 1991), *Euphorbia milli* (Sanoamuang, 1990),

Euphorbia pulcherrima (Holcomb & Brown, 1990), *Euphorbia heterophylla* (Barreto & Evans, 1998), *Euphorbia inarticulata* (Alwadie & Baka, 2003), *Acalypha hispida* and *Jatropha podagrica*

(Lima et al., 2008). Besides these reports of natural infection, artificial inoculations tests have shown that this pathogen has a wide host range within the Euphorbiaceae, including species of economic interest, e.g. cassava (*Manihot utilissima*) (Holcomb et al., 1989; Kumar et al., 2007; Lima et al., 2008). The sole report of natural infection by *B. ricini* outside the Euphorbiaceae family was made by Hansen & Bega (1955) on *Caladium bicolor* (Araceae) and, it is quite likely a case of fungus misidentification.

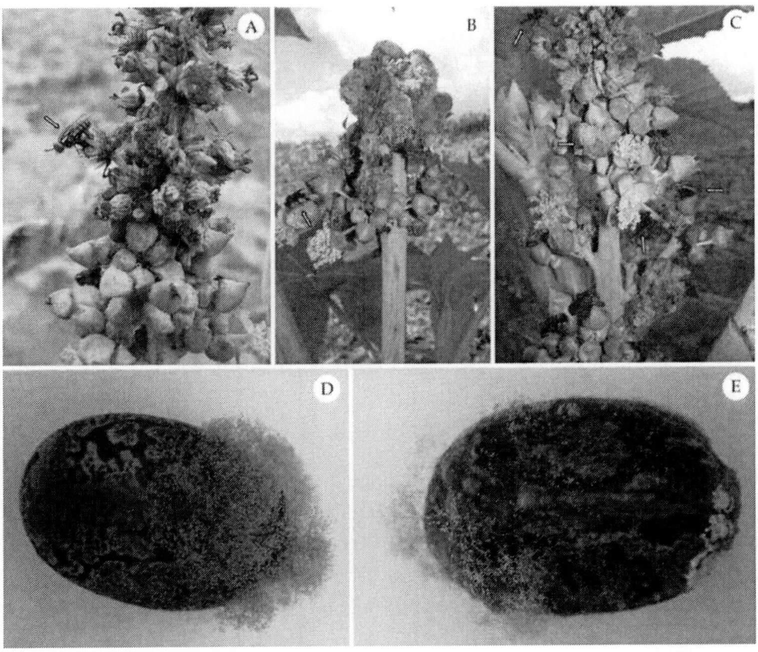

Figure 4. Visitor insects (flies and stingless bees) on castor inflorescence infected by *B. ricini* (A-C). Seeds with profuse gray mold growth and sporulation (D-E). Photos: D.J. Soares.

Hence, it is possible that the fungus can survive in the field on several other plant species belonging to Euphorbiaceae, and these can be an inoculum reservoir of the fungus in such a way that it would be ready to infect its preferential host, the castor plant, as soon as susceptible tissues become available.

2.6 Host resistance

The search for resistance to gray mold has been investigated ever since the disease was described. In his classic work, Godfrey (1923) pointed out general conclusions that clearly have been overlooked over the decades by plant breeders and plant pathologists alike. Among these, three are worth mentioning: "(1) Plants of more ornamental type, with stalk, foliage, and sometimes pods in different shades of red or reddish green were more resistant;

(2) All smaller, many branched plants, which by their yield indicated commercial possibilities, showed high susceptibility to the disease; and (3) Cross pollination in castor-bean fields probably occurs very extensively. It would require years of work to develop pure strains and then to select and breed for desirable qualities combined with resistance before permanent results could be secured" (Godfrey, 1923). Based on this information, it is clear that only minor progress has been achieved toward the development of resistant cultivars during the last century. As a result, breeding programs have failed to develop a resistant cultivar or hybrid until today.

Gonçalves (1936) reported that "spontaneous varieties" are highly resistant to the disease, since the fungus attacks only few capsules, while most of the capsules of the same raceme or from others raceme remain healthy. Although this behaviour is frequently observed in wild types, and even in some commercial cultivars, this statement must be viewed with caution since it has no scientific basis and probably such behaviour is a result of the wide genetic variation within castor plants.

According to several authors, varieties with more compact racemes, shorter internodes, and male flowers distributed all long the inflorescence are considered to be more susceptible to the pathogen (Batista et al., 1998; Costa et al., 2004; Dange et al., 2005; Milani et al., 2005; Thomas & Orellana, 1963b; Ueno et al., 2006; Zarzycka, 1958): a fact already noted by Godfrey (1923), and also by Esuruoso (1966), who concluded that the disease severity was more intense on short-stalked varieties than on those with longer stalks. Another relevant aspect is that, apparently, the presence of spines in the capsule predisposes them to pathogen attack (Alcântara et al., 2008; Cook, 1981; Lima & Soares, 1990).

Plants with capsules containing high soluble sugar concentrations are more susceptible to fungal development than plants with low soluble sugar concentrations (Orellana & Thomas, 1962). According to these authors, capsule resistance is intimately associated with its capacity of inactivation of pectic, cellulolytic and others hydrolytic enzymes through the products of the phenol oxidation (Thomas & Orellana, 1963a), as well as lower content of water-soluble pectin, higher content of calcium and magnesium and lower sodium and potassium contents (Thomas & Orellana, 1964).

Although several studies have been conducted to assess the resistance of castor bean genotypes, none of them deals with the inheritance of resistance to *B. ricini*, nor how it is governed. Probably, the resistance to gray mold is quantitative and possibly governed by several minor genes.

Quantitative resistance, also called horizontal resistance, is, according to Robinson (1976), universal and occurs in all plants against all parasites; it is also permanent and permits cumulative plant breeding. Horizontal resistance can be either passive or active, and usually is conferred by numerous and complex mechanisms and can include traits such as tolerance and disease-escape which are not strictly resistance mechanisms *per se* (Robinson, 1976). It is also important to highlight the fact that horizontal resistance usually is an efficient way to achieve disease control, and that in the recent years selection for horizontal resistance has become easier with the use of marker-assisted breeding (Keane, 2012).

In Brazil, several studies have been carried out aiming to select for a resistance source to gray mold. However, what has become clear is that whilst there are differences in susceptibility among the assayed genotypes, none has the desired resistance level (Batista et al., 1998; Costa et al., 2004; Lima & Soares, 1990; Milani et al., 2005; Rego Filho et al., 2007). Among the genotypes assessed in different countries, thus far, several distinct levels of susceptibility have been observed, but not immunity (Anjani et al. 2004; Esuruoso, 1996; Zarzycka, 1958). The way that the breeding programs are being conducted, using only information generated by field investigations, and which are usually affected by uncontrollable factors, is much more likely to select for "field resistance" rather than for genetic resistance, i.e., the plant and raceme architecture, together with the weather conditions, play a bigger role in the disease development, by inducing a micro-climate formation, which might be more or less favourable to the pathogen. This means that unless we redirect our thinking to incorporate horizontal resistance, through a careful marker-assisted breeding program, whether it be genetic or morphological, it will probably never be possible to obtain a variety with satisfactory levels of resistance.

Some castor plant varieties have a thick outer wax layer which can act as a constitutive barrier to pathogen infection. However, there is no information about the role of this wax layer in the infection process of *B. ricini* and, although the adhesion of the conidia and subsequently their penetration into the host tissue could be difficult, the wax could also act as an elicitor; being responsible for the initial recognition of the host-pathogen interaction. Thus, the wax layer could actually favour the pathogen rather than inhibit its development. Field observations lead us to hypothesize that the latter, is the most probable scenario in the present pathosystem.

2.6.1 Host resistance assessment

As commented previously, most of the research involving the assessment of genotypes resistance has been conducted under field conditions and without any standard to quantify the disease and, worst still, sometimes using very subjective assessment methods, like that adopted by Lima & Soares (1990). This has been reflected by the fact that it is almost impossible perform a comparison among the different already studies undertaken. However, a diagrammatic scale to assess gray mold severity in castor- was published recently (Figure 5) (Sussel et al., 2009; Sussel, 2009b). This was developed in order to standardize disease assessment in field experiments, and was constructed based on the Weber-Fechner law and divided into 10 levels. Although useful, such a scale faces two crucial factors which might constitute an impediment to its wider adoption: first, there is great variation in raceme architecture among the castor genotypes, and; secondly, it only considers the fully developed raceme. In the first case, as the diagrammatic scale was drawn based on long conical racemes, its use to assess disease severity in genotypes with more or less globose raceme will be difficult. In the second case, the scale does not take into account disease severity in inflorescences or even in immature racemes, where the disease is usually more severe and more difficult to estimate. Therefore, it is possible that the scale will underestimate disease severity. Nonetheless, the simple fact that such a diagrammatic scale is now available is a huge advance towards more reliable disease assessment and the possibility of comparing assays conducted by different researchers in different localities. Chagas et al. (2010) also developed a diagrammatic scale, with six levels, to assess the disease severity of gray mold on castor, but similar to the scale developed by Sussel et al. (2009), this scale was also developed based on long conical, full developed, racemes, and thus, will face the same problem mentioned above.

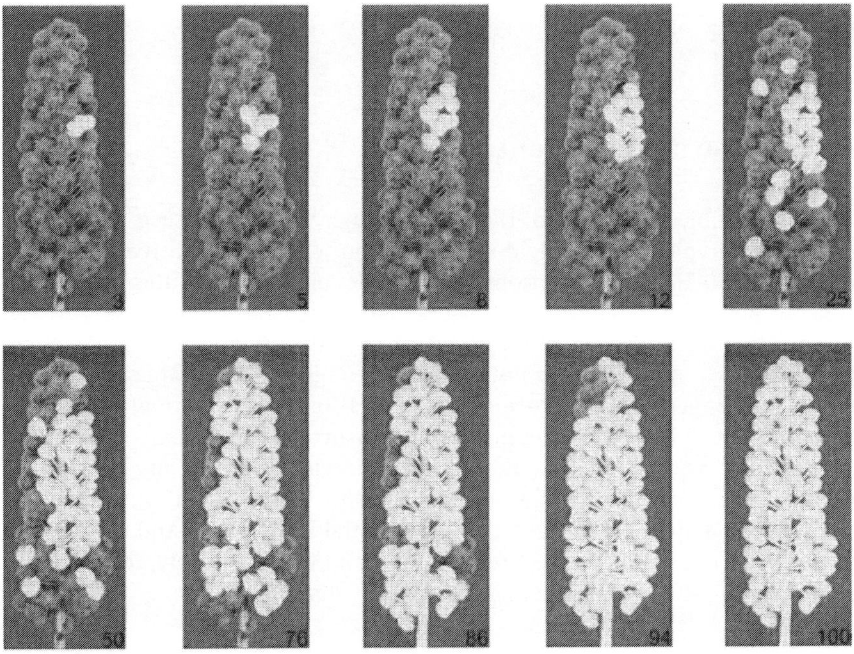

Figure 5. Diagrammatic scale to assess the gray mold severity on castor (Reprinted from: Sussel et al. *Tropical Plant Pathology*, Vol. 34, No.3, pp.186-191, 2009, by permission). Numbers represents the percent area affected by the disease.

The first attempt to evaluate disease resistance under controlled conditions was made by C.A. Thomas & R.G. Orellana in 1963 using a biochemical test (Thomas & Orellana, 1963b). This methodology was reproduced by O.F. Esuruoso in Nigeria, a few years later (Esuruoso, 1969). This last author concluded that the laboratory results were similar to previous observations from field tests, and none of the tested varieties were resistant to the disease (Esuruoso 1969). At the Embrapa Algodão Station - a branch of the Brazilian Agricultural Research Agency, responsible for developing research on castor - we have developed a controlled method to assess castor resistance to gray mold. The principal advantage of this method is to eliminate the unpredictable changes usually present under field evaluation, and then to select the castor plants based solely on their genetic background rather than select a phenotype. Based on such studies, it has been possible to obtain a clear difference among the so-called susceptible and resistant genotypes, which had been previously screened under field conditions. It has also been possible to stratify the genotypes within at least three ranks: highly susceptible, moderately susceptible and less susceptible (Silva et al., 2008; Soares et al., 2010), although none of the screened genotypes were regarded as resistant.

2.7 Disease management

Without doubt, protection of the inflorescences and immature capsules is crucial to avoid heavy yield losses when castor is cultivated under favourable disease conditions. Fully developed capsules are less susceptible to pathogen attack and the severity levels are usually lower when compared with infection in young capsules or on the inflorescence. However, it is important to note that under highly favourable conditions with high inoculum pressure, losses of 100% are relatively frequent (Anjani et al., 2004). There is no single measure to keep the disease under acceptable levels; as well as there is no knowledge about any acceptable disease level. As the pathogen has a very short incubation period, and is easily wind-dispersed, its destructive potential is very high and usually the growers do not want to take their chances and wait passively, they usually prefer to act before or at the first signs of the disease.

2.7.1 Cultural

Cultural practices are usually applied at aiming to prevent the introduction of inoculum into the field, reducing its survival, spread or build-up, or rendering the host less prone to disease attack (Palti & Rotem, 1983; Termorshuizen, 2001). The use of varietal resistance is regarded as the better method for disease management. However, as highlighted previously, there are no varieties with satisfactory resistance levels to gray mold (Anjani et al., 2003; Araújo et al., 2007; Cook, 1981; Dange et al., 2005; Kolte, 1995; Milani et al., 2005).

Several authors have recommended the use of healthy seeds, removal of plant debris, adequate choice of planting area and growing season, and use of less susceptible cultivars (Galli et al., 1968; Massola Jr. & Bedendo, 2005; Sussel, 2009a). It is also recommended to use plant spacing adjusted for maximum aeration (Kolte, 1995; Lima et al., 2005). The use of healthy seeds, including seed treatment with fungicides, is always a desirable practice, however, its practical benefits for the management gray mold are questionable, because as mentioned previously, it is unlikely that such seeds will serve as an inoculum source for that crop season, so this practice has much more value in promoting vigorous plant growth and avoiding the introduction of the pathogen into new areas. Elimination of alternate and reservoir hosts (euphorbiaceous hosts), as well as removal and destruction of inoculum persisting in plant residues, are welcome practices for management of gray mold, and usually result in lower disease levels. Nevertheless, this practice must be followed by rigorous field inspection since the fungus is easily wind-dispersed, and once established in an area, such practice will become unfeasible. Perhaps, among the recognized cultural practices, the choice of growing season, or sowing time in such a way that spike development and maturity occur during the dry season, consequently avoiding the long, wet periods favourable to disease development, should be the most efficient one (Dange et al., 2005; Kolte, 1995); as evidenced by the fact that in locations where dry weather prevails, the disease does not occur (Godfrey, 1923; Kolte, 1995). This situation is typically observed in the Bahia state of Brazil, as mentioned previously.

2.7.2 Chemical

Seed treatment has been the management strategy most frequently recommended (Araújo et al., 2007; Batista et al., 1996; Godfrey, 1923; Gonçalves, 1936; Massola Jr. & Bedendo, 2005; Milani et al., 2005; Sussel, 2009), mainly to avoid the introduction of the pathogen into new areas. However, as the pathogen is air-borne and since it is already reported in almost all countries where castor is cultivated, the efficacy of such measures needs to be corroborated, because doubts about the role of the seeds in this pathosystem, as a primary inoculum source, still need to be clarified.

After disease establishment, fungicide spraying is usually the only way to stop or reduce the disease progress. However, there are few studies on chemical control of gray mold, and, of the fungicides tested, none is registered for use on castor crops in Brazil. According to Araújo et al. (2007), spraying of systemic fungicides soon after the appearance of the first symptoms delayed the epidemic and reduced disease progress.

In India Dange et al. (2005) recommended two prophylactic sprays with carbendazim (0.05%): the first at 50% of flowering; and the second, when the first disease symptoms appear. Although, Anjani et al. (2004) considered that non-genetic management measures have failed to control gray mold.

In the last decades, significant progress has been achieved regarding the use of fungicides to control plant diseases. Several new active fungicides with distinct modes of action, and usually with high specificity, have provided satisfactory levels of control of many plant diseases. However, for gray mold of castor, the main issue is not the ineffectiveness of fungicidal products, but the lack of research on the most appropriate timing of fungicide application and optimum dose, including also cost-benefit analyses.

Preliminary studies under controlled conditions have shown that carbendazim and azoxystrobin are effective against the gray mold pathogen (Bezerra, 2007). According to Chagas (2009), however, azoxystrobin was ineffective against *B. ricini*, while carbendazin and several others fungicides, including tebuconazole, iprodione and procymidone, were highly effective.

A field study, conducted under highly favourable conditions for disease development and using susceptible varieties, has confirmed that procymidone and iprodione are effective in disease control, but only if applied at the beginning of the epidemic and at weekly intervals. Where the timing of the first spraying was lost, and the application intervals were longer (15

days), the use of the same fungicides could not stop the disease and the losses reached 100% (D.J. Soares unpublished data). Despite the limited studies dealing with chemical control of gray mold of castor, there are several fungicides known as "botryoticides" which are effective in protecting crops against *Botrytis* spp. (Leroux, 2007), and probably we can assume that these fungicides also are effective against *B. ricini.* In several *Botrytis* pathosystems, there is a high concern about the use of fungicides just before, or even after the harvest because of the toxicological risks of their residues (Leroux, 2007). In contrast, in castor crops, such a restriction is not a concern and fungicides can be applied just before harvest. However, if used indiscriminately, there must be concern about resistance phenomena associated with several major botryticide families, including benzimidazoles, phenylcarbamates and dicarboxymides (Leroux, 2007). Hence, control of gray mold of castor must consider several, rather than just one, management practices, as is currently recommended for the management of other *Botrytis* diseases (Leroux, 2007).

2.7.3 Biological

There are several studies dealing with the use of biological control agents, mainly *Trichoderma* spp. and *Clonostachys rosea* to control diseases caused by *Botrytis* spp. (Elad & Stewart, 2007). If we consider the fact that the genera *Botrytis* and *Amphobotrys* are biologically similar and, probably, phylogenetically related, we could expect that the use of such biological control agents could be applied as an effective strategy in the pathosystem *B. ricini* x *R. communis.* Actually, there are some studies conducted with *Trichoderma* and *Clonostachys rosea* for the control of gray mold of castor and promising results have been obtained (Bhattiprolu & Bhattiprolu,

2006; Chagas, 2009; Demant et al., 2006; Raoof et al., 2003; Tirupathi et al., 2006). However, it is clear that, although promising, these results are still experimental and much work needs to be done before permanent recommendations regarding the use of biological control agents can be secured for gray mold management.

3. FUTURE CHALLENGES

The major problem about castor gray mold remains the lack of basic knowledge about its causal agent, how the disease develops, which factors are conducive to epidemics, and how we can manage it. It is necessary to elucidate, for example, what is the role of the climatic variables over the monocyclic components of the disease and how they affect the development of epidemics in the field. A better understanding of such relationships will determine which areas are suitable to grow castor. It is also imperative to know whether sexual reproduction is occurring within *B. ricini* populations in order to prevent fungicide resistance developing, perhaps by recommending permutations of fungicides with distinct active molecules and to determine the role of ascospores during the beginning of epidemics. However, we must first determine which fungicides are effective against gray mold, as well as their timing and frequency of application. We must also determine how resistance to the disease is inherited and if there are any phenotypic or genetic markers associated with it, so that breeders can more effectively generate resistant varieties using marker-assisted programs. Additionally, economic and cost-benefit analyses should be conducted to determine which practices for disease management are worthwhile recommending. In others words, there is still much work to be done before we can better define the best strategies to avoid economic losses due to gray mold in castor crops.

4. CONCLUSION

Despite being one of the most important diseases of castor worldwide, causing severe losses on castor yield for almost a century, since its first report, gray mold is still poorly studied. The recent concern about renewable energy sources has proportioned a unique opportunity to draw attention back to this pathosystem. So, researchers involved with castor cultivation must now take this opportunity to try to elucidate the many still to be answered questions about this disease in order to mitigate the constant menace of gray mold to castor crops.

5. ACKNOWLEDGMENT

The author wishes to thank the CNPq (Proc. 472953/2009-5) and Petrobrás (TC 0050.0064181.10.9) for the financial support on research of castor gray mold. I also wish to thanks Dr. Harry C. Evans by his priceless suggestions to improve the text.

REFERENCES

1. Alcântara, M.J.; Nassur, R.C.M.R.; Sá, G.A.; Fraga, A.C. & Neto, P.C. (2008). Avaliação de materiais de mamoneira visando obtenção de resistência ao mofo cinzento,*Proccedings of 5th Congresso Brasileiro de Plantas Oleaginosas, Óleos, Gorduras e Biodiesel.* 04.06.2010. Available from http://oleo.ufla.br/anais_05/artigos/a5_464.pdf
2. Alwadie, H.M. & Baka, Z.A.M. (2003). New records of fungal pathogens of *Euphorbia inarticulate* from Aseer region. *Archives of Phytopathology and Plant Protection*, Vol.36, No.3/4, pp.195-209, ISSN 0323-5408
3. Amorim Neto, M. S.; Araújo, A.E.; Beltrão, N. E.M.; Silva, L.C. & Gomes, D.C. (2000) Zoneamento e época de plantio para a mamoneira - estado da Paraíba. Embrapa Algodão, [Comunicado Técnico 108], ISSN 0102-0099, Campina Grande, Brazil
4. Anjani, K.; Raoof, M.A.; Ashoka Vardhana Reddy, V. & Hanumanata Rao, C. (2004). Sources of resistance to major castor (*Ricinus communis* L.) diseases. *Plant Genetic Resources Newsletter*. Vol.137, (March 2004) pp.46-48, ISSN 0048-4334
5. Araújo, A.E.; Suassuna, N.D.; Bandeira, C.M. & Agra, K.N. (2003). Efeito da temperatura na germinação de esporos de *Amphobotrys ricini* (=*Botrytis ricini*). *Fitopatologia Brasileira*, Vol.28, pp.S200 [Abstract], ISSN 0100-4158
6. Araújo, A.E.; Suassuna, N.D. & Coutinho, W.M. (2007). *Doenças e seu Manejo*. In: *O Agronegócio da Mamona no Brasil*, D.M.P. Azevedo & N.E. de M. Beltrão, (Eds.), 283-303, Embrapa Informação Tecnológica, ISBN 978-85-7383-381-2, Brasília, Brazil
7. Barreto, R.W. & Evans H.C. (1998). Fungal pathogens of *Euphorbia heterophylla* and E. *hirta* in Brazil and their potential as weed biocontrol agents. *Mycopathologia*, Vol.141, pp.21-36, ISSN 0301-486X
8. Batista, F.A.S.; Lima, E.F.; Moreira, J.A.N.; Azevedo, D.M.P.; Pires, V.A.; Vieira, R.M. & Santos, J.W. (1998). Avaliação da resistência de genótipos de mamoneira, *Ricinus communis* L., ao mofo cinzento causado por *Botrytis ricini* Godfrey. Embrapa Algodão. [Comunicado Técnico 73], ISSN 0102-0099, Campina Grande, Brazil
9. Batista, F.A.S.; Lima, E.F.; Soares, J.J. & Azevedo, D.M.P. (1996). Doenças e pragas da mamoneira (*Ricinus communis* L.) e seu controle. Embrapa Algodão, [Circular Técnica 21], ISSN 0100-6460, Campina Grande, Brazil
10. Beever, R.E. & Weeds, P.L. (2007) *Taxonomy and Genetic Variation of* Botrytis *and* Botryotinia. In: Botrytis: *Biology, Pathology and Control*, Y. Elad; B. Williamson; P. Tudzynski & N. Delen (Eds), 29-52, Springer, ISBN 978-1-4020-6586-6, Dordrecht, The Netherlands

11. Bezerra, C.S. (2007). Estrutura genética e sensibilidade a fungicidas de *Amphobotrys ricini* agente causal do mofo cinzento da mamoneira. MSc Dissertation (Genetic and Molecular Biology), Universidade Federal do Rio Grande do Norte, Natal, Brazil
12. Bhattiprolu, S.L. & Bhattiprolu, G.R. (2006). Management of castor grey rot disease using botanical and biological agents. *Indian Journal of Plant Protection*, Vol.34, No.1, pp.101-104, ISSN 0253-4355
13. Buchwald, N. F. (1949). Studies in the Scletoriniaceae: I. Taxonomy of the Sclerotiniaceae.*Kongelige Veterinær- og Landbohøjskole, Aarsskrift*, Vol.32, pp.1-116. ISSN 03687171
14. Chagas, H.A. (2009). Controle de mofo-cinzento (*Amphobotrys ricini*) da mamoneira (*Ricinuscommunis* L.) por métodos químico, biológico e com óleos essenciais. MSc Dissertation, (Agronomy), Universidade Estadual Paulista "Júlio de Mesquita Filho", Botucatu, Brazil
15. Chagas, H.A.; Basseto, M.A.; Rosa, D.D.; Zanotto, M.D.; & Furtado E.L. (2010). Escala diagramática para avaliação de mofo cinzento (*Amphobotrys ricini*) da mamoneira (*Ricinus communis* L.). *Summa Phytopathologica*, Vol.36, No.2, pp.164-167, ISSN 0100-5405
16. Cook, A.A. (1981). *Disease of Tropical and Subtropical Field, Fiber and Oil Plants,* Macmillan, ISBN 0-02-949300-5, New York, United States
17. Costa, R. S.; Suassuna, T. M .F.; Milani, M.; Costa, M. N. & Suassuna, N. D. (2004). Avaliação de resistência de genótipos de mamoneira ao mofo cinzento (*Amphobotrys ricini*),*Proccedings of 1st Congresso Brasileiro de Mamona.* 04.06.2010. Available from: http://www.cnpa.embrapa.br/produtos/ mamona/publicacoes/trabalhos_cbm2/ 049.pdf
18. Dange, S.R.S; Desal, A.G. & Patel, S.I. (2005). *Diseases of castor.* In: G.S. Saharan; N. Mehta & M.S. Sangwan (Eds), 211-234, *Diseases of Oilseed Crops*, Indus Publishing Co, ISBN 81-7387-176-0, New Delhi, India
19. Demant, C.A.R.; Furtado, E.L.; Zanotto, M. & Chagas, A.A. Controle do mofo cinzento com o uso de *Trichoderma*, *Proccedings of 2nd Congresso Brasileiro de Mamona.* 04.06.2010. Available from: http://www.cnpa.embrapa.br/produtos /mamona/publicacoes/trabalhos_cbm2/ 043.pdf
20. Drumond, O. A. & Coelho S. J. (1981). Doenças da mamoneira. *Informe Agropecuário* Vol.7, No.82, pp.38-42, ISSN 0100-3364
21. Elad, Y & Stewart, A. (2007). *Microbial Control* of Botrytis *spp.* In: Botrytis: *Biology, Pathology and Control,* Y. Elad; B. Williamson; P. Tudzynski & N. Delen (Eds), 223-236, Springer, ISBN 978-1-4020-6586-6, Dordrecht, The Netherlands
22. Esuruoso, O.F. (1966). A preliminary study on the susceptibility of certain varieties of castor (*Ricinus communis* L.) to inflorescence blight disease caused by *Sclerotinia* (*Botrytis*) *ricini* (Godfrey) Whet. *The Nigerian Agricultural Journal*, Vol.3, No.1, pp.15-17, ISSN 0300-368X
23. Esuruoso, O.F. (1969). The reaction of the capsules of certain varieties of castor to a bio-chemical test for susceptibility to the inflorescence blight diseases. *The Nigerian Agricultural Journal*, Vol.6, No.1, pp.15-17, ISSN 0300-368X
24. Freire, E.C.; Lima, E.F.; Andrade, F.P.; Milani, M. & Nóbrega, M.B.M. (2007). *Melhoramento Genético.* In: *O Agronegócio da Mamona no Brasil,* D.M.P. Azevedo & N.E. de M. Beltrão, (Eds.), 171-194, Embrapa Informação Tecnológica, ISBN 978-85-7383-381-2, Brasília, Brazil
25. Galli, F.; Tokeshi, H.; Carvalho, P.C.T.; Balmer, E.; Cardoso, C.O.N. & Salgado, C.L. (1968).*Doenças da Mamoneira.* In: *Manual de Fitopatologia: Doenças das Plantas e seu Controle*, Galli, F.; Tokeshi, H.; Carvalho, P.C.T.; Balmer, E.; Cardoso, C.O.N. & Salgado, C.L., 292-197, Editora Agronomica Ceres, São Paulo, Brazil
26. Godfrey, G.H. (1919). *Sclerotinia ricini* n. sp. on the castor bean (*Ricinus communis*). *Phytopathology*, Vol.9, pp.565-567, ISSN 0031-949X
27. Godfrey, G. H. (1923). Gray mold of castor bean. *Journal* of *Agricultural Research*, Vol.23, No. 9, pp.679-715 + 13 plates, ISSN 0095-9758

28. Gonçalves, R.D. (1936). Mofo cinzento da mamoneira. *O Biológico*, Vol.2, No.7, pp.232-235, ISSN 0366-0567
29. Hansen H.N. & Bega, R.V. (1955). *Botrytis* rot of *Caladium* tubers. *Plant Disease Reporter*, Vol.39, No.3, pp.283, ISSN 0032-0811
30. Hennebert, G.L. (1973). *Botrytis* and *Botrytis*-like genera. *Persoonia*, Vol.7, No.2, pp.183-204, ISSN 0031-5850
31. Hoffmann, L.V.; Coutinho, T.C.; Duarte, E.A.A.; Bandeira, C.M. & Suassuna, N. D. (2004) Cultivo de *Amphobotrys ricini* e detecção das enzimas málica, superóxido dismutase e esterase, *Proccedings of 1st Congresso Brasileiro de Mamona*, 14.06.2010. Available from: http://www.cnpa.embrapa.br/produtos /mamona/publicacoes/trabalhos_cbm1/ 126.PDF
32. Holcomb, G.E. & Brown, W.L. (1990). Basal stem rot of cultivated *Poinsettia* caused by
33. *Amphobotrys ricini*. *Plant Disease*, Vol.74, pp.828, ISSN 0191-2917
34. Holcomb, G.E.; Jones, J.P. & Wells, D.W. (1989). Blight of prostate spurge and cultivated *Poinsettia* caused by *Amphobotrys ricini*. *Plant Disease*, Vol.73, pp.74-75, ISSN 0191-2917
35. Holz, G.; Coertze, S. & Williamson, B. (2007). *The Ecology of* Botrytis *on Plant Surfaces*. In: Botrytis: *Biology, Pathology and Control*, Y. Elad; B. Williamson; P. Tudzynski & N. Delen (Eds), 9-28, Springer, ISBN 978-1-4020-6586-6, Dordrecht, The Netherlands
36. Jarvis, W.R. (1978). Epidemiology. In: *The Biology of* Botrytis, J.R. Coley-Smith, K. Verhoeff & W.R. Jarvis (Eds), 219-250, Academic Press, ISBN 0-12-179850-X, London, UK
37. Kars, I. & van Kan, J.A.L. (2007) *Extracellular Enzymes and Metabolites Involved in Pathogenesis of* Botrytis. In: Botrytis: *Biology, Pathology and Control*, Y. Elad; B. Williamson; P. Tudzynski & N. Delen (Eds), 99-118, Springer, ISBN 978-1-4020-6586-6, Dordrecht, The Netherlands
38. Keane, P.J. (2012) *Horizontal or Generalized Resistance to Pathogens in Plants*. In: *Plant Pathology*, C.J.R. Cumagun (Ed), 317-352, InTech, ISBN 978-953-307-933-2, Viena, Austria
39. Kimati, H. (1980). *Doenças da Mamoneira*. In: *Manual de Fitopatologia: Doenças das Plantas Cultivadas*, F. Galli, (Ed), 347-351, Editora Agronomica Ceres, São Paulo, Brazil
40. Kirk, P.M.; Cannon, P.F.; Minter, D.W. & Stalpers, J.A. (2008). *Dictionary of the Fungi*, CABI Publishing, ISBN 978-0-85199-826-8, Wallingford, UK
41. Kolte, J.S. (1995). *Castor: Diseases and Crop Improvement*. Shipra Publications, ISBN 81-85402-54-X, Delhi, India
42. Kumar, A.; Reddy, P.N. & Rao, T.G.N. (2007). Host range studies of *Botrytis ricini*, the causal agent of castor grey mold. *Indian Journal of Plant Protection*, Vol.35, No.1, pp.140-141, ISSN 0253-4355
43. Leroux, P. (2007) *Chemical Control of* Botrytis *and its Resistance to Chemical Fungicides*. In: Botrytis: *Biology, Pathology and Control*, Y. Elad; B. Williamson; P. Tudzynski & N. Delen (Eds), 195-222, Springer, ISBN 978-1-4020-6586-6, Dordrecht, The Netherlands
44. Lima, B.V.; Soares, D.J.; Perreira, O.L. & Barreto, R.W. (2008). Natural infection of *Acalypha* hispida and *Jatropra podagrica* inflorescences by *Amphobotrys ricini* in brazil. *Australasian Plant Disease Notes* Vol.3, pp.5-7, ISSN 1833-928X
45. Lima, E.F.; Araújo, A.E. & Batista, F.A.S. (2001) *Doenças e seu Controle*. In: *O Agronegócio da Mamona no Brasil*, D.M.P. Azevedo & E.F.Lima (Eds), 192-212, Embrapa Informação Tecnológica, ISBN85-7383-116-2, Brasília, Brazil
46. Lima, E.F. & Soares, J.J. (1990). Resistência de cultivares de mamoneira ao mofo cinzento, causado por *Botrytis ricini*. *Fitopatologia Brasileira*, Vol.15, No.1, pp.96-98, ISSN 0100-4158
47. Lima, V.P.T.; Graça Leite, E.A.; Botrel, E.P.; Fraga, A.C. &Castro Neto, P. (2005). Avaliação de ataque de mofo cinzento da mamoneira, variedade Al Guarany 2002,

em diferentes espaçamentos. *Proccedings of 2nd Congresso Brasileiro de Plantas Oleaginosas, Óleos, Gorduras e Biodiesel.* 12.06.2010. Available from: http://oleo.ufla.br/anais_02/artigos/t183.pdf

48. Lyon, G.D.; Goodman, B.A. & Williamson, B. (2007). Botrytis cinerea *Perturbs Redox Processes as an Attack Strategy in Plants.* In: Botrytis: *Biology, Pathology and Control,* Y. Elad; B. Williamson; P. Tudzynski & N. Delen (Eds), 119-142, Springer, ISBN 978-1-4020-6586-6, Dordrecht, The Netherlands

49. Massola JR, N. S. & Bedendo, I. P. (2005). *Doenças da Mamoneira (*Ricinus communis *L.).* In:*Manual de Fitopatologia: Doenças das Plantas Cultivadas,* H. Kimati; L. Amorim; A. Bergamin Filho; L.E.A. Camargo & J.A.M. Rezende (Eds), 497-500,. Agronomica Ceres, ISBN 85-318-0008-0, São Paulo, Brazil

50. Maude, R.B. (1996). *Seedborne Diseases and their Control: Principles and Practice.* 978-0851989228, CABI Publishing, Kew, UK

51. Melhorança, A. L. & Staut, L.A. (2005). Informações técnicas para a cultura da mamona em Mato Grosso do Sul. Embrapa Agropecuária Oeste [Sistemas de Produção 8], ISSN 1679-1320, Dourados, Brazil

52. Milani M.; Nóbrega, M.B.M.; Suassuna, N.D. & Coutinho, W.M. (2005). Resistência da mamoneira (*Ricinus communis* L.) ao mofo cinzento causado por *Amphobotrys ricini.* Embrapa Algodão [Documentos 137], ISSN 0103-0205, Campina Grande, Brazil.

53. Moraes, W.B.; Souza, A.F.; Tomas, M.A.; Cecilio, R.A. & Jesus Junior, W.C. (2009). Zoneamento das áreas de risco da ocorrência de mofo cinzento da mamona no Brasil. *Proccedings of 13th Encontro Latino Americano de Iniciação Científica and 9th Encontro Latino Americano de Pós-Graduação,* 01.06.2010. Available from: http://www.inicepg.univap.br/cd/inic_2009/anais/arquivos/0821_1434_04.pdf

54. Orellana, R.G. & Thomas, C.A. (1962). Nature of predisposition of castor beans to *Botrytis.* I. Relation of leachable sugar and certain other biochemical constituents of the capsule to varietal susceptibility. *Phytopathology,* Vol.52, No.6, pp.533-538, ISSN 0031-949X

55. Palti, J. & Rotem, J. (1983). *Cultural Practices for the Control of Crop Diseases.* In: *Plant Pathologist's Pocketbook,* A. Johnston & C. Booth (Eds), 186-195, CMI, ISBN 0-85-198-460-6, Kew, England

56. Raoof, M.A.; Yasmeen, M. & Kausar, R. (2003). Potential of biocontrol agents for the management of castor grey mold, *Botrytis ricini* godfrey. *Indian Journal of Plant Protection,* Vol.31, No.2, pp.124-126, ISSN 0253-4355

57. Rego Filho, L.M.; Bezerra Neto, F.V. & Santos, Z.M. (2007). Avaliação da incidência de mofo-cinzento em genótipos de mamoneira no período de outono-inverno em campos dos Goytacazes-RJ. *Proccedings of the 2nd Congresso da Rede Brasileira de Tecnologia de Biodiesel 2006.* 27.07.2011. Available from: http://www.biodiesel.gov.br/docs/congresso2007/agricultura/3.pdf

58. Robinson, R.A. (1976). *Horizontal Pathosystem Analysis.* In: *Plant Pathosysytem (Advanced Series in Agricultural Sciences 3*), R.A. Robinson, 74-89, Springer-Verlag, ISBN 3-540-07712-X, Berlin, Germany

59. Russo, V.M. & Rossman, A.Y. (1991). Occurrence of *Amphobotrys ricini* on prostate spurge in Oklahoma. *Plant Disease,* Vol. 75, pp. 750, ISSN 0191-2917

60. Santos. R.F.; Kouri, J.; Barros, M.A.L.; Firmino, P.T. & Requião. L.E.G. (2007). *Aspectos Econômicos do Agronegócio da Mamoneira.* In: *O Agronegócio da Mamona no Brasil,* D.M.P. Azevedo & N.E. de M. Beltrão, (Eds.), 22-41, Embrapa Informação Tecnológica, ISBN 978-85-7383-381-2, Brasília, Brazil

61. Sanoamuang, N. (1996). First report of gray mold blight caused by *Amphobotrys ricini* on crown of thorns in Thailand. *Plant Disease,* Vol.80, pp.223, ISSN 0191-2917

62. Seifert, K.; Morgan-Jones, G.; Gams, W. & Kendrick, B. (2011). *The Genera of Hyphomycetes,* CBS-KNAW Fungal Biodiversity Centre, ISBN 978-90-70351-85-4, Utrecht, The Netherlands

63. Silva, J.A.; Suassuna, N.D.; Coutinho, W.M. & Milani, M. (2008). Esporulação de *Amphobotrys ricini* em frutos de mamoneira como pomponente de resistencia ao

mofo cinzento, *Proccedings of the 3th Congresso Brasileiro de Mamona*. 04.06.2010. Available from: http://www.cnpa.embrapa.br/produtos/mamona/publicacoes /cbm3/trabalhos/ FITOSSANIDADE/F%2012.pdf

64. Soares, D.J.; Fernandes J.N. & Araújo, A.E. (2010). Componentes monociclicos do mofo cinzento (*Amphobotrys ricini*) em diferentes genótipos de mamoneira, *Proceedings of 4th Congresso Brasileiro de Mamona,* 12.07.2011. Available from http://www.cbmamona.com.br/pdfs/FIT-01.pdf
65. Suassuna, N.D.; Araújo, A.E.; Bandeira, C.M.; Agra, K.N. (2003). Efeito de temperatura no crescimento e esporulação de *Amphobotrys ricini* (=*Botrytis ricini*). *Fitopatologia Brasileira* Vol.28, pp.S232 [Abstract], ISSN0100-4158
66. Sussel, A.A.B. (2008). Epidemiologia do mofo-cinzento (*Amphobotrys ricini* Buchw.) da mamoneira. PhD Thesis (Phytopathology), (June 2008)Universidade Federal de Lavras, Lavras, Brazil
67. Sussel, A.A.B. (2009a) Epidemiologia e Manejo do Mofo-cinzento-da-mamoneira. Embrapa Cerrados, [Documentos 241], ISSN 1517-5111, Brasília, Brazil
68. Sussel, A.A.B. (2009b) Escala Diagramática para Avaliação do Mofo-cinzento-da-mamoneira. Embrapa Cerrados, [Documentos 247], ISSN1517-5111, Brasília, Brazil
69. Sussel, A.A.B.; Pozza, E.A. & Castro, H.A. (2009). Elaboração e validação de escala diagramática para avaliação da severidade do mofo cinzento da mamoneira. *Tropical Plant Pathology*, Vol.34, No.3, pp.186-191, ISSN 1982-5676
70. Sussel, A.A.B.; Pozza, E.A. & Castro, H.A. & Lasmar, E.B.C. (2011). Incidência e severidade do mofo-cinzento-da-mamoneira sob diferentes temperaturas, períodos de molhamento e concentração de conídios. *Summa Phytopatologica* Vol.37, No.1, pp.30-34, ISSN0100-5405
71. Termorshuizen, A.J. (2001). *Cutural Control.* In: *Plant Pathologist's Pokcketbook*, J.M. Waller; J.M. Lenné & S.J. Waller (Eds), 318-327, CABI Publishing, ISBN 0851994598, Wallingford, UK
72. Thomas, C.A. & Orellana, R.G. (1963a) Nature of predisposition of castor beans to *Botrytis*. II. Raceme compactness, internode lenghs, position of staminate flowers and bloom in raltion to capsule susceptibility. *Phytopathology*, Vol.53, No.2, pp.249-251, ISSN 0031-949X
73. Thomas, C.A. & Orellana, R.G. (1963b) Biochemical tests indicative of reaction of castor bean to *Botrytis*. *Science*, Vol.139, pp.334-335, ISSN 0036-8075
74. Thomas, C.A. & Orellana, R.G. (1964) Phenols and pectins in relation to browning and maceration of castorbean capsules by *Botrytis*. *Phytopathologische Zeitschrift*, Vol.50, No.4, pp.359-366, ISSN 0931-1785
75. Tirupathi, J.; Kumar, C.P.C. & Reddy, D.R.R. (2006). *Trichoderma* as potential biocontrol agents for the management of grey mold of castor. *Journal of Research Angrau*, Vol.34, No.2, pp.31-36
76. Ueno, B.; Hellwig, T.C.; Nickel, G.; Silva, S.D.A. (2006). Resistência ao mofo cinzento em 15 genótipos de mamoneira cultivadas na região de Pelotas, RS, na safra 2004/2005, *Proceedings of 2nd Congresso Brasileiro de Mamona*. 12.06.2010. Available from: http://www.cnpa.embrapa.br/produtos/mamona/publicacoes/ trabalhos_cbm2/ 049.pdf
77. Whetzel, H.H. (1945). A synopsis of the genera and species of the Sclerotiniaceae, a family of stromatic inoperculate discomycetes. *Mycologia*, Vol.37, pp.648-714, ISSN 0027-5514
78. Whitney, N.G. & Taber, R.A. (1986). First report of *Amphobotrys ricini* infecting *Caperonia palustris* in the United States. *Plant Disease*, Vol.70, pp.892, ISSN 0191-2917
79. Zarzycka, H. (1958). Resistance of some varieties of castor bean (*Ricinus communis*) to the fungus *Botrytis cinerea* Pers. *Acta Agrobotanica*, Vol.7, pp.117-124, ISSN 0065-0951

Index